Area-Wide Management of Fruit Fly Pests

Area-Wide Management of Fruit Fly Pests

Edited by

Diana Pérez-Staples
Francisco Díaz-Fleischer
Pablo Montoya
María Teresa Vera

CRC Press
Taylor & Francis Group
6000 Broken Sound Parkway NW, Suite 300
Boca Raton, FL 33487-2742

First issued in paperback 2021

ISBN 13: 978-1-03-208429-9 (pbk)
ISBN 13: 978-1-138-47745-2 (hbk)

Library of Congress Cataloging-in-Publication Data

Names: Pérez-Staples, Diana, editor.
Title: Area-wide management of fruit fly pests / edited by Diana
Pérez-Staples, Francisco Díaz-Fleischer, Pablo Montoya, María Teresa
Vera.
Description: Boca Raton, FL : CRC Press, [2020] | Includes bibliographical references and index. |
Summary: "Fruit fly (Diptera: Tephritidae) pests have a profound impact on horticultural production and economy of many countries. It is fundamental to understand their biology and evaluate methods for their suppression, containment, or eradication. Area-Wide Management of Fruit Fly Pests comprises contributions from scientists from around the world on several species of tephritids working on diverse subjects with a focus on area-wide management of these pests. The first three sections of the book explore aspects of the biology, ecology, physiology, behavior, taxonomy, and morphology of fruit flies. The next two sections provide evidence on the efficacy of attractants, risk assessment, quarantine, and post-harvest control methods. The fifth and sixth sections examine biological control methods such as the Sterile Insect Technique and the use of natural enemies of fruit flies. The seventh section focuses on Area-Wide Integrated Pest Management and action programs. Finally, the eighth section examines social, economic, and policy issues of action programs aimed at involving the wider community in the control of these pests and facilitate the development of control programs. Features: Presents information on the biology of tephritid flies. Provides knowledge on the use of natural enemies of fruit flies for their biological control. Includes research results on models and diets used for the Sterile Insect Technique. Reports developments on the chemical ecology of fruit flies that contribute to make control methods more specific and efficient. Reviews subjects such as Holistic Pest Management and Area-Wide Management Programs including social, economic, and policy issues in various countries"-- Provided by publisher.
Identifiers: LCCN 2019040410 (print) | LCCN 2019040411 (ebook) | ISBN
9781138477452 (hardback) | ISBN 9780429355738 (ebook)
Subjects: LCSH: Fruit-flies--Control.
Classification: LCC SB945.F8 A743 2020 (print) | LCC SB945.F8 (ebook) |
DDC 634/.049774--dc23
LC record available at https://lccn.loc.gov/2019040410
LC ebook record available at https://lccn.loc.gov/2019040411

Visit the Taylor & Francis Web site at
http://www.taylorandfrancis.com

and the CRC Press Web site at
http://www.crcpress.com

Dedication

Dedicated to the Memory of Don McInnis, Rubén Leal Mubarqui, and Roger Vargas, leaders in area-wide management of fruit flies, and to Jorge Hendrichs, still a pillar in the fruit fly community

Contents

SECTION III Chemical Ecology and Attractants

SECTION IV Risk Assessment, Quarantine, and Post-Harvest

SECTION V Sterile Insect Technique

Preface

The *10th International Symposium on Fruit Flies of Economic Importance* was held in Tapachula, Chiapas, Mexico, from April 23 to 27, 2018. It was co-organized by El Colegio de la Frontera Sur (ECOSUR), the Mexican Fruit Fly program of the Servicio Nacional de Sanidad Inocuidad y Calidad Agroalimentaria (SENASICA), the Interamerican Institute for Cooperation in Agriculture (IICA), the Soconusco Association of Fruit Growers and the Joint Food and Agriculture Organization/ International Atomic Energy Agency (FAO/IAEA) Division of Nuclear Techniques in Food and Agriculture.

The symposium was attended by 289 fruit fly researchers, plant protection officials, fruit industry representatives, students, and exhibitors from 56 countries. There were 59 oral presentations and 134 posters. These were organized in 10 sessions: (1) Biology, Ecology, Physiology, and Behavior; (2) Taxonomy and Morphology; (3) Genetics and Biotechnology; (4) Chemical Ecology and Attractants; (5) Risk Assessment, Quarantine, and Post-Harvest; (6) Sterile Insect Technique; (7) Natural Enemies and Biological Control; (8) Other Control Methods and New Developments; (9) Area-Wide Integrated Pest Management (AW-IPM) and Action Programs; and (10) Social, Economic, and Policy Issues of Action Programs. Three field trips took place: (1) moscafrut mass-rearing facility in Metapa, (2) mango exporting process, release of sterile flies, and mango packing export center, and (3) surveillance of Mediterranean fruit fly, field operations, and coffee plantations.

Highlights of the symposium were new knowledge on microbial symbionts associations, the use of models to better understand and predict population dynamics, and new knowledge and developments regarding the chemical ecology of fruit flies that contribute to more specific and efficient control methods. The audience received research on social aspects regarding farmers' perceptions and education on fruit fly problems and management options.

Successful stories on the use of the sterile insect technique (SIT) were shared, including the US–Mexico–Guatemala Medfly program, Mediterranean fruit fly eradication in the Dominican Republic, and the Moscafrut program in Mexico.

A special session was devoted to honoring those that have left their print in the fruit fly community: Serge Quilici, Don McInnis, Rubén Leal-Mubarqui, and Jorge Gutiérrez-Samperio.

The poster sessions, coffee breaks, lunch, welcome cocktail reception, closing dinner, and field trips provided ample opportunity for participants to share their knowledge and experiences informally and establish new friendships and collaboration ties and demonstrated the camaraderie that exists in the fruit fly community, which undoubtedly has contributed to the various success stories.

Two videos were prepared for the symposium, one on the graphic history of the nine previous symposia (previous ISFFEIs) (https://www.youtube.com/watch?v=BJrBkfkDWrg) and another one with pictures from the 10th ISFFEI (10th ISFFEI) (https://www.youtube.com/watch?v=zEpvxK4eVN8).

This book represents the proceedings of the symposium, and continuing a long-lasting tradition, it is the 10th volume. It contains 31 contributions from 126 authors from all over the world. All these papers were peer reviewed. The editorial work was carried out by Diana Pérez-Staples, María Teresa Vera, Francisco Díaz-Fleischer, and Pablo Montoya. I highly appreciate and acknowledge their high level of commitment and the quality of their work. For the first time, this proceedings book will be open access, available online to a large audience through the CRC website. This was possible thanks to the registration fees of all the participants.

A few weeks after the symposium, we received the very sad news that our colleague and appreciated friend, Roger Vargas passed away in an unfortunate accident. We want to remember and honor him here. Roger's impact and influence on the community of fruit fly workers will never be diminished.

Pablo Liedo
Tapachula, Chiapas, Mexico

10 th International Symposium on Fruit Flies of Economic Importance
April 23 - 27, 2018 Tapachula, Chiapas México.

Acknowledgments

This book is a compilation of the *Proceedings of the 10th International Symposium on Fruit Flies of Economic Importance*, held in Tapachula, Chiapas, Mexico, from April 21 to 27, 2018. We thank Pablo Liedo and Rui Pereira for the invitation and the opportunity to edit the Proceedings.

Thank you Pablo and Rui!

The aim of this book is to highlight research on tephritid flies in different countries, showcasing research that would not necessarily be available in peer-reviewed journals. The book is organized in the following sections: Biology, Ecology, Physiology, and Behavior; Taxonomy and Morphology; Chemical Ecology and Attractants; Risk Assessment, Quarantine, and Post-Harvest; Sterile Insect Technique; Natural Enemies and Biological Control; Area-Wide Integrated Pest Management and Action Programs; and Social, Economic and Policy Issues of Action Programs. It also contains the plenary talk on 'Holistic Pest Management' by Dr. Barrera and reviews on subjects such as long distance flight of *Bactrocera dorsalis*, desiccation resistance, *Anastrepha* immature stage taxonomy, biological control of *Anastrepha*, area-wide management of *Anastrepha grandis* in Brazil, and natural host plants of the *Anastrepha fraterculus* complex, among others.

All chapters were peer reviewed by at least two experts in the field. Reviewers were from the following countries: Argentina, Australia, Austria, Belgium, Brazil, Chile, Czech Republic, France, Greece, Israel, Italy, Kenya, Malaysia, Mexico, Morocco, South Africa, Spain, Suriname, and the United States. As such we are in debt for thorough and insightful reviews by (in alphabetical order):

Isabel Arevalo-Vigne, Abdel Bakri, Ken Bloem, Carlos Cáceres, Jorge Cancino, Dong Cha, Jorge Luis Cladera, Des Conlong, Hugh Conway, Carol Cuashie-Williams, Francisco Devescovi, Maria Luisa Dindo, Bernie Dominiack, Sunday Ekesi, Salvador Flores, Flávio Roberto Mello Garcia, Yoav Gazit, Guy Hallman, Alvin Hee, Jorge Hendrichs, Martha Hendrichs, Michael K. Hennessey, Emilio Hernández, Vicente Hernández-Ortiz, Iara Joachim-Bravo, Nikos Kouloussis, Daniel Frías Lasserre, Aruna Manrakhan, Marc De Meyer, Salvador Meza, David Midgarden, Tahere Moadeli, Laura Moquet, Allies van Sauers Muller, Devaiah A. Muruvanda, Dori E. Nava, Vicente Navarro-Llopis, David Nestel, Allen Norrbom, Dina Orozco-Davila, Andrea Oviedo, Nikos T. Papadopoulos, Beatriz Jordao Parhanos, Jaime Piñero, Polychronis Rempoulakis, Jesús Reyes, Olivia Reynolds, Juan Rull, Mark Schutze, Diego Segura, Todd Shelly, Greg Simmons, Gary Steck, Karl Suiter, Donald Thomas, Jorge Toledo, Lucie Vaníčková, Venancio Vanoye, Marc Vreysen, Roberto Zucchi.

Thank you very much to all reviewers!

We are also thankful to Christian Rodriguez, Ricardo Macías Díaz del Castillo, and Helena Ajuria for help during the editorial process and especially Randy Brehm, Laura Piedrahita, Monica Felomina and Marsha Hecht of Taylor & Francis Group.

Last but not least, we would like to thank our families during this process for their invaluable support, in particular Dinesh, Maya and Lila Rao, Raquel Cervantes, Marcelo, Santiago and Andrés de la Vega.

Cover photographs (left to right): (1) *Ceratitis capitata* (photograph by Katja Schulz, licensed by Attribution [CC BY 2.0]), (2) *Bactrocera tryoni* mating (photograph by Ajay Narendra), (3) *Anastrepha ludens* ovipositing (photograph by Andrés Díaz Cervantes), (4) flies attracted to multi-lure trap (photograph by Pablo Montoya).

Editorial Team
Diana Pérez-Staples
Francisco Díaz-Fleischer
Pablo Montoya
María Teresa Vera

Editors

Diana Pérez-Staples is a faculty member at the Universidad Veracruzana in Xalapa, Veracruz, Mexico. Her research is focused on the sexual behavior of tephritid fruit flies and other insect pests and on improving current control methods with more than 50 published papers in scientific journals.

Francisco Díaz-Fleischer is a faculty member at the Universidad Veracruzana in Xalapa, Veracruz, Mexico. He is interested in the relationship between behavior and life history of tephritid fruit flies to improve control method, with more than 70 published papers in scientific journals.

Pablo Montoya is an expert on the use of biological control by augmentation (BCA) and the application of the sterile insect technique (SIT) against fruit fly pests, with more than 80 published papers in scientific journals. He is a researcher and head of the Unit of Methods Development in the Mexican Program against Fruit Flies, SENASICA-SADER.

María Teresa Vera is an expert on fruit fly reproductive biology and the assessment of sexual competitiveness for the implementation of the SIT with more than 50 published papers in scientific journals. Currently she is a researcher at the Consejo Nacional de Investigaciones Científicas y Técnicas (CONICET) and teaches at the Facultad de Agronomía y Zootecnia, Universidad Nacional de Tucumán, where she is member of the editorial board of the Revista Agronómica del Noroeste Argentino.

Contributors

Marysol Aceituno-Medina
Subdirección de Desarrollo de Métodos
Programa Moscafrut SENASICA-SADER
Chiapas, Mexico
Email: marysol.aceituno.i@senasica.gob.mx
ORCID ID: https://orcid.org/0000-0002-1236-9978

Reynaldo Aguilar-Laparra
Programa Moscamed
Acuerdo SADER-IICA
Chiapas, Mexico
Email: reynaldo.aguilar.i@senasica.gob.mx
ORCID ID: 0000-0001-7322-0272

Márcio Alves Silva
Entomology Laboratory
Piaui State University
Paranaíba, Piauí, Brazil
Email: silvamarcioalves@phb.uespi.br

Abdeljelil Bakri
Marrakech, Morocco
Email: bakri@uca.ac.ma

Nancy Barradas-Juanz
Red de Interacciones Multitróficas
Instituto de Ecología A.C.
Veracruz, Veracruz, Mexico
Email: eneida_juanz@hotmail.com
ORCID ID: https://orcid.org/0000-0003-4290-2304

César J. Barragán-Sol
Laboratorio de Ecología Química de Insectos
Centro de Desarrollo de Productos Bióticos (CEPROBIO)
Instituto Politécnico Nacional
Yautepec, Mexico
Email: rperaltafl600@alumno.ipn.mx
ORCID ID: https://orcid.org/0000-0002-2397-2211

Juan F. Barrera
Grupo Académico Ecología de Artrópodos y Manejo de Plagas
Departamento de Agricultura
Sociedad y Ambiente
El Colegio de la Frontera Sur (ECOSUR)
Chiapas, Mexico
Email: jbarrera@ecosur.mx
ORCID ID: https://orcid.org/0000-0002-8488-7782

Arturo Bello-Rivera
Subdirección de Operaciones de Campo Sur
Programa Nacional de Moscas de la Fruta
SENASICA-SADER
Mexico City, Mexico

Gerane Celly Dias Bezerra Silva
Brazilian Regional Faculty
Parnaíba, Piauí, Brazil
Email: gcdbezerra@gmail.com

Kenneth Bloem
US Department of Agriculture
Animal and Plant Health and Inspection Service
Plant Protection and Quarantine
Raleigh, North Carolina, USA
Email: kenneth.bloem@usda.gov

Anderson Bolzan
University of São Paulo
Piracicaba, São Paulo, Brazil
Email: ander_bolzan@hotmail.com

Naowarat Boonmee
Trok Nong Subdistrict Administrative Organization
Chanthaburi, Thailand
Email: saowchan@hotmail.com

Mirtha Borges-Soto
Instituto de Investigaciones en Fruticultura Tropical (IIFT)
Miramar, Playa C. Havana, Cuba
Email: ecologia1@iift.cu
ORCID ID: https://orcid.org/0000-0002-4163-7935

Clara A. Brandão
Agrihealth State Agency
State of Para (ADEPARA)
Belém, Pará, Brazil
Email: clara.angelica.brandao@gmail.com

Emilia Bustos-Griffin
North Carolina State University
Center for Integrated Pest Management
Raleigh, North Carolina, USA
Email: mbustos@ncsu.edu
ORCID ID: https://orcid.org/0000-0003-2504-3704

María V. Calvo
Departamento de Protección Vegetal
Facultad de Agronomía, UDELAR
Montevideo, Uruguay
Email: vcalvo@fagro.edu.uy

Eduardo Camacho-Bojórquez
Comité Estatal de Sanidad Vegetal del Estado de Sinaloa
CESAVESIN
Culiacán, Sinaloa, Mexico
Email: ecamacho@cesavesin.org.mx

Sergio Campos
Subdirección de Desarrollo de Métodos
Programa Moscafrut SENASICA-SADER
Chiapas, Mexico
Email: sergio.campos.i@senasica.gob.mx
ORCID ID: https://orcid.org/0000-0001-6478-8114

Jorge Cancino
Subdirección de Desarrollo de Métodos
Programa Moscafrut SENASICA-SADER
Chiapas, Mexico
Email: jorge.cancino.i@senasica.gob.mx
ORCID ID: https://orcid.org/0000-0003-3287-3060

Jesús Cárdenas-Lozano
Subdirección de Operaciones de Campo Norte
Programa Nacional de Moscas de la Fruta
SENASICA-SADER
Mexico City, Mexico
Email: jesus.cardenas@senasica.gob.mx

Chiou Ling Chang
USDA–ARS
Daniel K. Inouye US Pacific Basin Agricultural Research Center
Hilo, Hawaii, USA
Email: chiouling.chang@outlook.com

Thanat Chanket
Chanthaburi Provincial Agricultural Extension Office
Chanthaburi, Thailand
Email: thanatchan2502@gmail.com

Suksom Chinvinijkul
Irradiation for Pest Management Section
Department of Agricultural Extension
Ministry of Agriculture and Cooperatives
Bangkok, Thailand
Email: chinvinijkuls@gmail.com

Hugh Conway
US Department of Agriculture
Animal and Plant Health Inspection Service
S&T–CPHST–Mission Lab
Edinburg, Texas, USA
Email: Hugh.E.Conway@USDA.gov

Peter Crisp
South Australian Research and Development Institute
and
University of Adelaide
School of Agriculture, Food & Wine
Urrbrae, South Australia
Email: peter.crisp@sa.gov.au

Gabriela Costa de Sousa Cunha
Agrihealth State Agency
State of Pará (ADEPARA)
Belém, Pará, Brazil
Email: gabriela.adepara@hotmail.com

Joseph Jonathan Dantas de Oliveira
Science and Technology of Piauí
Federal Institute of Education
Cocal, Piauí, Brazil
Email: joseph@ifpi.edu.br

José Antonio De la Cruz-De la Cruz
Programa Moscamed
Acuerdo SADER-IICA
Chiapas, Mexico
Email: jose.delacruz.i@senasica.gob.mx
ORCID ID: 0000-0002-4770-9928

Soledad Delgado
Departamento de Protección Vegetal
Facultad de Agronomía
UDELAR
Montevideo, Uruguay
Email: soledaddelgadojorge@hotmail.com

Cecilia Díaz-Castelazo
Red de Interacciones Multitróficas
Instituto de Ecología A.C.
Xalapa, Veracruz, Mexico
Email: cecilia.diaz@inecol.mx
ORCID ID: https://orcid.
org/0000-0002-0185-6607

Francisco Díaz-Fleischer
Instituto de Biotecnología y Ecología Aplicada (INBIOTECA)
Universidad Veracruzana
Xalapa, Veracruz, Mexico
Email: fradiaz@uv.mx
ORCID ID: https://orcid.
org/0000-0003-2137-6587

Felicia Duarte
División Protección Agrícola
Dirección General de Servicios Agrícolas
Ministerio de Ganadería Agricultura y Pesca
and
Departamento de Protección Vegetal
Facultad de Agronomía
UDELAR
Montevideo, Uruguay
Email: fduarte@mgap.gub.uy

Ivonne Esmeralda Duarte Ubaldo
Escuela Superior de Ciencias Agropecuarias de la Universidad Autónoma de Campeche
Campeche, Mexico
Email: ieduarte@uacam.mx
ORCID ID: https://orcid.
org/000-0001-9683-1594

Vivian S. Dutra
Coordenação de Biodiversidade
Instituto Nacional de Pesquisas da Amazônia
Manaus, Amazonas, Brazil
Email: dutrasv@gmail.com
ORCID ID: https://orcid.
org/0000-0001-8499-8880

Esam Elghadi
Biotechnology Research Center
Tripoli, Libya
Email: elghadiesam@yahoo.com
ORCID ID: https://orcid.
org/0000-0003-1629-4288

Lisandro Encalada Mena
Escuela Superior de Ciencias Agropecuarias de la Universidad Autónoma de Campeche
Escárcega, Mexico

Walther Enkerlin
Insect Pest Control Section
Joint FAO/IAEA Division
Vienna, Austria
Email: W.R.Enkerlin@iaea.org

Arseny Escobar
Subdirección de Desarrollo de Métodos
Programa Moscafrut SENASICA-SADER
Chiapas, Mexico
Email: e_ariadne@hotmail.com

Evi R. Estévez Terrero
Instituto de Investigaciones en Fruticultura Tropical (IIFT)
Miramar, Playa C. Havana, Cuba
Email: ecologial0@iift.cu

Sunita Facknath
Faculty of Agriculture
University of Mauritius
Réduit, Mauritius
Email: sunif@uom.ac.mu

Salvador Flores
Subdirección de Desarrollo de Métodos,
Programa Moscafrut SENASICA-SADER
Chiapas, Mexico
Email: salvador.flores.i@senasica.gob.mx
ORCID ID: https://orcid.
org/0000-0002-9426-6824

Fredy Gálvez-Cárdenas
Campaña Moscas de la Fruta
CESAVECHIS
Chiapas, Mexico
Email: pf.tapachula@cesavechiapas.org.mx

Flávio M. García
Universidade Federal de Pelotas
Instituto de Biologia Departamento de Zoologia e Genética
Laboratório de Ecologia de Insetos
Pelotas, Rio Grande do Sul, Brazil
Email: flavio.garcia@ufpel.edu.br

Víctor García-Pérez
Campaña Moscas de la Fruta
CESVO
Oaxaca, Mexico
Email: vigape_uach@hotmail.com

María de Jesús García Ramírez
Escuela Superior de Ciencias Agropecuarias de la Universidad Autónoma de Campeche
Escárcega, Mexico
Email: mjgarcia@uacam.mx
ORCID ID: https://orcid.
org/0000-0002-2707-8081

Maria Julia S. Godoy
Department of Plant Health
Ministry of Agriculture
Livestock and Food Supply (MAPA)
Brasília, Brazil
Email: mariajuliasigodoy@gmail.com

Janisete Gomes Silva
Departamento de Ciências Biológicas
Universidade Estadual de Santa Cruz
Ilhéus, Bahia, Brazil
Email: jgs10@uol.com.br

Enoc Gómez
Subdirección de Validación de Tecnología
Programa Moscamed SENASICA-SADER
Chiapas, Mexico
Email: enoc.gomez.i@senasica.gob.mx

Enrique A. González Durán
Facultad de Ciencias Químico-Biológicas
Universidad Autónoma de Campeche
San Francisco de Campeche, Mexico

Guadalupe Gracia
US Department of Agriculture
Animal and Plant Health Inspection Service
Field Operations,
Harlingen, Texas, USA
Email: Guadalupe.Gracia@USDA.gov

Guy J. Hallman
Phytosanitation
Oceanside, California, USA
Email: n5551212@yahoo.com
ORCID ID: https://orcid.
org/0000-0002-0708-7618

Heather M. Hartzog
US Department of Agriculture
Animal and Plant Health and Inspection Service
Plant Protection and Quarantine
Raleigh, North Carolina, USA
Email: heather.m.hartzog@usda.gov

Jorge Hendrichs
Insect Pest Control Section (IPCS)
Joint FAO/IAEA Program
Vienna, Austria
Email: jorgehendrichs@gmail.com

Emilio Hernández
Subdirección de Desarrollo de Métodos
Programa Moscafrut SENASICA-SADER
Chiapas, Mexico
Email: emilio.hernandez.i@senasica.gob.mx
ORCID ID: https://orcid.org/0000-0002-1011-3637

Enrique Antonio Hernández
Escuela Superior de Ciencias Agropecuarias de la Universidad Autónoma de Campeche
Campeche, Mexico
Email: anastrephaproject@uacam.com

Refugio Hernández
Programa Moscamed SENASICA-SADER
Chiapas, Mexico
Email: refugio.hernandez.i@senasica.gob.mx

Vicente Hernández-Ortiz
Red de Interacciones Multitróficas
Instituto de Ecología A.C.
Xalapa, Veracruz, Mexico
Email: vicente.hernandez@inecol.mx
ORCID ID: https://orcid.org/0000-0003-0494-0895

Carol B. Hicks
US Department of Agriculture
Animal and Plant Health and Inspection Service
Plant Protection and Quarantine
Raleigh, North Carolina, USA
Email: carol.b.hicks@usda.gov

Nguyen T.T. Hien
Entomology Division
Plant Protection Research Institute
Hanoi, Viet Nam
Email: thanhhien1456@gmail.com

Atsushi Honma
Okinawa Prefectural Plant Protection Center
and
Ryukyu Sankei Co., Ltd
Naha, Japan
and
Faculty of Agriculture
University of the Ryukyus
Nishihara, Japan
Email: honma.tetrix@gmail.com

Yusuke Ikegawa
Okinawa Prefectural Plant Protection Center
and
Ryukyu Sankei Co., Ltd
Naha, Japan
and
Faculty of Agriculture
University of the Ryukyus
Nishihara, Japan
Email: y.ikegawa224@gmail.com

Weera Kimjong
Irradiation for Pest Management Section
Department of Agricultural Extension
Ministry of Agriculture and Cooperatives
Bangkok, Thailand
Email: chinvinijkuls@gmail.com

Phatchara Kumjing
Irradiation for Pest Management Section
Department of Agricultural Extension
Ministry of Agriculture and Cooperatives
Bangkok, Thailand
Email: chinvinijkuls@gmail.com

Rubén Leal Mubarqui
Servicios Aéreos Biológicos y Forestales Mubarqui
Cuidad Victoria, Mexico

Maximino Leyva-Castro
CESAVEGRO
Guerrero, Mexico
Email: cbiologico_gro@yahoo.com.mx

Ha K. Lien
Entomology Division
Plant Protection Research Institute
Hanoi, Viet Nam
Email: thanhhien1456@gmail.com

Wanitch Limohpasmanee
Thailand Institute of Nuclear Technology (Public Organization)
Nakhon Nayok, Thailand
Email: wanitch1@yahoo.co.th

Nicanor J. Liquido
US Department of Agriculture-
Animal and Plant Health Inspection Service-
Plant Protection and Quarantine-Science and Technology
Plant Epidemiology and Risk Analysis Laboratory
Honolulu, Hawaii, USA
Email: Nicanor.J.Liquido@usda.gov

Patricia López
Subdirección de Desarrollo de Métodos
Programa Moscafrut SENASICA-SADER
Chiapas, Mexico
Email: olga.lopez.i@senasica.gob.mx.
ORCID ID: https://orcid.org/0000-0002-5907-603X

Juan Heliodoro Luis
Subdirección de Desarrollo de Métodos
Programa Moscafrut SENASICA-SADER
Chiapas, Mexico
Email: juan.luis.i@senasica.gob.mx

Chanon Maneerat
Irradiation for Pest Management Section
Department of Agricultural Extension
Ministry of Agriculture and Cooperatives
Bangkok, Thailand
Email: chinvinijkuls@gmail.com

Rita Teresa Martínez-Salgado
Subdirección de Desarrollo de Métodos
Programa Moscafrut SENASICA-SADER
Chiapas, Mexico
Email: rita.mtz@hotmail.com

Grant T. McQuate
US Department of Agriculture-
Agricultural Research Service
Daniel K. Inouye US Pacific Basin Agricultural Research Center
Hilo, Hawaii, USA
Email: gmmcquate@hawaii.rr.com

Pablo Montoya
Subdirección de Desarrollo de Métodos
Programa Moscafrut SENASICA-SADER
Chiapas, Mexico
Email: pablo.montoya.i@senasica.org.mx.
ORCID ID: https://orcid.org/0000-0002-8415-3367

Jorge Luis Morales-Marin
Comité Estatal de Sanidad Vegetal de Tamaulipas
Tamaulipas, Mexico
Email: yaqui1952@hotmail.com

Elindinalva Antônia Nascimento
Plant Health Service
MAPA
Boa Vista, Roraima, Brazil
Email: elindinalva.nascimento@agricultura.br

Allen L. Norrbom
US Department of Agriculture
Agricultural Research Service Systematic Entomology Laboratory
Washington, District of Columbia, USA
Email: Allen.Norrbom@ARS.USDA.GOV
ORCID ID: https://orcid.org/0000-0002-5854-089X

Godshen R. Pallipparambil
North Carolina State University
NSF Center for Integrated Pest Management
Raleigh, North Carolina, USA
Email: godshen.r.pallipparambil@usda.gov
ORCID ID: https://orcid.org/0000-0002-6423-9649

Nausheen A. Patel
Entomology Division
Agricultural Services
Ministry of Agro Industry & Food Security
Réduit, Mauritius
Email: npatel@govmu.org

Ricardo Peralta-Falcón
Laboratorio de Ecología Química de Insectos
Centro de Desarrollo de Productos Bióticos
(CEPROBIO) Instituto Politécnico Nacional
Yautepec, Mexico
Email: rperaltafl600@alumno.ipn.mx
ORCID ID: https://orcid.org/0000-0001-6403-9324

Rui Pereira
Insect Pest Control Section
Join FAO/IAEA Division of Nuclear
Techniques in Food and Agriculture
Vienna, Austria
Email: R.Cardoso-Pereira@iaea.org

Diana Pérez-Staples
Instituto de Biotecnología y Ecología Aplicada
(INBIOTECA)
Universidad Veracruzana
Xalapa, Veracruz, Mexico
Email: diperez@uv.mx
ORCID ID: https://orcid.org/0000-0002-6804-0346

Puttipong Phopanit
Khlung District Agricultural Extension Office
Chanthaburi, Thailand
Email: puttipong_nich1818@hotmail.com

Luzia Picanço
Plant Health Service
MAPA
Macapá, Amapá, Brazil
Email: luzia.picanco@agricultura.gov.br

Wilda S. Pinto
Plant Protection Service
MAPA
Belém, Pará, Brazil
Email: wilda.silveira@agricultura.gov.br

José M. Pires
Plant Protection Service
MAPA
Macapá, Amapá, Brazil
Email: josemacdowellpiresfilho@gmail.com

Gordon Port
School of Natural and Environmental Sciences
Newcastle University
United Kingdom
Email: gordon.port@newcastle.ac.uk
ORCID ID: https://orcid.org/0000-0002-1409-4576.

Carol Quashie-Williams
Entomologist
CSIRO STEM Professionals in Schools-Farrer
Primary School
Canberra, ACT, Australia
Email: Carol.Quashie-Williams@agriculture.gov.au

Maria Eliana Queiroz
Plant Health Service
MAPA
Macapá, Amapá, Brazil
Email: maria.eliana@agricultura.gov.br

Francisco Ramírez y Ramírez
Dirección General de Sanidad Vegetal
SENASICA-SADER
Mexico City, Mexico
Email: francisco.ramirez@senasica.gob.mx

Milton Arturo Rasgado-Marroquín
Subdirección de Producción
Programa Moscamed acuerdo SADER-IICA
Chiapas, Mexico
Email: milton.rasgado.i@senasica.gob.mx
ORCID ID: 0000-0003-4115-8851

Pedro Rendón
IAEA–Technical Cooperation-Latin America & Caribbean Section
Guatemala City, Guatemala
Email: Pedro.Rendon@USDA.gov

Norma R. Robledo-Quintos
Laboratorio de Ecología Química de Insectos
Centro de Desarrollo de Productos Bióticos (CEPROBIO) Instituto Politécnico Nacional
Yautepec, Mexico
Email: rperaltafl600@alumno.ipn.mx
ORCID ID: https://orcid.org/0000-0002-8988-9875

Erick J. Rodriguez
Department of Entomology and Nematology
University of Florida
Gainesville, Florida, USA
Email: erick.rodriguez@ufl.edu
ORCID ID: https://orcid.org/0000-0001-8132-0863

Maylin Rodríguez Rubial
Instituto de Investigaciones en Fruticultura Tropical (IIFT),
Miramar, Playa C. Havana, Cuba
Email: ecologia1@iift.cu

Beatriz Ronchi-Teles
Coordenação de Biodiversidade
Instituto Nacional de Pesquisas da Amazônia
Manaus, Amazonas, Brazil
Email: ronchi@inpa.gov.br
ORCID ID: https://orcid.org/0000-0001-8840-2757

Beatriz Sabater-Munoz
Smurfit Institute of Genetics
Trinity College of Dublin, College Green
Dublin, Ireland

and

Institute of Plant Molecular and Cellular Biology (IBMCP) of the Spanish National Research Council (CSIC) and Polytechnic University of Valencia (UPV)
Valencia, Spain
Email: b.sabater.munyoz@gmail.com
ORCID ID: https://orcid.org/0000-0002-0301-215X

Iris B. Scatoni
Departamento de Protección Vegetal
Facultad de Agronomía, UDELAR
Montevideo, Uruguay
Email: iscatoni@gmail.com

Emiliano Segura-Bailon
Campaña Moscas de la Fruta
CESAVEGRO
Guerrero, Mexico
Email: segurabailon@yahoo.com

Mohammad Sabbir Siddiqui
South Australian Research and Development Institute
Urrbrae, Adelaide, South Australia

and

Department of Biological Sciences
Macquarie University
New South Wales, Australia
Email: mohammad.siddiqui@mq.edu.au

Luis Cristóbal Silva Villareal
Programa Moscamed acuerdo SADER-IICA
Chiapas, Mexico
Email: luis.silva.i@senasica.gob.mx
ORCID ID: 0000-0003-2481-866X

Preeaduth Sookar
Entomology Division, Agricultural Services
Ministry of Agro Industry & Food Security
Réduit, Mauritius
Email: psookar@govmu.org

Gary J. Steck
Florida Department of Agriculture and Consumer Services
Division of Plant Industry
Gainesville, Florida, USA
Email: gary.steck@freshfromflorida.com
ORCID ID: https://orcid.org/0000-0003-3714-0560

Karl A. Suiter
Center for Integrated Pest Management
North Carolina State University
Raleigh, North Carolina, USA
Email: Karl_Suiter@cipm.info

Weerawan Sukamnouyporn
Irradiation for Pest Management Section
Department of Agricultural Extension
Ministry of Agriculture and Cooperatives
Bangkok, Thailand
Email: chinvinijkuls@gmail.com

Phillip Taylor
Department of Biological Sciences
Macquarie University
New South Wales, Australia
Email: phil.taylor@mq.edu.au
ORCID ID: https://orcid.org/0000-0002-7574-7737

Marco Tulio Tejeda
Subdirección de Producción
Programa Moscamed acuerdo SADER-IICA
Chiapas, Mexico
Email: marco.tejeda@programamoscamed.mx
ORCID ID: 0000-0003-0691-6960

Dang Đ. Thang
Entomology Division
Plant Protection Research Institute
Hanoi, Viet Nam
Email: thanhhien1456@gmail.com

Vu V. Thanh
Entomology Division
Plant Protection Research Institute
Hanoi, Viet Nam
Email: thanhhien1456@gmail.com

John Thomas
National Horticultural Research Institute
Ibadan, Nigeria
Email: vumeha@yahoo.com

Vu T.T. Trang
Entomology Division
Plant Protection Research Institute
Hanoi, Viet Nam
Email: thanhhien1456@gmail.com

Luiz Carlos Trassato
Plant Health Service
Boa Vista, Roraima, Brazil
Email: luiz.trassato@agricultura.gov.br

Vincent Umeh
National Horticultural Research Institute,
Ibadan, Nigeria
Email: vumeha@yahoo.com

Vivian Umeh
Department of Crop Protection and Environmental Biology
University of Ibadan
Ibadan, Nigeria
Email: vivianumeh37@gmail.com

Alongkot Uthaitanakit
Chanthaburi Provincial Agricultural Extension Office
Chanthaburi, Thailand
Email: alongkot.1976@gmail.com

Marvel del Carmen Valencia Gutiérrez
Facultad de Ciencias Químico-Biológicas
Universidad Autónoma de Campeche
Campeche, Mexico
Email: mcvalenc@uacam.mx

Juan José Vargas Magaña
Escuela Superior de Ciencias Agropecuarias de la Universidad Autónoma de Campeche
Campeche, Mexico
Email: jjvargas@uacam.mx
ORCID ID: https://orcid.org/000-0002-9218-3259

Clóvis V. Vasconcelos
Agrihealth State Agency
State of Para (ADEPARA)
Belém, Pará, Brazil
Email: clovissantarem@hotmail.com

Emmanuel Velázquez-Dávila
Programa Moscamed, Acuerdo SADER-IICA
Planta Moscamed
Chiapas, Mexico
Email: emmanuel.velazquez.i@senasica.gob.mx
ORCID ID: 0000-0002-3127-7753

Jorge Vélez
Campaña Moscas de la Fruta
CESAVETAM
Tamaulipas, Mexico
Email: velezaguilucho@hotmail.com

Carmen Ventura
Subdirección de Desarrollo de Métodos
Programa Moscafrut SENASICA-SADER
Chiapas, Mexico
Email: ibtventura@gmail.com

María Teresa Vera
Facultad de Agronomía y Zootecnia
Universidad Nacional de Tucumán
and
Consejo Nacional de Investigaciones Científicas y Técnicas (CONICET)
Tucumán, Argentina
Email: teretina@hotmail.com
ORCID ID: https://orcid.org/0000-0002-1446-6042

Christopher Vitek
Department of Biology
University of Texas Rio Grande Valley
Edinburg, Texas, USA
Email: Christopher.Vitek@UTRGV.edu

Christopher W. Weldon
Department of Zoology and Entomology
University of Pretoria
Pretoria, South Africa
Email: cwweldon@zoology.up.ac.za
ORCID ID: https://orcid.org/0000-0002-9897-2689

Le T. Xuyen
Entomology Division
Plant Protection Research Institute
Hanoi, Viet Nam
Email: thanhhien1456@gmail.com

Wee L. Yee
US Department of Agricutlure-Agricultural Research Service
Temperate Tree Fruit and Vegetable Research Unit
Wapato, Washington, USA
Email: Wee.Yee@ARS.USDA.GOV

Section I

Biology, Ecology, Physiology, and Behavior

1 Identification of the Profile of Cuticular Hydrocarbons of *Anastrepha curvicauda* (Diptera: Tephritidae)

Ricardo Peralta-Falcón[*], *Norma R. Robledo-Quintos, and César J. Barragán-Sol*

CONTENTS

Abstract Cuticular hydrocarbons (CHCs) are constituents of the epicuticle of insects, which have the function of preventing dehydration and are signs of inter- and intraspecific recognition. Because CHCs vary between species and according to adulthood, sex, and mating status, they have been studied in species of economic importance such as *Ceratitis capitata* Wiedemann, *C. anonae* Graham, *C. rosa* Karsch, and *Anastrepha fraterculus* Wiedemann as an effective means of taxonomic identification. However, there are no studies of its intervention in chemical communication, and they have not been studied in *Anastrepha curvicauda* Gerstaecker, an insect pest of *Carica papaya* Linnaeus. In this work, we studied the CHC profile of virgin males and females of different ages of *A. curvicauda*. The extraction was done with hexane and was injected into a gas chromatograph coupled with a mass spectrometer. The identification of compounds was performed considering retention times, retention index, and spectral evaluation through a comparison with the NIST mass spectra library. The CHC profile of *A. curvicauda* consists of long chains of 20–29 carbons, and four major compounds were identified: 2-methyloctacosane, 1-heptacosanol, (Z)-14-tricosenyl formate, and a (Z)-14-tricosenyl formate isomer. 1-heptacosanol was the main compound in females and (Z)-14-tricosenyl formate in males. 1-heptacosanol in females increased in abundance at 5–7 days, a period that coincides with their sexual maturity. The obtained CHC profile is specific to this species. The compounds are sex-specific, too, and their differences are apparent at 7 days of age when abundance is higher in females than males.

[*] Corresponding author.

1.1 INTRODUCTION

The cuticle of insects has several functions, such as protecting against environmental conditions, pathogens, and other insects, as well as supporting the body. The epicuticle is the external layer of the cuticle and consists of two layers: the first one, composed of chitin, and the second one, composed of hydrocarbons, which helps the insect avoid dehydration and damage by ultraviolet (UV) rays (Hadley 1984; Vrkoslav et al. 2010). These hydrocarbons are inter- and intraspecific recognition signals because they vary in species, age, sex, and physiological stage. Nevertheless, cuticular hydrocarbons (CHC) have been studied to taxonomically identify pest insects, such as *Ceratitis capitata, Ceratitis anonae, Ceratitis fasciventris* Bezzi, *Ceratitis rosa*, and *Anastrepha fraterculus* (Vaníčková et al. 2014). However, our interest focuses on inter- and intraspecific recognition (Blomquist 2010). In this work, we studied the CHC profile of virgin males and females of different ages of *Anastrepha curvicauda*, formerly *Toxotrypana* (Diptera: Tephritidae) (Norrbom et al. 2018).

1.2 MATERIALS AND METHODS

Insects were collected from a small plantation at CeProBi, IPN, Yautepec, Morelos, Mexico, between 18° 05′ N latitude and 99° 03′ W longitude. Larvae were collected from infested fruit and deposited in plastic containers with soil from the original collection site for pupation. Once emerged, adults were individually separated according to sex into 9 × 4-cm acrylic containers (NX2185C, Daiger, Vernon Hills, IL) and provided with 10% sugar water as food.

We used virgin flies of 1 (24 hours after the fly emerged), 3, 5, 7, 9, and 11 days of age, and a weight interval of 0.0469 g ± 0.0050 for males and 0.0510 g ± 0.0055 for females. Flies were kept for 30 minutes in a freezer (–20°C) (Norlake Scientific). They were then placed for 15 min in a desiccator with silica gel granules. The extraction was done with 1 mL of hexane in a glass vial, which was concentrated to 150 μL under a nitrogen flow. The extracts were stored at –20°C until chemical analysis. Two μL of extract were injected into a gas chromatograph (GC; HP6890) coupled with a mass spectrometer (MS; HP 5972) (Agilent, USA). The samples were analyzed using a nonpolar column SLB-5ms (30-m long, 250-μm internal diameter, and 0.25-μm film thickness, SUPELCO Analytical). The initial oven temperature was 150°C for 2 min, increasing to 5°C/min until reaching 308°C. The carrier gas was hydrogen at a constant flow of 2 mL/min. The injector temperature was 250°C and the auxiliary was 280°C; the injector functioned in split mode 1:25. The MS functioned by electronic ionization (70 EV) in SCAN mode and at a mass interval of 29 to 400 AMU (modified from Vaníčková et al. 2014).

The identification of compounds was performed considering retention times, retention index (Clarke 1978), and spectral evaluation through a comparison with the NIST mass spectra library (NIST/EPA/NIH 2002).

Linear regressions were performed to determine the influence of the flies' weight on the abundance of CHCs. Results of the regressions showed a relation between these variables. Therefore, CHC abundance was divided by weight for each fly.

Comparisons of CHC abundance between virgin females and males were performed with a *t*-student test; natural logarithm (α), logarithm base 10 (β), square root (σ), and reciprocal (γ) transformations were applied to some data to fulfill the requirements of normality and equality of variance. A Mann–Whitney test was used for data that could not be normalized or homogenized by a transformation. For all cases, mean ± standard error of mean (SEM) is reported, even for nontransformed data. All analyses were performed using SigmaPlot 12.5, and the rejecting error was 0.05.

1.3 RESULTS

Our results show that the profiles for virgin flies included a mix of large-chain hydrocarbons of 20-31 carbons for both sexes. We identified the following CHCs: 2-methylactosane, 1-heptacosanol, (Z)-14-tricosenyl formate, and a (Z)-14-tricosenyl formate isomer.

For females, 1-heptacosanol and (Z)-14-tricosenyl formate were compounds that were highly abundant in all samples. For males, (Z)-14-tricosenyl formate was highly abundant (Table 1.1).

There were no qualitative differences in CHC profiles between virgin males and females. The abundance of 2-methyloctacosane in 5-day-old virgin males was higher than in 5-day-old virgin females ($t = 7.072$, $df = 24$, $P < 0.001$). This CHC also had a higher abundance in 7-day-old virgin females than in virgin males of the same age ($T = 256$, $df = 24$, $P < 0.001$, respectively). There were no significant differences between virgin females and males, which were 1, 3, 9, and 11 days old ($t = 2.014$, $df = 24$, $P = 0.028$; $t = 0.395$, $df = 24$, $P = 0.348$, $t = 0.463$, $df = 24$, $P = 0.324$; $t = 0.598$, $df = 24$, $P = 0.278$, respectively) (Figure 1.1).

The abundance of 1-heptacosanol was higher in virgin males than in virgin females at 5 days of age ($T = 238$; $df = 24$; $P < 0.001$). However, the abundance of this CHC was higher in virgin females than in virgin males of 7 and 11 days of age ($t = 6.732$, $df = 24$, $P < 0.001$; $t = 2.118$, $df = 24$, $P = 0.022$, respectively). There were no differences in this CHC among sexes in flies of 1, 3, and 9 days of age ($T = 153$, $df = 24$, $P = 0.885$; $t = 1.485$, $df = 24$, $P = 0.075$; $t = 1.47$, $df = 24$, $P = 0.324$, respectively) (Figure 1.1).

The abundance of (Z)-14-tricosenyl formate was higher in virgin females than in virgin males of 7 days of age ($t = 3.102$, $df = 24$, $P = 0.002$). However, its abundance was similar between sexes at 1, 3, 5, 9, and 11 days of age ($T = 134$, $df = 24$, $P = 0.470$; $t = 0.330$, $df = 24$, $P = 372$; $t = 0.330$, $df = 24$, $P = 0.372$; $t = 0.835$, $df = 24$, $P = 206$; $t = 1.042$, $df = 24$, $P = 0.154$, respectively) (Figure 1.1).

Virgin females showed a higher abundance of (Z)-14-tricosenyl formate than virgin males at 7 days of age ($t = 2.881$, $df = 24$, $P = 0.004$). There were no differences between sexes in the abundance of this compound for the rest of the evaluated ages.

Ethanol, 2-(Z) (octadecene-9-enoxy) traces were detected in both sexes; thus, no analysis was performed.

TABLE 1.1
Cuticular Hydrocarbon Compounds in Virgin Males and Females of *Anastrepha curvicauda*

Compound[a]	Rt (min)	Formula[b]	MW (g/mol)	CAS	RI	F%	M%
2-methyloctacosane	24.981	$C_{29}H_{60}$	408.4695	1560-98-1	2868	6.84	8.80
1-heptacosanol	27.784	$C_{27}H_{56}O$	396.4331	2004-39-9	3086	**36.17**	**28.93**
(Z)-14-tricosenyl formate	29.960	$C_{24}H_{46}O_2$	366.5776	77899-10-6	3246	**36.14**	**41.04**
A (Z)-14-tricosenyl formate isomer	30.088	—	278.48	—	3255	18.34	18.55
Ethanol, 2-(Z) (octadecene-9-enoxy)	32.310	$C_{20}H_{40}O_2$	312.4928	5353-25-3	3412	2.48	2.67

CAS, Chemical Abstract Service; F%, female percentage; IUPAC, International Union of Pure and Applied Chemistry; M%, male percentage; MW, molecular weight; RI, retention index; Rt, retention time.

[a] IUPAC name

[b] Condensed formula.

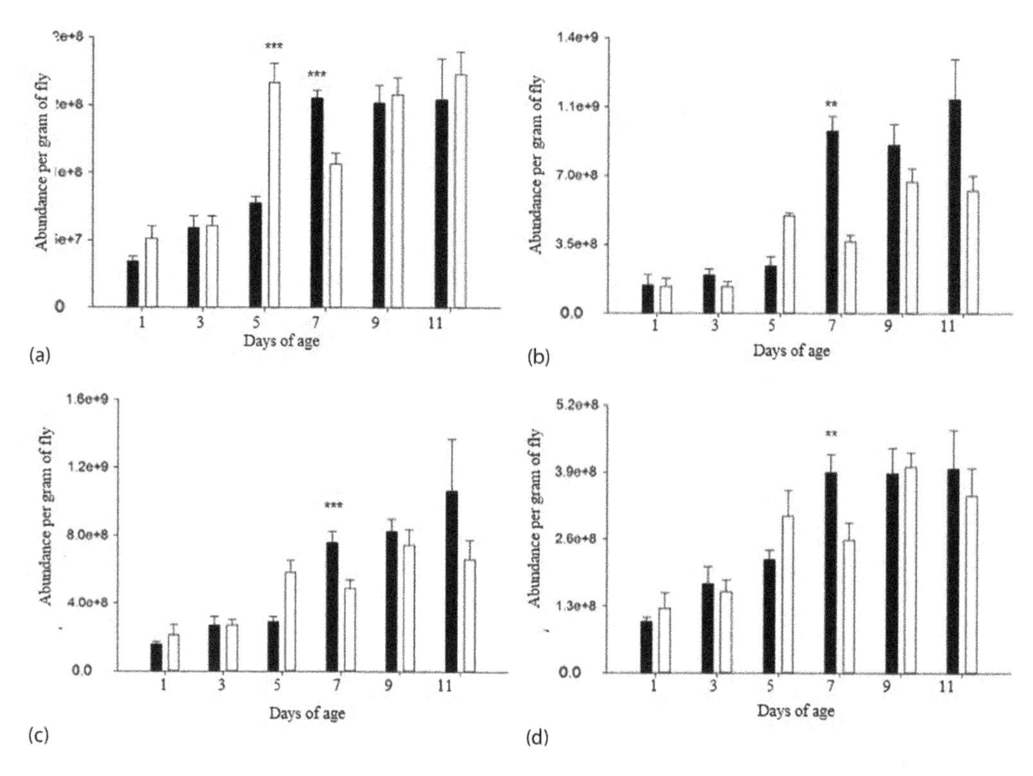

FIGURE 1.1 Abundance (mean + SEM) of (a) 2-methyloctacosane, (b) 1-heptacosanol, (c) (Z)-14-tricosenyl formate, and (d) an isomer of (Z)-14-tricosenyl formate present in virgin females (black bar) and males (white bar) (n = 13). For the 2-methyloctacosane compound analysis, a α data transformation was applied, except for 7 days of age; for the 1-heptacosanol compound, a α transformation was applied for 3, 7, and 11 days of age and no transformation was used for 1, 5, and 9 days of age; for (Z)-14-tricosenyl, a α transformation was applied, except for 1 and 7 days of age; and for the (Z)-14-tricosenyl formate isomer, a α transformation was applied, except for 1 and 3 days of age ($P < 0.05$; $^{**}P < 0.01$; $^{***}P < 0.001$).

1.4 DISCUSSION

In profiles of virgin males and females of *A. curvicauda*, as well as in both sexes of the dipteran species *Drosophila birchii*, Ayala, *Drosophila serrata* Ayala, and *Drosophila suzukii* Matsumura (Howard et al. 2003; Snellings et al. 2018), in species such as *Aldrichina grahami* Aldrich, *Achoetandrus rufifacies* Macquart, *Chrysomya megacephala* Fabricius, and *Lucilia sericata* Meigen (Diptera: Calliphoridae) (Ye et al. 2007), in *A. fraterculus* (Vaníčková et al. 2012), *C. capitata*, *C. anonae*, *C. fascivertis*, and in female *C. rosa* flies 2-methyloctacosane has been detected (Vaníčková et al. 2014, 2015). However, this is the first study that reports 1-heptacosanol, (Z)-14-tricosenyl formate, and a 14-tricosenyl formate isomer as part of CHC profiles. Therefore, the results of this study suggest a profile specificity for *A. curvicauda*.

Compound profiles were the same for both sexes, which is known as a monomorphic profile and has been observed in *D. serrata* (Howard et al. 2003), *D. suzukii* (Snellings et al. 2018), *Drosophila persimilis* Dobzhansky and Epling, *Drosophila pseudobscura* Frolova, and *Drosophila takahashii* Sturtevant, which are cases that reported similar results as this study (Shirangi et al. 2009).

Abundance differences between sexes have been reported in profiles of different species; for example, in *D. suzukii*, 7-tricosene was more abundant in virgin females than in virgin males but only in 1-day-old flies (Shirangi et al. 2009). Abundance of all compounds has been observed to be higher in females than in males in similar species such as *D. birchii*, *D. serrata* (Howard et al. 2003), *A. fraterculus* (Vaníčková et al. 2012), and *C. rosa* (Vaníčková et al. 2014, 2015). Such differences are associated with sex differentiation because they are related to recognition between the sexes (Blomquist 2010).

The difference is observed at 7 days of age, when females showed a high CHC abundance. Substantial differences between sexes of advanced ages were reported for *Anopheles gambiae* Giles (Diptera: Culicidae, Caputo et al. 2005), where the abundance of n-alkanes increased with age.

Quantitative variability of co mpound profiles between sexes is related to a specific role of recognition between sexes that can be involved in selection or discrimination of a fly of the same sex or of the opposite sex (Blomquist 2010). Thus, in studies with *D. serrata* and *Drosophila melanogaster* Meigen, CHCs can be used as a signal for pheromonal communication (Grillet et al. 2006; Thomas and Simmons 2010), or intraspecific recognition (Blomquist 2010). Currently, bioassays on intraspecific recognition are in process, taking as a starting point patterns that occur in agonistic and courtship behavior.

ACKNOWLEDGMENTS

N. Robledo is a COFAA and EDI, IPN fellow. R. Peralta and Barragán C. would like to thank CONACyT and IPN for a scholarship. We are also very grateful to SIP Projects (20160834 and 20171338).

REFERENCES

Blomquist, G.J. 2010. Structure and analysis of insect hydrocarbons. In: *Insect Hydrocarbons Biology, Biochemistry, and Chemical Ecology*, Ed. Blomquist G. J., and A. G. Bagnères, pp. 19–34. New York: Cambridge University Press.

Caputo, B., F. R. Dani, G. L. Horne, V. Petrarca, S. Turillazi, M. Coluzzi, A. A. Priestman, and A. Torre. 2005. Identification and composition of cuticular hydrocarbons of the major Afrotropical malaria vector *Anopheles gambiae* s.s. (Diptera: Culicidae): Analysis of sexual dimorphism and age-related changes. *Journal of Mass Spectrum* 40:1595–1604.

Clarke, E.G. 1978. Isolation and identification of drugs. *The Pharmaceutical Press* 2:927–928.

Grillet, M., L. Dartevelle, and J. F. Ferveur. 2006. A *Drosophila* male pheromone affects female sexual receptivity. *Proceedings of the Royal Society B* 273:315–323.

Hadley, F. N. 1984. Cuticle: Ecological significance. In: *Biology of the Integument Vol.1*, Ed. Beretier, H. J., A. G. Matoltsy, and S. K. Richards, pp. 685–693. Berlin, Germany: Springer-Verlag.

Howard, R. W., L. L Jackson, H. Banse, and M.W. Blows. 2003. Cuticular hydrocarbons of *Drosophila birchii* and *D. serrata*: Identification and role in mate choice in *D. serrata*. *Journal of Chemical Ecology* 29(4):961–976.

NIST/EPA/NIH. 2002. Mass spectral library. Mass Spectral Library with Search Program (Data Version: NIST05, Software Version 2.0).

Norrbom, A. L., N. B. Barr, P. Kerr, X. Mengual, N. Nolazco et al. 2018. Synonymy of *Toxotrypana* Gerstaecker with *Anastrepha* Shiner (Diptera: Tephritidae). *Proceedings of the Entomological Society of Washington* 120(4):834–841.

Shirangi, T. R., H. D. Dufour, T. M. Williams, and S. B. Carroll. 2009. Rapid evolution of sex pheromone-producing enzyme expression in *Drosophila*. *PLOS Biology* 7:e1000168.

Snellings, Y., B. Herrera, B. Wildemann, M. Beelen, L. Zwarts, T. Wenseleers, and P. Callaerts. 2018. The role of cuticular hydrocarbons in mate recognition in *Drosophila suzukii*. *Scientific Reports* 8:4996.

Thomas, M. L. and L. W. Simmons. 2010. Cuticular hydrocarbons influence female attractiveness to males in the Australian field cricket, *Teleogryllus oceanicus*. *Journal of Evolutionary Biology* 23:707–714.

Vaníčková, L., A. Svatoš, J. Kroiss, M. Kaltenpoth, R. Nascimento, M. Hoskovec, R. Břízová, and B. Kalinová. 2012. Cuticular hydrocarbons of the South American fruit fly *Anastrepha fraterculus*: Variability with sex and age. *Journal of Chemical Ecology* 38:1133–1142.

Vaníčková, L., M. Virgilio, A. Tomčala, R. Břízová, S. Ekesi, M. Hoskovec, and M. De Meyer. 2014. Resolution of three cryptic agricultural pests (*Ceratitis fasciventris, C. anonae, C. rosa*, Diptera: Tephritidae) using cuticular hydrocarbon profiling. *Bulletin of Entomological Research* 104:631–638.

Vaníčková, L., R. Břízová, A. Pompeiano, S. Ekesi, and M. De Meyer. 2015. Cuticular hydrocarbons corroborate the distinction between lowland and highland natal fruit fly (Tephritidae, *Ceratitis rosa*) populations. *ZooKeys* 540:507–524.

Vrkoslav, V., A. Muck, J. Cvačka, and A. Svatoš. 2010. MALDI imaging of neutral cuticular lipids in insects and plants. *Journal of the American Society for Mass Spectrometry* 21:220–231.

Ye, G., K. Li, J. Zhu, G. Zhu, and C. Hu. 2007. Cuticular hydrocarbon composition in pupal exuviae for taxonomic differentiation of six necrophagous flies. *Journal of Medical Entomology* 44(3):450–456.

2 Reported Long-Distance Flight of the Invasive Oriental Fruit Fly and Its Trade Implications

Carol B. Hicks, Kenneth Bloem, Godshen R. Pallipparambil, and Heather M. Hartzog*

CONTENTS

Abstract Online biological databases are a popular method of summarizing and storing scientific information. Invasive species databases are used by risk analysts and policy makers from many counties as their main source of scientific data. A majority of the information found in invasive species databases is useful, but data can be oversimplified and errors do exist. For example, the statement "Many *Bactrocera* spp. can fly 50–100 km (Fletcher 1989)" is found in multiple invasive species databases and has been repeated in phytosanitary documents written in different countries. This broad statement has been presented to the United States as evidence that they should extend the radii of quarantine areas placed around new detections of the Oriental fruit fly, *Bactrocera dorsalis* (Hendel), in California from 7.5 to 150 km. In reviewing the available literature, this work finds that the flight distance of 50–100 km for *Bactrocera* spp. as summarized in invasive species databases cannot be attributed to Brian S. Fletcher. A review of 17 publications describing mark-release-recapture studies or field observations on *B. dorsalis* showed that long distance (>20 km) captures of the flies do occur, but such captures are atypical and occur only rarely. Dispersal distances up to 2 km are much more typical and commonly reported. Data summarized in invasive species databases on fruit flies may not be precise. Therefore, consideration

* Corresponding author.

of biological evidence found in the original source material and other scientific publications is also necessary when developing pest-management strategies or phytosanitary policies. Mark-release-recapture studies clearly show that flight capacity differs among *Bactrocera* species. Although flight capacity is a major factor in determining the size of quarantine areas for fruit flies, host availability, climate suitability, potential pathways, and community demographics are also important risk factors. In California and Florida, *B. dorsalis* have been captured on a number of occasions in their respective state fruit fly detection trapping networks. However, despite these detections, rapidly delimiting an outbreak, establishing quarantine areas, and when needed implementing additional eradication measures have successfully prevented establishment of this invasive pest and has prevented the export of any infested host fruits from quarantine areas to other countries for more than 30 years. Therefore, trading partners should also consider whether rigorous, established trapping programs are in place and proven response protocols exist when determining the radii of required quarantine zones rather than simply setting standards based on the most distant recapture of a fruit fly species.

2.1 INTRODUCTION

Scientists and policy makers routinely refer to pest databases and other scientific literature to gather facts when developing phytosanitary documents, formulating trade policies, and implementing management strategies for harmful exotic pest species. Online biological databases are popular repositories for sharing scientific information. The rising incidence of detection, introduction, and establishment of invasive species has prompted the development of new databases that contain biological information necessary to prevent, detect, manage, and develop policies on invasive species (Katsanevakis and Roy 2015). There are more than 250 comprehensive open-sourced databases that contain information on invasive species that are listed on the Global Invasive Species Information Network (GISIN; http://www.gisin.org).

Overwhelmingly, the scientific information that biological databases provide is valuable and helpful; however, databases are not free of errors. In organizational databases, between 1% and 10% of data items are estimated to be inaccurate (Klein et al. 1997). Errors in invasive species lists have been reported by McGeoch et al. (2012) and Jacobs et al. (2017). Misidentification of a species can lead to incorrect distribution records in databases (Emig et al. 2015).

In addition, information that originates in one reference database is often replicated in related databases with the information spreading throughout both printed and online reference material. An example of a controversial statement that originated in an invasive species database is the conclusion that "Many *Bactrocera* spp. can fly 50–100 km (Fletcher 1989)." This exact statement has been repeated in multiple databases. It has been presented as evidence to implement 150-km pest-free areas beyond the established standard quarantine of 7.2-km (4.5 mile) radius surrounding new detections of *Bactrocera dorsalis* (Hendel) in California (FreshPlaza 2015). *Bactrocera dorsalis* is an invasive species to the continental United States that has been introduced, quarantined, and then eradicated on a number of occasions in California (CDFA 2008) and Florida (Weems et al. 2016).

The flight ability of a fruit fly species to move or disperse is a major factor that influences the size of imposed quarantine areas and boundaries for pest-free zones. Agricultural-based countries spend millions of dollars to detect, delimit, quarantine, and eradicate new fruit fly introductions. The imposition of unreasonably large quarantine areas can result in unnecessary trade restrictions and loss of export markets for growers, as well as unnecessary pesticide applications and use of monetary resources. In this chapter, a review and an analysis of published data indicates that the statement "Many *Bactrocera* spp. can fly 50–100 km" does not reflect the typical dispersal distance and should be replaced with more specific flight details for each *Bactrocera* species. In addition, this review on the flight capability of *B. dorsalis* will provide support for decisions on phytosanitary measures, quarantine restrictions, and integrated pest management (IPM) for this invasive species.

2.2 METHODS

A search of the World Wide Web was performed July 27, 2016 for the statement "Many *Bactrocera* spp. can fly 50–100 km (Fletcher 1989)." Multiple searches were performed while including or excluding the citation "(Fletcher 1989)."

Literature searches for articles by Brian S. Fletcher and other authors on fruit fly biology and movement were performed through scientific search engines available through Academic Search™ Premier (https://www.ebscohost.com/academic/academic-search-premier), Web of Science™ (http:/ clarivate.com/scientific-and-academic-research/research-discovery/web-of-science/), and US Department of Agriculture's National Agricultural Library (https://www.nal.usda.gov/). A generic search of the World Wide Web was also performed. Literature searches for biology and flight data for *Bactrocera dorsalis* were inclusive of all taxonomic synonyms including *Bactrocera papayae*, Drew & Hancock, *Bactrocera invadens*, Drew, Tsuruta & White, and *Bactrocera philippinensis*, Drew & Hancock, which were recently declared synonyms of *B. dorsalis* (Schutze et al. 2015). Titles and abstracts of articles were examined and relevant articles on fruit fly biology and movement were reviewed. References cited within each article were screened and pertinent articles were selected for further review. Thirty-five articles or book chapters by Fletcher focusing on fruit fly biology, population dynamics, and movement were reviewed (Table 2.1). Two books, Shelly et al.

TABLE 2.1
List of Publications by Brian S. Fletcher on *Bactrocera* Ecology or Biology

Bellas, T. E., and B. S. Fletcher. 1979. Identification of the major components in the secretion from the rectal pheromone glands of the Queensland fruit flies *Dacus tryoni* and *Dacus neohumeralis* (Diptera: Tephritidae). *Journal of Chemical Ecology* 5:795–803.

Comins, H. N., and B. S. Fletcher. 1988. Simulation of fruit fly population-dynamics with particular reference to the olive fruit-fly, *Dacus oleae*. *Ecological Modelling* 40:213–231.

Dorji, C., A. R. Clarke, R. A. I. Drew et al. 2006. Seasonal phenology of *Bactrocera minax* (Diptera: Tephritidae) in Western Bhutan. *Bulletin of Entomological Research* 96:531–538.

Fletcher, B. S. 1968. Storage and release of a sex pheromone by Queensland fruit fly *Dacus tryoni* (Diptera: Tephritidae). *Nature* 219:631–632.

Fletcher, B. S. 1969. Structure and function of sex pheromone glands of male Queensland fruit fly, *Dacus tryoni*. *Journal of Insect Physiology* 15:1309.

Fletcher, B. S. 1973. The ecology of a natural population of the Queensland fruit fly, *Dacus tryoni*. IV. The immigration and emigration of adults. *Australian Journal of Zoology* 21:541–556.

Fletcher, B. S. 1974. The ecology of a natural population of the Queensland fruit fly, *Dacus tryoni*. V. The dispersal of adults. *Australian Journal of Zoology* 22:189–202.

Fletcher, B. S. 1974. The ecology of a natural population of the Queensland fruit fly, *Dacus tryoni*. VI. Seasonal changes in fruit fly numbers in the areas surrounding the orchard. *Australian Journal of Zoology* 22:353–363.

Fletcher, B. S. 1975. Temperature-regulated changes in ovaries of overwintering females of Queensland fruit fly, *Dacus tryoni*. *Australian Journal of Zoology* 23:91–102.

Fletcher, B. S. 1979. Overwintering survival of adults of the Queensland Fruit fly, *Dacus tryoni*, under natural conditions. *Australian Journal of Zoology* 27:403–411.

Fletcher, B. S. 1987. The biology of dacine fruit flies. *Annual Reviews of Entomology* 32:115–144.

Fletcher, B. S. 1989. Life history strategies of tephritid fruit flies. In *World Crop Pests. Fruit Flies: Their Biology, Natural Enemies and Their Control*, Vol. 3B, ed. A. S. Robinson and G. Hooper, pp. 195–208. Amsterdam, the Netherlands: Elsevier Science Publishers.

Fletcher, B. S. 1989. Movements of tephritid fruit flies. In *World Crop Pests. Fruit Flies: Their Biology, Natural Enemies and Their Control*, Vol. 3B, ed. A. S. Robinson and G. Hooper, pp. 209–219. Amsterdam, the Netherlands: Elsevier Science Publishers.

Fletcher, B. S. 1998. Dacine fruit flies collected during the dry season in the lowland rainforest of Madang Province, Papua New Guinea (Diptera: Tephritidae). *Australian Journal of Entomology* 37:315–318.

(Continued)

TABLE 2.1 (*Continued*)
List of Publications by Brian S. Fletcher on *Bactrocera* Ecology or Biology

Fletcher, B. S., M. A. Bateman, N. K. Hart et al. 1975. Identification of a fruit fly attractant in an Australian plant, *Zieria smithii*, as O-methyl eugenol. *Journal of Economic Entomology* 68:815–816.

Fletcher, B. S., and A. P. Economopoulos. 1976. Dispersal of normal and irradiated laboratory strains and wild strains of the olive fly *Dacus oleae* in an olive grove. *Entomologia Experimentalis et Applicata* 20:183–194.

Fletcher, B. S., and A. Giannakakis. 1973. Factors limiting response of females of Queensland fruit fly, *Dacus tryoni*, to sex pheromone of male. *Journal of Insect Physiology* 19:1147–1155.

Fletcher, B. S., and A. Giannakakis. 1973. Sex pheromone production in irradiated males of *Dacus (Strumeta) tryoni*. *Journal of Economic Entomology* 66:62–64.

Fletcher, B. S., and E. Kapatos. 1981. Dispersal of the olive fly, *Dacus oleae*, during the summer period on Corfu. *Entomologia Experimentalis et Applicata* 29:1–8.

Fletcher, B. S., E. Kapatos, and T. R. E. Southwood. 1981. A modification of the Lincoln index for estimating the population densities of mobile insects. *Ecological Entomology* 6:397–400.

Fletcher, B. S., and E. Kapatos. 1983. The influence of temperature, diet and olive fruits on the maturation rates of female olive flies at different times of the year. *Entomologia Experimentalis et Applicata* 33:244–252.

Fletcher, B. S., S. Pappas, and E. Kapatos. 1978. Changes in ovaries of olive flies *Dacus oleae* (*Gmelin*) during summer, and their relationship to temperature, humidity and fruit availability. *Ecological Entomology* 3:99–107.

Fletcher, B. S., and C. A. Watson. 1974. Ovipositional response of Tephritid fruit fly, *Dacus tryoni*, to 2-Chloro-ethanol in laboratory bioassays. *Annals of the Entomological Society of America* 67:21–23.

Fletcher, B. S., and G. Zervas. 1977. Acclimation of different strains of olive fly, *Dacus oleae*, to low temperatures. *Journal of Insect Physiology* 23:649–653.

Giannakakis, A., and B. S. Fletcher. 1974. Production and release of sex-pheromone in *Dacus tryoni* males sterilized with aziridine derivative HMAC. *Journal of Economic Entomology* 67:3–4.

Giannakakis, A., and B. S. Fletcher. 1978. Improved bioassay technique for sex-pheromone of male *Dacus tryoni* (Diptera: Tephritidae). *Canadian Entomologist* 110:125–129.

Giannakakis, A, and B. S. Fletcher. 1981. Ablation studies related to the location of the sex-pheromone receptors of the Queensland fruit fly, *Dacus tryoni* (Froggatt) (Diptera, Tephritidae). *Journal of the Australian Entomological Society* 20:9–12.

Giannakakis, A., and B. S. Fletcher. 1985. Morphology and distribution of antennal sensilla of *Dacus tryoni* (Froggatt) (Diptera: Tephritidae). *Journal of the Australian Entomological Society* 24:31–35.

Hendrichs, J., B. S. Fletcher., and R. J. Prokopy. 1993. Feeding behavior of *Rhagoletis pomonella* flies (Diptera: Tephritidae): Effect of initial food quantity and quality on food foraging, handling costs, and bubbling. *Journal of Insect Behavior* 6:43–64.

Kapatos, E., and B. S. Fletcher. 1983. Seasonal changes in the efficiency of McPhail traps and a model for estimating olive fly densities from trap catches using temperature data. *Entomologia Experimentalis et Applicata* 33:20–26.

Kapatos, E., B. S. Fletcher, S. Pappas, and Y. Laudeho. 1977. Release of *Opius concolor* and *Opius concolor* var. *siculus* (Hymenoptera: Braconidae) against spring generation of *Dacus oleae* (Diptera: Tephritidae) on Corfu. *Entomophaga* 22:265–270.

Kapatos, E. T., and B. S. Fletcher. 1984. The phenology of the olive fly, *Dacus oleae* (Gmel) (Diptera: Tephritidae), in Corfu. *Zeitschrift Fur Angewandte Entomologie. Journal of Applied Entomology* 97:360–370.

Kapatos, E. T., and B. S. Fletcher. 1986. Mortality factors and life-budgets for immature stages of the olive fly, *Dacus oleae* (Gmel) (Diptera: Tephritidae), in Corfu. *Zeitschrift Fur Angewandte Entomologie*. 102:326–342.

Prokopy, R. J., and B. S. Fletcher. 1987. The role of adult learning in the acceptance of host fruit for egg laying by the Queensland fruit fly, *Dacus tryoni*. *Entomologia Experimentalis et Applicata* 45:259–263.

Tychsen, P. H., and B. S. Fletcher. 1971. Studies on rhythm of mating in Queensland fruit fly, *Dacus tryoni*. *Journal of Insect Physiology* 17:2139–2156.

(2014) and Robinson and Hooper (1989), on fruit flies were searched in their entirety. Flight-distance records for *B. dorsalis* were collected from the scientific articles and tabulated.

The classification of the genus *Bactrocera* Macquart has been revised recently based on phylogenetic studies of the tribe Dacini (Doorenweerd et al. 2018). As a result, several species classified as *Bactrocera* in the searched literature of flight studies are now included in the genus *Zeugodacus*

Hendel. The revised names reported in Doorenweerd et al. (2018) are used here, but the *Bactrocera* and *Zeugodacus* species treated as one group when comparing trends to the older classification of *Bactrocera* in the literature search.

2.3 RESULTS AND DISCUSSION

2.3.1 "Many *Bactrocera* spp. Can Fly 50–100 Km" and Its Frequency on the World Wide Web

A Web search for the statement "Many *Bactrocera* spp. can fly 50–100 km (Fletcher 1989)" returned 66 results, which included links to databases, Websites, and trade-related phytosanitary documents. The results included 20 unique sources, and the remainder were duplicates from databases that had information on numerous species of *Bactrocera*. When the citation, "(Fletcher 1989)," was not included in the search, the results increased to 76.

Databases or Websites that contain the flight distance of 50–100 km and cite Fletcher (1989) include the CABI Invasive Species Compendium, EPPO Global Database, the Pests and Diseases Image Library (PaDIL) Plant Biosecurity Toolbox (http://www.padil.gov.au/), EcoPort (http://epf.ecoport.org/), and DiscoverLife (http://www.discoverlife.org/). The CABI Invasive Species Compendium is a comprehensive database with datasheets on 42 different *Bactrocera* spp., and 10 of these datasheets contained the identical statement "Many *Bactrocera* spp. can fly 50–100 km (Fletcher 1989)." In addition to the online databases, phytosanitary documents from Australia (Plant Health Australia 2010), Iran (Bureau of Plant Pest Surveillance and Pest Risk Analysis 2013), Kenya (KEPHIS n.d.), and Malaysia (CAB International Southeast Asia 2013) were listed in the search results.

2.3.2 Review of Scientific Publications on *Bactrocera* Movement by Brian S. Fletcher

Fletcher's 1989 book chapter "Life History Strategies of Tephritid Fruit Flies" (Fletcher 1989a) is often referenced in databases as the source for the statement "Many *Bactrocera* spp. can fly 50–100 km." However, there is no evidence in "Life History Strategies of Tephritid Fruit Flies" (Fletcher 1989a) that supports the flight range of 50–100 km. The closest reference to the flight distance of *Bactrocera* spp. in this chapter is a general description that polyphagous, multivoltine tephritids have "high mobility" and a "high capacity for dispersal" without any quantification for these descriptors.

Although Fletcher, an expert on fruit fly biology, never implied that *Bactrocera* spp. have a flight range of 50–100 km in any of his 35 articles (Table 2.1), the article "Movement of Tephritid Fruit Flies" (Fletcher 1989b) is most likely the intended citation for the "50–100 km" conclusion found in biological databases for three reasons. First, "Movement of Tephritid Fruit Flies" is found in the same book, *World Crop Pests: Fruit Flies: Their Biology, Natural Enemies and Control* (Robinson and Hooper 1989), as the often miscited chapter "Life History Strategies of Tephritid Fruit Flies." Given that both chapters are in the same book and written by Fletcher, the misidentification of chapter title is plausible. Second, some resources including (EPPO 2018) cite "Movement of Tephritid Fruit Flies" (Fletcher 1989b) as the source for the "50–100 km" flight capability statement. Third, this chapter by Fletcher includes dispersal information for several species of *Bactrocera* along with some quantitative information. Although our reasoning finds the "Movement of Tephritid Fruit Flies" (Fletcher 1989b) as the most logical source in support of the statement "Many *Bactrocera* spp. can fly 50–100 km," we conclude that this chapter lacks sufficient documentation to indicate that in fact "many *Bactrocera* spp. can fly 50–100 km" using the following rationale:

2.3.2.1 Flight Data for Many *Bactrocera* Species Are Not Found in the Chapter

Foremost, the phrase "many *Bactrocera* spp." does not accurately reflect the data presented in the review "Movement of Tephritid Fruit Flies" (Fletcher 1989b). There are 657 described species in the genera *Bactrocera* and *Zeugodacus* (Doorenweerd et al. 2018), and the book chapter by Fletcher (1989b) provides dispersal information on exactly 7 species and only 3 of the 7 had information indicating that their dispersal distances were 50 km or greater (Table 2.2). Flight data are not available for many species because research is conducted most often on the economically important *Bactrocera* spp. (Aluja 1993). Furthermore, movement by fruit flies is influenced in part by their life history

TABLE 2.2
Movements of *Bactrocera* spp. as Described in "Movements of Tephritid Fruit Flies"

Species Common Name	Movement Recorded	Sex and Numbers Trapped	Citation Used by Fletcher
Z. cucurbitae[1] (Coquillett)	Left field; some traveled long distances	Some adults	Nishida and Bess (1957)
Melon fly	Move from host to surrounding vegetation before nightfall with diurnal pattern of movement	Mature females	
	Up to 65 km in Mariana Islands	Sterile marked	Steiner et al. (1962)
	34–64 km away on adjacent islands	Small number of marked males	Kawai et al. (1978)
	200 km away on Okinoerabi Island	1 sterile male	Miyahara and Kawai (1979)
	Less than 0.2 km on average	Released mature males 2–3 weeks old	Hamada (1980), Nakamori and Soemori (1981) Soemori and Kuba (1983)
Z. diversus[1] (Coquillett) Three striped fruit fly	Seek sheltered refuges – flight distance not provided	Adults	Syed (1968)
B. dorsalis (Hendel)	Up to 65 km in Mariana Islands	Sterile marked	Steiner et al. (1962)
Oriental fruit fly	"Must have flown at least 50 km, mostly over open ocean"	9 marked males	Iwahashi (1972)
	Considerable amount, between islands	Marked males	
	Moved toward host trees	Sterile adults	Yao et al. (1977)
	0.6 km mean 0.33 km	Sterile adults	Chiu (1983)
	2 km mean 0.94 km	Some sterile adults	
	Moved toward host trees	Sterile adults	Yao et al. (1977)
B. oleae (Rossi) Olive fruit fly	0.017–0.018 km, mean dispersal rate	Very few flies left the grove, lab reared, wild males and females	Fletcher and Economopoulos (1976)
	4 km	Small number of males and females	Economopoulos et al. (1978)
	10 km	Small number of released flies	Brnetic (1981)
	Movement increases in the absence of hosts	Adults	Michelakis and Neuenschwander (1981)
	up to 0.02 km	Males and females	Katsoyannos (1983)

(*Continued*)

TABLE 2.2 (*Continued*)
Movements of *Bactrocera* spp. as Described in "Movements of Tephritid Fruit Flies"

Species Common Name	Movement Recorded	Sex and Numbers Trapped	Citation Used by Fletcher
Z. scutellaris[1] (Bezzi) Cucurbit fruit fly	Seek sheltered refuges – flight distance not provided	Adults	Syed (1968)
B. tryoni (Froggatt) Queensland fruit fly	Remain on and around hosts (non-dispersive)	Mature adults	Sonleitner and Bateman (1963), Bateman and Sonleitner (1967)
	Distance not provided	75% of released males emigrated	Fletcher (1973)
	Estimated 3–4 km in 2 to 3 weeks	Male flies	Fletcher (1974)
	12–13 km overall	Male flies	
	24 km	Few marked males	
	"Circumstantial evidence suggested" migration of 25–35 km	Some gravid females	
	Distance not provided	Overwintering adults	Fletcher (1979) and unpublished data
	High rate	Males	Drew and Hooper (1983)
	1.5 km	Most released males	MacFarlane et al. (1987)
	80–94 km	Some males	
B. zonata (Saunders) Peach fruit fly	40 km	Few sterile males	Qureshi et al. (1975)
	Considerable amount of dispersal occurred	Sterile males	

[1] Recent phylogenetic work places this species in the genus *Zeugodacus*. (Doorenweerd et al. 2018)

strategies (univoltine vs. multivoltine, monophagous vs. polyphagous), intrinsic capabilities (flight capacity, polymorphism), physiology (age, nutrition), and sex and body traits (wing shape, size) (Aluja 1993), which makes it difficult to generalize flight capabilities among different species of *Bactrocera*.

2.3.2.2 Case Studies on *Bactrocera* spp. Lack Necessary Quantifiable Dispersal Data

In "Movement of Tephritid Fruit Flies," Fletcher (1989b) includes flight information from field trials, mainly mark-release-recapture experiments of *Bactrocera* species. The pertinent data from these case studies, which include both qualitative and quantitative records, are summarized in Table 2.2. Other genera of fruit flies are discussed in the chapter but are not relevant to this analysis and are therefore not included.

The 22 case studies as described in the chapter include 31 observations (Table 2.2, column 2) on the movements of *Bactrocera* spp. in mark-release-recapture or field studies. Almost 40% of these comments do not provide a calculated distance that the flies traveled; instead, the conclusion are generalizations such as "seeked sheltered refuges," "some adults traveled long distances," or "moved toward host trees." These vague comments emphasize long flight distances over shorter distances and leave the reader to define the term "long."

Quantitative flight distances for *Bactrocera* spp. are provided for 13 case studies (Table 2.2). Five of the 13 studies resulted in recaptures of one to nine flies at distances that ranged from 50 to 200 km. Fletcher does not provide the number of flies that are captured at distances less than

50 km but indicates that more flies are captured at distances less than 4 km by using descriptive terms such as "most males," "male flies," or "very few left the grove." The lack of case study details presented in this chapter have led to different interpretation of the results. For example, when citing "Movement of Tephritid Fruit Flies" (Fletcher 1989b), Peck et al. (2005) came to the conclusion, "Many studies on *Bactrocera* have reported that these flies do not move far," whereas the EPPO Global Database (EPPO 2018) states nearly the opposite: "Many *Bactrocera* spp. can fly 50–100 km." An article published in the EFSA Journal (European Food Safety Authority 2007) addressing the pest risk of *B. zonata* (Saunders) clarifies the findings in the chapter by stating, "Although many *Bactrocera* spp. can fly 50–100 km (Fletcher 1989), the maximum reported for *B. zonata* is 40 km (Qureshi et al. 1975)." Overall, Fletcher (1989b) provides a good historical review of fruit fly movement, but the evidence from the case studies is insufficient to speculate generally about *Bactrocera* spp. flight capacity. Most studies simply do not place traps out more than a few kilometers from the release sites because of increased need for resources and diminished likelihood of captures.

2.3.3 Additional Published Reviews on Fruit Fly Movement

Historically, the majority of scientific publications on the movement of fruit flies focus on the species in the tribe Dacini that are of economic importance, especially *Zeugodacus cucurbitae* (Coquillett), *B. dorsalis*, *Bactrocera oleae* (Rossi), and *Bactrocera tryoni* (Froggatt) (Aluja 1993). Published reviews on tephritid fruit fly biology that, in whole or part, discuss fruit fly movements date back to Christenson and Foote (1960) and include articles by Aluja (1993); Bateman (1972, 1977); Díaz-Fleischer and Aluja (2000); Dominiak (2012); Fletcher (1987, 1989b); Prokopy and Roitberg (1984); and Zwolfer (1983).

Weldon et al. (2014) provide an excellent review of tephritid movement. They redefine types of movement, discuss dispersal and its implications for pest management, and include mark-release-recapture studies, molecular methods, and remote sensing as tools for measuring movement. Key results from published recapture studies on 12 tephritid species are summarized in a table and evaluated in depth. The consolidation of the recapture data required a significant effort by Weldon et al. (2014), and the data are a useful resource when comparing movement among fruit fly species.

Although mark-and-recapture studies have some limitations, they are the most practical means for studying movement, especially the long-distance movement of organisms (Southwood and Henderson 2000). In the future, new tracking technologies could provide more accurate estimates of long-distance dispersal (Nathan et al. 2003). Trap array design and the strength of the trap attractant can influence recapture rates (Weldon et al. 2014). The maximum flight distance studied can be limited by the costs associated with setting up, maintaining, and checking the trap array. Overcrowding because of large numbers of released flies in a small area may result in a higher occurrence of long-distance dispersal. Despite the limitations, the trapping results from 38 recapture studies on tephritid fruit flies that are reviewed by Weldon et al. (2014) indicate that some species are more mobile than others and that long-distance recapture is a rare event. Mean dispersal distance of tephritid flies is usually well below 1 km (Weldon et al. 2014).

The recapture results indicate that the most mobile economically important fruit fly species are in the genus *Bactrocera* (Weldon et al. 2014). Among the *Bactrocera* species studied, *B. oleae*, the olive fruit fly, is the least mobile. *B. oleae* adults from the first two generations emerge during a time when olive fruit is ripe and abundant (Fletcher 1989a). When suitable fruit is available, dispersal is low. In a recapture study with eight replications, *B. oleae* adults moved a distance of only 0.019 to 0.068 km on average (Weldon et al. 2014). Other studies not included in Weldon et al. (2014) also indicate that *B. oleae* does not move far. Laboratory-reared olive fruit flies took 12 to 14 days to travel a mean distance of 0.018 to 0.020 km (Fletcher and Economopoulos 1976). When wild and artificially reared *B. oleae* flies were released outside an olive grove, fewer than 15 flies moved 2 km to reach the grove, and just a few flies were trapped 4 km away (Economopoulos et al. 1978). During a 4-year study, an average of 99.45% of the recaptured flies were trapped within 1 km and only a few flies were found 10 km away (Brnetic 1981).

In contrast, other studied species have ranges beyond 10 km. *Zeugodacus cucurbitae*, *B. dorsalis*, and *B. tryoni* male adults were captured at 56, 50, and 94 km, respectively, from their release sites (Weldon et al. 2014). It is not possible to tell from these studies how common such dispersal distances were given the fact that relatively few of the total number of released flies were recaptured at any distance (0.3% or less when recorded) and the fact that, by necessity, trap density decreased the further out from the release site. In the study where a single *B. tryoni* fly was captured at 94 km, the mean distance for all captured flies (approximately 800 of the 400,000 released), was 1.1 km (MacFarlane et al. 1987; Weldon et al. 2014). Although most recapture studies do not report mean distance traveled, it is apparent that dispersive movements can vary greatly among species of dacine fruit flies.

2.3.4 Flight Capacity of *Bactrocera dorsalis*

Shortly after *Bactrocera dorsalis* was discovered in Hawaii, research on this invasive pest became a top priority to the Hawaiian islands and to the mainland United States. Research began on biological and chemical control, area-wide control, ecology, biology, and commodity treatments (Carter 1950). Reinfestation of areas by *B. dorsalis* was perceived as a problem; therefore, data were collected on the movements of Oriental fruit flies during area-wide control projects. The published results from these initial studies, which include the earliest knowledge of movements by *B. dorsalis* in Hawaii and subsequent research on the flight capabilities of *B. dorsalis*, are summarized in Table 2.3. There are 17 primary source publications (Table 2.3). Published flight distances from field studies on *B. dorsalis* can be grouped into two broad categories:

1. Observations reported without experimental methods
2. Experimental results, including scientific methods

Seven out of the 17 publications listed in Table 2.3 do not provide methodology to demonstrate how the flight data were collected. If materials and methods are furnished in a listed publication, details for the release sites are given. In addition, environmental factors that are considered to affect flight, including topography, host availability, and wind currents, are noted.

From their inception, the reported results on the flight of *B. dorsalis* emphasized the movement of longer distances by a few flies. For example, Carter (1950) does not include experimental methods for the ecology and control studies in Hawaii and summarizes the results without population densities and seasonal population trends. In his article, Carter (1950) states,

> Data were therefore obtained on fly movements into and out of the study areas, and from this data detailed studies were made of population densities and seasonal fluctuations. During these studies marked male flies have been recovered 20 miles [32.2 km] from their original point of liberation. Other data on fly movement acquired throughout the investigations have confirmed the migratory habit of the fly or at least its great capacity for dispersal. It has crossed an ocean strait 9 miles [14.5 km] wide; it evidently moves back and forth over each island and possibly over more than one island; and it can be carried on the outside of fast-moving vehicles for long distances.

Four subsequent articles (Christenson and Foote 1960; Porter and Christenson 1960; Steiner 1957; Steiner et al. 1962) also generalize findings on flight capacity of *B. dorsalis* as observed in the Hawaiian islands (Table 2.3). These articles recorded flight distances spanning 6 to 42 km. Two articles include the numbers trapped, with just 1 fly at 38.6 km and 133 flies at 15 to 17 km. Steiner et al. (1962) combine capture data for three fruit fly species and report distances of 19–72 km. Unfortunately, detailed information regarding the total number of flies that were released and recaptured at different distances from the release sites during these studies is not indicated. Certainly, some flies must have been captured closer to the release sites than 19 km, and excluding this data can

TABLE 2.3
Field Research on the Flight Capability of *Bactrocera dorsalis*

Distance (km)	Number Trapped (% trapped)†	Release Area and Trapping Details and Reference
14.5	Not provided	Cited as "It has crossed an ocean strait 9 miles wide; it evidently moves back and forth over each island and possibly over more than one island."
32.1	Marked male flies	Cited as "recovered 20 miles from their original point of liberation." No materials and methods provided for study. Carter (1950)
38.6	1	Cited as "One marked male has been recovered 24 miles from its release point, and many others have been taken in methyl eugenol traps far removed from any known breeding sites-even at an elevation of 7600 feet on Mauna Loa volcano." No materials and methods provided for study. Steiner (1957)
6–24	Not provided	Cited as "The males of *D. dorsalis* have frequently been found to travel 4 to 15 mi. from the point of their release and to cross 9 mi. of open sea between islands. Occasionally this species may fly several miles within a few days (United States Department of Agriculture, unpublished data)." Christenson and Foote (1960)
42	Not provided	Cited as "Marked Oriental fruit flies have been retaken in traps as far as 26 miles from a liberation point, having crossed at least
15 over water	Not provided	9 miles of water in the course of their flight. In one recent experiment, 133 Oriental fruit flies were recovered in traps
15–17	133	located 9 to 11 miles from the release site." Porter and Christenson (1960)
16 (256 km²)	Not provided	Cited as "Oriental fruit flies from a single release site have spread over 100 square miles of surrounding mountain and coastal areas" (Hawaii–no materials and methods).
40–72 after emergence 19–64 over water	Not provided	Cited as "These flies have moved distances ranging from 25 to 45 miles from their point of emergence and made sustained overwater flights of 12–40 miles" (collectively referring to *B. dorsalis, B. cucurbitae* and *Ceratitis capitata* (Wiedemann). No materials and methods provided.
Not measured		Aerial drop boxes of marked sterile flies ½ mile [0.8 km] apart over length of Rota Island. Releases averaging 6 million Oriental fruit flies weekly. "It is apparent from the results that a predominant downward drift of flies, both sterile and wild, is taking place." Steiner et al. (1962)
2–5	202 (87%)	A total of 4,831 marked males released on 4 different islets in the Ogasawara Islands, Japan. Islets in China Jima group are 2 km
7–8	26 (11%)	apart. All traps set at least 2 km from the release sites. Host availability varied among release sites. Overall recapture rate
15–16	5 (2%)	was 4.8%.
50*	9 (0.3%)	3,000 flies were released on the islet Haha Jima, approximately 50 km from Chichi Jima. *3 months later, 9 flies were caught in 3 traps; flies were apparently aided by prevailing winds. Iwahashi (1972)

(Continued)

TABLE 2.3 (*Continued*)
Field Research on the Flight Capability of *Bactrocera dorsalis*

Distance (km)	Number Trapped (% trapped)†	Release Area and Trapping Details and Reference
0.2–2.8	Not provided	Taiwan Yao et al. (1977) [cited in Chiu (1983)]
0.60 avg. 0.33	Not provided	Release site in host plant areas. Traps set in 4 directions at 150, 300, 450 and 600 m. Note: Furthest trap set 0.60 km from release site.
2.0 avg. 0.94		Release site in non-host plant areas 0.5, 1 m, 1.5 and 2 km Note: Furthest trap set 2 km from release site. Chiu (1983)
0.5	16 (4 - N, 12 - S)	Wild flies marked and released in town with hosts plants present. Traps placed to north, south and west of release sites. Traps 1 km from town did not trap any marked flies.
1	None	Tan (1985)
Within villages	Avg of 7.4%–11.9% of released flies	Released total of 6,838 marked wild flies in 2 villages, Batu Uban and Sungei Dua. Both village had host plants, but differed in kind and quantity. Villages were 1 km apart.
Outside villages	1	Tan and Jaal (1986)
Within respective ecosystem	Most marked flies	Three different ecosystems (each 2.25 ha) about 1 km apart. Most wild marked flies stayed within each ecosystem.
1 km	37 (0.20%)	Only 0.20% of flies emigrated (based on 18,624 total released flies). 13 out of the 37 flies moved to forest ecosystem, where host plants were scarce. Tan and Serit (1988)
27 max	Few males	50,000 sterilized males were released in a favorable habitat in western Taiwan. A few flies were captured on Lambay Island. During releases on Lambay Island (6.8 km^2), *B. dorsalis* flies moved to more favorable habitats within the island Chu and Chiu (1989).
13	1 marked	Cited as "observed a marked *B. dorsalis* traverse an upwind distance of 13 km in a single 24-h period in response to a methyl eugenol–baited trap (unpublished data)" Peck et al. (2005)
0.2	Most marked flies	Studied diurnal movement in orchard. Released flies moved from guava orchards to other fruit orchards up to 0.2 km away. Chen et al. (2006)
13–34**	30	All marked flies released from 1 site. Numbers released is not apparent. Set traps at 4 cardinal directions. No marked flies
63–82**	8	were trapped to the east or west due obstruction by mountain ranges.
97**	5	**Under suitable climatic conditions, longer movement inside the Nujiang valley from south to north could be attributed to southern air currents.
8–15 max	17	Captured at southern sites. Chen et al. (2007)

(*Continued*)

TABLE 2.3 (*Continued*)
Field Research on the Flight Capability of *Bactrocera dorsalis*

Distance (km)	Number Trapped (% trapped)†	Release Area and Trapping Details and Reference
<0.5	571 (30%)	In Hawaii, 217,560 total sterile male fliers were released at 4 different sites located 5 km (release 1), 8 km (release 2), 10 km (release 3) and 2 km (release 4) from a 51 km^2 study grid with 2 traps/ km^2. Five days after the 4th release, traps were placed within 0.1 and 0.5 km of the release site. Recapture rates are as follows:
0.5–2	1310 (68%)	1st release - 0.0005%
>2–5	6	2nd release – 0.0%
> 5–10	7	3rd release – 0.10%
>10–11.7	17 (0.8%)	4th release – 0.98%
		Host fruits were found around all 4 release sites, but were distributed heterogeneously. The study grid contained agricultural crops, secondary vegetation or native/disturbed forests with abundant preferred host plants. Wind speed and direction patterns were consistent.
		Froerer et al. (2010)
0.3 or less	1820 (71.1%)	Released 14-day-old sterilized males in 6 grids, each 2.6 km^2 with hosts available. Flies were released in 4 compass directions at 25, 50, 75, 100, 200 and 300 m away from a central trap within the grid.
Within grid	316 (12.4%)	The 6 grids were 3.2 km or more apart in Orange County, CA, and within the *Bactrocera* detection program that spans 6,400 km^2 in the Los Angeles area.
Adjoining grid	304 (11.9%)	Overall 2,558 flies were recaptured (22.9%).
1.6–19.2 in detached grids	96 (3.7%)	Shelly et al. (2010)
in Orange Co.	22 (0.9%)	
<19 in LA Co. and San Bernadino Co. Exact distances not measured	None	
>19.2		

† Percent tapped = (numbers trapped within specified distance/ total trapped)*100

unintentionally mislead readers in regard to which are the expected dispersal patterns for these species, especially if most of the flies were captured close to release sites. During the 1950s and 1960s, discovery of efficacious control methods was of utmost importance; thus, results from research on flight activity, although directly related to control, were rarely published in their entirety. Thus, specific questions about the flight habits of *B. dorsalis*, such as the timing of captures, the likelihood of multiple flights, and the occurrence of short versus long flights, were not recorded.

Mark-release-recapture studies are labor and financially intensive. Therefore, there are more investigations on movement of fruit flies within a habitat, a grove, or between closely aligned fields than between islands or from urban to rural agricultural production areas that involve monitoring long distances. *Bactrocera dorsalis* has been the subject of multiple experiments monitoring movement in host-rich habitats versus poor host or nonhost environments (See Table 2.3: Chen, et al. 2006; Chiu 1983; Chu and Chiu 1989; Tan 1985; Tan and Jaal 1986; Tan and Serit 1988). Although these studies concluded that released *B. dorsalis* adults stay within host-rich habitats, traps beyond 2 km were not monitored.

Osamu Iwahashi (1972) was the first researcher to design a recapture experiment to specifically measure the long-distance movement (2 km or greater) of *B. dorsalis*. Iwahashi demonstrated that marked males could move from one islet to another in the Ogasawara Islands; each islet is about 2 km apart. A majority, 87%, of the flies were captured 2–5 km away from four release sites

(Table 2.3). Thirteen percent of captured flies were trapped at distances from 7–16 km. Differences were found among recapture rates for the four releases. It was hypothesized that the greatest recapture rates were associated with the lack of suitable hosts at the release sites; however, 43 flies moved from favorable habitat conditions to less-ideal conditions.

A fifth release took place on the distant island of Haha Jima. Nine flies were recaptured three months later on another island 50 km away. This recapture study receives the most attention in the literature. Multiple authors cite Iwahashi's discovery of a 50-km flight but do not mention the possibility of prevailing winds influencing the results (Chen et al. 2006; Fletcher 1989b; MacFarlane et al. 1987; Shelly et al. 2010). Findings from the 1972 study by Iwahashi in the Ogasawara Islands is controversial because the study did not differentiate between active and passive transport (Froerer et al. 2010); however, for the release at Haha Jima, Iwahashi (1972) suggests that prevailing winds were likely a factor in the dispersal of the nine flies that were trapped 50 km away.

In addition to the 1972 paper by Iwahashi, there are three other published articles on recapture studies that investigated the long distance movement of *B. dorsalis* (Chen et al. 2007; Chu and Chiu 1989; Froerer et al. 2010). Research on the capture probability of *B. dorsalis* in the southern California grid system conducted by T. Shelly et al. (2010) provides additional trap data. Summaries for these four publications are found in Table 2.3. All four studies report trap captures at distances greater than 10 km. The four releases occurred in different climatic and geographical environments, but host plants were available at all sites. In two environments, the movement of *B. dorsalis* was likely influenced by wind, as noted by the authors (Chen et al. 2007; Chu and Chiu 1989). In one study, out of 50,000 released flies, a few male flies moved from mainland Taiwan to Lambay Inlet, a maximum of 27 km away (Chu and Chiu 1989). The farthest distance traveled by *B. dorsalis* was recorded in the Nujiang Valley, where five flies were captured 97 km away when aided by southerly winds (Chen et al. 2007). In contrast, in Hawaii and southern California releases when wind was not likely a factor, *B. dorsalis* were trapped at a maximum distance of 11.7 km (Froerer et al. 2010) and 19.2 km, respectively, from their release point (Shelly et al. 2010). Results from these release studies (Chen et al. 2007; Chu and Chiu 1989; Froerer et al. 2010; Iwahashi 1972; Shelly et al. 2010) suggest that wind is a factor in the dispersal of *B. dorsalis* flies and that trap recaptures at distances greater than 20 km are not typical (less than 1% of all flies captured). Dispersion distances of 2 km are common (Froerer et al. 2010).

2.4 CONCLUSIONS

Biological information in invasive species databases provides data for integrated pest management (IPM) and policy decisions for many countries. Data errors associated with an invasive species can cause unnecessary disagreements or confusion among farmers, policy makers, and international trade partners. The statement "Many *Bactrocera* spp. can fly 50–100 km (Fletcher 1989)," which is found in multiple databases and phytosanitary documents, has been presented as evidence to extend the radii of quarantine areas for new detections of *B. dorsalis* in California from 7.5 to 150 km. The tabulated summary of scientific literature on the movements of *Bactrocera* spp. contained herein provides evidence that the statement does not accurately represent the material presented by Fletcher in his 1989 publications (Fletcher 1989a, 1989b) or any of his publications on *Bactrocera* biology or ecology (Table 2.1). Initially, long-distance movement was emphasized in research findings about fruit flies. Generalizations about flight distances from field observations and recapture studies have led readers to conclude that long-distance flights of 50 km or greater are common occurrences. In fact, few *Bactrocera* species have actually been studied regarding their dispersal capabilities and patterns, and for those that have been studied, dispersal distances more than 50 km, although it does occur, is atypical.

Results from recapture studies can support decision makers in area-wide control, survey, and quarantine activities. Interpreting the results from recapture studies can be challenging because of complex interactions of abiotic and biotic forces on fruit fly flight. Most published results from recapture studies of fruit flies include the maximum dispersal distance, but do not communicate the mean dispersal distance and variance, which could be important in defining quarantine areas (Weldon et al. 2014). Currently, there are no universal standards among trading partners for quarantine distances based solely on the biology or dispersal habits of a fruit fly species. Dominiak (2012) points out that quarantine distances imposed for *B. tryoni* varies among trading countries and a consensus based on scientific principles is needed to harmonize trade.

Like *B. tryoni*, there is not a universal quarantine size area defined for *B. dorsalis*. Quarantine radii up to 150 km have been imposed based on the statement "Many *Bactrocera* spp. can fly 50–100 km (Fletcher 1989a, 1989b)" found in invasive species databases. This review of scientific publications on *B. dorsalis* reported distances of 0.2–97 km; however very few *B. dorsalis* flies moved beyond 20 km during the duration of recapture studies and the majority of released Oriental fruit flies moved 2 km or less. Outbreaks of *B. dorsalis* in California and Florida have been successfully contained with a quarantine radius of 7.4 km for more than 30 years. Therefore, quarantine distances and trade regulations for *B. dorsalis* (and other invasive species) should be based on the overall behavior of the species as well as the effectiveness of the trapping grid that is in place and the level of preparedness a trading partner is to implement eradication tactics. Although, long-distance dispersal for *B. dorsalis* is atypical, the longer it takes to detect an outbreak and initiate control actions, the more likely it may occur.

ACKNOWLEDGMENTS

This material was made possible, in part, by the Technical Assistance for Specialty Crops (TASC) Grant from the US Department of Agriculture's Foreign Agricultural Service (FAS). The findings and conclusions in this preliminary publication have not been formally disseminated by the US Department of Agriculture and should not be construed to represent any agency determination or policy. We are grateful to John Stewart, James W. Smith, Walter A. Gutierrez and Norman B. Barr with USDA–APHIS–PPQ–CPHST for their valuable comments on this manuscript.

REFERENCES

Aluja, M. 1993. The study of movement in Tephritid flies: Review of concepts and recent advances. In *Fruit Flies: Biology and Management*, ed. M. Aluja, and P. Liedo, pp. 105–113. New York: Springer-Verlag.

Bateman, M. A. 1972. The ecology of fruit flies. *Annual Review of Entomology* 17:493–518.

Bateman, M. A. 1977. Dispersal and species interaction as factors in the establishment and success of tropical fruit flies in new areas. *Paper presented at the Proceedings of the Ecological Society of Australia*. The Society Canberra, Australia.

Bateman, M. A., and F. J. Sonleitner. 1967. The ecology of a natural population of the Queensland fruit fly, *Dacus tryoni*, I. The parameters of the pupal and adult populations during a single season. *Australian Journal of Zoology* 15:303–335.

Brnetic, D. 1981. Biological control of olive fly by means of sterile male technique and by *Opius concolor*. Report of Project YO-ARS-9-JBS, Institut za Jadranske Kulture I Melioraciju KRSA, Split, Croatia.

Bureau of Plant Pest Surveillance and Pest Risk Analysis. 2013. A guide for diagnosis and detection of quarantine pests: Oriental fruit fly. Islamic Republic of Iran, Ministry of Jihad-e-Agriculture Plant Protection Organization, Tehran, Iran.

CAB International Southeast Asia. 2013. Towards Improved Market Access for ASEAN Agricultural Commodities. Selangor, Malaysia. https://www.idrc.ca/en/project/toward-improved-market-access-asean-agricultural-commodities. Accessed April 24, 2017.

California Department of Food and Agriculture (CDFA). 2008. Oriental fruit fly factsheet. https://www.cdfa.ca.gov/plant/factsheets/OFF_FactSheet.pdf. Accessed November 2, 2016.

Carter, W. 1950. The Oriental fruit fly: Progress on research. *Journal of Economic Entomology* 43:677–683.

Chen, C.-C., Y.-J. Dong, C.-T. Li et al. 2006. Movement of the Oriental fruit fly, *Bactrocera dorsalis* (Hendel) (Diptera: Tephritidae), in a guava orchard with special reference to its population changes. *Formosan Entomology* 26:143–159.

Chen, P., H. Ye, and Q.-A. Mu. 2007. Migration and dispersal of the Oriental fruit fly, *Bactrocera dorsalis* in regions of Nujiang River based on fluorescence mark. *Acta Ecologica Sinica* 27:2468–2476.

Chiu, H. 1983. Movements of Oriental fruit flies in the field. *Chinese Journal of Entomology* 3:93–102.

Christenson, L. D., and R. H. Foote. 1960. Biology of fruit flies. *Annual Review of Entomology* 5:171–192.

Chu, Y. I., and H. T. Chiu. 1989. The re-establishment of *Dacus dorsalis* Hendel (Diptera: Tephitidae) after the eradication on Lambay Island. *Chinese Journal of Entomology* 9: 217–230.

Díaz-Fleischer, F., and M. Aluja. 2000. Behavior of tephritid flies: A historical perspective. In *Fruit Flies (Tephritidae): Phylogeny and Evolution of Behavior*, ed. M. Aluja and A. Norrbom, pp. 39–68. Boca Raton, FL: CRC Press.

Dominiak, B. C. 2012. Review of dispersal, survival, and establishment of *Bactrocera tryoni* (Diptera: Tephritidae) for quarantine purposes. *Annals of the Entomological Society of America* 105: 434–446.

Doorenweerd, C., L. Lublanc, A. L. Norrbom et al. 2018. A global checklist of the 932 fruit fly species in the tribe Dacini (Diptera, Tephritidae). *ZooKeys* 730:19–56.

Drew, R. A. I., and G. H. S. Hooper. 1983. Population studies of fruit flies (Diptera: Tephritidae) in south-east Queensland. *Oecologia* 56:153–159.

Economopoulos, A. P., G. E. Haniotakis, J. Mathioudis, N. Missis, and P. Kinigakis. 1978. Long-distance flight of wild and artificially-reared *Dacus oleae* (Gmelin) (Diptera: Tephritidae). *Zeitschrift für Angewandte Entomologie* 87:101–108.

Emig, C. C., M. A. Bitner, and F. Alvarez. 2015. Scientific death-knell of databases? Errors induced by database manipulations and its consequences. *Carnets De Geologie* 15:231–238.

EPPO 2018. Data sheets on quarantine pests: *Bactrocera dorsalis*. European Plant Protection Organization, EPPO Global Database. https://gd.eppo.int https://gd.eppo.int. Accessed April 4, 2018.

European Food Safety Authority (EFSA). 2007. Opinion of the scientific panel on plant health on a request from the commission on pest risk assessment made by Spain on *Bactrocera zonata*. *The EFSA Journal* 5:1–25.

Fletcher, B. S. 1973. The ecology of a natural population of the Queensland fruit fly, *Dacus tryoni*. IV. The immigration and emigration of adults. *Australian Journal of Zoology* 21:541–556.

Fletcher, B. S. 1974. The ecology of a natural population of the Queensland fruit fly, *Dacus tryoni*. V. The dispersal of adults. *Australian Journal of Zoology* 22:189–202.

Fletcher, B. S. 1979. Overwintering survival of adults of the Queensland fruit fly, *Dacus tryoni*, under natural conditions. *Australian Journal of Zoology* 27:403–411.

Fletcher, B. S. 1987. The biology of dacine fruit flies. *Annual Review of Entomology* 32: 115–144.

Fletcher, B. S. 1989a. Life history strategies of tephritid fruit flies. In *World Crop Pests. Fruit Flies: Their Biology, Natural Enemies and Their Control*, ed. A. S. Robinson and G. Hooper, Vol. 3B, pp. 195–208. Amsterdam, the Netherlands: Elsevier Science Publishers.

Fletcher, B. S. 1989b. Movements of tephritid fruit flies. In *World Crop Pests: Fruit Flies—Their Biology, Natural Enemies and their Control*, ed. A. S. Robinson and G. Hooper, Vol. 3B, pp. 209–219. Amsterdam, the Netherlands: Elsevier Science Publishers.

Fletcher, B. S., and A. P. Economopoulos. 1976. Dispersal of normal and irradiated laboratory strains and wild strains of the olive fly *Dacus oleae* in an olive grove. *Entomologia Experimentalis et Applicata* 20:183–194.

FreshPlaza. 2015. Dominican Republic bans imports of certain fruit, veg to fight fruit fly. http://www.freshplaza.com/article/147640/Dominican-Republic-bans-imports-of-certain-fruit,-veg-to-fight-fruit-fly. Accessed November 19, 2016.

Froerer, K. M., S. L. Peck, G. T. McQuate et al. 2010. Long-distance movement of *Bactrocera dorsalis* (Diptera: Tephritidae) in Puna, Hawaii: How far can they go? *American Entomologist* 56:88–94.

Hamada, R. 1980. Studies on the dispersal behavior of melon flies, *Dacus cucurbitae* Coquillett (Diptera: Tephritidae), and the influence of gamma-irradiation on dispersal. *Applied Entomology and Zoology* 15:363–371.

Iwahashi, O. (1972). Movement of the oriental fruit fly adults among islets of the Ogasawara Islands. *Environmental Entomology* 1:176–179.

Jacobs, L. E. O., D. M. Richardson, B. J. Lepsch, and J. R. U. Wilson. 2017. Quantifying errors and omissions in alien species lists: The introduction status of *Melaleuca* species in South Africa as a case study. *NeoBiota* 32:89–105.

Katsanevakis, S., and H. E. Roy. 2015. Alien species related information systems and information management. *Management of Biological Invasions* 6:115–117.

Katsoyannos, B. J. 1983. Captures of *Ceratitis capitata* and *Dacus oleae* (Diptera: Tephritidae) by McPhail and Rebell color traps suspended on citrus, fig and olive trees on Chios, Greece. In *Fruit Flies of Economic Importance*, ed. R. Cavalloro, pp. 451–456. Rotterdam, the Netherlands: Balkema.

Kawai, A., O. Iwahashi, and Y. Ito. 1978. Movement of the sterilized melon fly from Kume Is. to adjacent islets. *Applied Entomology and Zoology* 13:314–315.

KEPHIS. n.d. Fact sheets of pests of phytosanitary significance to Kenya: *Bactrocera dorsalis*. http://www.kephis.org/.

Klein, B. D., D. L. Goodhue, and G. B Davis. 1997. Can humans detect errors in data? Impact of base rates, incentives, and goals. *MIS Quarterly* 21:169–194.

MacFarlane, J. R., R. W. East, R. A. I. Drew, and G. A. Betlinski. 1987. Dispersal of irradiated Queensland fruit flies, *Dacus tryoni* (Froggatt) (Diptera: Tephritidae), in southeastern Australia. *Australian Journal of Zoology* 35:275–281.

McGeoch, M. A., D. Spear, E. J. Kleynhans, and E. Marais. 2012. Uncertainty in invasive alien species listing. *Ecological Applications* 22:959–971.

Michelakis, S., and P. Neuenschwander. 1981. Etude des deplacements de la population imaginale de *Dacus oleae* (Gmel.) (Diptera: Tephritidae) en Crete, Greece. *Acta Oecologica Oecologia Applicata* 2:127–137.

Miyahara, Y., and A. Kawai. 1979. Movement of sterilized melon fly from Kume Is. to the Amami Islands. *Applied Entomology and Zoology* 144:496–497.

Nakamori, H., and H. Soemori. 1981. Comparison of dispersal ability and longevity of wild and mass-reared melon flies, *Dacus curcubitae* Coquillett (Diptera: Tephritidae), under field conditions. *Applied Entomology and Zoology* 16:321–327.

Nathan, R., G. Perry, J. T. Cronin, A. E. Strand, and M. L. Cain. 2003. Methods for estimating long-distance dispersal. *Oikos* 103:261–273.

Nishida, T., and H. A. Bess, 1957. Studies on the ecology and control of the melon fly *Dacus (Strumeta) cucurbitae. Hawaii Agricultural Experiment Station Technical Bulletin* 34:44.

Peck, S. L., G. T. McQuate, R. I. Vargas, D. C. Seager, H. C. Revis, and E. B. Jang. 2005. Movement of sterile male *Bactrocera cucurbitae* (Diptera: Tephritidae) in a Hawaiian agroecosystem. *Journal of Economic Entomology* 98:1539–1550.

Plant Health Australia. 2010. Industry Biosecurity Plan for the Apple and Pear Industry (Version 2.01). http:/ www.planthealthaustralia.com.au/. Accessed March 27, 2017.

Porter, B. A., and L. D. Christenson. 1960. A review of the fruit fly research program of the United States Department of Agriculture. *The Florida Entomologist* 43:163–169.

Prokopy, R. J., and B. D. Roitberg. 1984. Foraging behavior of true fruit flies: Concepts of foraging can be used to determine how tephritids search for food, mates, and egg-laying sites and to help control these pests. *American Scientist* 72:41–49.

Qureshi, Z. A., M. Ashraf, A. R. Bughio, and Q. H. Siddiqui. 1975. Population fluctuation and dispersal studies of the fruit fly, *Dacus zonatus* Saunders. Paper presented at the International Atomic Energy Agency; Food and Agriculture Organization. *Proceedings of the Symposium on the Sterility Principle for Insect Control*. IAEA and the FAO of the United Nations, Innsbruck, Austria.

Robinson, A. S., and G. Hooper, ed. 1989. *World Crop Pests. Fruit Flies: Their Biology, Natural Enemies and Control*, Vol. 3B. Amsterdam, the Netherlands: Elsevier.

Schutze, M. K., N. Aketarawong, N. W. Amornsak et al. 2015. Synonymization of key pest species within the *Bactrocera dorsalis* species complex (Diptera: Tephritidae): Taxonomic changes based on a review of 20 years of integrative morphological, molecular, cytogenetic, behavioural and chemoecological data. *Systematic Entomology* 40:456–471.

Shelly, T., J. Nishimoto, F. Díaz-Fleischer et al. 2010. Capture probability of released males of two *Bactrocera* species (Diptera: Tephritidae) in detection traps in California. *Journal of Economic Entomology* 103:2042–2051.

Shelly, T., N. Epsky, E. B Jang, J. Reyes-Flores, and R. Vargas. 2014. *Trapping and the Detection, Control, and Regulation of Tephritid Fruit Flies: Lures, Area-Wide Programs, and Trade Implications*. New York: Springer.

Soemori, H., and H. Kuba, H. 1983. Comparison of dispersal ability among two mass-reared and one wild strain of the melon fly, *Dacus cucurbitae* Coquillett (Diptera: Tephritidae), under the field conditions. *Bulletin of the Okinawa Agricultural Experiment Station* 8:37–41.

Sonleitner, F. J., and M. A. Bateman. 1963. Mark-recapture analysis of a population of Queensland fruit fly, *Dacus tryoni* in an orchard. *Journal of Animal Ecology* 32: 259–269.

Southwood, R., and P. A. Henderson. 2000. *Ecological Methods*, 3rd ed. Malden, MA: Blackwell Science.

Steiner, L. F. 1957. Field evaluation of oriental fruit fly insecticides in Hawaii. *Journal of Economic Entomology* 50:16–24.

Steiner, L. F., W. C. Mitchell, and A. H. Baumhover. 1962. Progress of fruit-fly control by irradiation sterilization in Hawaii and the Marianas Islands. *International Journal of Applied Radiation and Isotopes* 13:427–434.

Syed, R. A. 1969. Studies on the ecology of some important species of fruit flies and their natural enemies in West Pakistan. Pakistan Commonwealth Institute of Biology Control Station Report. Slough, UK: Commonwealth Agricultural Bureau.

Tan, K. H. 1985. Estimation of native populations of male *Dacus* spp. by Jolly's stochastic method using a new designed attractant trap in a village ecosystem. *Journal of Plant Protection in the Tropics* 2:87–95.

Tan, K. H., and M. Serit. 1988. Movements and population density comparisons of native male adult *Dacus dorsalis* and *Dacus umbrosus* (Diptera: Tephritidae) between three ecosystems in Tanjong Bungah, Penang. *Journal of Plant Protection in the Tropics* 5:17–21.

Tan, K. H., and Z. Jaal. 1986. Comparison of male adult population densities of the Oriental and *Artocarpus* fruit flies, *Dacus* spp. (Diptera: Tephritidae), in two nearby villages in Penang, Malaysia. *Researches on Population Ecology* 28:85–89.

Weems, H. V., J. B. Heppner, J. L. Nation, and G. J. Steck. 2016. Oriental fruit fly, *Bactrocera dorsalis* (Hendel) (Insecta: Diptera: Tephritidae) (EENY-08). University of Florida, IFAS Extension, Department of Entomology and Nematology. http://edis.ifas.ufl.edu/in240. Accessed April 17, 2017.

Weldon, C. W., M. K. Schutze, and M. Karsten. 2014. Trapping to monitor tephritid movement: Results, best practice, and assessment of alternatives. In *Trapping and the Detection, Control, and Regulation of Tephritid Fruit Flies: Lures, Area-Wide Programs, and Trade Implications*, ed. T. Shelly, N. Epsky, E. B. Jang, J. Reyes-Flores, and R. Vargas, pp. 175–217. New York: Springer.

Yao, A., Y. Hsu, and W. Lee. 1977. Moving abilities of sterile Oriental fruit fly (Diptera: Tephritidae). *National Science Council Monthly (Taiwan)* 5: 668–673 (In Chinese with English abstract).

Zwolfer, H. 1983. Life systems and strategies of resource exploitation in tephritids. In *Fruit Flies of Economic Importance*, ed. R. Cavalloro, pp. 16–30. Rotterdam, the Netherlands: Balkema.

3 Desiccation Resistance of Tephritid Flies

Recent Research Results and Future Directions

Christopher W. Weldon, Francisco Díaz-Fleischer, and Diana Pérez-Staples*

CONTENTS

Abstract The ability of organisms to withstand water stress is a fundamental determinant of their abundance and distribution. The ability to survive periods of water loss ("desiccation resistance") is also related to the ability of species to become invasive pests. This is equally true of the true fruit flies (Diptera: Tephritidae), which include highly invasive, damaging pests of fruit and vegetable production. This chapter describes current knowledge of the desiccation resistance of tephritid species. Patterns of whole-organism desiccation resistance are summarized for the egg, larval, pupal, and adult life stages. Associations of desiccation resistance in tephritids with body size, body water content, and lipid content are explained. Artificial selection for desiccation resistance as a means to improve the performance of sterile

* Corresponding author.

males in sterile insect technique (SIT) programs is examined. With few exceptions, desiccation resistance of the adult and pupal stages of tephritid species is best studied. Adult desiccation resistance is much higher than anticipated. At least in *Ceratitis capitata* (Wiedemann), this may result from the ability to use water released by lipid catabolism. However, there is also considerable evidence of variation in adult desiccation resistance within species in relation to environmental variability, sex, age, and genetic background. As a consequence, there is capacity for adult tephritid desiccation resistance to be improved by artificial selection through water stress in the laboratory. It is necessary to improve taxonomic, life stage, and life history coverage of studies on tephritid desiccation resistance. Building on results available to date will improve our understanding of the genetic and physiological mechanisms conferring desiccation resistance and develop tephritid strains that can tolerate water stress. By doing so, a better understanding of responses by nonpest tephritids to a changing world and an improved management of pest tephritids will be achieved.

3.1 WHAT IS DESICCATION RESISTANCE AND WHY IS IT IMPORTANT?

Water is an important nutrient for the growth and survival of insects, so its regulation is essential. For blood-feeding, xylem-feeding, and freshwater insects, water is in abundance, at least during part of their life cycle. As a result, they face the problem of removing excess water to regulate body mass and solute concentrations in their tissues (Benoit and Denlinger 2010; Le Caherec et al. 1997). However, most insects are small and terrestrial, so they are at risk of losing water ("dehydration") because of their high surface area-to-volume ratio and the relative scarcity of water in their environment. Additionally, insects living in saline waters (Bradley 2008) or feeding on diets with high osmotic pressure (Douglas 2006) are also susceptible to osmotic water loss. The consequences of dehydration for insects include changes in membrane potential and enzyme function as cellular or hemolymph concentrations increase, decreased circulatory transport of nutrients and hormones as hemolymph volume declines, and retarded movement (and ultimately feeding and predator avoidance) in insects with hydrostatic skeletons (Harrison et al. 2012). The length of time that an organism can survive in a dehydrating environment is referred to as "desiccation resistance." This is distinct from "dehydration tolerance," which is the proportion of body water that can be lost before death (Gibbs et al. 1997).

Environmental stress resistance, and the ability of species to adapt to novel or variable environments, is of central interest in biology because it contributes to niche partitioning and biogeographic patterns. For example, across *Drosophila* species, there are clear correlations between average minimum temperature and critical thermal minimum (lower threshold for muscular function) and between annual precipitation and desiccation resistance (Kellermann et al. 2012). Environmental adaptations have particular importance for invasion biology because the ability of invasive species to survive variable environmental conditions has been suggested as a key trait that contributes to their dispersal and potential to invade new habitats (Lee 2002). Invasive species are introduced species that become established, spread, and cause negative impacts on the environment, human activities, or human health (Lee 2002); as such, they are regarded as major global threats and their management is an international research priority. Studies estimate the monetary costs for the control of invasive species to be enormous (Olson 2006). As climate change increases global temperatures, the threat of invasive insect species will increase as tropical and subtropical insects expand their range into temperate areas (Chown et al. 2007). In this context, the desiccation resistance of insects is likely to contribute to their invasive potential because, as they move away from the wetter tropical and subtropical regions, they will be exposed to reduced rainfall and increasingly dry environments (Chown et al. 2007). This has been illustrated in the dengue mosquito, *Aedes aegypti* (Linnaeus in Hasselquist) (Diptera: Culicidae), in which desiccation resistance of the eggs was a key determinant of its potential current and future distribution in Australia (Kearney et al. 2009).

The first hurdle in the success of an insect pest to colonize new environments is its capacity to adapt to stressful abiotic factors, mainly to those considered critical, such as temperature and water deficit, which can both lead to desiccation. Rapid physiological responses through processes of acclimation and acclimatization allow organisms to display reproductive behaviors temporally and spatially and to establish in new areas (Meats 1989a, 1989b, 1989c; Weldon et al. 2016). Acclimation is a rapid and reversible change in phenotype (be it physiological, biochemical, or anatomical) in response to chronic exposure under controlled conditions to a new environmental condition within the lifetime of an individual (Bowler 2005; Woods and Harrison 2002). Acclimatization is similar, except that it occurs under natural conditions (Chown and Terblanche 2006). Both acclimation and acclimatization are examples of phenotypic plasticity, which permits organism performance under varying natural conditions without any change to the genetic architecture of the individual (Arendt 2015). However, if fitness is enhanced by the capacity to exhibit phenotypic plasticity, its expression over many generations may originate differences in life history and basal tolerance among populations (Chown and Terblanche 2006). Both basal adaptations and plastic responses of individuals and populations determine their bioclimatic potential, or ability to colonize and persist in a particular environment (Meats 1989b).

To manage invasive insect pests, ecologically and socially sustainable tactics are continually being developed and optimized. However, the efficacy of tactics relying on living control agents may be reduced if released organisms are sensitive to local environmental conditions (Sørensen et al. 2012). Some evidence exists of a mismatch of environmental stress tolerance between pest insects and parasitoids released as biological control agents (Hance et al. 2007; Mutamiswa et al. 2018), although studies have focused on thermal tolerance rather than on desiccation resistance. This difference in tolerance of pests and their parasitoids is further exacerbated by changes in the tolerance of parasitoids reared under and adapted to controlled environments for classical biological control or augmentative releases (Colinet and Boivin 2011). The same considerations apply to genetic control tactics, such as various forms of strain replacement or the sterile insect technique (SIT), where live insects must be released into the field, disperse, survive, and ultimately mate with their wild counterparts. In the case of sterile insects, the capacity to achieve these goals is often diminished as a consequence of adaptation to mass-rearing conditions, inbreeding depression, direct rearing and handling effects, irradiation, and transport and release methods. However, environmental tolerance traits, including desiccation resistance, of insects reared for SIT programs are relatively understudied. This is concerning because poor survival of sterile insects in the field under current environmental conditions may be expected to worsen as climates change (Chidawanyika et al. 2012). Because of this, steps need to be taken to ensure the continued success of biologically based control tactics that rely on the release of live insects.

Some notable exceptions to the paucity of studies on the environmental tolerance traits of insects used in SIT programs include the Mexican fruit fly, *Anastrepha ludens* (Loew) (Tejeda et al. 2017); Queensland fruit fly, *Bactrocera tryoni* (Froggatt) (Weldon et al. 2013); and the Mediterranean fruit fly, *Ceratitis capitata* (Wiedemann) (Nyamukondiwa et al. 2013). These three species are representatives of the true fruit flies (Diptera: Tephritidae), a family with about 4300 species from more than 420 genera (White and Elson-Harris 1992). Nearly all tephritids lay eggs inside of plants and the larvae feed on stems, flowers, or fruit before pupating in the soil or inside galls. However, most agriculturally important tephritids are frugivorous, with larval feeding making fruit unmarketable. Fruit fly pests also pose a significant quarantine risk, which affects international trade (De Meyer and Ekesi 2016). Pest tephritids are found throughout the tropical and temperate regions of the world, and some, including *C. capitata*, have become globally invasive pests (Malacrida et al. 2007). The success of some tephritid flies to invade new regions of the world has been attributed to their wide host range and basal developmental response and tolerance to physical environmental variables (e.g., temperature and water availability) (Hill and Terblanche 2014; Malacrida et al. 2007; Vera et al. 2002). However, it is clear that they are also highly adaptable, with flexible responses to environmental conditions through phenotypic plasticity and genetic adaptation (Diamantidis et al. 2011a, 2011b, 2008, 2009; Malacrida et al.

2007; Weldon et al. 2018). Consequently, it is important to understand the abilities of tephritid species to tolerate the physical environment. This includes determining the genetic and epigenetic architecture, biochemical pathways, and physiological processes that lead to variation in desiccation resistance within and between tephritid species.

This review describes current knowledge of the desiccation resistance of tephritid species. Whole-organism desiccation resistance is summarized in the egg, larval, pupal, and adult life stages before the proposed mechanisms underlying these patterns are explained. Because of limited knowledge of these patterns and processes in tephritids, the review is limited to pest fruit fly species within the genera *Anastrepha*, *Bactrocera*, *Ceratitis*, and *Rhagoletis*, as well as the non-pest, stem-galling genus *Eurosta*. As such, it represents species and populations scattered across Africa, Australia, and Central and North America. Thereafter, recent selection experiments to produce desiccation resistant fruit flies are reviewed. Artificial selection for desiccation resistance may offer a means to improve the performance of sterile males in SIT programs. Finally, future directions for fundamental and applied research on desiccation resistance in tephritids are identified.

3.2 DESICCATION RESISTANCE IN TEPHRITIDS

3.2.1 Eggs and Larvae

It has been proposed that eggs and larvae of frugivorous tephritid flies are usually not exposed to desiccating conditions (Meats 1989c). This is because of eggs and larvae being located inside fruits, in a moist environment. Exceptions may occur when fruits drop from a tree prematurely or become hyperosmotic during dry conditions (Bateman 1968), increasing the potential for eggs and larvae to suffer osmotic stress resulting from high fruit solute concentration. However, we are not aware of any studies that have directly compared the osmolality of the cells and hemolymph or osmotic regulation of frugivorous tephritid larvae in relation to their host fruit.

In contrast to the assumptions made for frugivorous tephritids, larvae of the stem-feeding Goldenrod gall fly, *Eurosta solidaginis* (Coquillett), survive extremely cold and dry conditions. This has led to *E. solidaginis* becoming an important model for understanding programmed responses to the cold in Nearctic region, and the link between cold temperature tolerance and desiccation resistance (Sinclair et al. 2003). Larvae of *E. solidaginis* experience extremely low temperatures and humidity during winter inside galls that they induce in their goldenrod host plants, *Solidago canadensis* L., *Solidago gigantea* Ait., and *Solidago altissima* L. Despite this, they maintain a relatively constant water content throughout the winter (e.g., Nelson and Lee 2004).

The high desiccation resistance of *E. solidaginis* larvae relates to the impermeability of their cuticular lipids to water (Ramløv and Lee 2000). Impermeability of the cuticle is acquired as hydrocarbons, mainly 2-methyltriacontane, are deposited during the third instar from September to January (Nelson and Lee 2004). During this period, total hydrocarbons increase from 122 ng/larva to 4900 ng/larva (Nelson and Lee 2004). The permeability of the cuticular lipids to water remains low up to a temperature of 40°C, at which point permeability abruptly increases (Ramløv and Lee 2000). This temperature-dependent change in cuticular permeability likely represents the melting point of hydrocarbons coating the cuticle (Gibbs 1998). To further illustrate the water-proofing properties of the cuticular lipids deposited during the third instar, studies have shown that larvae treated with a chloroform and methanol solution to remove lipids experience high water loss rates (Nelson and Lee 2004; Ramløv and Lee 2000).

Larvae of *E. solidaginis* also exhibit high levels of dehydration tolerance, which results from the movement of solutes and cryoprotectants (glycerol and sorbitol) from the hemolymph to the cells (Williams and Lee 2011). This reduces the osmotic gradient for water to leave the cells and likely maintains cellular water volume during desiccation (Williams and Lee 2011). Expression of pathways leading to stress resistance in larval *E. solidaginis* are triggered after as few as two hours of desiccation and a loss of less than 1% of fresh mass (Gantz and Lee 2015). Larvae experience these conditions during the senescence of host plants. Evaporative water loss is also reduced through depressed metabolic rate (Williams and Lee 2005).

3.2.2 Pupae

The effects of relative humidity and immersion in water on the survival of tephritid pupae were determined by Duyck et al. (2006). Pupae of *C. capitata*, *Ceratitis catoirii* Guérin-Mèneville, the Cape fruit fly, *Ceratitis quilicii* De Meyer, Mwatawala & Virgilio (formerly identified as the Natal fly, *Ceratitis rosa* Karsch), and the peach fruit fly, *Bactrocera zonata* (Saunders), exhibited high survival at 100% relative humidity (Duyck et al. 2006). Pupae of *C. catoirii* and *C. quilicii* were the most susceptible to reduced relative humidity. In contrast, *C. capitata* suffered from even short durations of immersion in water, which may be experienced during periods of flooding. *Bactrocera zonata* pupae were the most tolerant to both low humidity and flooding (Duyck et al. 2006). These patterns of pupal desiccation resistance and flooding tolerance were aligned with the identified niche differentiation of these indigenous and invasive tephritids in the island of La Reunion (Duyck et al. 2006).

Exposure to low relative humidity has also been implicated in poor pupal survival in *Rhagoletis* species but is dependent on their origin. Results from snowberry maggot, *Rhagoletis zephyria* Snow, populations sampled from regions along a rainfall gradient in the state of Washington, United States, suggest local adaptation of pupae to dry conditions (Hill 2016). Exposure of *R. zephyria* pupae sampled from a wet, coastal location to 43% relative humidity for 8 days led to a greater than 60% reduction in adult emergence after diapause in comparison with pupae from a drier, inland location. Along the rainfall gradient, from high to low, pupal mass increased and the proportion of body water remaining after desiccation also increased (Hill 2016). Analyses of gene expression found greater differences between populations than among humidity treatments (Kohnert 2017), supporting the role of local adaptation suggested by phenotypic observations. In particular, there was an upregulation of oxidioreductases in a desiccation resistant population of *R. zephyria* (Kohnert 2017), which are important for the production of long-chain cuticular hydrocarbons (Qiu et al. 2012). In the apple and hawthorn host-races of *Rhagoletis pomonella* (Walsh), the proportion of water remaining in pupae after desiccation was higher in the hawthorn host-race (Hill 2016). This was despite pupae from the apple host-race being 33% heavier than those from the hawthorn race. In both *R. pomonella* and *R. zephyria*, pupal diapause does not contribute to improved resistance to desiccation (Kohnert 2017).

3.2.3 Adults

Meats (1989c) noted that no study had been performed to determine how adult tephritids maintain water reserves within viable limits. In the absence of empirical data, he used a physiological model based on the fundamental relationships between temperature, relative humidity, rate of water loss, and an assumption of the water reserve required for survival (i.e., dehydration tolerance) to predict adult tephritid survival. By assuming that most terrestrial insects can only replace lost water by drinking or eating food with adequate water content, Meats (1989c) predicted that a fly weighing 15 mg with 10 μg of water would be able to survive 24 hours when held at 25°C and 10% relative humidity. Declining temperature or increasing relative humidity would lead to lower water loss rates and higher survival time, and vice versa, in the absence of a source of ingestible water. At 100% relative humidity, no evaporative water loss was anticipated.

When desiccation resistance at 25°C and relative humidity below 10% was evaluated in a range of tephritid species, the survival times expected by Meats (1989c) were exceeded by a considerable margin (Figure 3.1). In *B. tryoni*, the species on which the predictions of Meats (1989c) were based, median survival time was longer than 24 hours in most cases (Weldon and Taylor 2010; Weldon et al. 2013). This was regardless of fly origin (wild or mass-reared), sex, or age, which all have significant effects on desiccation resistance in *B. tryoni* (Weldon and Taylor 2010; Weldon et al. 2013). Only in mass-reared *B. tryoni* that were 20 days old did desiccation resistance decline to a median of

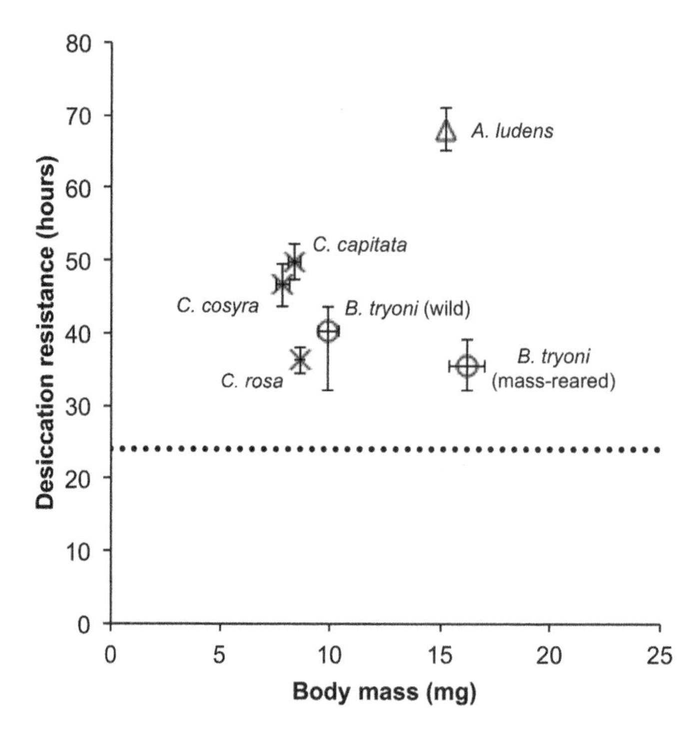

FIGURE 3.1 Relationship between mean body mass and mean desiccation resistance in adult tephritid flies. Different marker styles represent data sources: triangles (Tejeda et al. 2014); crosses (Weldon et al. 2013); circles (Weldon et al. 2016). With the exception of *Anastrepha ludens*, all body mass and desiccation resistance values are for ten-day-old flies. Body mass and desiccation resistance of *A. ludens* were determined on adult emergence. Error bars represent ± 1 standard error except for *Bactrocera tryoni*, where they indicate 95% confidence intervals. The dotted line shows desiccation resistance predicted for *B. tryoni* by Meats (1989c).

25 hours (Weldon et al. 2013). In a comparison of *Ceratitis* species, even the least desiccation resistant species, 10-day-old *C. rosa*, survived an average of 36 hours at 25°C and less than 10% relative humidity. Desiccation resistance of *C. capitata* and marula fruit flies, *Ceratitis cosyra* (Walker), of the same age was even higher, at 50 and 47 hours, respectively (Weldon et al. 2016). This is despite *C. capitata* and *C. cosyra* generally being smaller with less body water than *C. rosa* (Weldon et al. 2016). In *A. ludens*, which are large flies with a body mass within the range of 11–20 mg, mean desiccation resistance was approximately 68 hours at 25°C and 22% relative humidity (Tejeda et al. 2014).

Increasing empirical evidence shows that desiccation resistance of adult tephritid flies varies within species. There are strong effects of age, sex, and their interaction that may relate to differences in life history or senescence. Desiccation resistance of a laboratory-adapted strain of *C. cosyra* tested on the day of adult emergence was higher than when they were 10 days old (Weldon et al. 2019). In a laboratory-adapted strain of *B. tryoni*, desiccation resistance of adults declined in a continuous and regular manner over the first 20 days after adult eclosion (Weldon and Taylor 2010; Weldon et al. 2013). However, in the same species, there was no consistent effect of age on desiccation resistance over 20 days among flies derived from field-collected fruit (loquats, *Eriobotrya japonica* (Thunb.) Lindl.) and their first- and second-generation offspring (Weldon et al. 2013). Desiccation resistance of adult female *B. dorsalis*, *C. capitata*, *C. cosyra*, and *A. ludens* is generally lower than that of males (Weldon et al. 2013, 2016; Weldon and Taylor 2010). This observation correlates with lower dehydration tolerance but higher total body water in females, which suggests that the water contained in eggs represents an inaccessible pool of total body water (Weldon et al. 2013, 2016, 2019; Weldon and Taylor 2010).

Desiccation resistance of tephritid flies may also vary within individuals due to prior experience of various environmental conditions. In *C. capitata*, desiccation resistance varied among individuals based on exposure for 5 days to temperatures of 20°C, 25°C, or 30°C, although the direction of the change in desiccation resistance differed among populations (Weldon et al. 2018). Larval diet can also affect desiccation resistance, with adult *C. cosyra* reared on a standard, high yeast larval diet exhibiting lower desiccation resistance on the day of eclosion when compared to those reared on a low yeast larval diet with a protein content akin to the preferred host (Weldon et al. 2019).

As already mentioned, the origin of adult tephritid fly populations also affects desiccation resistance within a species. In populations of *C. capitata* sampled from parts of southeastern Africa and tested under common conditions (25°C, <10% relative humidity), there are considerable differences in desiccation resistance between populations (Weldon et al. 2018). These differences appear to be associated with the bioclimatic conditions prevailing in the region where the populations were sampled. Relevant to this discussion, desiccation resistance of adult *C. capitata* was weakly but negatively affected by growing degree-days, which may suggest a role of resource availability coupled with plant productivity (Weldon et al. 2018). This was despite evidence from microsatellite markers of little genetic differentiation of populations, which suggests that *C. capitata* has a high potential for evolutionary responses to environmental conditions (Weldon et al. 2018). Rapid adaptation to prevailing conditions among tephritid flies is also evident in the laboratory. Intentional application of water stress has successfully selected for improved desiccation resistance in *A. ludens* (Tejeda et al. 2016). In this case, significantly improved desiccation resistance relative to the parental generation was observed within as little as two generations and was double of that of a control population after 10 generations of selection (112 vs. 56 hours) (Tejeda et al. 2016). Unintentional selection for differences in desiccation resistance also results from mass-rearing, with wild *B. tryoni* exhibiting greater desiccation resistance than their mass-reared counterparts (Weldon et al. 2013). This was particularly evident in older adults, where median desiccation resistance of wild female and male *B. tryoni* was approximately 12 hours longer than mass-reared *B. tryoni* (25 vs. 37 hours) (Weldon et al. 2013). Whether recorded population differences in desiccation resistance in the field and laboratory represent genetic differences or potential heritable epigenetic effects is yet to be determined.

3.3 WHAT ENABLES DESICCATION RESISTANCE?

There are three physiological mechanisms by which insects can reduce the risk of water stress: increased water storage, reduced water loss rates, and enhanced dehydration tolerance (Gibbs et al. 1997). Body water content can be improved by increasing body size and hemolymph volume (Folk and Bradley 2003). Increased body size can reduce the surface area-to-volume ratio and, thereby, reduce evaporative water losses. More importantly, insect hemolymph may often serve as a reservoir that buffers insect tissues during periods of water stress (Gibbs et al. 1997). For example, laboratory-selected, desiccation-resistant *Drosophila* populations exhibited a striking increase in hemolymph volume (~330 nl, a >6-fold increase) (Folk and Bradley 2003). Body water reserves can also be elevated by increased food intake and not only due to the free water content of food but also through the release of metabolic water from the breakdown of ingested nutrients. Metabolic reserves are thought to play a large role in adaptation to desiccation resistance (Djawdan et al. 1998). Oxidative phosphorylation of glucose leads to the production of water and carbon dioxide (CO_2) as by-products, with water retained and CO_2 expired into the environment (Djawdan et al. 1998). Lipids also play a part in desiccation resistance as they represent the main form of energy storage in insects as triglycerides in body fat (Arrese and Soulages 2010) and act as a source of water upon oxidation in some species (Kleynhans and Terblanche 2009; Naidu 2001; Naidu and Hattingh 1988; Nicolson 1980). Reduced water loss rates are achieved by closing spiracles to reduce respiratory water loss, increasing rectal water reabsorption to reduce excretory water loss, or by enhancing the water-proofing properties of the cuticle to reduce cuticular evaporative water loss (Gibbs 1998; Harrison et al. 2012). Improved dehydration tolerance in insects is largely associated with osmotic properties of cells and the hemolymph. Carbohydrates, particularly trehalose, are known to play a large role in insect osmoregulation by binding water molecules (Djawdan et al. 1998; Gefen et al. 2006). In addition, proteins also help insects to tolerate dry conditions, with heat shock protein expression upregulated, presumably to protect water stressed cells from changes in pH and solute concentrations (Hayward et al. 2004; Tammariello et al. 1999). In addition, expression of late embryogenic abundant (LEA) proteins is also associated with dehydration (Kikawada et al. 2006) and may have a role in DNA protection during desiccation (Ryabova et al. 2016).

In the case of tephritid flies, the mechanisms that confer improved desiccation resistance are yet to be fully explored. However, as will be discussed, results to date suggest some of the mechanisms associated with improved water stress in other insects, particularly body size, water content, and lipid reserves, are also correlated with enhanced desiccation resistance in tephritid species that have been studied.

3.3.1 Body Size

In the model organism *Drosophila melanogaster* Meigen, body size has not been a good indicator of individual desiccation tolerance. Some studies have reported a positive relationship between size and stress resistance (Chippindale et al. 1996; Gibbs and Matzkin 2001; Telonis-Scott et al. 2006), whereas others have not found this relationship (Gibbs et al. 1997; Hoffmann and Harshman 1999). For *B. tryoni*, as in *Drosophila*, mixed patterns have been observed among size, sex, and stress resistance. When wing length was used as a proxy for size, resistance under desiccation was not correlated with size, and both sexes showed similar resistance (Weldon and Taylor 2010). However, further studies demonstrated that body mass correlated positively with desiccation resistance (Weldon et al. 2013). In the case of *A. ludens*, it was observed that desiccation resistant individuals were heavier and exhibited higher lipid and water content than unselected, control flies (Tejeda et al. 2014). Larger pupae of *R. pomonella* and *R. zephyria* also survive better during exposure to dry conditions than smaller ones (Hill 2016). Nevertheless, despite often being smaller than females, males are more desiccation resistant (see above). This pattern was observed in *B. tryoni* and *A. ludens* and may be explained by differences in life history between the sexes. In addition, males have higher lipid and water content than females at emergence (Tejeda et al. 2014; Weldon et al. 2013). Thus, size by itself does not explain improved desiccation resistance in tephritids studied to date.

3.3.2 Lipid and Water Content and Lipid Catabolism

Ambiguous results have been observed among studies on the role of lipid reserves and water content in insects. Higher lipid reserves have correlated with higher desiccation resistance in some *Drosophila* species (Telonis-Scott et al. 2006). However, in studies involving artificial selection for desiccation resistance, invariable (Hoffmann and Parsons 1989) or even reduced (Djawdan et al. 1998) lipid content has been observed. This inconsistency is also observed in the role of water reserves. For example, some studies in *Drosophila* report the canteen strategy, with an increase of body water content and water storage in the hemolymph (Folk et al. 2001; Gibbs et al. 1997). However, more recently, it has been observed that individuals with a high water content presented reduced desiccation resistance if associated with a low level of desaturated cuticular hydrocarbons (CHCs) (Ferveur et al. 2018).

In tephritids, lipid and water contents seem to play a decisive role in the desiccation resistance of some species. Individuals with higher desiccation resistance exhibit higher levels of both lipid and water contents (Tejeda et al. 2014). Moreover, *A. ludens* flies artificially selected for desiccation resistance exhibited higher levels of water and lipids than nonselected flies (Tejeda et al. 2016). Desiccation resistance is greatest among adults with high body water content in *B. tryoni* (Weldon et al. 2013), *C. capitata*, *C. cosyra*, and *C. rosa* (Weldon et al. 2016). In *C. capitata*, body lipid content was higher than in other tested *Ceratitis* species, and lipids were selectively catabolized during a short period of dehydration (Weldon et al. 2016). However, the use of lipids as a source of metabolic water was not apparent in *C. cosyra* or *C. rosa* (Weldon et al. 2016).

3.3.3 Cuticular Lipids

Transpiration through the cuticle is the main route of water loss in insects (Gibbs and Rajpurohit 2010). For this reason, cuticular water loss by insects in dry environments is minimized through changes in the quantity and composition of water-proofing epicuticular lipids (Gibbs 1998). For example, different desiccation resistant lines of *D. melanogaster* have reduced cuticular water permeability as a consequence of altered lipid composition, in addition to increased hemolymph volume, higher extracellular carbohydrate storage that increases hemolymph osmolality (Gibbs et al. 1997), and elevated tolerance to water loss and lipid storage (Telonis-Scott et al. 2006).

The epicuticular layer, which covers almost the entire surface of an insect, is mainly responsible for protecting insects from desiccation (Downer and Matthews 1976; Drijfthout et al. 2010). A large component of the lipids that comprise the epicuticle are long-chain hydrocarbons, generally known as CHCs that, aside from protection, also play an important role in sexual selection (Chung and Carroll 2015; Ferveur et al. 2018). By using artificial selection in *D. melanogaster*, it was observed that desiccation resistance was positively linked to the proportion of desaturated CHCs, which considerably reduce transpiration (Ferveur et al. 2018). Furthermore, once selected, this proportion is kept even after several generations without the stressor (Ferveur et al. 2018).

In tephritids, CHC profiles have been employed mainly for taxonomic purposes. For example, in "lowland" and "highland" populations of *C. rosa* (now *C. rosa s.s.* and *C. quilicii*, respectively), statistical analyses of CHC composition showed distinct interspecific identities, with several CHC specific to each of the lowland and highland populations (Vaníčková et al. 2015). Whether this difference confers differences in desiccation tolerance is not known, but the two species are associated with different bioclimatic regions (Mwatawala et al. 2015; Tanga et al. 2018). Furthermore, it has been reported that two stenophagous species, *C. cosyra* and *C. rosa*, lost water at significantly higher rates under hot, dry conditions and do not catabolize lipids or other sources of metabolic water during water stress compared to the polyphagous *C. capitata* (Weldon et al. 2018). Thus, it is possible that differences in the CHC profiles of the populations also confer differences in their desiccation resistance. Furthermore, Weldon et al. (2019) postulated that changes in desiccation resistance as adult tephritids age may also relate to abrasion or qualitative changes of the epicuticular lipids.

3.4 DEVELOPMENT OF DESICCATION RESISTANT STRAINS FOR SIT

The development of desiccation resistant strains can be a powerful tool that improves the efficiency of SIT through increased longevity of released males in environments causing water stress. Thus, a promising method to improve SIT is to produce and release strains that are resistant to desiccation. So far, desiccation resistant strains have been produced only in *A. ludens*. These strains, derived from the already adapted mass-rearing strains, would provide SIT with an added advantage over regular bisexual strains in terms of increased longevity of sterile males under environmentally stressful situations. The increased survival of adults from the selected strain has been attributed to the fact that they are heavier and store 20% more water than control strains (Tejeda et al. 2014). Also, they store considerably higher lipid reserves than control strains.

Nevertheless, although these strains have increased longevity for males, they are not without disadvantages because certain trade-offs have been detected. Furthermore, these strains are still at the experimental stage; the added detrimental effect of mass rearing has not been documented, and they have not been tested under field conditions. For example, even though no detrimental effects are predicted, their ability to detect protein baited traps has not been tested. Thus, we review the effects on fecundity, fertility, and pre- and postmating competitiveness of the desiccation resistant strain of *A. ludens*.

The desiccation resistant strain of *A. ludens* was developed from the standard bisexual mass-rearing strain produced in the Moscafrut facility, Metapa de Dominguez, Chiapas, Mexico. Ten experimental families were separated from the main colony. Five of these were selected for experiments using directional selection for fly longevity under desiccation stress, and the other five families served as unselected, control populations. Each family or population was comprised of 200 males and 200 females and placed in separate plexiglass cages from emergence. Selection for desiccation resistance was carried out by placing each selected population in a cage with three containers of silica gel, covered with mesh to avoid direct contact but no food or water. Cages were then sealed with plastic film. Humidity inside each cage was thus reduced to 20%–30%. When 12% of the population remained alive, the survivors were transferred to another cage with water and food. Control cages were handled in the same way except they were not subjected to low humidity. For each of the 10 populations, 25 fly pairs were used to produce the following generation. This was repeated for 10 generations (Tejeda et al. 2016).

3.4.1 Fecundity/Fertility/Longevity

Changes in the mean survival time of flies was observed as early as the second (F_2) generation (Control: 62.01 ± 1.31 [mean ± s.e.]; Resistant: 74.12 ± 1.34). After only 10 generations, selected populations exposed to low humidity without food or water lived twice as long as unselected populations (Tejeda et al. 2016). Individuals also had a higher life expectancy than control flies. The average number of eggs that females laid was not significantly different between selected and control lines (Control: 56.38 ± 2.69; Resistant: 112.33 ± 5.50) (Tejeda et al. 2016).

However, some interesting trade-offs were observed. For example, the mean age of females at which reproduction started was significantly delayed. Control flies started reproducing (laying eggs) at an average of 25 days of age in comparison with selected females, which started approximately 10 days later. This could have detrimental effects in terms of logistics and costs of mass rearing because flies would need to be kept and fed for longer periods of time before they start producing eggs. Thus, this is something that warrants further research.

Also, another trade-off between longer survival and reproduction was observed, as selected females, on average, had a lower daily egg production compared to control females (34 eggs/female vs. 25 eggs/female). The intervals between generations were shorter for control compared to desiccation-selected lines. Pupal stage duration was approximately 40 hours longer for the selected populations compared to the control (Tejeda et al. 2016). Again, this may imply considerable costs for mass rearing in terms of diet, personnel, and oviposition devices allocated to the colony. For example, the size of the colony would probably need to be increased to meet production standards. However, a longer pupal stage duration could be advantageous if pupae need to be transported between production and prerelease holding facilities or release sites.

3.4.2 Sexual Behavior

Sexual behavior can be grossly divided into precopulatory and postcopulatory behaviors. For the desiccation resistant line of *A. ludens*, male sexual competitiveness was evaluated in field cages, where desiccation resistant and control males competed for matings with wild females. There was no significant difference in the frequency of matings between selected or control males, indicating that selected males were just as competitive as control nonresistant males in obtaining matings with wild females (Tejeda et al. 2016). This is an important result because the main objective of SIT is for males to mate with a wild female and render her infertile. No detrimental effects on mating performance indicate that, at least for this component of behavior, there are no apparent trade-offs between desiccation resistance and sexual performance.

One further aspect of sexual behavior that needed to be evaluated is the effect of irradiation on sexual performance. For this, the resistant and nonresistant strains were sterilized 48 hours before emergence and in hypoxia to the standard sterilization dose of 80 Gy using a ^{60}Co irradiator (model GB-127, Nordion International Inc., Ottawa, ON, Canada). Selected and nonselected males competed for matings with wild females in field cages. Again, as with fertile males, there was no significant difference in mating competitiveness between selected and control lines (Tejeda et al. 2017).

Compatibility tests were carried out also in field cages between sterile males and females of the selected strain against the standard bisexual mass-reared strain of *A. ludens* from Moscafrut and against wild males for matings with wild females. The Relative Sterility Index (RSI) and the Male Relative Performance Index (MRPI) (Cayol et al. 1999) were the same between both the selected and the Moscafrut strain. Wild males were more likely to obtain copulations than males from either of these strains. There was no significant difference in the Index of Sexual Isolation (ISI) between strains, indicating that both strains were compatible with the wild population. The Female Relative Performance Index (FRPI) (Cayol et al. 1999) indicated for both strains a higher participation of sterile females in obtaining matings compared to wild females. Despite slightly higher FRPI values for the Moscafrut strain compared to the resistant selected strain, there was no statistically significant difference between these two strains (Tejeda et al. 2017).

A further test of previously stressed sterile males (no water or food and 30%–40% humidity for 24 hours before observations) compared with nonstressed sterile males (food and water ad libitum at 60%–80% humidity) of both strains (Moscafrut vs. selected) revealed that desiccation-selected males obtained 88% of matings with wild females. When males were reared in the nonstressed environment, both strains had similar RSI, as opposed to when males were reared in the stressful environment, where RSI was significantly different from the expected value, indicating that the selected strain outperformed the Moscafrut strain (Tejeda et al. 2017). This suggests that if the desiccation-selected males were released in a dry environment, their capacity to mate with and induce sterility in wild females would likely be better than the currently used strain.

3.4.3 Postcopulatory Behaviors and Mechanisms

There is now widespread recognition that sexual selection and behavior do not end with mating but, rather, continue during and after copulation. During mating, males transfer sperm and secretions from the male accessory glands. These are products from the male ejaculate and have important effects on female remating (Abraham et al. 2016; Radhakrishnan and Taylor 2007, 2008). Thus, any strain developed for SIT should not only exhibit suitable precopulatory behavior but should also have adequate postcopulatory behaviors, including sperm and accessory gland product transfer.

Recent studies have found that there are also biological trade-offs between resistance to desiccation and the male ejaculate. Control females mating with the selected strain of *A. ludens* had less sperm stored in their spermathecae than control females mating with nonselected males (Pérez-Staples et al. 2017). Also, resistant males were found to have smaller accessory glands and seminal vesicles (organ where males store mature sperm; Martínez and Hernández-Ortiz 1997) than nonselected males (Pérez-Staples et al. 2017). These results suggest that the evolution of desiccation resistance comes at a cost for the male, in this case manifested through smaller accessory glands and lower amount of sperm stored by females, which probably indicates lower sperm transfer as well.

Control females that mated with resistant males also suffered a cost in terms of lower fecundity compared to control females mating with nonresistant males. One of the peptides that are produced in the male accessory glands is ovulin, which in *D. melanogaster* affects female oogenesis, fecundity, and the egg-laying process (Avila et al. 2011). Although the production and function of specific peptides in the ejaculate is unknown in *A. ludens* and, indeed for most tephritids of economic importance, a lower fecundity for females mating with resistant males suggests that some components of the male ejaculate may be compromised during the evolution of the ability to withstand water stress.

Desiccation-selection did not affect all components of the ejaculate equally. For example, no effect on protein content of the male accessory glands or testes were found, and no detrimental effects were found in the male ability to inhibit female remating. Wild *A. ludens* females were just as likely to remate after mating with a desiccation-selected or control male.

3.5 FUTURE DIRECTIONS

3.5.1 Improved Taxonomic, Life Stage, and Life History Coverage

To date, desiccation resistance data are available only for *A. ludens*, *B. tryoni*, *B. zonata*, *C. capitata*, *C. catoirii*, *C. cosyra*, *C. rosa*, *E. solidaginis*, *R. pomonella*, and *R. zephyria*. Of these, *C. capitata* is the only species with published desiccation resistance data for more than one life stage: pupae (Duyck et al. 2006) and adults (Weldon et al. 2016; Weldon et al. 2018). There is clearly a need to study the water stress experienced and desiccation resistance exhibited by a wider range of tephritid species throughout all life stages. As noted previously, eggs and larvae may experience water stress in some fruits owing to osmotic gradients between the insect and the osmolality of fruits of varying species and conditions. The pupal stage may also be susceptible to dehydration or drowning under some circumstances, and the tolerance of species, and even populations, to these conditions may vary (Duyck et al. 2006; Hill 2016). Development of a wider database of tephritid desiccation resistance will assist with identifying the potential invasiveness of pest species. It will also enable predictions of the effects of global change on the majority of tephritid species that are not of economic concern but are rather important components of ecosystems throughout the world. In relation to this, it is important to obtain environmental tolerance data not only for multivoltine frugivorous pest tephritids but also for the full range of life histories and host use patterns that are encompassed by this diverse insect group. Here we have already discussed how *E. solidaginis*, a univoltine, stem-galling host specialist with no economic importance, has profoundly shaped our understanding of the role of water loss in insect cold tolerance (e.g., Gantz and Lee 2015; Williams et al. 2004). It may be that other understudied species in this group will have a similar contribution.

3.5.2 Understanding Mechanisms of Desiccation Resistance

Most studies indicate that the invasive potential of polyphagous flies, such as *C. capitata* and *A. ludens*, is related to a rapid capacity for adaptation to abiotic stressors. Polyphagous insects have the ability to face heterogeneous environments of host plants and abiotic conditions but keep the reproductive potential that allows them to persist in those variable habitats (Gilchrist et al. 2008; Weldon et al. 2016). Thus, studying complex interactions among different types of trade-offs is necessary for a better understanding of the response to stressors (i.e., specialist–generalist trade-offs, allocation trade-offs, and acquisition trade-offs) (Angilletta et al. 2003). For example, future studies could include the voltinism of species as a character of their life history to determine their bioclimatic potential. Comparative genomic, epigenetic, and transcriptomic data are also required to understand the mechanisms leading to desiccation resistance and any associated trade-offs.

3.5.3 Selection for Desiccation Resistant, Mass-Reared Tephritids

In their recent review, Hoffmann and Ross (2018) showed that, in general, the evolutionary response of laboratory-adapted lines to stress was negative, although studies for Diptera were particularly promising. Thus, it is important to carry out these selection experiments on different tephritid species. For programs that use SIT and that will release sterile males in dry and arid conditions, it would be desirable to examine how their mass-rearing strains can develop a higher resistance to desiccation. Because laboratory-adapted lines tend to be more sensitive to stress (Hoffmann and Ross 2018), this should be countered in mass-rearing programs. Potentially, all mass-rearing strains that require release in arid environments can be improved in terms of response to water stress.

Further studies are needed on some key issues of desiccation resistant lines. Although field cage studies for *A. ludens* have demonstrated that selected lines exposed to stress before mating perform as well or better than control lines (Tejeda et al. 2017), it would be ideal to test calling and mating activities of these males in arid conditions. As biological trade-offs were detected in certain life history components of the selected line, it is also important to study the effect of resistance to water stress on the age of sexual maturation for males. This can have important consequences for holding males during the prerelease period.

3.6 CONCLUSIONS

Knowledge of the desiccation resistance of tephritid flies is limited, which is surprising considering their importance as economic pests of fruit and vegetable production, as well as biological control agents for invasive plants. The research reviewed here highlights that the adults of some tephritid fly species, particularly those that are widespread pests, are able to tolerate water stress to a greater extent than predicted by physiological models. Furthermore, selection for improved desiccation resistance is possible, which may lead to the development of strains better able to survive when released in SIT programs. However, much still needs to be done to identify the mechanisms underlying the ability of tephritids to tolerate water stress. The in-depth understanding of the ability of the larval stages of *E. solidaginis*, a nonpest, univoltine, host specialist, to tolerate extreme desiccation, illustrates how the study of this trait in a wider range of tephritid species and life stages can contribute to fundamental knowledge of insect physiology. But furthermore, this kind of knowledge can help to predict how species distributions may be affected in a changing world, in particular given the scenarios of climate change that consistently predict higher temperatures and reduced rainfall in many temperate areas. Tephritid species particularly tolerant of desiccation may also pose the next threat to global food security through their effect on fruit and vegetable production.

REFERENCES

Abraham, S., L. A. Lara-Pérez, C. Rodríguez et al. 2016. The male ejaculate as inhibitor of female remating in two tephritid flies. *Journal of Insect Physiology* 88:40–47.

Angilletta, M. J., R. S. Wilson, C. A. Navas, and R. S. James. 2003. Tradeoffs and the evolution of thermal reaction norms. *Trends in Ecology & Evolution* 18:234–240.

Arendt, J. D. 2015. Effects of dispersal plasticity on population divergence and speciation. *Heredity* 115:306–311.

Arrese, E. L., and J. L. Soulages. 2010. Insect fat body: Energy, metabolism, and regulation. *Annual Review of Entomology* 55:207–225.

Avila, F. W., L. K. Sirot, B. A. LaFlamme, C. D. Rubinstein, and M. F. Wolfner. 2011. Insect seminal fluid proteins: Identification and function. *Annual Review of Entomology* 56:21–40.

Bateman, M. A. 1968. Determinants of abundance in a population of the Queensland fruit fly. *Symposia of the Royal Entomological Society of London* 4:119–131.

Benoit, J. B., and D. L. Denlinger. 2010. Meeting the challenges of on-host and off-host water balance in blood-feeding arthropods. *Journal of Insect Physiology* 56:1366–1376.

Bowler, K. 2005. Acclimation, heat shock and hardening. *Journal of Thermal Biology* 30:125–130.

Bradley, T. J. 2008. Saline-water insects: Ecology, physiology and evolution. In *Aquatic Insects: Challenges to Populations*, Ed. J. Lancaster and R. A. Briers, Wallingford, UK: CAB International.

Cayol, J. P., J. Vilardi, E. Rial, and M. T. Vera. 1999. New indices and method to measure the sexual compatibility and mating performance of *Ceratitis capitata* (Diptera: Tephritidae) laboratory-reared strains under filed cage conditions. *Journal of Economic Entomology* 92:140–145.

Chidawanyika, F., P. Mudavanhu, and C. Nyamukondiwa. 2012. Biologically based methods for pest management in agriculture under changing climates: Challenges and future directions. *Insects* 3:1171–1189.

Chippindale, A. K., T. J. F. Chu, and M. R. Rose. 1996. Complex trade-offs and the evolution of starvation resistance in *Drosophila melanogaster. Evolution* 50:753–766.

Chown, S. L., and J. S. Terblanche. 2006. Physiological diversity in insects: Ecological and evolutionary contexts. *Advances in Insect Physiology* 33:50–152.

Chown, S. L., S. Slabber, M. A. McGeoch, C. Janion, and H. P. Leinaas. 2007. Phenotypic plasticity mediates climate change responses among invasive and indigenous arthropods. *Proceedings of the Royal Society of London, Series B: Biological Sciences* 274:2531–2537.

Chung, H., and S. B. Carroll. 2015. Wax, sex and the origin of species: Dual roles of insect cuticular hydrocarbons in adaptation and mating. *Bioessays* 37:822–830.

Colinet, H., and G. Boivin. 2011. Insect parasitoids cold storage: A comprehensive review of factors of variability and consequences. *Biological Control* 58:83–95.

De Meyer, M., and S. Ekesi. 2016. Exotic invasive fruit flies (Diptera: Tephritidae): In and out of Africa. In *Fruit Fly Research and Development in Africa*, Eds. S. Ekesi, S. A. Mohamed, and M. De Meyer, Cham, Switzerland: Springer International Publishing.

Diamantidis, A. D., J. R. Carey, C. T. Nakas, and N. T. Papadopoulos. 2011a. Ancestral populations perform better in a novel environment: Domestication of Mediterranean fruit fly populations from five global regions. *Biological Journal of the Linnean Society* 02:334–345.

Diamantidis, A. D., J. R. Carey, C. T. Nakas, and N. T. Papadopoulos. 2011b. Population-specific demography and invasion potential in medfly. *Ecology and Evolution* 1:479–488.

Diamantidis, A. D., N. T. Papadopoulos, and J. R. Carey. 2008. Medfly populations differ in diel and age patterns of sexual signalling. *Entomologia Experimentalis et Applicata* 128:389–397.

Diamantidis, A. D., N. T. Papadopoulos, C. T. Nakas, S. Wu, H.-G. Müller, and J. R. Carey. 2009. Life history evolution in a globally invading tephritid: Patterns of survival and reproduction in medflies from six world regions. *Biological Journal of the Linnean Society* 97:106–117.

Djawdan, M., A. K. Chippindale, M. R. Rose, and T. J. Bradley. 1998. Metabolic reserves and evolved stress resistance in *Drosophila melanogaster. Physiological Zoology* 71:584–594.

Douglas, A. E. 2006. Phloem-sap feeding by animals: Problems and solutions. *Journal of Experimental Botany* 57:747–754.

Downer, R. G. H., and J. R. Matthews. 1976. Patterns of lipid distribution and utilisation in insects. *American Zoologist* 16:733–745.

Drijfthout, F. P., R. Kather, and S. J. Martin. 2010. The role of cuticular hydrocarbons in insects. In *Behavioural and Chemical Ecology*, Eds. W. Zhang and H. Liu, New York: Nove Science Publishers.

Duyck, P.-F., P. David, and S. Quilici. 2006. Climatic niche partitioning following successive invasions by fruit flies in La Réunion. *Journal of Animal Ecology* 75:518–526.

Ferveur, J. F., J. Cortot, K. Rihani, M. Cobb, and C. Everaerts. 2018. Desiccation resistance: Effect of cuticular hydrocarbons and water content in *Drosophila melanogaster* adults. *Peer J* 6:e4318.

Folk, D. G., and T. J. Bradley. 2003. Evolved patterns and rates of water loss and ion regulation in laboratory-selected populations of *Drosophila melanogaster. Journal of Experimental Biology* 206:2779–2786.

Folk, D. G., C. Han, and T. J. Bradley. 2001. Water acquisition and partitioning in *Drosophila melanogaster*: Effects of selection for desiccation-resistance. *Journal of Experimental Biology* 204:3323–3331.

Gantz, J. D., and R. E. Lee, Jr. 2015. The limits of drought-induced rapid cold-hardening: Extremely brief, mild desiccation triggers enhanced freeze-tolerance in *Eurosta solidaginis* larvae. *Journal of Insect Physiology* 73:30–36.

Gefen, E., A. J. Marlon, and A. G. Gibbs. 2006. Selection for desiccation resistance in adult *Drosophila melanogaster* affects larval development and metabolite accumulation. *Journal of Experimental Biology* 209:3293–3300.

Gibbs, A. 1998. Water-proofing properties of cuticular lipids. *American Zoologist* 38:471–482.

Gibbs, A. G., A. K. Chippindale, and M. R. Rose. 1997. Physiological mechanisms of evolved desiccation resistance in *Drosophila melanogaster. Journal of Experimental Biology* 200:1821–1832.

Gibbs, A. G., and L. M. Matzkin. 2001. Evolution of water balance in the genus *Drosophila. Journal of Experimental Biology* 204:2331–2338.

Gibbs, A. G., and S. Rajpurohit. 2010. Cuticular lipids and water balance. In *Insect Hydrocarbons: Biology, Biochemistry and Chemical Ecology*, Eds. G. J. Blomquist and A.-G. Bagnères, Cambridge, UK: Cambridge University Press.

Gilchrist, G. W., L. M. Jeffers, B. West, D. G. Folk, J. Suess, and R. B. Huey. 2008. Clinal patterns of desiccation and starvation resistance in ancestral and invading populations of *Drosophila subobscura. Evolutionary Applications* 1:513–523.

Hance, T., J. van Baaren, P. Vernon, and G. Boivin. 2007. Impact of extreme temperatures on parasitoids in a climate change perspective. *Annual Review of Entomology* 52:107–126.

Harrison, J. F., H. A. Woods, and S. P. Roberts. 2012. *Ecological and Environmental Physiology of Insects.* Oxford, UK: Oxford University Press.

Hayward, S. A. L., J. P. Rinehart, and D. L. Denlinger. 2004. Desiccation and rehydration elicit distinct heat shock protein transcript responses in flesh fly pupae. *Journal of Experimental Biology* 207:963–971.

Hill, J. L. 2016. *Adaptive Variation in Dessication Resistance in Rhagoletis.* Bellingham, WA: Western Washington University.

Hill, M. P., and J. S. Terblanche. 2014. Niche overlap of congeneric invaders supports a single-species hypothesis and provides insight into future invasion risk: Implications for global management of the *Bactrocera dorsalis* complex. *PLoS One* 9:e90121.

Hoffmann, A. A., and L. G. Harshman. 1999. Desiccation and starvation resistance in *Drosophila*: Patterns of variation at the species, population and intrapopulation levels. *Heredity* 83:637–643.

Hoffmann, A. A., and P. A. Parsons. 1989. An integrated approach to environmental stress tolerance and life-history variation: Desiccation tolerance in *Drosophila. Biological Journal of the Linnean Society* 37:117–136.

Hoffmann, A. A., and P. A. Ross. 2018. Rates and patterns of laboratory adaptation in (mostly) insects. *Journal of Economic Entomology* 111:501–509.

Kearney, M., W. P. Porter, C. Williams, S. Ritchie, and A. A. Hoffmann. 2009. Integrating biophysical models and evolutionary theory to predict climatic impacts on species' ranges: The dengue mosquito *Aedes aegypti* in Australia. *Functional Ecology* 23:528–538.

Kellermann, V., V. Loeschcke, A. A. Hoffmann et al. 2012. Phylogenetic constraints in key functional traits behind species' climate niches: Patterns of desiccation and cold resistance across 95 *Drosophila* species. *Evolution* 66:3377–3389.

Kikawada, T., Y. Nakahara, Y. Kanamori et al. 2006. Dehydration-induced expression of LEA proteins in an anhydrobiotic chironomid. *Biochemical and Biophysical Research Communications* 348:56–61.

Kleynhans, E., and J. S. Terblanche. 2009. The evolution of water balance in *Glossina* (Diptera: Glossinidae): Correlations with climate. *Biology Letters* 5:93–96.

Kohnert, C. M. 2017. *Physiological Mechanisms of Desiccation Resistance in Fruit-Parasitic Rhagoletis Flies.* Bellingham, WA: Western Washington University.

Le Caherec, F., M.-T. Guillam, F. Beuron et al. 1997. Aquaporin-related proteins in the filter chamber of homopteran insects. *Cell and Tissue Research* 290:143–151.

Lee, C. E. 2002. Evolutionary genetics of invasive species. *Trends in Ecology and Evolution* 17:386–391.

Malacrida, A. R., L. M. Gomulski, M. Bonizzoni, S. Bertin, G. Gasperi, and C. R. Guglielmino. 2007. Globalization and fruit fly invasion and expansion: The medfly paradigm. *Genetica* 131:1–9.

Martínez, I., and Hernández-Ortiz. 1997. Anatomy of the reproductive system in six *Anastrepha* species and comments regarding their terminology in Tephritidae (Diptera). *Proceedings of the Entomological Society of Washington* 99:727–743.

Meats, A. 1989a. Acclimation, activity levels and survival. In *Fruit Flies: Biology, Natural Enemies and Control*, Eds. A. S. Robinson and G. H. S. Hooper, Vol. 3A, pp. 231–239. Rotterdam, the Netherlands: Elsevier World Crop Pest Series.

Meats, A. 1989b. Bioclimatic potential. In *Fruit Flies: Biology, Natural Enemies and Control*, Eds. A. S. Robinson and G. H. S. Hooper, Vol. 3A, pp. 241–252. Rotterdam, the Netherlands: Elsevier World Crop Pest Series.

Meats, A. 1989c. Water relations of Tephritidae. In *Fruit Flies: Biology, Natural Enemies and Control*, Eds. A. S. Robinson and G. H. S. Hooper, Vol. 3A, pp. 241–246. Rotterdam, the Netherlands: Elsevier World Crop Pest Series.

Mutamiswa, R., H. Machekano, F. Chidawanyika, and C. Nyamukondiwa. 2018. Thermal resilience may shape population abundance of two sympatric congeneric *Cotesia* species (Hymenoptera: Braconidae). *PLoS One* 13:e0191840.

Mwatawala, M., M. Virgilio, J. Joseph, and M. De Meyer. 2015. Niche partitioning among two Ceratitis rosa morphotypes and other *Ceratitis* pest species (Diptera, Tephritidae) along an altitudinal transect in Central Tanzania. *Zookeys* 429–442.

Naidu, S. G. 2001. Water balance and osmoregulation in *Stenocara gracilipes*, a wax-blooming tenebrionid beetle from the Namib Desert. *Journal of Insect Physiology* 47:1429–1440.

Naidu, S. G., and J. Hattingh. 1988. Water balance and osmoregulation in *Physademia globosa*, a diurnal tenebrionid beetle from the Namib Desert. *Journal of Insect Physiology* 34:911–917.

Nelson, D. R., and R. E. Lee, Jr. 2004. Cuticular lipids and desiccation resistance in overwintering larvae of the goldenrod gall fly, *Eurosta solidaginis* (Diptera: Tephritidae). *Comparative Biochemistry and Physiology B* 138:313–320.

Nicolson, S. W. 1980. Water balance and osmoregulation in *Onymacris plana*, a tenebrionid beetle from the Namib Desert. *Journal of Insect Physiology* 26:315–320.

Nyamukondiwa, C., C. W. Weldon, S. L. Chown, P. C. le Roux, and J. S. Terblanche. 2013. Thermal biology, population fluctuations and implications of temperature extremes for the management of two globally significant insect pests. *Journal of Insect Physiology* 59:1199–1211.

Olson, L. J. 2006. The economics of terrestrial invasive species: A review of the literature. *Agricultural and Resource Economics Review* 35:178–194.

Pérez-Staples, D., S. Abraham, M. Herrera-Cruz et al. 2017. Evolutionary consequences of desiccation resistance in the male ejaculate. *Evolutionary Biology* 45:56–66.

Qiu, Y., C. Tittiger, C. Wicker-Thomas et al. 2012. An insect-specific P450 oxidative decarbonylase for cuticular hydrocarbon biosynthesis. *Proceedings of the National Academy of Sciences of the United States of America* 109:14858–14863.

Radhakrishnan, P., and P. W. Taylor. 2007. Seminal fluids mediate sexual inhibition and short copula duration in mated female Queensland fruit flies. *Journal of Insect Physiology* 53:741–745.

Radhakrishnan, P., and P. W. Taylor. 2008. Ability of male Queensland fruit flies to inhibit receptivity in multiple mates, and the associated recovery of accessory glands. *Journal of Insect Physiology* 54:421–428.

Ramløv, H., and R. E. Lee, Jr. 2000. Extreme resistance to desiccation in overwintering larvae of the gall fly *Eurosta solidaginis* (Diptera: Tephritidae). *Journal of Experimental Biology* 203:783–789.

Ryabova, A., A. Cherkasov, R. Yamaguchi, R. Cornette, T. Kikawada, and O. Gusev. 2016. LEA4 protein is likely to be involved in direct protection of DNA against external damage. *BioNanoScience* 6:554–557.

Sinclair, B. J., P. Vernon, C. Jaco Klok, and S. L. Chown. 2003. Insects at low temperatures: An ecological perspective. *Trends in Ecology & Evolution* 18:257–262.

Sørensen, J. G., M. F. Addison, and J. S. Terblanche. 2012. Mass-rearing of insects for pest management: Challenges, synergies and advances from evolutionary physiology. *Crop Protection* 38:87–94.

Tammariello, S. P., J. P. Rinehart, and D. L. Denlinger. 1999. Desiccation elicits heat shock protein transcription in the flesh fly, *Sarcophaga crassipalpis*, but does not enhance tolerance to high or low temperatures. *Journal of Insect Physiology* 45:933–938.

Tanga, C. M., F. M. Khamis, H. E. Z. Tonnang et al. 2018. Risk assessment and spread of the potentially invasive *Ceratitis rosa* Karsch and *Ceratitis quilicii* De Meyer, Mwatawala & Virgilio sp. Nov. using life-cycle simulation models: Implications for phytosanitary measures and management. *PLoS One* 13:e0189138.

Tejeda, M. T., J. Arredondo, D. Pérez-Staples, P. Ramos-Morales, P. Liedo, and F. Díaz-Fleischer. 2014. Effects of size, sex and teneral resources on the resistance to hydric stress in the tephritid fruit fly *Anastrepha ludens*. *Journal of Insect Physiology* 70:73–80.

Tejeda, M. T., J. Arredondo, P. Liedo, D. Pérez-Staples, P. Ramos-Morales, and F. Díaz-Fleischer. 2016. Reasons for success: Rapid evolution for desiccation resistance and life-history changes in the polyphagous fly *Anastrepha ludens*. *Evolution* 70:2583–2594.

Tejeda, M. T., J. Arredondo-Gordillo, D. Orozco-Davila, L. Quintero-Fong, and F. Díaz-Fleischer. 2017. Directional selection to improve the sterile insect technique: Survival and sexual performance of desiccation resistant *Anastrepha ludens* strains. *Evolutionary Applications* 10:1020–1030.

Telonis-Scott, M., K. M. Guthridge, and A. A. Hoffmann. 2006. A new set of laboratory-selected *Drosophila melanogaster* lines for the analysis of desiccation resistance: Response to selection, physiology and correlated responses. *Journal of Experimental Biology* 209:1837–1847.

Vaníčková, L., R. Břízová, A. Pompeiano, S. Ekesi, and M. De Meyer. 2015. Cuticular hydrocarbons corroborate the distinction between lowland and highland Natal fruit fly (Tephritidae, *Ceratitis rosa*) populations. *Zookeys* 540:507–524.

Vera, M. T., R. Rodriguez, D. F. Segura, J. L. Cladera, and R. W. Sutherst. 2002. Potential geographical distribution of the Mediterranean fruit fly, *Ceratitis capitata* (Diptera: Tephritidae), with emphasis on Argentina and Australia. *Environmental Entomology* 31:1009–1022.

Weldon, C. W., and P. W. Taylor. 2010. Desiccation resistance of adult Queensland fruit flies *Bactrocera tryoni* decreases with age. *Physiological Entomology* 35:385–390.

Weldon, C. W., C. Nyamukondiwa, M. Karsten, S. L. Chown, and J. S. Terblanche. 2018. Geographic variation and plasticity in climate stress resistance among southern African populations of *Ceratitis capitata* (Wiedemann) (Diptera: Tephritidae). *Scientific Reports* 8:9849.

Weldon, C. W., L. Boardman, D. Marlin, and J. S. Terblanche. 2016. Physiological mechanisms of dehydration tolerance contribute to the invasion potential of *Ceratitis capitata* (Wiedemann) (Diptera: Tephritidae) relative to its less widely distributed congeners. *Frontiers in Zoology* 13:15.

Weldon, C. W., S. Mnguni, F. Démares et al. 2019. Adult diet of a tephritid fruit fly does not compensate for impact of a poor larval diet on stress resistance. *Journal of Experimental Biology* 222: jeb192534.

Weldon, C. W., S. Yap, and P. W. Taylor. 2013. Desiccation resistance of wild and mass-reared *Bactrocera tryoni* (Diptera: Tephritidae). *Bulletin of Entomological Research* 103:690–699.

White, I. M., and M. M. Elson-Harris. 1992. *Fruit Flies of Economic Significance: Their Identification and Bionomics*. Wallingford, UK: CAB International.

Williams, J. B., and R. E. Lee, Jr. 2005. Plant senescence cues entry into diapause in the gall fly *Eurosta solidaginis*: Resulting metabolic depression is critical for water conservation. *Journal of Experimental Biology* 208:4437–4444.

Williams, J. B., and R. E. Lee, Jr. 2011. Effect of freezing and dehydration on ion and cryoprotectant distribution and hemolymph volume in the goldenrod gall fly, *Eurosta solidaginis*. *Journal of Insect Physiology* 57:1163–1169.

Williams, J. B., N. C. Ruehl, and R. E. Lee, Jr. 2004. Partial link between the seasonal acquisition of cold-tolerance and desiccation resistance in the goldenrod gall fly *Eurosta solidaginis* (Diptera: Tephritidae). *Journal of Experimental Biology* 207:4407–4414.

Woods, H. A., and J. F. Harrison. 2002. Interpreting rejections of the beneficial acclimation hypothesis: When is physiological plasticity adaptive? *Evolution* 56:1863–1866.

4 Mating Compatibility between Two Populations of *Anastrepha fraterculus* (Wiedemann) (Diptera: Tephritidae) from Argentina and Uruguay

Felicia Duarte, María V. Calvo, Soledad Delgado, María Teresa Vera, Flávio M. García, and Iris B. Scatoni*

CONTENTS

Abstract *Anastrepha fraterculus* (Wiedemann) has been reported to show extensive morphological variation along its geographic distribution and is currently recognized as a complex of cryptic species composed of at least eight different morphotypes. The Brazilian-1 morphotype includes the Argentinean and southern Brazilian populations. To contribute with basic information on the distribution of *A. fraterculus* morphotypes, the sexual compatibility between a Uruguayan and an Argentinean population was evaluated. Mating compatibility was evaluated in field cages under semi-natural conditions. The Argentinean population was obtained from a colony of the laboratory of the Instituto de Genética "E. A. Favret" (INTA Castelar), Buenos Aires, established in 2007. The Uruguayan population came from infested fruits of *Acca sellowiana* (Berg. 1855) Burret 1941 (Myrtaceae). At the moment of the trials, Argentinean flies were between 11 and 17 days old and Uruguayan flies were between 16 and 26 days old. Sexual compatibility was established using the index of sexual isolation (ISI), the male and female relative performance indices (MRPI, FRPI), and a Kruskal-Wallis one-way analysis of variance with subsequent pairwise comparison tests of the four types of pairs formed according to male and female origin. Latency, mating duration, and location of the couples were also recorded. The ISI value was significantly different from zero because

* Corresponding author.

of a greater performance of the Argentinean adults. There were no significant differences between the frequency of homotypic Uruguayan couples and heterotypic couples, whereas the frequency of Argentinean homotypic couples was significantly higher than the rest. No significant differences were found for the other evaluated parameters. Results suggest that Uruguayan populations belong to the Brazilian-1 morphotype considering that the greater performance of Argentinean flies is probably because of faster sexual maturation rates and an inherent greater mating propensity rather than to reproductive isolation.

4.1 INTRODUCTION

Anastrepha fraterculus (Wiedemann), the South American fruit fly, is widely distributed from the Rio Grande Valley in northern Mexico to central Argentina (Malavasi et al. 1999). With more than 100 plants reported as hosts (Norrbom 2004), it is a species of major economic importance in many countries in South America, not only because of its destructive potential but also because of quarantine restrictions imposed on fruit export (Steck 1999).

The South American fruit fly has long been reported to show extensive morphological variation along its geographic distribution (Lima 1934, Stone 1942, Steck 1999, Hernández-Ortíz et al. 2004, 2015). Many studies confirm that this morphological variation is associated with differences in host use, the presence and degree of reproductive isolation, karyotypic differences, isozyme divergence, and DNA sequence divergence. The existence of this morphological variation has been revealed by multivariate morphometric analyses (Hernández-Ortiz et al. 2012). An extensive list of bibliography reviewing this research can be consulted in Rull et al. (2013), Cladera et al. (2014), Hernández-Ortíz et al. (2015), and Manni et al. (2015). In consequence, *A. fraterculus* is currently recognized as a complex of cryptic species composed of at least eight different morphotypes clustered into three phenotypic lineages. The Mesoamerican-Caribbean lineage consists of Mexican and Venezuelan morphotypes. The Andean lineage consists of the Andean, the Peruvian, and the Ecuadorian morphotypes. Finally, the Brazilian lineage is composed of three Brazilian morphotypes: Brazilian-1, Brazilian-2, and Brazilian-3 (Hernández-Ortiz et al. 2015).

In the Brazilian lineage, the Brazilian-1 morphotype includes the Argentinean populations (Alberti et al. 2002, Petit-Marty et al. 2004, Vera et al. 2006) and southern Brazilian populations (Smith-Caldas et al. 2001, Alberti et al. 2002, Basso et al. 2003, Hernández-Ortíz et al. 2004, Rull et al. 2013, Vaníčková et al. 2015, Dias et al. 2016). In the state of Sao Paulo, Brazilian-1 overlaps with Brazilian-2, where both morphotypes maintain their genetic integrity despite sympatry and partial reproductive compatibility (Selivon et al. 2005). The Brazilian-3 morphotype was also found in sympatry with Brazilian-1 and Brazilian-2 in the coastal areas of the state of Sao Paulo and in the inland plateau of southeastern and southern Brazil (Selivon et al. 2004).

The main practical reason that makes the delimitation of the *A. fraterculus* morphotypes essential is that it is a basic requirement for the implementation of the sterile insect technique (SIT). SIT is a method of pest control that consists of inundative releases of sterile insects into a wide area to reduce reproduction in a field population of the same species (FAO-IPPC 2016). A great research effort is being made to gather basic knowledge to enable the adjustment of this technique for the control of *A. fraterculus* (Cladera et al. 2014). In addition to the control of the pest, the key benefits of SIT derive mostly from a reduction in pesticide use, minimizing environmental and health costs. Furthermore, SIT is an environmentally friendly strategy for use in area-wide pest management (AWPM) because it can be applied not only in commercial orchards but also in backyards and urban areas not usually protected by insecticides (FAO/IAEA 2005, Dias and García 2014).

Although there is a great deal of research attempting to elucidate how the complex of cryptic species under the name of *A. fraterculus* is composed, none of it includes populations from Uruguay. Rull et al. (2013) suggest a geographical range of the Brazilian-1 morphotype from

Monte Alegre do Sul in southeastern Brazil to Buenos Aires, Argentina; the southern limit of this morphotype has been determined up to the present day. To contribute with basic information on the distribution of *A. fraterculus* morphotypes and to prepare for an eventual application of the SIT in Uruguay, a study on the sexual compatibility between a population from Uruguay and a population from Argentina (Brazilian-1 morphotype) of *A. fraterculus* was carried out.

4.2 MATERIALS AND METHODS

4.2.1 Biological Material

The Argentinean population of *A. fraterculus* was obtained from a laboratory colony. This population was derived from an experimental colony kept at the Estación Experimental Agroindustrial Obispo Colombres (EEAOC), Tucumán, Argentina, which was originally established with pupae recovered from infested guavas (*Psidium guajava* L., Myrtaceae) at the vicinity of Tafi Viejo (Tucuman) in 1997, (Jaldo et al. 2001, Vera et al. 2007).

The Uruguayan population was obtained directly from infested fruits of *Acca sellowiana* (Berg. 1855) Burret 1941 (Myrtaceae) collected from a commercial orchard located in Montevideo (34° 44′S; 56° 16′ W). Fruits were placed in sandboxes covered with voile and checked daily to separate the pupae.

Pupae and adult flies from both origins were maintained under the same controlled conditions (temperature [T]: 23°C, relative humidity [RH]: 60%–70%, photoperiod: 10L–14D) until the day of the trials. Pupae were first placed separately in emergence cages. After emergence, flies were separated and placed in 1-L containers with 25 adult flies each and sorted according to date, sex, and origin. In each container, adult flies were supplied with water and a diet composed of bee honey, hydrolyzed yeast (in a 3:1 ratio), and food dye (Laboratorio Fleibor S.R.L., Buenos Aires, Argentina). The Argentinean population was colored red and the Uruguayan population was colored blue.

4.2.2 Mating Compatibility Test

The test took place in the campus of the Facultad de Agronomía, Universidad de la República, Uruguay. Three field cages made of screen fabric, measuring 3 meters in diameter by 2 meters in height, were set inside a greenhouse to obtain suitable temperature conditions at the moment of release. Six potted plants of *Citrus limon* (Rutaceae) var. "Limon criollo" were placed inside each cage to provide perching places.

Between June 9 and 22, 2018, eight replicates of the trial were carried out: three on June 9, one on June 13, three on June 20, and one on June 22. The dates of the replications depended on the availability of sexually mature adults. At the moment of the trials, Argentinean flies were between 11 and 17 days old and Uruguayan flies were between 16 and 26 days old.

Twenty-five males and 25 females of each origin were released inside of each field cage. Because mating occurs mainly at sunrise (Malavasi et al. 1983, Vera et al. 2006) and given that the experiments were carried out in winter (where the average temperature is 10.4°C), the photoperiod in the rearing room was adjusted so the lights were turned on at 10 am to allow the greenhouse to reach temperatures higher than 18°C at the moment when the flies were released. The maximum temperature recorded during the tests was 32°C and the mean relative humidity (RH) was of 75%.

Most releases started at 9:45 am with males and finished at 10:00 am with females. Only on June 13, due to prevailing low temperatures, fruit flies were kept in the dark and males and females were released at 10:45 am and 11:00 am, respectively. One observer remained inside each cage until the end of the trials, until 2 hours after the last male callings occurred and sexual activity ceased. Each observed mating pair was collected in a 50 mL vial and placed in the shade until the pair disengaged. Male and female color, time of start and end of copulation, and location were recorded for each mating pair. Location was recorded as either net, ground, stem, abaxial-adaxial side of leaf, height on tree, and cardinal point (FAO/IAEA/USDA 2014).

4.2.3 Data Analysis

The percentage of mating inside each cage was calculated to corroborate that all replicates had reached at least 20% of potential couples on the plants (FAO/IAEA/USDA 2014).

Mating compatibility was assessed by means of the Index of Sexual Isolation (ISI). An ISI value of 1 indicates complete assortative mating, an ISI value of −1 indicates complete outbreeding, and an ISI estimate of 0 indicates random mating. Mating competitiveness was assessed by means of the Male and Female Relative Performance Indices (MRPI and FRPI). MRPI varies from 1, when males of one of the tested populations engage in all the copulations, to −1, when males from the other population are present in all copulations. The range for FRPI is the same as that of MRPI. Values close to zero indicate similar participation from males (MRPI) or females (FRPI) independently of the origin of the population (Cayol et al. 1999). To verify random mating, confidence intervals at 95% were estimated to assess departures from zero (Rull et al. 2013). Differences among mating combination frequencies were tested using a Kruskall–Wallis one-way analysis of variance by ranks, followed by pairwise comparisons (Conover 1999). Statistical analyses were performed with InfoStat/Libre (Infostat 2018).

Data recorded for latency to first mating and mating duration were log transformed and compared between mating combinations by means of a one-way ANOVA followed by Tukey multiple comparison tests. Differences between homotypic couples in location of mating on plant height and cardinal point were tested using a Kruskall–Wallis one-way ANOVA by ranks and pairwise comparisons.

4.3 RESULTS

The percentage of mating exceeded 20% in all cages, with a mean value of 28.8% ± 2.59%, indicating that the environment was suitable for mating. The ISI, MRPI, and FRPI were statistically different from zero (Figure 4.1). The sexual isolation index was 0.29 ± 0.13, a positive value that indicates that there were more homotypic than heterotypic mating couples. The mating performance index was 0.23 ± 0.07 for males and 0.18 ± 0.12 for females, which implies that males and females from the Argentinean population were engaged in most of the copulations, which indicates a greater activity of the Argentinean population in both sexes (Figure 4.1). In addition, when we compared the number of couples per type of mating combination, there were no statistically significant differences among the heterotypic mating combinations and the Uruguayan homotypic combination, whereas the frequency of the homotypic combination of the Argentinean population was about twice the frequency of the homotypic combination of the Uruguayan population (Figure 4.2).

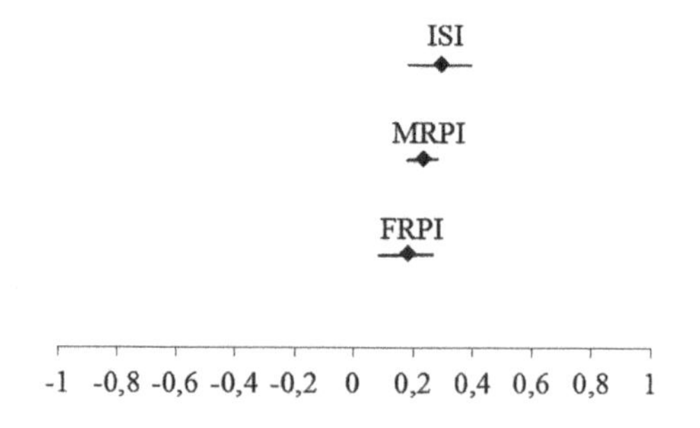

FIGURE 4.1 Index of sexual isolation (ISI) and relative performance indices for males (MRPI) and females (FRPI) with associated 95 Cis.

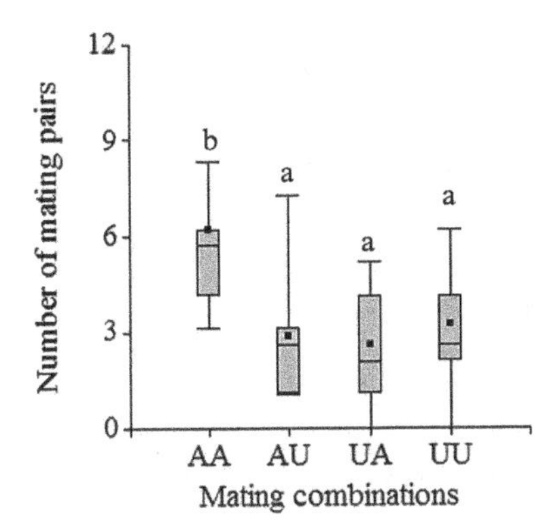

FIGURE 4.2 Mean, median, and quartiles of each mating combination. Capital letters indicate the origin of fruit flies: A refers to fruit flies from Argentina and U refers to fruit flies from Uruguay. The first letter indicates the origin of males and the second letter the origin of females. Data with the same lowercase letter are not significantly different according to the Kruskal–Wallis test ($P < 0.02$) and pairwise comparisons tests ($P < 0.05$). (From Conover, W.J., *Practical Nonparametric Statistics*, John Wiley & Sons, New York, 1999).

TABLE 4.1
Mean ± SE Mating Duration and Latency in Minutes

Mating Combination	Mating Duration	Latency
AA	59.5 ± 6.8	45.9 ± 5.4
AU	55.6 ± 6.5	34.5 ± 6.7
UA	45.1 ± 8.5	51.0 ± 14.4
UU	60.5 ± 6.8	43.6 ± 7.7

n = 8; Data were log transformed prior to the ANOVA and no statistical differences were found for either mating duration ($P = 0.23$) or latency ($P = 0.4$). Capital letters indicate the origin of fruit flies: A refers to fruit flies from Argentina and U refers to fruit flies from Uruguay. The first letter indicates the origin of males and the second letter the origin of females.

There were no statistically significant differences ($P = 0, 13$) in location on plant height, and just 1 out the 113 couples that mated on the leaves did so on the adaxial side. Matings outside the tree involved 17.2% of those recorded, with 9% taking place on the floor of the cage and 8.2% on the net walls and the ceiling. Matings on the ceiling were all homotypic mating couples from Uruguay, whereas matings on the floor were all homotypic mating couples from Argentina. Mating duration and latency were not statistically different among mating combinations (Table 4.1).

All copulations occurred mostly during the first hour after flies were released (Figure 4.3), and no sexual activity was observed after 3 hours.

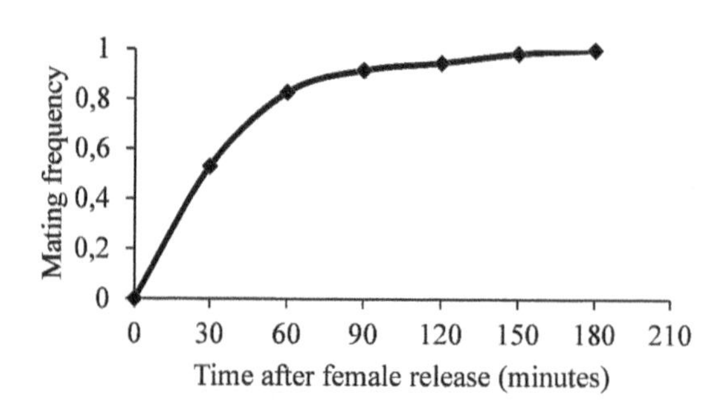

FIGURE 4.3 Accumulated mating frequency as a function of time.

4.4 DISCUSSION

In previous studies where *A. fraterculus* populations from southern South America showed full sexual compatibility, ISI values ranged between −0.01 and −0.14. Pairwise comparisons included populations from Argentina (Petit-Marti et al. 2004), southern Brazil (Dias et al. 2016), and Argentina and southern Brazil (Rull et al. 2012). In comparisons among populations with strong sexual incompatibility, such as the Mexican, Peruvian, and Brazilian-1 linages, ISI values varied between 0.74 and 0.92 (Vera et al. 2006, Cáceres et al. 2009, Rull et al. 2012), whereas populations from southeastern Brazil mating with one Argentinean population and one Peruvian population showed partial incompatibility, with ISI values of 0.43 and 0.55, respectively (Vera et al. 2006, Dias et al. 2016).

The ISI value obtained during this study (0.29) indicates a tendency toward random mating, being closer to 0 than to 1 or −1. Comparisons among different mating combinations showed similar frequencies for the Uruguayan homotypic combination and both heterotypic combinations; thus, the slight deviance from 0 (random mating) could be attributed to a greater propensity to mate of the Argentinean laboratory population. Liedo et al. (2002) observed that laboratory-reared females of *Ceratitis capitata* were more prone to mate than wild females, and that laboratory flies, both females and males, matured much earlier than wild flies. Calkins (1984) stated that it takes a few generations for insects to thrive under artificial rearing conditions and that the rearing regime probably selects for simpler courtship and mating behavior and also for earlier reproduction than in wild populations. That is considering that long and intense courtships have a risk of being interrupted by other males adapted to crowded conditions. In addition, because under artificial conditions all females become sexually mature at about the same time and earlier than wild females, an intense competition between females for fit males is promoted, favoring females that require fewer criteria for male selection (Calkins 1984, Calcagno et al. 1999).

Argentinean males had a better mating performance than Uruguayan males. Although it is expected that males from artificial-rearing conditions are less competitive than wild males when they are competing for wild females (Liedo et al. 2002), the better performance of Argentinean males was as a result of their mating with females coming from the same artificial-rearing conditions. In addition, the evaluated laboratory colony of *A. fraterculus* has shown good competitiveness in previous studies, particularly when given the same adult diet as wild males (Rull et al. 2013, Vera et al. 2007). If we consider only Uruguayan females, Uruguayan and Argentinean males showed a similar performance. It is possible that the Argentinean population had a greater proportion of sexually mature individuals because of artificial selection during rearing and, in some way, it favored a greater mating performance in this population.

Regarding the location on the plant and cardinal point, we did not find significant differences. However, it is important to consider that to interpret the results on location on the plant, we have to take into account that the plants were relatively small, although they provided enough foliage for flies to perch on, and it was not possible to clearly delimit the top, middle, and bottom. A similar situation occurred with the cardinal points.

Most of the copulations occurred during the first 2 hours after artificial sunrise, which agrees with the results of Petit-Marty et al. (2004). Also, there were no differences in latency, indicating that there is no temporal isolation between these populations.

Different studies agree with the fact that the morphotype present in Argentina and southern Brazil is the Brazilian-1 morphotype (Hernández-Ortíz 2012). Petit-Marty et al. (2004) evaluated mating compatibility among four populations from Argentina, including a population from Tucumán, the same region where the Argentinean flies used in the present study came from, and confirmed full compatibility among populations. Rull et al. (2013) evaluated the reproductive compatibility among two strains from southern Brazil (Pelotas and Vacaria) and one strain from Argentina (Tucumán) and also found that they were fully compatible. These results were later confirmed by a study with a multidisciplinary approach that included wild flies of four populations from southern Brazil (Bento Gonçalvez, Pelotas, Vacaria and São Joaquim) (Dias et al. 2016) and by a recent courtship-behavior study including different populations of *A. fraterculus* (Roriz et al. 2019).

The distribution suggested by Rull et al. (2013) puts Montevideo, Uruguay, in the distribution limits of the Brazilian-1 morphotype. The relatively small size and geographical location of Uruguay, with relatively homogeneous climate conditions and a geographical landscape with a topography without great variation in elevation (INIA 2018), suggest that Brazilian-1, reported as the only morphotype currently present in Argentina and also present in the south of Brazil, would be the morphotype that is present in Uruguay. Although our results are not entirely conclusive, if we consider only the ISI value, the fact that the mating combination of flies that deviated from the expected behavior were those coming exclusively from artificial breeding could be indicating that the observed behavioral differences were mainly as a result of breeding conditions and origin rather than to interpopulation differences. In any case, it is important to consider that the three morphotypes of the Brazilian lineage are reported to coexist in Vale do Paraiba in the state of Sao Paulo (Selivon et al. 2004, 2005), and hence, it is still not clear whether the distribution of the Brazilian morphotypes is related to lower and higher altitudes or to a north/south differentiation (Selivon et al. 2005, Vaníčková et al. 2015).

This is the first study intending to begin to clarify the taxonomic situation of *A. fraterculus* in Uruguay. Further studies using molecular and morphometric approaches need to be carried out before a specific morphotype can be conclusively assigned. Comparing populations of wild origin from Argentina, Uruguay, and Brazil would be useful to resolve this issue.

ACKNOWLEDGMENTS

We are very grateful to Diego Segura from INTA Castelar, Buenos Aires, Argentina, for providing constructive comments; to Angelo Turra, Evelyn Pechi, Nicolas Yakimik, and Christian Inzaurralde for assistance during the field cage trials; to Cristina Mori from the UdelaR, Uruguay, for loaning the greenhouse; and to Carlos Caceres from the FAO/IAEA Agriculture and Biotechnology Laboratories in Seibersdorf, Austria for supplying the pupae.

This study was financed by Fondo de Promoción de Tecnología Agropecuaria of the Instituto Nacional de Investigación Agropecuaria.

REFERENCES

Alberti, A. C., M. S. Rodriguero, P. G. Cendra, B. O. Saidman, and J. C. Vilardi. 2002. Evidence indicating that Argentine populations of *Anastrepha fraterculus* (Diptera: Tephritidae) belong to a single biological species. *Annals of the Entomological Society of America* 95:505–512.

Basso, A., A. Sonvico, L. A. Quesada-Allue, and F. Manso. 2003. Karyotypic and molecular identification of laboratory stocks of the South American fruit fly *Anastrepha fraterculus* (Wied) (Diptera: Tephritidae). *Journal of Economic Entomology* 96:1237–1244.

Cáceres C., D. F. Segura, M. T. Vera et al. 2009. Incipient speciation revealed in *Anastrepha fraterculus* (Diptera; Tephritidae) by studies on mating compatibility, sex pheromones, hybridization, and cytology. *Biological Journal of the Linnean Society* 97:152–165.

Calcagno, G. E., M. T. Vera, F. Manso, S. A. Lux, F. M. Norry, and F. N. Munyiri. 1999. Courtship behavior of wild and mass-reared Mediterranean fruit fly (Diptera: Tephritidae) males from Argentina. *Journal of Economic Entomology* 92:373–379.

Calkins, C. O. 1984. The importance of understanding fruit fly mating behavior in sterile male release Programs (Diptera, Tephritidae). *Folia Entomológica Mexicana* 61:205–213.

Cayol, J. P., J. C. Vilardi, E. Rial, and T. Vera. 1999. New Indices and method to measure the sexual compatibility and mating performance of medfly (Diptera: Tephritidae) laboratory reared strains under field cage conditions. *Journal of Economic Entomology* 92:140–145.

Cladera, J. L., J. C. Vilardi, M. Juri et al. 2014. Genetics and biology of *Anastrepha fraterculus*: Research supporting the use of the sterile insect technique (SIT) to control this pest in Argentina. *BMC Genetics* 15(2):1–14.

Conover, W. J. 1999. *Practical Nonparametric Statistics.* New York: John Wiley & Sons.

Dias, N., and F. R. M. Garcia. 2014. Fundamentos da técnica do inseto estéril (TIE) para o controle de Moscas das frutas (Diptera, Tephritidae). *O Biologico* 76:58–62.

Dias, V. S., J. G. Silva, K. M. Lima et al. 2016. An integrative multidisciplinary approach to understanding cryptic divergence in Brazilian species of the *Anastrepha fraterculus* complex (Diptera: Tephritidae). *Biological Journal of the Linnean Society* 117:725–746.

FAO/IAEA. 2005. *Environmental Benefits of Medfly Sterile Insect Technique in Madeira and Their Inclusion in a Cost–Benefit Analysis.* IAEA-TECDOC-1475. Vienna, Austria: International Atomic Energy Agency.

FAO/IAEA/USDA. 2014. *Product Quality Control for Sterile Mass-Reared and Released Tephritid Fruit Flies*, Version 6.0. Vienna, Austria: International Atomic Energy Agency.

FAO-IPPC. 2016. *International Standards for Phytosanitary Measures. Glossary of Phytosanitary Terms.* Roma, Italy: Food and Agriculture Organization.

Hernández-Ortiz, V, A. F. Bartolucci, P. Morales-Valles, D. Frías, and D. Selivon. 2012. Cryptic Species of the *Anastrepha fraterculus* Complex (Diptera: Tephritidae): A Multivariate Approach for the Recognition of South American Morphotypes. *Annals of the Entomological Society of America* 105(2):305–318.

Hernández-Ortiz, V., J. A. Gómez-Anaya, A. Sánchez, B. A. McPheron, and M. Aluja. 2004. Morphometric analysis of Mexican and South American populations of the *Anastrepha fraterculus* complex (Diptera: Tephritidae), and recognition of a distinct Mexican morphotype. *Bulletin of Entomological Research* 94:487–499.

Hernández-Ortiz, V., N. A. Canal, J. O. Tigrero Salas, F. M. Ruíz-Hurtado, and J. F. Dzul-Cauich. 2015. Taxonomy and phenotypic relationships of the *Anastrepha fraterculus* complex in the Mesoamerican and Pacific Neotropical dominions (Diptera, Tephritidae). *ZooKeys* 540:95–124.

Infostat. 2018. Infostat/Libre. Version 2018. http://www.infostat.com.ar. Accessed November 2018.

INIA-Instituto Nacional de Investigación Agropecuaria. 2018. Características Geográficas y Socioeconómicas del Uruguay. http://www.inia.org.uy/disciplinas/agroclima/uruguay_gral.htm. Accessed November 2018.

Jaldo H. E., M. C. Gramajo, and E. Willink. 2001. Mass rearing of *Anastrepha fraterculus* (Diptera: Tephritidae): A preliminary strategy. *Florida Entomologist* 84:716–718.

Liedo, P., E. De Leon, M. I. Barrios, J. F. Valle-Mora, and G. Ibarra. 2002. Effect of age on the mating propensity of the Mediterranean fruit fly (Diptera: Tephritidae). *Florida Entomologist* 85(1):94–101.

Lima, A. C. 1934. Moscas de frutas do genero *Anastrepha* Schiner, 1868 (Diptera, Trypetidae). *Memórias do Instituto Oswaldo Cruz* 28:487–575.

Malavasi, A., J. S. Morgante, and R. J. Prokopy. 1983. Distribution and activities of *Anastrepha fraterculus* (Diptera: Tephritidae) flies on host and non-host trees. *Annals of the Entomological Society of America* 76:286–292.

Malavasi, A., R. A. Zucchi, and R. L. Sugayama. 1999. Biogeografía. In: *Moscas-das-frutas de importancia económica no Brasil. Conhecimento básico e aplicado*, ed. A. Malavasi and R. A. Zucchi, pp. 93–98. Ribeirao Preto, Brazil: Holos.

Manni, M., K. M. Lima, C. R. Guglielmino et al. 2015. Relevant genetic differentiation among Brazilian populations of *Anastrepha fraterculus* (Diptera, Tephritidae). *ZooKeys* 540:157–173.

Norrbom, A. L. 2004. Host plant database for *Anastrepha* and *Toxotrypana* (Diptera: Tephritidae: Toxotrypanini). Diptera Data Dissemination Disk 2 (CD-Rom). http://www.sel.barc.usda.gov:591/diptera/Tephritidae/TephIntro.html.

Petit-Marty, N., M. T. Vera, G. Calcagno et al. 2004. Sexual behavior and mating compatibility among four populations of *Anastrepha fraterculus* (Diptera: Tephritidae) from Argentina. *Annals of the Entomological Society of America* 97:1320–1327.

Roriz A. K. P., H. F. Japyassú, C. Cáceres, M. T. Vera, and I. Joachim-Bravo. 2019. Pheromone emission patterns and courtship sequences across distinct populations within *Anastrepha fraterculus* (Diptera-Tephritidae) cryptic species complex. *Bulletin of Entomological Research* 109:408–417.

Rull, J, S. Abraham, A. Kovaleski et al. 2012. Random mating and reproductive compatibility among Argentinean and southern Brazilian populations of *Anastrepha fraterculus* (Diptera: Tephritidae). *Bulletin of Entomological Research* 102:435–443.

Rull, J., S. Abraham, A. Kovaleski et al. 2013. Evolution of pre-zygotic and post-zygotic barriers to gene flow among three cryptic species within the *Anastrepha fraterculus* complex. *Entomologia Experimentalis et Applicata* 148:213–222.

Selivon, D., A. L. P. Perondini, and J. S. Morgante. 2005. A genetic-morphological characterization of two cryptic species of the *Anastrepha fraterculus* complex (Diptera: Tephritidae). *Annals of the Entomological Society of America* 98:367–381.

Selivon, D., C. Vretos, L. Fontes, and A. L. P. Perondini. 2004. New variant forms in the *Anastrepha fraterculus* complex (Diptera, Tephritidae). In *Proceedings 6th International Symposium on Fruit Flies of Economic Importance*, ed. B. N. Barnes, pp. 253–258. Stellenbosch: Isteg Scientific Publications.

Smith-Caldas, M. R. B., B. A. McPheron, J. G. Silva, and R. A. Zucchi. 2001. Phylogenetic relationships among species of the *fraterculus* group (*Anastrepha*: Diptera: Tephritidae) inferred from DNA sequences of mitochondrial cytochrome oxidase I. *Neotropical Entomology* 30:565–573.

Steck, G. J. 1999. Taxonomic status of *Anastrepha fraterculus*. In *The South American Fruit Fly, Anastrepha fraterculus (Wied.): Advances in Artificial Rearing, Taxonomic Status and Biological Studies*, ed. International Atomic Energy Agency, IAEA 13–20. Vienna, Austria: Tech-Doc 1064.

Stone, A. 1942. *The Fruit Flies of the Genus Anastrepha*. Washington, DC: United State Department of Agriculture.

Vaníčková, L., V. Hernández-Ortiz, I. Sordi Joachim Bravo et al. 2015. Current knowledge of the species complex *Anastrepha fraterculus* (Diptera, Tephritidae) in Brazil. *ZooKeys* 540:211–237.

Vera, T., C. Cáceres, V. Wornoayporn et al. 2006. Mating incompatibility among populations of the South American fruit fly *Anastrepha fraterculus* (Diptera: Tephritidae). *Annals of the Entomological Society of America* 99:387–397.

Vera, T., S. Abraham, A. Oviedo, and E. Willink. 2007. Demographic and quality control parameters of *Anastrepha fraterculus* (Diptera: Tephritidae) maintained under artificial rearing. *Florida Entomologist* 90(1):53–57.

Section II

Taxonomy and Morphology

5 Review of *Anastrepha* (Diptera: Tephritidae) Immature Stage Taxonomy

Gary J. Steck, Erick J. Rodriguez, Allen L. Norrbom, Vivian S. Dutra, Beatriz Ronchi-Teles, and Janisete Gomes Silva*

CONTENTS

Abstract Taxonomic study of fruit fly immature stages is important for developing identification keys, especially of pest species, and understanding phylogeny of the Tephritidae. A review of the entomological literature revealed 78 publications describing one or more of the immature stages (egg, larvae, pupa) of the genus *Anastrepha* dating from 1909 to the present. Descriptions of varying quality exist for larvae of 27 species and eggs of 49 species. A table listing 74 species of *Anastrepha* with corresponding publications (or lack of), and annotations on their descriptive content is provided. A diagnosis of *Anastrepha* larvae distinguishes them from other genera. Synapomorphies to distinguish larvae of the species groups of *Anastrepha* have not been found, except for the *curvicauda* group. Taxonomic study of immature stages will advance with new collections, detailed scanning electron microscopic observation, and development of multi-entry keys.

5.1 INTRODUCTION

Invasive fruit fly pests move to new areas primarily through human transport of infested fruits bearing the immature stages. Much of the morphological study of fruit fly immature stages has been driven by the need to identify larvae at all ports of entry where infested fruits are frequently intercepted. In the United States, for example, commodities infested with fruit flies arrive on a daily basis at one or more of the many ports of entry. During the period from 1984 to 2015, more than 122,000 such instances were recorded, and the diversity of host material is very large, comprising at least 328 genera of plants; approximately 97% of the intercepted material arrived as "baggage" (AQAS:PestID 2016). Prior analysis of airline-baggage interception data between 1984 and 1999 suggested that air travel could be an important pathway for fruit fly invasion (Liebhold et al. 2006).

* Corresponding author.

Although traditional commercial shipments of produce have fewer total numbers of interceptions than passenger baggage, it is important to note that larvae could be present in other forms of transport such as e-commerce (Humair et al. 2015). The chronic invasions of fruit flies into California, Florida, and elsewhere may frequently originate from these various sources. Clearly, there is a need to understand the threat posed by this vast human-mediated movement of economically injurious pests by knowing the identity of the insects in transit. Knowledge of the sources, host plants, seasonality, and pathways of pest species allows development of mitigations to reduce their threat to agriculture.

From the 122,000+ documented interceptions of items infested with flies referenced previously, approximately 82,000 (67%) originated in the Americas (South and Central America, Mexico, and the West Indies) and were identified as genus *Anastrepha*. The genus is indigenous to the Americas, where it is the most speciose of the fruit-infesting tephritids and the most economically important because of its numerous major and minor pests of commercial agricultural and dooryard crops. Only about 300 (0.4%) of the *Anastrepha* interceptions were identified to species level, mostly as *Anastrepha ludens* (Loew), the Mexican fruit fly.

5.2 TAXONOMIC STUDY OF *ANASTREPHA* IMMATURE STAGES

To date, 305 species of *Anastrepha* have been described (Norrbom et al. 2012, 2015, 2018). Of these, larvae of 27 species have been described (9% of the total) with varying levels of detail. Similarly, eggs of 49 species have been described (16% of the total). The history and breadth of descriptive morphological work on *Anastrepha* immature stages can be seen in Table 5.1, which lists the published (and some unpublished) scientific papers and a brief summary of their contents on a species-by-species basis.

Fifty-one species of *Anastrepha* are considered to be of economic interest. Designation as major or minor pests is somewhat arbitrary, but in general, those species that are important to commercial agriculture are considered major pests, and those species that primarily infest edible, noncommercial dooryard or forest hosts are considered minor pests. Most of these species are included in White and Elson-Harris (1992). Larvae have been described for 23 such pest species (43% of total pest species).

Published descriptions vary greatly in detail, quality, and originality. Table 5.1 summarizes this information by noting which developmental stage was described (egg, any of the three instars, pupa) denoted by an *x* in the table. An "original description" included some novel descriptive or measurement data. Numerous publications duplicated or assembled data or figures from previous authors into keys or comparative tables without adding original observations. Descriptions may be supplemented by photomicrographs, drawings, and scanning electron micrographs (SEMs), also denoted by an *x* in the table. In the Comments column, we further characterize the descriptions by their thoroughness. For example, a "thorough description" provided details for all major body features including head (sensory structures and associated lobes, oral ridges, cephalopharyngeal skeleton), anterior and posterior spiracles (internal and external portions), cuticle (spinules and creeping welts), usually including imagery (photomicrographs or SEM) and comparison with other species; a "partial description" lacked details for one or more major body features; a "basic description" lacked comparison to other species; and a "superficial description" had insufficient detail for identification or classification purposes.

The earliest larval descriptions beginning with Froggatt (1909) were of a few major pests, such as *A. ludens*, *Anastrepha fraterculus* (Wiedemann), *Anastrepha serpentina* (Wiedemann), *Anastrepha striata* Schiner, and *Anastrepha curvicauda* (Gerstaecker), with no attempt to describe larvae carefully enough to distinguish them from other pest species. Greene (1929) was the first to describe larvae in a comparative manner and provide an identification key to these same *Anastrepha* species plus those of other major pests: *Ceratitis capitata* (Wiedemann), *Bactrocera oleae* (Rossi), *Rhagoletis cingulata* (Loew), *Rhagoletis pomonella* (Walsh), and *Zeugodacus cucurbitae* (Coquillett). Over time, the number of species described and the general quality of descriptions increased. Notable

TABLE 5.1
Literature Describing Immature Stages of *Anastrepha* Species

Species/Species Group/Pest Status Publication	Original Description	Key	Photo	Drawing	SEM	Egg	1st	2nd	3rd	Pupa	Comments
***Anastrepha acris* Stone/*fraterculus*/minor**											Not described or keyed.
***Anastrepha alveata* Stone/*spatulata*/not a pest**											
Figueiredo et al. (2011, 2017)	Yes		x		x	x					Thorough description, comparisons among 17 *Anastrepha* spp.
***Anastrepha alveatoides* Blanchard/*spatulata*/not a pest**											
Norrbom et al. (1999)	Yes					x					Comparative table of egg dimensions, shape, and sculpture of 26 species.
***Anastrepha amita* Zucchi/*fraterculus*/not a pest**											
Norrbom et al. (1999)	Yes					x					Comparative table of egg dimensions, shape, and sculpture of 26 species.
Figueiredo et al. (2011, 2017)	Yes		x		x	x					Thorough description, comparisons among 17 *Anastrepha* spp.
Dutra et al. (2018b)	Yes		x		x			x	x		Thorough description.
***Anastrepha amplidentata* Norrbom/*fraterculus*/minor**											Not described or keyed.
***Anastrepha annonae* Norrbom/not assigned/minor**											Not described or keyed.
***Anastrepha antunesi* Lima/*fraterculus*/minor**											
Dutra et al. (2011a)	Yes		x		x	x					Thorough description.
***Anastrepha atrox* (Aldrich)/*mucronota*/minor**											
Norrbom (1985)	Yes					x					Comparative table of egg length, width, lobe length of 11 species.
Norrbom et al. (1999)	Yes					x					Comparative table of egg dimensions, shape, and sculpture of 26 species.

(Continued)

TABLE 5.1 (*Continued*)
Literature Describing Immature Stages of *Anastrepha* Species

Species/Species Group/Pest Status Publication	Original Description	Key	Photo	Drawing	SEM	Egg	1st	2nd	3rd	Pupa	Comments
***Anastrepha bahiensis* Lima/*fraterculus*/minor**											
Dutra et al. (2011a)	Yes		x		x	x					Thorough description.
Dutra et al. (2012)	Yes		x	x	x			x	x		Thorough description.
***Anastrepha barbiellinii* Lima/not assigned/not a pest**											
Norrbom et al. (1999)	Yes					x					Comparative table of egg dimensions, shape, and sculpture of 26 species.
***Anastrepha bella* Norrbom & Korytkowski/not assigned/not a pest**											
Norrbom and Korytkowski (2009)	Yes					x					Basic description.
***Anastrepha bezzii* Lima/*mucronota*/minor**											Not described or keyed.
***Anastrepha bistrigata* Bezzi/*striata*/minor**											
Steck and Malavasi (1988)	Yes			x		x	x	x	x	x	Thorough description, but lacking imagery. Comparison with larvae of other *Anastrepha* spp. that infest guava.
Steck et al. (1990)	Yes	Yes							x		Key to larvae of 13 *Anastrepha* spp. based on novel data.
White and Elson-Harris (1992)		Yes							x		Description based on Steck and Malavasi (1988).
Norrbom et al. (1999)						x					Comparative table of egg dimensions, shape, and sculpture of 26 species. Data from Steck and Malavasi (1988).
Carroll et al. (2004)		Yes		x							Description and drawings taken from Steck and Malavasi (1988).
Figueiredo et al. (2011, 2017)	Yes		x		x	x					Thorough description, comparisons among 17 *Anastrepha* spp.
***Anastrepha carreroi* Canal/*fraterculus*/minor**											Not described or keyed.
***Anastrepha consobrina* (Loew)/*pseudoparallela*/minor**											

(*Continued*)

TABLE 5.1 (*Continued*)
Literature Describing Immature Stages of *Anastrepha* Species

Species/Species Group/Pest Status Publication	Original Description	Key	Photo	Drawing	SEM	Egg	1st	2nd	3rd	Pupa	Comments
Lima (1930)	Yes		x						x		Superficial description, but clear images of CPS, anterior and posterior spiracles.
Norrbom (1985)									x		Comparative table of various larval characters of 14 species; data from various authors.
Figueiredo et al. (2011, 2017)	Yes		x		x	x					Thorough description, comparisons among 17 *Anastrepha* spp.
***Anastrepha cordata* Aldrich/*cryptostrepha*/not a pest**											
Norrbom (1985)	Yes				x	x					Comparative table of egg length, width, lobe length of 11 species.
Norrbom and Foote (1989)			x			x					Limited comparison with other *Anastrepha* species.
Norrbom et al. (1999)	Yes					x					Comparative table of egg dimensions, shape, and sculpture of 26 species.
Norrbom and Korytkowski (2009)	Yes					x					Basic description.
***Anastrepha coronilli* Carrejo & González/*fraterculus*/not a pest**											
Dutra et al. (2011a)	Yes		x		x	x					Thorough description.
Dutra el al. (2012)	Yes		x	x	x			x	x		Thorough description.
***Anastrepha curitis* Stone/*pseudoparallela*/minor**											
Dutra et al. (2013)	Yes		x		x	x					Thorough description.
Dutra et al. (2018a)	Yes		x		x			x	x		Thorough description.
***Anastrepha curvicauda* (Gerstaecker)/*curvicauda*/major**											
Knab and Yothers (1914)				x		x			x	x	As *Toxotrypana curvicauda*. Partial, basic description.
Greene (1929)	Yes	Yes		x					x	x	As *Toxotrypana curvicauda*. Keys to larvae and pupae. Partial description.
Baker et al. (1944)	Yes			x					x		As *Toxotrypana curvicauda*. Partial description (posterior spiracle).

(*Continued*)

TABLE 5.1 (*Continued*)
Literature Describing Immature Stages of *Anastrepha* Species

Species/Species Group/Pest Status Publication	Original Description	Key	Photo	Drawing	SEM	Egg	1st	2nd	3rd	Pupa	Comments
Phillips (1946)	Yes	Yes		x					x		As *Toxotrypana curvicauda*. Thorough description but lacking imagery.
Peterson (1951)	Yes			x					x		As *Toxotrypana curvicauda*. Partial description with good detail.
Deputy (1957)		Yes		x					x		As *Toxotrypana curvicauda*. Key to larvae of pest fruit flies. Superficial drawings modified from Greene (1929).
Berg (1979)		Yes		x					x		As *Toxotrypana curvicauda*. Key to larvae of pest fruit flies. Superficial drawing of caudal segment.
Heppner (1986)	Yes			x					x		As *Toxotrypana curvicauda*. Partial description but good detail and comparisons with other fruit fly larvae.
Foote (1991)									x		As *Toxotrypana curvicauda*. Drawing duplicated from Peterson (1951).
Carroll (1992)	Yes				x				x		As *Toxotrypana curvicauda*. Partial description (lacks CPS, cuticular spinules).
White and Elson-Harris (1992)		Yes							x		As *Toxotrypana curvicauda*. Partial description based on Phillips (1946) and Heppner (1986).
Norrbom et al. (1999)						x					As *Toxotrypana curvicauda*. Comparative table of egg dimensions, shape, and sculpture of 26 species. Data from Knab and Yothers (1914).
Carroll et al. (2004)	Yes	Yes			x				x		As *Toxotrypana curvicauda*. Thorough description based on Heppner (1986), Phillips (1946), unpublished data.
Frías et al. (2006)	Yes	Yes		x	x				x		As *Toxotrypana curvicauda*. Key to genus of pest New World fruit flies. Thorough description (except lacking cuticular spinules).
Frías et al. (2008)		Yes	x	x	x				x		As *Toxotrypana curvicauda*. General comparison with larvae of other fruit flies.
***Anastrepha* sp./*curvicauda*/not a pest**											
Frías et al. (2006)	Yes			x	x				x		As *Toxotrypana* sp. Thorough description (except lacking cuticular spinules).

(*Continued*)

TABLE 5.1 (*Continued*)
Literature Describing Immature Stages of *Anastrepha* Species

Species/Species Group/Pest Status Publication	Original Description	Key	Photo	Drawing	SEM	Egg	1st	2nd	3rd	Pupa	Comments
***Anastrepha daciformis* Bezzi/*daciformis*/ not a pest**											Not described or keyed.
***Anastrepha distincta* Greene/*fraterculus*/ minor**											
Norrbom (1985)	Yes								x		Comparative table of various larval characters of 14 species; data from various authors.
Steck et al. (1990)	Yes	Yes							x		Key to larvae of 13 *Anastrepha* spp. based on novel data.
Carroll (1992)	Yes				x				x		Partial description (lacks CPS and cuticular spinules).
Carroll et al. (2004)	Yes	Yes			x				x		Thorough description based on Steck et al. (1990), unpublished data.
Dutra et al. (2011a)	Yes		x		x	x					Thorough description.
***Anastrepha ethalea* (Walker)/*pseudoparallela*/minor**											Not described or keyed.
***Anastrepha fraterculus* (Wiedemann)—Andean morphotype/*fraterculus*/major**											
Canal et al. (2015)	Yes		x	x					x		Geometric and linear morphometric comparisons of Andean, Brazilian-1, Ecuadorian, Mexican, and Peruvian morphotypes.
Canal et al. (2018)	Yes					x			x		Morphometric study of adults, larvae, and eggs of nine Colombian populations.
***Anastrepha fraterculus* (Wiedemann)—Brazil-1 morphotype/*fraterculus*/major**											
Pruitt (1953)	Yes								x		As *A. fraterculus* Brazil form. Partial description with good detail (lacks spinules, facial features).
Selivon et al. (1997)	Yes					x					As *A. fraterculus* type I. Egg and embryo characteristics compared with those of *A. fraterculus* type II.
Selivon and Perondini (1998)	Yes		x		x	x					As *A.* sp.1 *aff. fraterculus*. Comparison with eggs of Brazil-1 morphotype (as *A.* sp.2 *aff. fraterculus*).

(*Continued*)

TABLE 5.1 (*Continued*)
Literature Describing Immature Stages of *Anastrepha* Species

Species/Species Group/Pest Status Publication	Original Description	Key	Photo	Drawing	SEM	Egg	1st	2nd	3rd	Pupa	Comments
Perondini and Selivon (1999)	Yes		x		x	x					As *A. fraterculus* sp. 1. Includes transmission electron micrographs and comparison with eggs of other genera.
Norrbom et al. (1999)						x					As *A. fraterculus* II in comparative table of egg dimensions, shape, and sculpture of 26 species. Data from Selivon et al. (1997).
Selivon and Perondini (2000)	Yes		x		x	x					As *A.* sp.1 *aff. fraterculus*. Comparisons with numerous *Anastrepha* spp.
Delprat et al. 2001	Yes		x	x			x	x	x		Thorough description.
Carroll et al. (2004)		Yes	x						x		As *A. fraterculus* (Brazil). Partial description based on Steck et al. (1990) and unpublished data, lacking imagery.
Frías et al. (2006)	Yes			x	x		x	x	x		Presumed Brazil-1 morphotype based on geographical source. Key to genus of pest New World fruit flies.
Frías et al. (2008)									x		Generic comparison with other fruit flies.
Figueiredo et al. (2011, 2017)						x					As *A.* sp.1 *aff. fraterculus*. Thorough description and comparisons among 17 *Anastrepha* spp.
Canal et al. (2015)			x	x					x		Geometric and linear morphometric comparisons of Andean, Brazil-1, Ecuadorian, Mexican, and Peruvian morphotypes.
***Anastrepha fraterculus* (Wiedemann)—Brazil-2 morphotype/*fraterculus*/major**											
Selivon et al. (1997)	Yes					x					As *A. fraterculus* type II. Egg and embryo characteristics compared with those of *A. fraterculus* type I.
Selivon and Perondini (1998)	Yes		x		x	x					As *A.* sp.2 *aff. fraterculus*. Comparison with eggs of Brazil-1 morphotype (as *A.* sp.1 *aff. fraterculus*).
Norrbom et al. (1999)						x					As *A. fraterculus* I in comparative table of egg dimensions, shape, and sculpture of 26 species. Data from Selivon et al. (1997).
Selivon and Perondini (2000)	Yes		x		x	x					As *A.* sp.2 *aff. fraterculus*. Comparisons with numerous *Anastrepha* spp.

(*Continued*)

TABLE 5.1 (*Continued*)
Literature Describing Immature Stages of *Anastrepha* Species

Species/Species Group/Pest Status Publication	Original Description	Key	Photo	Drawing	SEM	Egg	1st	2nd	3rd	Pupa	Comments
Selivon et al. (2003)					x	x					As *Anastrepha* sp. 2 *aff fraterculus*. Compared egg morphology of eggs taken from females preserved in ethanol.
Frías et al. (2008)			x						x		Generic comparison with other fruit flies.
Figueiredo et al. (2011, 2017)					x	x					As *fraterculus* (sp.2), comparisons among 17 *Anastrepha* spp.
***Anastrepha fraterculus* (Wiedemann)—Brazil-3 morphotype/*fraterculus*/major**											
Selivon et al. (2004)	Yes				x	x					As *A.* sp. 3. Comparison with eggs of Brazil-1 and Brazil-2 morphotypes (as *A.* sp. 1 and *A.* sp. 2, respectively).
Figueiredo et al. (2011, 2017)						x					As *A.* sp.3 *aff. fraterculus*. Thorough description and comparisons among 17 *Anastrepha* spp.
Frías et al. (2008)	Yes				x				x		Generic comparison with other fruit flies.
***Anastrepha fraterculus* (Wiedemann)—Ecuadorian morphotype/*fraterculus*/major**											
White and Elson-Harris (1992)	Yes	Yes			x				x		As *A. fraterculus* species complex. Thorough description (but limited detail of CPS) based on specimens from Ecuador: Pichincha: Perucho.
Elson-Harris (1992)	Yes	Yes		x	x				x		As *A. fraterculus*. Specimens from Ecuador: Pichincha: Perucho. Thorough description (but limited detail of CPS).
Carroll et al. (2004)		Yes		x					x		Based on White and Elson-Harris (1992).
Canal et al. (2015)	Yes		x	x					x		Geometric and linear morphometric comparisons of Andean, Brazil-1, Ecuadorian, Mexican, and Peruvian morphotypes.
***Anastrepha fraterculus* (Wiedemann)—Mexican/Mesoamerican morphotype/*fraterculus*/major**											

(*Continued*)

TABLE 5.1 (*Continued*)
Literature Describing Immature Stages of *Anastrepha* Species

Species/Species Group/Pest Status Publication	Original Description	Key	Photo	Drawing	SEM	Egg	1st	2nd	3rd	Pupa	Comments
Greene (1929)	Yes	Yes		x					x	x	Morphotype presumed from geographic origin of specimens. Key to larvae and pupae. Partial description.
Emmart (1933)	Yes			x		x					A mis-identification of *A. obliqua* (see Norrbom & Foote 1989).
Baker et al. (1944)	Yes			x					x		Partial description (posterior spiracles).
Baker (1945)	Yes			x		x	x	x	x		Partial description (mandibles).
Pruitt (1953)	Yes								x		As *A. fraterculus* Mexican form. Partial description with good detail (lacks spinules, facial features).
Weems (1980)				x		x			x		Descriptions and drawings taken from Greene (1929), Emmart (1933), and Baker et al. (1944).
Anonymous 1982	Yes								x		Character table: spiracles, anal lobes, tubercles.
Carroll (1992)	Yes				x				x		Partial description (lacks CPS and distribution of cuticular spinules).
Murillo and Jiron (1994)	Yes		x		x	x					Comparison with egg of *Anastrepha obliqua*.
Norrbom et al. (1999)						x					Comparative table of egg dimensions, shape, and sculpture of 26 species. Data from Murillo and Jiron (1994).
Carroll (2004)	Yes	Yes		x	x				x		Thorough description based on Steck et al. (1990) and unpublished data.
Frías et al. (2008)									x		Generic comparison with other fruit flies.
Canal et al. (2015)			x	x					x		Geometric and linear morphometric comparisons of Andean, Brazil-1, Ecuadorian, Mexican, and Peruvian morphotypes.
***Anastrepha fraterculus* (Wiedemann); Peruvian morphotype/*fraterculus*/ major**											
Selivon et al. (2004)	Yes				x	x					As *A.* sp. 4. Comparison with eggs of Brazil-1 and Brazil-2 morphotypes (as *A.* sp. 1 and *A.* sp. 2, respectively).

(*Continued*)

TABLE 5.1 (*Continued*)
Literature Describing Immature Stages of *Anastrepha* Species

Species/Species Group/Pest Status Publication	Original Description	Key	Photo	Drawing	SEM	Egg	1st	2nd	3rd	Pupa	Comments
Figueiredo et al. (2011, 2017)						x					As *A*. sp.4 *aff. fraterculus*. Thorough description and comparisons among 17 *Anastrepha* spp.
Frías et al. (2008)									x		As *A fraterculus* Ecuador. Presumed Peruvian morphotype based on Guayaquil origin. Generic comparison with other fruit flies.
Canal et al. (2015)			x	x					x		Geometric and linear morphometric comparisons of Andean, Brazil-1, Ecuadorian, Mexican, and Peruvian morphotypes.
***Anastrepha fraterculus* (Wiedemann); Venezuelan morphotype/*fraterculus*/ major**											Not described or keyed.
***Anastrepha fraterculus* (Wiedemann)—morphotype not specified/*fraterculus*/ major**											
Bezzi (1913)										x	Superficial description.
Deputy (1957)		Yes		x					x		Key to larvae of pest fruit flies. Superficial drawings modified from Greene (1929).
Berg (1979)		Yes							x		Key to larvae of pest fruit flies.
Norrbom (1985)									x		Comparative table of various larval characters of 14 species; data from various authors.
Steck et al. (1990)		Yes			x				x		Novel data taken from larvae representing five different morphotypes. Key to larvae of 13 *Anastrepha* spp.
***Anastrepha grandis* (Macquart)/*grandis*/ major**											
Fischer (1932)	Yes			x					x	x	Partial description, detailed drawings.
Pruitt (1953)	Yes								x		Partial description with good detail (lacks spinules, facial features).
Norrbom (1985)									x		Comparative table of various larval characters of 14 species; data from various authors.

(*Continued*)

TABLE 5.1 (*Continued*)
Literature Describing Immature Stages of *Anastrepha* Species

Species/Species Group/Pest Status Publication	Original Description	Key	Photo	Drawing	SEM	Egg	1st	2nd	3rd	Pupa	Comments
Whittle and Norrbom (1987)				x					x	x	Description and drawings taken from Fischer (1932).
Steck and Wharton (1988)	Yes			x		x		x	x	x	Thorough description but lacking imagery.
Norrbom and Foote (1989)									x		Limited comparison with larvae of other *Anastrepha* spp.
Steck et al. (1990)	Yes	Yes							x		Key to larvae of 13 *Anastrepha* spp. based on novel data.
White and Elson-Harris (1992)		Yes							x		Description based on Steck and Wharton (1988).
Norrbom et al. (1999)						x					Comparative table of egg dimensions, shape, and sculpture of 26 species. Data from Steck and Wharton (1988).
Carroll et al. (2004)		Yes		x					x		Description and drawings taken from Steck and Wharton (1988).
Figueiredo et al. (2011, 2017)	Yes		x		x	x					Thorough description, comparisons among 17 *Anastrepha* spp.
***Anastrepha haywardi* Blanchard/*spatulata*/not a pest**											
Norrbom et al. (1999)	Yes					x					Comparative table of egg dimensions, shape, and sculpture of 26 species.
***Anastrepha interrupta* Stone/*spatulata*/ not a pest**											
Norrbom (1985)	Yes			x	x				x		Comparative table of various larval characters of 14 species; data from various authors.
Steck and Wharton (1988)	Yes			x					x		Thorough description but lacking imagery.
Norrbom and Foote (1989)									x		Limited comparison with larvae of other *Anastrepha* species.
Steck et al. (1990)	Yes	Yes							x		Key to larvae of 13 *Anastrepha* spp. based on novel data.
Heppner (1990)	Yes			x					x		Partial description but good detail and comparisons with other *Anastrepha* larvae.
Carroll et al. (2004)		Yes		x					x		Description and drawings taken from Steck and Wharton (1988).

(*Continued*)

TABLE 5.1 (*Continued*)
Literature Describing Immature Stages of *Anastrepha* Species

Species/Species Group/Pest Status Publication	Original Description	Key	Photo	Drawing	SEM	Egg	1st	2nd	3rd	Pupa	Comments
***Anastrepha leptozona* Hendel/*leptozona*/major**											
Norrbom (1985)	Yes				x	x					Comparative table of egg length, width, lobe length of 11 species.
Norrbom and Foote (1989)			x			x					Limited comparison with eggs of other *Anastrepha* spp.
Steck et al. (1990)	Yes	Yes							x		Key to larvae of 13 *Anastrepha* spp. based on novel data.
Carroll (1992)	Yes				x				x		Partial description (lacks CPS and cuticular spinules).
Norrbom et al. (1999)					x	x					Comparative table of egg dimensions, shape, and sculpture of 26 species. Data and SEM from Norrbom (1985).
Carroll et al. (2004)	Yes	Yes			x				x		Thorough description based on Steck et al. (1990) and unpublished data.
Frías et al. (2008)									x		Generic comparison with other fruit flies.
Frias et al. (2009)	Yes		x		x				x		Thorough description.
Dutra et al. (2013)	Yes		x		x	x					Thorough description.
***Anastrepha limae* Stone/*pseudoparallela*/minor**											
Norrbom (1985)	Yes			x	x				x		Comparative table of various larval characters of 14 species; data from various authors.
Steck and Wharton (1988)	Yes			x					x	x	Thorough description but lacking imagery.
Norrbom and Foote (1989)									x		Limited comparison with larvae of other *Anastrepha* spp.
Steck et al. (1990)	Yes	Yes			x				x		Key to larvae of 13 *Anastrepha* spp. based on novel data.
Carroll et al. (2004)		Yes		x							Description and drawings taken from Steck and Wharton (1988).
***Anastrepha ludens* (Loew)/*fraterculus*/major**											
Froggatt (1909)				x					x	x	Superficial drawings duplicated from report of California State Board of Horticulture 1905.

(*Continued*)

TABLE 5.1 (*Continued*)

Literature Describing Immature Stages of *Anastrepha* Species

Species/Species Group/Pest Status Publication	Original Description	Key	Photo	Drawing	SEM	Egg	1st	2nd	3rd	Pupa	Comments
Banks (1912)	Yes			x					x		Superficial description.
Greene (1929)	Yes	Yes		x					x	x	Key to larvae and pupae. Partial description.
Emmart (1933)	Yes			x		x					Detailed drawings and comparison with eggs of other pest species in Mexico.
Baker et al. (1944)	Yes		x	x		x			x		Partial description with good detail and biology. Egg drawings from Emmart (1933).
Phillips (1946)	Yes								x		Probably a misidentification of *A. obliqua* (see Carroll & Wharton 1989).
Peterson (1951)	Yes			x					x		Partial description with good detail.
Pruitt (1953)	Yes								x		Partial description with good detail (lacks most facial features).
Foote (1991)									x		Drawings duplicated from Peterson (1951).
Deputy (1957)		Yes		x					x		Key to larvae of pest fruit flies. Superficial drawings modified from Greene (1929).
Kandybina (1977)	Yes			x					x		Thorough description but lacking imagery.
Berg (1979)		Yes		x					x		Key to larvae of pest fruit flies.
Anonymous (1982)	Yes			x					x		Partial data (oral ridges, anterior spiracle, caudal papillae); superficial drawings.
Heppner (1984)	Yes	Yes		x					x		Partial description, but good detail and comparison with *A. suspensa*.
Norrbom (1985)	Yes				x	x			x		Comparative table of egg dimensions of 11 species and various larval characters of 14 species; data from various authors.
Norrbom and Foote (1989)			x			x					Limited comparison with other *Anastrepha* spp.
Carroll and Wharton (1989)	Yes			x	x	x	x	x	x	x	Thorough description.
Steck et al. (1990)	Yes	Yes							x		Key to larvae of 13 *Anastrepha* spp. based on novel data.
Carroll (1992)	Yes			x	x						Thorough description and discussion of identification vs. other pest fruit fly larvae.

(*Continued*)

TABLE 5.1 (*Continued*)
Literature Describing Immature Stages of *Anastrepha* Species

Species/Species Group/Pest Status Publication	Original Description	Key	Photo	Drawing	SEM	Egg	1st	2nd	3rd	Pupa	Comments
White and Elson-Harris (1992)		Yes							x		Description based on Carroll and Wharton (1989).
Norrbom et al. (1999)	Yes				x	x					Comparative table of egg dimensions, shape, and sculpture of 26 species, including data from previous authors.
Carroll et al. (2004)	Yes	Yes		x	x				x		Thorough description based on Carroll and Wharton (1989) and unpublished data.
Frías et al. (2006)	Yes			x					x		Key to genus of pest New World fruit flies. Partial description.
Frías et al. (2008)	Yes		x						x		Generic comparison with other fruit flies.
***Anastrepha macrura* Hendel/*daciformis*/not a pest**											Not described or keyed.
***Anastrepha manihoti* Lima/*spatulata*/minor**											
Norrbom et al. (1999)	Yes					x					Comparative table of egg dimensions, shape, and sculpture of 26 species.
Dutra et al. (2011b)	Yes				x	x					Thorough description.
***Anastrepha margarita* Caraballo/*panamensis*/minor**											Not described or keyed.
***Anastrepha montei* Lima/*spatulata*/minor**											
Dutra et al. (2011b)	Yes				x	x					Thorough description.
***Anastrepha mucronota* Stone/*mucronota*/minor**											
Steyskal (1977)	Yes			x					x		As *A. nunezae*. Superficial description with drawing of posterior spiracle.
Norrbom (1985)									x		As *A. nunezae*. Comparative table of various larval characters of 14 species; data from various authors.
Carroll et al. (2004)		Yes		x					x		Description and drawing taken from Steyskal (1977).
***Anastrepha nigrifascia* Stone/*robusta*/not a pest**											

(*Continued*)

TABLE 5.1 (*Continued*)
Literature Describing Immature Stages of *Anastrepha* Species

Species/Species Group/Pest Status Publication	Original Description	Key	Photo	Drawing	SEM	Egg	1st	2nd	3rd	Pupa	Comments
Norrbom (1985)	Yes			x		x					Comparative table of egg length, width, lobe length of 11 species.
Norrbom and Foote (1989)						x					Limited comparison with eggs of other *Anastrepha* species.
Norrbom et al. (1999)						x					Comparative table of egg dimensions, shape, and sculpture of 26 species. Data from Norrbom (1985).
Norrbom and Korytkowski (2009)	Yes					x					Basic description.
***Anastrepha nolazcoae* Norrbom & Korytkowski/*mucronota*/minor**											Not described or keyed.
***Anastrepha obliqua* (Macquart)/*fraterculus*/major**											
Sein (1933)	Yes		x			x			x	x	As *A. fraterculus* var. *mombinpraeoptans*. Partial descriptions, detailed figures.
Emmart (1933)	Yes					x					As *Anastrepha fraterculus* (Wiedemann) (misidentification). Detailed drawings and comparison with other pest species in Mexico.
Baker et al. (1944)	Yes			x		x			x		As *Anastrepha mombinpraeoptans* Sein. Partial description (posterior spiracles). Egg drawings from Emmart (1933).
Phillips (1946)	Yes	Yes		x							As *Anastrepha ludens* (Loew), probable misidentification (Carroll & Wharton 1989). Thorough description, but lacking imagery.
Peterson (1951)	Yes			x					x		As *Anastrepha fraterculus* prob. var. *mombinpraeoptans* Sein. Partial description with good detail.
Pruitt (1953)	Yes								x		As *A. mombinpraeoptans* Sein. Partial description with good detail (lacks most facial features).
Deputy (1957)		Yes		x					x		As *A. mombinpraeoptans*. Key to larvae of pest fruit flies. Superficial drawings modified from Greene (1929).

(*Continued*)

TABLE 5.1 (*Continued*)
Literature Describing Immature Stages of *Anastrepha* Species

Species/Species Group/Pest Status Publication	Original Description	Key	Photo	Drawing	SEM	Egg	1st	2nd	3rd	Pupa	Comments
Berg (1979)		Yes		x					x		As *A. mombinpraeoptans*. Key to larvae of pest fruit flies.
Jiron and Zeledon (1979)	Yes			x					x		Drawings only of anterior and posterior spiracles.
Anonymous (1982)	Yes								x		Character table: spiracles, anal lobes, tubercles.
Norrbom (1985)	Yes				x	x			x		Comparative table of egg dimensions of 11 species and various larval characters of 14 species; data from various authors.
Norrbom and Foote (1989)			x			x					Limited comparison with other *Anastrepha* spp.
Steck et al. (1990)	Yes	Yes							x		Key to larvae of 13 *Anastrepha* spp. based on novel data.
Heppner (1991)	Yes								x		Partial description but good detail and comparisons with other *Anastrepha* larvae.
Boleli and Teles (1992)	Yes					x					Egg dimensions.
Carroll (1992)	Yes				x				x		Partial description (lacks CPS, cuticular spinules).
White and Elson-Harris (1992)	Yes	Yes		x	x				x		Thorough description (but limited detail of CPS).
Murillo and Jiron (1994)	Yes		x		x	x					Comparison with egg of *Anastrepha obliqua*.
Norrbom et al. (1999)	Yes				x	x					Comparative table of egg dimensions, shape, and sculpture of 26 species including data from previous authors.
Selivon and Perondini (2000)	Yes		x			x					Comparisons with numerous *Anastrepha* species.
Carroll et al. (2004)	Yes	Yes		x	x				x		Thorough description based on Steck et al. (1990), White and Elson-Harris (1992), and unpublished data.
Frías et al. (2006)	Yes			x	x			x	x		Key to genus of pest New World fruit flies. Thorough description except lacking cuticular spinules.
Figueiredo et al. (2011, 2017)	Yes		x		x	x					Thorough description, comparisons among 17 *Anastrepha* spp.
Anastrepha ocresia **(Walker)/*serpentina*/minor**											Not described or keyed.
Anastrepha ornata **Aldrich/*striata*/minor**											Not described or keyed.
Anastrepha pallens **Coquillett/*daciformis*/not a pest**											

(*Continued*)

TABLE 5.1 (*Continued*)
Literature Describing Immature Stages of *Anastrepha* Species

Species/Species Group/Pest Status Publication	Original Description	Key	Photo	Drawing	SEM	Egg	1st	2nd	3rd	Pupa	Comments
Baker et al. (1944)	Yes			x					x		As *Pseudodacus pallens* Coq. Partial description (posterior spiracle).
Phillips (1946)	Yes	Yes		x					x		Thorough description but lacking imagery.
Peterson (1951)	Yes			x					x		As *Pseudodacus pallens*. Partial description.
Norrbom (1985)	Yes			x		x			x		Comparative table of egg dimensions of 11 species and various larval characters of 14 species; data from various authors.
Norrbom and Foote (1989)									x		Limited comparison with larvae of other *Anastrepha* species.
Steck et al. (1990)	Yes	Yes							x		Key to larvae of 13 *Anastrepha* species based on novel data.
Norrbom et al. (1999)						x					Comparative table of egg dimensions, shape, and sculpture of 26 species. Data from Norrbom (1985).
Carroll et al. (2004)		Yes		x					x		Description taken from Steck et al. (1990). Drawings taken from Phillips (1946).
***Anastrepha pallidipennis* Greene/*pseudoparallela*/minor**											Not described or keyed.
***Anastrepha panamensis* Greene/*panamensis*/minor**											Not described or keyed.
***Anastrepha parishi* Stone/not assigned/ minor**											Not described or keyed.
***Anastrepha passiflorae* Greene/*pseudoparallela*/minor**											Not described or keyed.
***Anastrepha pastranai* Blanchard/*pseudoparallela*/not a pest**											
Norrbom et al. (1999)	Yes					x					Comparative table of egg dimensions, shape, and sculpture of 26 species.
***Anastrepha perdita* Stone/*fraterculus*/ minor**											Not described or keyed.

(*Continued*)

TABLE 5.1 (*Continued*)
Literature Describing Immature Stages of *Anastrepha* Species

Species/Species Group/Pest Status Publication	Original Description	Key	Photo	Drawing	SEM	Egg	1st	2nd	3rd	Pupa	Comments
***Anastrepha pickeli* Lima/*spatulata*/not a pest**											
Norrbom et al. (1999)	Yes					x					Comparative table of egg dimensions, shape, and sculpture of 26 species.
Dutra et al. (2011b)					x	x					Thorough description.
Figueiredo et al. (2011, 2017)	Yes		x		x	x					Thorough description, comparisons among 17 *Anastrepha* spp.
Dutra et al. (2018a)	Yes		x		x				x		Thorough description.
***Anastrepha pittieri* Caraballo/*robusta*/not a pest**											
Norrbom (1985)	Yes				x	x					Comparative table of egg length, width, lobe length of 11 species.
Norrbom and Foote (1989)			x			x					Limited comparison with other *Anastrepha* spp.
Norrbom et al. (1999)					x	x					Comparative table of egg dimensions, shape, and sculpture of 26 species. Data from Norrbom (1985).
Norrbom and Korytkowski (2009)	Yes				x	x					Basic description.
***Anastrepha pseudoparallela* (Loew)/*pseudoparallela*/minor**											
Norrbom et al. (1999)	Yes					x					Comparative table of egg dimensions, shape, and sculpture of 26 species.
Figueiredo et al. (2011, 2017)	Yes		x		x	x					Thorough description, comparisons among 17 *Anastrepha* spp.
***Anastrepha psidivora* Norrbom/not assigned/minor**											Not described or keyed.
***Anastrepha pulchra* Stone/*serpentina*/not a pest**											
Dutra et al. (2018a)	Yes		x		x				x		Thorough description.

(*Continued*)

TABLE 5.1 (*Continued*)
Literature Describing Immature Stages of *Anastrepha* Species

Species/Species Group/Pest Status Publication	Original Description	Key	Photo	Drawing	SEM	Egg	1st	2nd	3rd	Pupa	Comments
***Anastrepha punctata* Hendel/*punctata*/minor**											Not described or keyed.
***Anastrepha rheediae* Stone/not assigned/minor**											Not described or keyed.
***Anastrepha sagitatta* (Stone)/*dentata*/minor**											
Baker et al. (1944)	Yes			x					x		As *Lucumaphila sagitatta* Stone. Partial description (posterior spiracle, CPS).
Norrbom (1985)				x					x		Comparative table of various larval characters of 14 species; data and drawing from various authors.
Norrbom and Foote (1989)									x		Limited comparison with larvae of other *Anastrepha* species.
Carroll et al. (2004)		Yes							x		Description taken from Steck et al. (1990) and unpublished data.
***Anastrepha schultzi* Blanchard/*fraterculus*/minor**											Not described or keyed.
***Anastrepha serpentina* (Wiedemann)/*serpentina*/major**											
Greene (1929)	Yes	Yes		x					x	x	Key to larvae and pupae. Partial description.
Emmart (1933)	Yes			x		x					Detailed drawings and comparison with eggs of other pest species in Mexico.
Baker et al. (1944)	Yes			x		x			x		Partial description (posterior spiracles). Egg drawings from Emmart (1933).
Shaw and Starr (1946)	Yes						x	x	x		Drawings and sizes of mandibles only.
Phillips (1946)	Yes	Yes		x					x		Thorough description but lacking imagery.
Pruitt (1953)	Yes			x					x		Partial description with good detail (lacks most facial features).
Burgers (1953)				x					x		Superficial description. Drawing of lateral view of the larva.

(*Continued*)

TABLE 5.1 (*Continued*)
Literature Describing Immature Stages of *Anastrepha* Species

Species/Species Group/Pest Status Publication	Original Description	Key	Photo	Drawing	SEM	Egg	1st	2nd	3rd	Pupa	Comments
Deputy (1957)		Yes		x					x		Key to larvae of pest fruit flies. Superficial drawings modified from Greene (1929).
Weems (1969)									x		Brief description from Phillips (1946).
Berg (1979)		Yes		x					x		Key to larvae of pest fruit flies.
Jiron and Zeledon (1979)	Yes			x					x		Drawings only of anterior and posterior spiracles.
Anonymous (1982)	Yes								x		Character table: spiracles, anal lobes, tubercles.
Norrbom (1985)						x			x		Comparative table of egg dimensions of 11 species and various larval characters of 14 species; data from various authors.
Steck et al. (1990)	Yes	Yes			x				x		Key to larvae of 13 *Anastrepha* spp. based on novel data.
Carroll (1992)	Yes				x				x		Partial description (lacks CPS and distribution of cuticular spinules).
White and Elson-Harris (1992)	Yes	Yes		x	x				x		Thorough description (but limited detail of CPS).
Selivon and Perondini (1999)	Yes				x	x					Thorough description.
Norrbom et al. (1999						x					Comparative table of egg dimensions, shape, and sculpture of 26 species, data from previous authors.
Selivon and Perondini (2000)	Yes		x			x					Comparisons with numerous *Anastrepha* spp.
Carroll et al. (2004)	Yes	Yes		x	x				x		Thorough description based on Steck et al. (1990), White and Elson-Harris (1992), and unpublished data. Drawings from Phillips (1946) and others.
Figueiredo et al. (2011, 2017)						x					Thorough description, comparisons among 17 *Anastrepha* spp.
***Anastrepha shannoni* Stone/*grandis*/not a pest**											
Norrbom (1991)	Yes			x		x					Comparison with eggs of *A. grandis*.
Norrbom et al. (1999)						x					Comparative table of egg dimensions, shape, and sculpture of 26 species. Data from Norrbom (1991).
***Anastrepha sororcula* Zucchi/*fraterculus*/minor**											

(*Continued*)

TABLE 5.1 (*Continued*)
Literature Describing Immature Stages of *Anastrepha* Species

Species/Species Group/Pest Status Publication	Original Description	Key	Photo	Drawing	SEM	Egg	1st	2nd	3rd	Pupa	Comments
Selivon and Perondini (1999)	Yes				x	x					Thorough description.
Norrbom et al. (1999)						x					Comparative table of egg dimensions, shape, and sculpture of 26 species. Data from Selivon and Perondini (1999).
Selivon and Perondini (2000)	Yes		x			x					Positions of larvae before eclosion; egg dimensions.
Figueiredo et al. (2011, 2017)						x					Thorough description, comparisons among 17 *Anastrepha* spp.
Dutra et al. (2018b)	Yes		x		x				x		Thorough description.
***Anastrepha speciosa* Stone/*speciosa*/not a pest**											
Norrbom and Korytkowski (2009)	Yes				x	x					Basic description.
***Anastrepha steyskali* Korytkowskii/*leptozona*/not a pest**											Not described or keyed.
***Anastrepha striata* Schiner/*striata*/major**											
Keilin and Picado (1920)	Yes			x					x	x	Detailed description and drawings, also internal anatomy.
Picado (1920)	Yes			x					x	x	Superficial description.
Greene (1929)	Yes	Yes		x					x	x	Key to larvae and pupae. Partial description.
Emmart (1933)	Yes			x		x					Detailed drawings and comparison with eggs of other pest species in Mexico.
Baker et al. (1944)	Yes			x		x			x		Partial description (posterior spiracles). Egg drawings from Emmart (1933).
Pruitt (1953)	Yes			x					x		Partial description with good detail (lacks some facial features).
Deputy (1957)		Yes		x					x		Key to larvae of pest fruit flies. Superficial drawings modified from Greene (1929).
Berg (1979)		Yes		x					x		Key to larvae of pest fruit flies.
Jiron and Zeledon (1979)	Yes			x					x		Drawings only of anterior and posterior spiracles.
Anonymous (1982)	Yes								x		Character table: spiracles, anal lobes, tubercles.

(*Continued*)

TABLE 5.1 (*Continued*)
Literature Describing Immature Stages of *Anastrepha* Species

Species/Species Group/Pest Status Publication	Original Description	Key	Photo	Drawing	SEM	Egg	1st	2nd	3rd	Pupa	Comments
Weems 1982				x					x		Description from Pruitt (1953) and others. Drawings from Berg (1979).
Norrbom 1985	Yes					x			x		Comparative table of egg dimensions of 11 species and various larval characters of 14 species; data from various authors.
Jones and Kim (1987)	Yes		x						x		Dimensions of posterior spiracles and abnormality.
Steck et al. (1990)	Yes	Yes							x		Key to larvae of 13 *Anastrepha* spp. based on novel data.
Carroll (1992)	Yes				x				x		Partial description (lacks CPS and distribution of cuticular spinules)
White and Elson-Harris (1992)	Yes	Yes		x	x				x		Thorough description (but limited detail of CPS).
Norrbom et al. (1999)						x					Comparative table of egg dimensions, shape, and sculpture of 26 species. Data from Emmart (1933).
Carroll et al. (2004)	Yes	Yes		x	x				x		Thorough description based on Steck et al. (1990), White and Elson-Harris (1992), and unpublished data.
Figueiredo et al. (2011, 2017)	Yes		x		x	x					Thorough description, comparisons among 17 *Anastrepha* spp.
***Anastrepha suspensa* (Loew)/*fraterculus*/ major**											
Sein (1933)	Yes					x			x	x	As *A. unipuncta*. Partial descriptions, detailed figures.
Pruitt (1953)	Yes			x					x		Partial description with good detail (lacks some facial features).
Lawrence (1979)	Yes			x			x	x	x		Partial descriptions, especially CPS and anterior spiracles.
Berg (1979)		Yes		x					x		Key to larvae of pest fruit flies.
Anonymous (1982)	Yes								x		Character table: spiracles, anal lobes, tubercles.
Heppner (1984)	Yes			x					x		Partial description but good detail and comparison *with A. ludens*.

(*Continued*)

TABLE 5.1 (*Continued*)
Literature Describing Immature Stages of *Anastrepha* Species

Species/Species Group/Pest Status Publication	Original Description	Key	Photo	Drawing	SEM	Egg	1st	2nd	3rd	Pupa	Comments
Norrbom (1985)						x			x		Comparative table of egg dimensions of 11 species and various larval characters of 14 species; data from various authors.
Steck et al. (1990)	Yes	Yes			x				x		Key to larvae of 13 *Anastrepha* spp. based on novel data.
Carroll (1992)	Yes				x				x		Partial description (lacks CPS and distribution of cuticular spinules)
White and Elson-Harris (1992)		Yes		x	x				x		Thorough description (but limited detail of CPS).
Norrbom et al. (1999)						x					Comparative table of egg dimensions, shape, and sculpture of 26 species. Data from previous authors.
Carroll et al. (2004)	Yes	Yes		x	x				x		Thorough description based on Steck et al. (1990), White and Elson-Harris (1992), and unpublished data.
Figueiredo et al. (2011, 2017)	Yes		x		x	x					Thorough description, comparisons among 17 *Anastrepha* spp.
***Anastrepha tehuacana* Norrbom/*tripunctata*/not a pest**											
Norrbom et al. (2014)	Yes		x			x					Compared to egg of *A. cordata*.
***Anastrepha tumida* Stone/not assigned/not a pest**											
Norrbom et al. (1999)	Yes					x					Comparative table of egg dimensions, shape, and sculpture of 26 species.
***Anastrepha turpiniae* Stone/*fraterculus*/minor**											
Dutra et al. (2011a)	Yes		x		x	x					Thorough description.
Dutra et al. (2012)	Yes		x	x	x			x	x		Thorough description.
***Anastrepha zenildae* Zucchi/*fraterculus*/minor**											

(*Continued*)

TABLE 5.1 (*Continued*)

Literature Describing Immature Stages of *Anastrepha* Species

Species/Species Group/Pest Status Publication	Original Description	Key	Photo	Drawing	SEM	Egg	1st	2nd	3rd	Pupa	Comments
Dutra et al. (2011a)	Yes		x		x	x					Thorough description.
Figueiredo et al. (2011, 2017)	Yes		x		x	x					Thorough description, comparisons among 17 *Anastrepha* spp.
Dutra et al. (2018b)	Yes		x		x				x		Thorough description.

X indicates feature is present in cited publication.

CPS, cephalopharyngeal skeleton; SEM, scanning electron micrograph.

contributors to further comparative larval studies of *Anastrepha* pest species were Baker et al. (1944, 1945; seven species), Phillips (1946; four species), Peterson (1951; four species), and Pruitt (1953; eight species). Berg's key (1979) did not show much progress over that of Greene (1929), as it incorporated only two additional *Anastrepha* species (*Anastrepha obliqua* [Macquart] and *Anastrepha suspensa* [Loew]) and did not add any original data. Around that time, it was generally agreed that there was low confidence in the reliability of any of the existing keys to provide accurate identifications of *Anastrepha* larvae to species level (Anonymous 1982).

A new generation of researchers began work on *Anastrepha*, including its immature stages, in the 1980s. Norrbom (1985) compiled existing and new comparative data on larvae of 14 species. Steck and Malavasi (1988) and Steck and Wharton (1988) described larvae of four *Anastrepha* species in excellent detail but without benefit of photomicroscopy. Carroll and Wharton (1989) first used SEMs of *Anastrepha* larvae in their very thorough descriptions of all immature stages of *A. ludens*. Steck et al. (1990) explicitly addressed the question of geographic variation in larval characters by examining specimens from multiple localities when available and created a reliable key to larvae of 13 *Anastrepha* species. Note, however, that some closely related species (*striata/bistrigata*, and *fraterculus/obliqua/suspensa*) could not be separated reliably based on the key character data set, and the concept of *A. fraterculus* at that time did not include the multiple morphotypes/putative species recognized today (Hernández-Ortiz et al. 2004, 2012, 2015). Carroll (1992), Elson-Harris (1992), and White and Elson-Harris (1992) greatly expanded the use of SEM in describing larvae, which revealed the fine detail and utility of facial mask characters such as oral ridges, their margins and accessory plates, and the stomal organ and its associated lobes that help to distinguish among species. Currently the most comprehensive public data set of comparative larval morphological characters is that on which the identification tool of Carroll et al. (2004) is based. It incorporates all of the character data included in Carroll's and Elson-Harris's previous publications with the addition of a large number of novel SEM images. Frías et al. (2006, 2008) provided some novel data on larvae of several *Anastrepha* species with the aim of improving the generic diagnosis of *Anastrepha* and keys to major genera of fruit fly pests. The most recent and highly detailed alpha-taxonomic studies of *Anastrepha* larvae are those of Dutra et al. (2011a, 2011b, 2012, 2013, 2018a, 2018b) and Dutra (2012). Canal et al. (2015, 2018) applied morphometric techniques to the study of larvae of several morphotypes of *A. fraterculus* to augment similar studies done on the adult stages.

Based on cumulative studies to date, *Anastrepha* larvae can be recognized at the genus level and separated from other fruit fly pest genera based on mandible with a single primary tooth (without secondary tooth) and basally truncate (absence of a neck); dental sclerite apparently absent, not visible in lateral view; preoral teeth (stomal guards) lacking; accessory plates to oral ridges present and short; stomal organ at apex of large elongate-rounded primary lobe that lacks secondary lobes; anterior spiracle usually concave centrally; and caudal ridge absent (White and Elson-Harris 1992; Carroll et al. 2004; Frías et al. 2006, 2008; Balmès and Mouttet 2017).

At the intrageneric level, however, potentially diagnostic information is sparse. At present 26 species groups are recognized (Norrbom et al. 2012; Mengual et al. 2017; see Table 5.2). Larvae have been described for at least 1 member of 11 of the species groups, and larvae of 15 species groups are entirely unknown. With the exception of the *fraterculus* species group, in which larvae of 11 species have been described, the overall level of descriptive coverage of *Anastrepha* larvae is very low: 9 of the species groups are represented by only 1 or 2 species with described larvae. Therefore, it is difficult to say whether informative synapomorphies exist to help define species groups. Even with this limited data, however, there appears to be a disappointing lack of larval character states useful in recognizing species groups of *Anastrepha*. For example, even the most character-rich feature of tephritid larvae, the facial mask, tends to be rather uniform across species groups. There is some variation in the appearance of the oral ridges and accessory plates, being either smooth or serrate, but both character states are shared among species groups. The one exception is the *curvicauda* group, which until recently was classified as

TABLE 5.2
***Anastrepha* Species Groups, Numbers of Included Species, and Numbers of Species for Which Larvae Have Been Described**

Species Group	# spp.	No. Species Described to Date
benjamini	10	—
binodosa	3	—
caudata	3	—
cryptostrepha	4	—
curvicauda	7	2
daciformis	14	1
dentata	11	1
doryphoros	4	—
fraterculus	40+[a]	11
grandis	10	1
hastata	3	—
leptozona	6	1
mucronota	52	1
nigrina	2	—
panamenis	3	—
pseudoparallela	25	3
punctata	4	—
ramosa	3	—
raveni	2	—
robusta	15	—
schausi	5	—
serpentina	9	2
spatulata	16	2
striata	3	2
speciosa	3	—
tripunctata	4	—
Unassigned	41	—
TOTAL		27

[a] plus unnamed members of the *A. fraterculus* complex.

a distinct genus, *Toxotrypana* (Norrbom et al. 2018). All of the major keys to fruit fly pest genera clearly distinguish this group from the remainder of other *Anastrepha* larvae based on several characters such as shape of mandible and reduction in tubercles on caudal segment (see especially Frías et al. 2008).

The study of eggs has outpaced that of larvae because fully developed eggs can be extracted from gravid females collected in the field and from museum specimens, and their identity is known from the associated adult specimens. The first comparative study of *Anastrepha* eggs was that of Emmart (1933), who described eggs of four pest species in Mexico. Norrbom (1985) compiled existing and new comparative data on eggs of 11 species and was the first to use SEMs to observe eggs. Norrbom et al. (1999) increased comparative data for *Anastrepha* eggs to 26 species. Selivon and Perondini (1999, 2000) and Selivon et al. (2004) provided detailed descriptions of eggs of several *Anastrepha* species and especially contributed to the growing realization that *A. fraterculus s.l.* comprises numerous cryptic species. Norrbom and Korytkowski (2009), Figueiredo et al. (2011), and Dutra et al. (2011a, 2011b, 2013) have further expanded the list to 49 *Anastrepha* species with described eggs. As is mostly the case with larvae, the egg stage does not present synapomorphies linking members of any of the proposed species groups together (Figueiredo et al. 2011), although some related species share some derived features.

5.3 FUTURE WORK

Further needs for study of *Anastrepha* immature stages include (1) alpha taxonomy, (2) development of identification tools, and (3) phylogenetic analysis.

The first order of business is to acquire specimens for study. There are no research specimens available for more than half of the recorded *Anastrepha* pest species. Larvae are troublesome because they must be collected from fruits infested in the field and a subset reared to adult stage for positive identification. However, even reared, associated adults do not guarantee a definitive identification of a given larva because field-collected fruits can be multiply infested by two or more different species. Alternatively, a pure laboratory colony can be established as a source of the immature stages.

Good alpha taxonomy is greatly facilitated by high-quality optical microscopes, digital cameras, and a scanning electron microscope. Optical microscopy has limited capability of imaging surface features of larvae and eggs, therefore thorough observation requires scanning electron microscopy, which is not readily available to many researchers and identifiers. However, it should be noted that skilled and observant optical microscopists are capable of detecting fine detail. For example, Keilin and Picado's (1920) drawings show amazing detail of the facial mask, including the stomal organ and lobes of *Anastrepha striata*, and the drawings of Phillips (1946) and Kandybina (1977) are excellent. Good standards are already in place for ongoing comparative morphological studies.

The future of fruit fly larvae identification based on morphology lies in digital multi-entry keys such as that of Carroll et al. (2004). The species-level key of Steck et al. (1990) relies on minor variation in numerous "trivial" characters such as presence or absence of dorsal spinules on various body segments, the number of lobes on the anterior spiracles, numbers of posterior spiracular processes and branches, etc. At present, the paucity of intrageneric synapomorphies makes species-level identification difficult when there are many possible outcomes. For example, all of Dutra et al.'s recent descriptions of *Anastrepha* larvae fail to easily incorporate into the binomial key of Steck et al. (1990). It will be relatively easy to build upon the key of Carroll et al. (2004), which includes 18 *Anastrepha* taxa, by adding new taxa and perhaps additional characters and imagery.

The current phylogeny of Tephritidae is largely based on adult morphology and molecular data, but larval characters provide useful corroboration and additional insight. For example, the affinities of the tribe Gastrozonini were questioned for many years, and the presence of a caudal ridge on their larvae strongly supports its inclusion in the Dacinae (Kovac et al. 2006). We expect that further taxon sampling within *Anastrepha* will reveal useful species group synapomorphies (unpublished data).

Finally, molecular identification tools offer great promise in addressing the problem of identifying fruit fly immature stages. We already know that morphological identification is difficult, and the larvae of closely related species may be effectively indistinguishable, just as is the case for adults (males, cryptic species complexes). Molecular data already are often used to support identifications in difficult cases. However, it should be noted that identifications based on molecular data suffer similar limitations as morphological keys (e.g., missing taxa, limited geographic sampling, lack of validation, and lack of in-house equipment and expertise by identifiers).

ACKNOWLEDGMENTS

We thank Steve Bullington (USDA-APHIS-PPQ) for providing fruit fly port interception data from AQAS:Pest ID and Norman Barr (USDA-APHIS-PPQ) for input on fruit fly invasion studies. The work was supported by USDA Farm Bill project "Enhancement of fruit fly immature stage identification and taxonomy" #3.0342 (2012), #13-8131-0291-CA (2013), #3.0295.01 (2014), #3.0281(2015); and "Development of more rapid and reliable diagnostic tools for all life stages of *Anastrepha* and other pest fruit flies" #3.0520 (2017). We thank the Florida Department of Agriculture and Consumer Services—Division of Plant Industry for their support on this contribution. Mention of trade names or commercial products in this publication is solely for the purpose of providing specific information and does not imply recommendation or endorsement.

REFERENCES

Anonymous. 1982. Workshop on identifying *Anastrepha* larvae. United States Department of Agriculture/ Animal and Plant Health Inspection Service/Plant Protection and Quarantine Memo, November 18, 1982.

AQAS:PestID. 2016. Agricultural Quarantine Activity System: Pest Interception Database (AQAS:PestID). U.S. Department of Agriculture, Animal and Plant Health Inspection Service, Plant Protection and Quarantine. https://aqas.aphis.usda.gov. Accessed June 1, 2016.

Baker, A. C., W. C. Stone, C. C. Plummer, and M. McPhail. 1944. A review of studies on the Mexican-fruit fly and related Mexican species. US Department of Agriculture Miscellaneous Publication Number 531, 155 p.

Baker, E. W. 1945. Studies on the Mexican fruit fly known as *Anastrepha fraterculus*. *Journal of Economic Entomology* 38: 95–100.

Balmès, V. and R. Mouttet. 2017. Development and validation of a simplified morphological identification key for larvae of tephritid species most commonly intercepted at import in Europe. *Bulletin OEPP/EPPO Bulletin* 47: 91–99.

Banks, N. 1912. The structure of certain dipterous larvae with particular reference to those in human foods. *United States Department of Agriculture Bureau of Entomology Technical Series* 22: 1–44.

Berg, G. H. 1979. *Pictorial Key to Fruit Fly Larvae of the Family Tephritidae*. Organismo Internacional Regional de Sanidad Agropecuaria (OIRSA), San Salvador, Rep. de El Salvador. 36 p.

Bezzi, M. 1913. Indian trypaneids (fruit-flies) in the collection of the Indian Museum, Calcutta. *Memoirs of the Indian Museum* 3: 53–175.

Boleli, I. C., and M. M. D. C. Teles. 1992. Egg length of *Anastrepha obliqua* Macquart (Diptera: Tephritidae) according to oviposition rate and maternal age. *Revista Brasileira de Zoologia* 9: 215–221.

Burgers, A. C. J. 1953. The fruitfly *Anastrepha serpentina* in Curacao. In P. Wegenaar Hummelinck, ed., *Studies on the Fauna of Curacao and Other Caribbean Islands* Vol. IV, No. 21. The Hague: *Natuurwetenschappelijke Studiekring voor Suriname en de Nederlandse Antillen* No. 8: 149–153.

Canal, N. A., V. Hernández-Ortiz, J. O. Tigrero Salas, and D. Selivon. 2015. Morphometric study of third-instar larvae from five morphotypes of the *Anastrepha fraterculus* cryptic species complex (Diptera, Tephritidae). In *Resolution of Cryptic Species Complexes of Tephritid Pests to Enhance SIT Application and Facilitate International Trade*, ed. M. De Meyer, A. R. Clarke, T. M. Vera, and J. Hendrichs. *ZooKeys* 540: 41–59. Sofia, Bulgaria: Pensoft Publishers.

Canal, N. A., P. E. Galeano-Olaya, and M. Rosario Castañeda. 2018. Phenotypic structure of colombian populations of Anastrepha fraterculus complex (Diptera: Tephritidae). *Florida Entomologist* 101: 299–310.

Carroll, L. E. 1992. Systematics of fruit fly larvae (Diptera: Tephritidae). PhD thesis. College Station, TX: Texas A&M University. xi + 330 p.

Carroll, L. E., A. L. Norrbom, M. J. Dallwitz, and F. C. Thompson. 2004 et seq. Pest fruit flies of the world—larvae. Version: 8. December 2006. http://delta–intkey.com

Carroll, L. E., and R. A. Wharton. 1989. Morphology of the immature stages of *Anastrepha ludens* (Diptera: Tephritidae). *Annals of the Entomological Society of America* 82: 201–214.

Delprat, M. A., F. C., Manso, and J. L. Cladera. 2001. Larval morphology of the pure strain Arg294 of *Anastrepha fraterculus* (Wiedeman) (Diptera:Tephritidae) from Argentina and a comparison with *Ceratitis capitata*. *Revista Chilena de Entomología* 28: 39–46.

Deputy, O. D. 1957. Diptera larvae. In *A Hand-Book on the Principles of Plant Quarantine Enforcement for Indonesia*, ed. O. D. Deputy, and M. S. Harahap, pp. 133–152. Bogor, Indonesia: Pemberitaan Balai Besar Penjelidikan Pertanian No. 146: 165 p.

Dutra, V. S. 2012. Caracterização de ovos e larvas de espécies de *Anastrepha* (Diptera: Tephritidae) utilizando análises morfológicas e moleculares. PhD thesis. Manaus, Brazil: Instituto Nacional de Pesquisas da Amazônia.

Dutra, V. S., B. Ronchi-Teles, G. J. Steck, and J. G. Silva. 2011a. Egg morphology of *Anastrepha* spp. (Diptera: Tephritidae) in the *fraterculus* group using scanning electron microscopy. *Annals Entomological Society of America* 104: 16–24.

Dutra, V. S., B. Ronchi-Teles, G. J. Steck, and J. G. Silva. 2011b. Description of eggs of *Anastrepha* spp. (Diptera: Tephritidae) in the *spatulata* group using scanning electron microscopy. *Annals of the Entomological Society of America* 104: 857–862.

Dutra, V. S., B. Ronchi-Teles, G. J. Steck, and J. G. Silva. 2012. Description of larvae of *Anastrepha* spp. (Diptera: Tephritidae) in the *fraterculus* group. *Annals of the Entomological Society of America* 105: 529–538.

Dutra, V. S., B. Ronchi-Teles, G. J. Steck, and J. G. Silva. 2013. Description of eggs of *Anastrepha curitis* and *Anastrepha leptozona* (Diptera: Tephritidae) using SEM. *Annals of the Entomological Society of America* 106: 13–17.

Dutra, V. S., B. Ronchi-Teles, G. J. Steck, and J. G. Silva. 2018a. Description of third instar larvae of *Anastrepha curitis*, *A. pickeli*, and *A. pulchra* (Diptera: Tephritidae). *Proceedings of the Entomological Society of Washington* 120: 9–24.

Dutra, V. S., B. Ronchi-Teles, G. J. Steck, E. L. Araujo, M. F. Souza-Filho, A. Raga, and J. G. Silva. 2018b. Description of larvae of three *Anastrepha* species in the *fraterculus* group (Diptera: Tephritidae). *Proceedings of the Entomological Society of Washington* 120: 708–724.

Elson-Harris, M. 1992. A systematic study of Tephritidae (Diptera) based on the comparative morphology of larvae. PhD thesis. Brisbane, Australia: University of Queensland. xvi + 366 p + 98 plates.

Emmart, E. W. 1933. The eggs of four species of fruit flies of the genus *Anastrepha*. *Proceedings of the Entomological Society of Washington* 35: 184–191.

Figueiredo, J. V. A., A. L. P. Perondini, E. M. Ruggiro, L. F. Prezzoto, and D. Selivon. 2011. External eggshell morphology of *Anastrepha* fruit flies (Diptera: Tephritidae). *Acta Zoologica* 00: 1–9 [in print (2013) 94: 125–133].

Figueiredo, J. V. A., A. L. P. Perondini, E. M. Ruggiro, and D. Selivon. 2017. Patterns of inner chorion structure in *Anastrepha* (Diptera: Tephritidae) eggs. *Arthropod Structure & Development* 46: 236–245.

Fischer, C. R. 1932. Nota taxonomica e biologica sobre *Anastrepha grandis* Macq. (Dipt., Trypetidae). *Revista de Entomologia* (Rio de Janeiro) 2: 302–310.

Foote, B. A. 1991. Tephritidae (Tephritoidea). In *Immature Insects*, ed. F. W. Stehr, Vol. 2, 809–815. Dubuque, IA: Kendall/Hunt Publishing.

Frías, D., D. Selivon, and V. Hernández-Ortiz. 2008. Taxonomy of immature stages: New morphological characters for Tephritidae larvae identification. In *Fruit flies of economic importance: From basic to applied knowledge. Proceedings of the 7th International Symposium on Fruit Flies of Economic Importance, September 10–15, 2006, Salvador, Brazil*, ed. R. L. Sugayama, R. A. Zucchi, S. M. Ovruski, and J. Sivinski,pp. 29–44. São Paulo, Brazil: Sociedade Brasileira para o Progresso de Ciência.

Frías, D., V. Hernández-Ortiz, and L. López. 2009. Description of the third-instar of *Anastrepha leptozona* Hendel (Diptera: Tephritidae). *Neotropical Entomology* 38: 491–496.

Frías, D., V. Hernández-O., N. C. Vaccaro, A. F. Bartolucci, and L. A. Salles. 2006. Comparative morphology of immature stages of some frugivorous species of fruit flies (Diptera: Tephritidae). *Israel Journal of Entomology* 35-36: 423–457.

Froggatt, W. W. 1909. Notes on fruit flies. *Report of the Estacion Central Agronomica. Havana* 2: 117–121, pl. XXII.

Greene, C. T. 1929. Characters of the larvae and pupae of certain fruit flies. *Journal of Agricultural Research* 38: 489–504.

Heppner, J. B. 1984. Larvae of fruit flies. I. *Anastrepha ludens* (Mexican fruit fly) and *Anastrepha suspensa* (Caribbean fruit fly) (Diptera: Tephritidae). *Florida Department of Agriculture and Consumer Services, Division of Plant Industry, Entomology Circular* 260: 4 p.

Heppner, J. B. 1986. Larvae of fruit flies. III. *Toxotrypana curvicauda* (papaya fruit fly) (Diptera: Tephritidae). *Florida Department of Agriculture and Consumer Services, Division of Plant Industry, Entomology Circular* 282: 2p.

Heppner, J. B. 1990. Larvae of fruit flies. 6. *Anastrepha interrupta* (*Schoepfia* fruit fly). *Florida Department of Agriculture and Consumer Services, Division of Plant Industry, Entomology Circular* 327: 2 p.

Heppner, J. B. 1991. Larvae of fruit flies, 7. *Anastrepha obliqua* (West Indian fruit fly) (Diptera: Tephritidae). *Florida Department of Agriculture and Consumer Services, Division of Plant Industry, Entomology Circular* 339: 2 p.

Hernández-Ortiz, V., A. F. Bartolucci, P. Morales-Valles, D. Frías, and D. Selivon. 2012. Cryptic species of the *Anastrepha fraterculus* complex (Diptera: Tephritidae): A multivariate approach for the recognition of south American morphotypes. *Annals of the Entomological Society of America* 105: 305–318.

Hernández-Ortiz, V., J. A. Gómez-Anaya, A. Sánchez, B. A. McPheron, and M. Aluja. 2004. Morphometric analysis of Mexican and South American populations of the *Anastrepha fraterculus* complex (Diptera: Tephritidae) and recognition of a distinct Mexican morphotype. *Bulletin of Entomological Research* 94: 487–499.

Hernández-Ortiz, V., N. A. Canal, J. O. Tigrero Salas, F. M. Ruíz-Hurtado, and J. F. Dzul-Cauich. 2015. Taxonomy and phenotypic relationships of the *Anastrepha fraterculus* complex in the Mesoamerican and Pacific Neotropical dominions (Diptera, Tephritidae). *Zookeys* 540: 95–124.

Humair, F., L. Humair, F. Kuhn, and C. Kueffer. 2015. E-commerce trade in invasive plants. *Conservation Biology* 29: 1658–1665.

Jiron, L. F., and R. Zeledon. 1979. El genero *Anastrepha* (Diptera; Tephritidae) en las principales frutas de Costa Rica y su relacion con pseudomyiasis humana. *Revista de Biología Tropical* 27: 155–161.

Jones, S. R., and K. C. Kim. 1987. A spiracular abnormality in *Anastrepha striata* (Diptera: Tephritidae) from Costa Rica. *Entomological News* 98: 217–220.

Kandybina, M. N. 1977. Lichinki plodovykh mukh-pestrokrylok (Diptera, Tephritidae). [Larvae of fruit-infesting fruit flies (Diptera, Tephritidae)]. *Opredeliteli po faune S.S.S.R.* 114: 1–210. [In Russian; unpublished English translation, 1987, produced by National Agricultural Library, Beltsville, MD.]

Keilin, D., and C. Picado. 1920. Biologie et morphologie larvaires d'*Anastrepha striata* Schiener [sic], mouche des fruits de l'Amerique centrale. *Bulletin scientifique de la France et de la Belgique* 48: 423–441.

Knab, F., and W. W. Yothers. 1914. Papaya fruit fly. *Journal of Agricultural Research* 2: 447–453.

Kovac, D., P. Dohm, A. Freidberg, and A. L. Norrbom, 2006. Catalog and revised classification of the Gastrozonini (Diptera: Tephritidae: Dacinae). *Israel Journal of Entomology* 35–36: 163–196.

Lawrence, P. O. 1979. Immature stages of the Caribbean fruit fly, *Anastrepha suspensa. Florida Entomologist* 62: 214–219.

Liebhold, A. M., T. T. Work, D. G. McCullough, and J. F. Cavey. 2006. Airline baggage as a pathway for alien insect species invading the United States. *American Entomologist* 52: 46–54.

Lima, A. M. da Costa. 1930. Sobre insectos que vivem em maracujas (*Passiflora* spp.). *Memórias do Instituto Oswaldo Cruz* 23: 159–162.

Mengual, X., P. Kerr, A. L. Norrbom, N. B. Barr, M. L. Lewis, A. M. Stapelfeldt, S. J. Scheffer, et al. 2017. Phylogenetic relationships of the tribe Toxotrypanini (Diptera: Tephritidae) based on molecular characters. *Molecular Phylogenetics and Evolution* 113: 84–112.

Murillo, T., and L. F. Jiron. 1994. Egg morphology of *Anastrepha obliqua* and some comparative aspects with eggs of *Anastrepha fraterculus* (Diptera: Tephritidae). *Florida Entomologist* 77: 342–348.

Norrbom, A. L., E. J. Rodriguez, G. J. Steck, B. A. Sutton, and N. Nolazco. 2015. New species and host plants of *Anastrepha* (Diptera: Tephritidae) primarily from Peru and Bolivia. *ZooTaxa* 4041: 1–94.

Norrbom, A. L. 1985. Phylogenetic analysis and taxonomy of the *cryptostrepha*, *daciformis*, *robusta* and *schausi* species groups of *Anastrepha* Schiner (Diptera; Tephritidae). PhD thesis. University Park, PA: Pennsylvania State University.

Norrbom, A. L. 1991. The species of *Anastrepha* (Diptera: Tephritidae) with a *grandis*-type wing pattern. *Proceedings of the Entomological Society of Washington* 93: 101–124.

Norrbom, A. L., A. L. Castillo-Meza, J. H. García-Chávez, M. Aluja, and J. Rull. 2014. A new species of *Anastrepha* (Diptera: Tephritidae) from *Euphorbia tehuacana* (Euphorbiaceae) in Mexico. *Zootaxa* 3780: 567–576.

Norrbom, A. L., and C. A. Korytkowski. 2009. A revision of the *Anastrepha robusta* species group (Diptera: Tephritidae). *Zootaxa* 2182: 1–91.

Norrbom, A. L., and R. H. Foote. 1989. The taxonomy and zoogeography of the genus *Anastrepha* (Diptera: Tephritidae). In *Fruit Flies, Their Biology, Natural Enemies and Control*, ed. A. S. Robinson and G. Hooper, W. Helle, pp. 15–26, World Crop Pests, Vol. 3(A). Amsterdam, the Netherlands: Elsevier Science Publishers. xii 372 p.

Norrbom, A. L., C. A. Korytkowski, R. A. Zucchi, K. Uramoto, G. L. Venable, J. McCormick, and M. J. Dallwitz. 2012 et seq. *Anastrepha* and *Toxotrypana*: Descriptions, illustrations, and interactive keys. Version: September 28, 2013. http://delta–intkey.com.

Norrbom, A. L., N. B. Barr, P. Kerr, X. Mengual, N. Nolazco, E. J. Rodriguez, G. J. Steck, B. D. Sutton, K. Uramoto, and R. A. Zucchi. 2018. Synonymy of *Toxotrypana* Gerstaecker with *Anastrepha* Schiner (Diptera: Tephritidae). *Proceedings of the Entomological Society of Washington* 120: 834–841.

Norrbom, A. L., R. A. Zucchi, and V. Hernandez-Ortiz. 1999. Phylogeny of the genera *Anastrepha* and *Toxotrypana* (Tryptetinae: Toxotrypanini) based on Morphology. In *Fruit Flies (Tephritidae): Phylogeny and Evolution of Behavior*, ed. M. Aluja, and A. L. Norrbom, pp. 299–342. Boca Raton, FL: CRC Press. 944 p.

Perondini, A. L. P., and D. Selivon. 1999. The structure of the chorion of *Anastrepha fraterculus* (Wiedemann, 1830) eggs (Diptera, Tephritidae). *Revista Brasileira de Entomologia* 43: 243–248.

Peterson, A. 1951. *Larvae of Insects.* Part II. Colombus, OH, 416 p.

Phillips, V. T. 1946. The biology and identification of trypetid larvae (Diptera: Trypetidae). *Memoirs of the American Entomological Society* 12: 1–161.

Picado, C. T. 1920. Hístoría del gusano de la guayaba. *Publicaciones del Colegio de Señoritas, Serie A* No. 2. 5–28.

Pruitt, J. H. 1953. Identification of fruit fly larvae frequently intercepted at ports of entry of the United States. MS thesis. Gainesville, FL: University of Florida. 69 p.

Sein, F., Jr. 1933. *Anastrepha* (Trypetydae [sic], Diptera) fruit flies in Puerto Rico. *The Journal of the Department of Agriculture of Porto Rico.* 17: 183–196.

Selivon, D., J. S. Morgante, and A. L. P. Perondini. 1997. Egg size, yolk mass extrusion and hatching behavior in two cryptic species of *Anastrepha fraterculus* (Wiedemann) (Diptera: Tephritidae). *Brazilian Journal of Genetics* 91: 471–478.

Selivon, D., and A. L. P. Perondini. 1998. Eggshell morphology in two cryptic species of the *Anastrepha fraterculus* complex (Diptera: Tephritidae). *Annals of the Entomological Society of America* 91: 473–478.

Selivon, D., and A. L. P. Perondini. 1999. Description of *Anastrepha sororcula* and *A. serpentina* (Diptera: Tephritidae) eggs. *Florida Entomologist* 82: 347–353.

Selivon, D., and A. L. P. Perondini. 2000. Morfologia dos ovos de moscas-das-frutas do genero *Anastrepha.* In *Moscasdas-frutas de importancia economica no Brasil: conhecimento basico e aplicado,* ed. A. Malavasi and R. A. Zucchi, 49–54. Ribeirao Preto, Brazil: Holos Editora.

Selivon, D., C. Vretos, and A. L. P. Perondini. 2003. Evaluation of egg morphology from ethanol preserved females of *Anastrepha* sp. 2 *aff. fraterculus* (Diptera: Tephritidae). *Neotropical Entomology* 32: 527–529.

Selivon, D., C. Vretos, L. Fontes, and A. L. P. Perondini. 2004. New variant forms in the *Anastrepha fraterculus* complex. In *Proceedings of the 6th International Fruit Flies Symposium,* ed. B. Barnes, 253–258. Irene: Isteg Scientific Publications.

Shaw, J. G., and D. F. Starr. 1946. Development of the immature stages of *Anastrepha serpentina* in relation to temperature. *Journal of Agricultural Research* 72: 265–276.

Steck, G. J., and A. Malavasi. 1988. Description of immature stages of *Anastrepha bistrigata* (Diptera: Tephritidae). *Annals of the Entomological Society of America* 81: 1004–1006.

Steck, G. J., and R. A. Wharton. 1988. Description of immature stages of *Anastrepha interrupta, A. limae,* and *A. grandis* (Diptera: Tephritidae). *Annals of the Entomological Society of America* 81: 994–1003.

Steck, G. J., L. E. Carroll, H. Celedonio H., and J. C. Guillen A. 1990. Methods for identification of *Anastrepha* larvae (Diptera: Tephritidae), and key to 13 species. *Proceedings of the Entomological Society of Washington* 92: 356–369.

Steyskal, G. C. 1977. Two new neotropical fruitflies of the genus *Anastrepha,* with notes on generic synonymy (Diptera, Tephritidae). *Proceedings of the Entomological Society of Washington* 79: 75–81.

Weems, H. V., Jr. 1969. *Anastrepha serpentina* (Wiedemann) (Diptera: Tephritidae). *Florida Department of Agriculture and Consumer Services, Division of Plant Industry, Entomology Circular* 91: 2 p.

Weems, H. V., Jr. 1980. *Anastrepha fraterculus* (Wiedemann) (Diptera: Tephritidae). *Florida Department of Agriculture and Consumer Services, Division of Plant Industry, Entomology Circular* 217: 2 p.

Weems, H. V., Jr. 1982. *Anastrepha striata* Schiner (Diptera: Tephritidae), *Florida Department of Agriculture and Consumer Services, Division of Plant Industry, Entomology Circular* 245: 2 p.

White, I. M., and M. M. Elson-Harris. 1992. *Fruit Flies of Economic Significance: Their Identification and Bionomics.* Wallingford, UK: CAB International.

Whittle, K. A., and A. L. Norrbom. 1987. A fruit fly *Anastrepha grandis* (Macquart). Pests not known to occur in the United States or of limited distribution, no. 82, 8 p.

6 A Review of the Natural Host Plants of the *Anastrepha fraterculus* Complex in the Americas

Vicente Hernández-Ortiz, Nancy Barradas-Juanz, and Cecilia Díaz-Castelazo*

CONTENTS

Abstract There is now enough support for the hypothesis that nominal *Anastrepha fraterculus* is a complex of cryptic species that are currently recognized, using morphometric procedures, as eight morphotypes that probably correspond to different biological species. In addition to this variability, there is also evidence that this nominal species presents important variation in its range of preferential host use. The aim of this chapter is to provide a comprehensive understanding of the natural host plants used by the nominal *A. fraterculus* under natural field conditions. This was accomplished through a bibliographic examination of information from the original sources of host plants recorded for this fly species. A total of 200 references from all regions of the Americas were examined. Data useful to the analysis were captured in a database incorporating information pertaining to host identity, original source of data, and location of distribution, where available. The list of host plants for the *A. fraterculus* complex comprised 177 species belonging to 40 plant families, which together accounted for 1,622 documented reports. The most highly represented families were Myrtaceae (27.1%), Rosaceae (11.9%), and Rutaceae (8.5%). The Myrtaceae exhibited a high percentage (>90%) of native species in contrast to the higher proportions of exotic species presented in the other families. Guava was the only common host shared by different populations throughout the tropical and subtropical landscapes of the Americas. The highest number of hosts was recorded in Brazil (121), followed by Argentina (40), Ecuador (40), Colombia (38), Venezuela (24), and Mexico (19). The landscapes occupied by different populations of this nominal species presented some preferential patterns in terms

* Corresponding author.

of resource use. This reinforces the hypothesis of distinct taxonomic entities because most of the plants are present throughout the range but are not found to be common hosts to all of the fly populations. In this context, the potential application of the sterile insect technique (SIT) in certain geographical areas requires knowledge of the particular hosts consumed by the target species.

6.1 INTRODUCTION

The South American fruit fly (SAFF), *Anastrepha fraterculus* (Wiedemann, 1830), is widely distributed throughout the Americas; from the southern United States (Texas), through Mexico, Central America and South America to Argentina (Hernández-Ortiz & Aluja 1993). Based on previous research under different approaches, including the use of karyotypes, molecular DNA sequences, reproductive isolation, pheromone profiles, or even integrative approaches (reviewed in Selivon et al. 2004, 2005a, 2005b, Vera et al. 2006, Cáceres et al. 2009, Vaníčková et al. 2015, Dias et al. 2016, and others), there is now enough support for the hypothesis that this nominal species in fact comprises a complex of cryptic species that are currently recognized using morphometric procedures as eight morphotypes, most likely corresponding to different biological species (Hernández-Ortiz et al. 2004, 2012, 2015).

There is also evidence that some populations exhibit important differences in the host range used, using preferential hosts at the regional scale. Moreover, their pest status can differ depending on the geographical area in which they occur; for instance, Baker et al. (1944) showed that citrus fruits such as oranges were unsuitable hosts for Mexican SAFF females and Aluja et al. (2003a) later confirmed that fruits of *Citrus sinensis* and *Citrus paradisi* in Mexico are not infested, either in the field or under laboratory-induced conditions. In contrast, studies in South America revealed that citrus fruits are common hosts for Brazilian populations of the SAFF (Malavasi et al. 1980, Zucchi 2007, and others). In this regard, there are some highly questionable reports indicating that *A. fraterculus* is able to infest citrus fruits in Guatemala (Eskafi & Cunningham 1987, Eskafi 1990). This variation in the range of host plants consumed in different regions of the neotropics presents a problem in terms of enforcing effective quarantine procedures.

Historical accounts indicate that nearly 150 plant species host larvae feeding on their fruits across the Americas (Norrbom 2004). However, many of these records were derived from observations under laboratory conditions or copied from previous literature reports, and some fail to specify an explicit location, or even country, or there is some uncertainty regarding the fly or host plant identity. Various studies at the regional level provide information on the population dynamics of the SAFF derived from specimens caught in traps baited with food attractants. These traps are usually hung on fruit trees of commercial importance; however, this does not necessarily demonstrate that these fly larvae infest such fruits under natural conditions. This information has led to the identification of three key problems of the phytosanitary measures that must be applied: (1) the presence of erroneous records featuring wrongly identified or unconfirmed hosts; (2) misidentification of the taxonomic species because other wild fly species are also attracted to the food baits; and (3) misinterpretation of distribution patterns based on records with inaccurate locations. As a consequence, implementation of quarantine measures and methods of integrated pest management (IPM) based on such erroneous or ambiguous information can lead to poor planning of control strategies based on concepts of fly-free or low prevalence areas, or even the application of the sterile insect technique (SIT) in specific geographical areas, which requires accurate identification of the hosts consumed by the target species.

This study therefore aims to provide a comprehensive understanding of the host plants used as natural food resources by *A. fraterculus*. The review focuses particularly on records produced under field conditions to conduct an analysis of botanical families and host species using current botanical nomenclature and homogenizing past and current species names, as well as examining their occurrence in different countries using the information available in the literature produced over nearly a century. This is a critical first step to elucidate the host range of the SAFF across several regions of the neotropics.

6.2 METHODS AND ANALYSIS

Information was sought in the original published sources of recorded host plants of the nominal species *A. fraterculus*. The search was largely conducted in articles published in scientific journals, books, and unpublished dissertations available online. The main requirements of credibility for each record were inclusion of the full identity of the host plant, as well as the occurrence of fruit infestation under natural conditions in the field, with no ambiguity regarding the recorded location (to country level at least). We therefore excluded doubtful reports, those obtained under laboratory conditions, and those in which the accurate identity of the fruit fly species could not be verified.

Nearly 170 references as primary sources of information from all regions of the Americas were examined. All information useful for the analysis was captured in a database, including the host identity, botanical family, original source of data, country, state or province, locality, coordinates and collection date, where available. If a single publication reported the same host in several locations, each report was considered a separate record in the database. Furthermore, each host plant was classified as native or exotic based on its origin because many were species that had been introduced to the Americas.

A few records were omitted from the analysis, even though they were obtained under natural conditions. This was the case of two historical records for *Annona cherimola* Mill. in Arica (Chile) because *A. fraterculus* does not occur in that country since its eradication in 1964. In addition, other reports for several countries of *Rubus* spp., as well as *Ravenia wampi* Oliv (Rutaceae) and *Cyphomandra betacea* (Cav.) Stendtn. (Solanaceae) for Brazil were also excluded from the analysis because their status as unresolved names prevented confirmation of their identity and origin.

Because some original names have changed because publication in the primary source, the attached list presents a single identity for each host plant. The nomenclature used for the scientific names of plants was updated following the classification of "*The Plant List*" (2013) to avoid duplication or potential synonymies among species, as well as to facilitate their correct assignment to the corresponding family. The list of host plants of *A. fraterculus* (*sensu lato*) is organized and presented by host family and specific binomial name, with the authorship, origin, key code by host species, total number of records and countries where this fruit fly has been reported, as well as the original source of information (Table 6.1).

To analyze host plant trends, interaction networks depicting the associations of *A. fraterculus* and its host plant species across the Americas were evaluated, considering the origin of the hosts (native or exotic) and their plant families. A cluster analysis was also conducted based on the frequency of records of hosts throughout the countries. All analyses were executed using the "plotweb function" of the "bipartite" package (Dormann & Gruber 2009) in "R software" (R Core Team 2014).

6.3 RESULTS

6.3.1 Diversity of Host Plants

As a result of the examination of natural host plants of *A. fraterculus* throughout the Americas, 177 host plants belonging to 40 plant families were recorded, which together accounted for 1,622 documented reports in different localities. The plant families with highest species richness were Myrtaceae (48), Rosaceae (20), Rutaceae (14), Leguminosae (11), Sapotaceae (10), Anacardiaceae (8), and Annonaceae (7), which together represented nearly 66% of all recorded hosts. For instance, the family Myrtaceae was represented by at least 10 different genera, having the highest species richness. The highest frequency of records was found for *Psidium guajava* (298), *Eugenia uniflora* (48), *Syzygium jambos* (35), *Acca sellowiana* (31), and *Psidium cattleianum* (22). Myrtaceae is therefore the most important food plant family widespread throughout the distribution range of the SAFF.

TABLE 6.1

List of Host Plants for the Nominal Species *Anastrepha fraterculus* Reported in the Literature as Food Plants Infested in Nature

Host Family	Valid Scientific Name	Host Origin	Host-Code	Total records	Countries	References
Actinidiaceae	*Actinidia deliciosa* (A Chev) CS Liang & AR Ferguson	Exotic	*Ac-del*	2	Bra	Hickel and Schuck (1993)
Anacardiaceae	*Anacardium occidentale* L.	Native	*An-occ*	2	Bra, Gua	Eskafi and Cunningham (1987), Jesus-Barros et al. (2012)
Anacardiaceae	*Byrsonima crassifolia* (L.) Kunth	Native	*By-cra*	1	Bra	Jesus-Barros et al. (2012)
Anacardiaceae	*Mangifera indica* L.	Exotic	*Ma-ind*	46	Arg, Bra, Col, Ecu, FrG, Gua, Mex, Per, Ven	Aluja et al. (1987), Alvarenga et al. (2009), Araujo (2015), Boscán de Martínez and Godoy (1996), Campos (1960), Caraballo (1981), Eskafi (1990), Eskafi and Cunningham (1987), Ferreira et al. (2003), Gonzalez Mendoza (1952), Guagliumi (1966), Hernández-Ortiz et al. (2012), Korytkowski and Ojeda Peña (1968), Korytkowski and Ojeda Peña. (1969), Malavasi et al. (1980), Marchiori et al. (2000), Marinho (2004), Molineros et al. (1992), Núñez Bueno (1981), Ovruski et al. (2003), Putruele et al. (1996), Raga et al. (2011), Sá et al. (2008), Silva (1993), Silva et al. (1996), Stone (1942b), Tigrero (2009), Uramoto (2002), Uramoto et al. (2004), Vayssières et al. (2013), Veloso et al. (2000), Zucchi (2000)
Anacardiaceae	*Spondias dulcis* Parkinson	Exotic	*Sp-dul*	5	Bra	Araujo (2015), Souza-Filho et al. (2000), Uramoto (2002), Uramoto et al. (2004)
Anacardiaceae	*Spondias mombin* L.	Native	*Sp-mom*	12	Bra, Col, Mex, Per, Ven	Aguiar-Menezes et al. (2007), Almeida (2016), Boscán de Martínez and Godoy (1996), Gonzalez Mendoza (1952), Jesus-Barros et al. (2012), Korytkowski and Ojeda Peña (1968), Korytkowski and Ojeda Peña (1969), Lemos (2014), Lemos et al. (2017), McPhail and Bliss (1933), Sarmiento et al. (2012)
Anacardiaceae	*Spondias purpurea* L.	Native	*Sp-pur*	23	Bra, Ecu, Pan, Per	Aguiar-Menezes et al. (2001), Aguiar-Menezes et al. (2007), Alvarenga et al. (2000, 2009), Bressan and Da Costa Teles (1991), Campos (1960), Korytkowski and Ojeda Peña (1968), Korytkowski and Ojeda Peña. (1969), Leal et al. (2009), Lemos (2014), Lemos et al. (2017), Malavasi et al. (1980), Pirovani (2011), Pirovani et al. (2010), Raga et al. (2011), Sá et al. (2008), Silva et al. (2010), Souza-Filho et al. (2000), Stone (1942b), Tigrero (2009), Zucchi (2000)

(*Continued*)

TABLE 6.1 (*Continued*)
List of Host Plants for the Nominal Species *Anastrepha fraterculus* Reported in the Literature as Food Plants Infested in Nature

Host Family	Valid Scientific Name	Host Origin	Host-Code	Total records	Countries	References
Anacardiaceae	*Spondias* sp.	Native	*Sp-sp*	7	Bra, Col, Ecu, Mex	Aluja (1984), Molineros et al. (1992), Núñez Bueno (1981), Sá et al. (2008), Santos (2003), Silva et al. (2010), Zucchi (2000)
Anacardiaceae	*Spondias tuberosa* Arruda	Native	*Sp-tub*	2	Bra	Alvarenga et al. (2009), Sá et al. (2008)
Annonaceae	*Annona cherimola* Mill.	Native	*An-che*	21	Arg, Bra, Col, Ecu, Per	Gonzalez Mendoza (1952), Greene (1934), Hernández-Ortiz et al. (2012), Korytkowski and Ojeda Peña (1968), Korytkowski and Ojeda Peña. (1969), Molineros et al. (1992), Núñez Bueno (1981), Ovruski et al. (2003), Rust (1918), Stone (1942b), Tigrero (2009), White and Elson-Harris (1992), Wille (1941)
Annonaceae	*Annona crassiflora* Mart.	Native	*An-cra*	2	Bra	Veloso et al. (2000), Zucchi (2000)
Annonaceae	*Annona muricata* L.	Native	*An-mur*	4	Col	Gonzalez Mendoza (1952)
Annonaceae	*Annona rugulosa* (Schltdl.) H. Rainer	Native	*An-rug*	2	Bra	Garcia and Norrbom (2011), Marsaro Júnior (2014)
Annonaceae	*Rollinia emarginata* Schltdl.	Native	*Ro-ema*	2	Bra	Souza-Filho et al. (2000), Zucchi (2000)
Annonaceae	*Rollinia laurifolia* Schltdl.	Native	*Ro-lau*	2	Bra	Uramoto (2007), Uramoto et al. (2008)
Annonaceae	*Rollinia sericea* (R. E. Fr.) R. E. Fr.	Native	*Ro-ser*	3	Bra	Raga et al. (2011), Souza-Filho et al. (2000), Zucchi (2000)
Arecaceae	*Butia eriospatha* (Mart. ex Drude) Becc.	Native	*Bu-eri*	1	Bra	Savaris et al. (2013)
Arecaceae	*Syagrus romanzoffiana* (Cham.) Glassman	Native	*Sy-rom*	2	Bra	Araujo (2015), Uramoto (2002)
Calophyllaceae	*Mammea americana* L.	Exotic	*Ma-ame*	2	Col, Ecu	Campos (1960), Gonzalez Mendoza (1952)
Cannabaceae	*Celtis iguanaea* (Jacq.) Sarg.	Native	*Ce-igu*	1	Bra	Garcia and Norrbom (2011)
Caricaceae	*Carica papaya* L.	Native	*Ca-pap*	2	Arg, Col	Gonzalez Mendoza (1952)

(*Continued*)

TABLE 6.1 (*Continued*)
List of Host Plants for the Nominal Species *Anastrepha fraterculus* Reported in the Literature as Food Plants Infested in Nature

Host Family	Valid Scientific Name	Host Origin	Host-Code	Total records	Countries	References
Celastraceae	*Peritassa campestris* (Cambess.) A. C. Sm.	Native	*Pe-cam*	2	Bra	Veloso et al. (2000), Zucchi (2000)
Chrysobalanaceae	*Chrysobalanus icaco* L.	Native	*Ch-ica*	2	Bra, Col	Deus et al. (2013), Gonzalez Mendoza (1952)
Clusiaceae	*Garcinia brasiliensis* Mart.	Native	*Ga-bra*	2	Bra	Pirovani (2011), Raga et al. (2011)
Combretaceae	*Terminalia catappa* L.	Exotic	*Te-cat*	32	Bra, CR, Ecu, Mex, Ven	Aguiar-Menezes et al. (2007), Aluja et al. (1987, 2000), Araujo et al. (2000), Araujo (2012), Boscán de Martínez and Godoy (1996), Boscán de Martínez et al. (1980), Caraballo (1981), Hernández-Ortiz and Morales-Valles (2004), Hernández-Ortiz et al. (2012), Katiyar et al. (1995), Mascarenhas (2007), Raga et al. (2011), Silva et al. (1996), Souza-Filho et al. (2000), Steck (1991), Steck et al. (1990), Tigrero (2009), Zucchi (2000)
Ebenaceae	*Diospyros kaki* L.	Exotic	*Di-kak*	12	Arg, Bra	Garcia and Norrbom (2011), Malavasi et al. (1980), Marsaro Júnior (2014), Nasca et al. (1996), Ovruski et al. (2003), Segura et al. (2004, 2006), Souza-Filho et al. (2000), Zucchi (2000)
Euphorbiaceae	*Alchornea latifolia* Sw.	Native	*Al-lat*	2	Mex	Aluja (1984), Aluja et al. (1987)
Euphorbiaceae	*Manihot esculenta* Crantz	Native	*Mn-esc*	1	Ven	Boscán de Martínez and Godoy (1996)
Juglandaceae	*Juglans australis* Griseb.	Native	*Ju-aus*	6	Arg	Ovruski et al. (2003, 2004)
Juglandaceae	*Juglans neotropica* Diels	Native	*Ju-neo*	4	Ecu, Per	Korytkowski and Ojeda Peña (1969), Molineros et al. (1992), Tigrero (2009)
Lauraceae	*Cryptocarya aschersoniana* Mez	Native	*Cr-asc*	2	Bra	Marinho et al. (2009), Raga et al. (2011)
Lauraceae	*Endlicheria paniculata* (Spreng.) J. F. Macbr.	Native	*En-pan*	1	Bra	Pirovani (2011)
Lauraceae	*Persea americana* Mill.	Native	*Pe-ame*	4	Arg, Bra	Araujo (2015), Putruele et al. (1996), Rust (1918), Uramoto (2002)
Leguminosae	*Andira humilis* Mart. ex Benth.	Native	*An-hum*	2	Bra	Veloso et al. (2000), Zucchi (2000)

(*Continued*)

TABLE 6.1 (*Continued*)
List of Host Plants for the Nominal Species *Anastrepha fraterculus* Reported in the Literature as Food Plants Infested in Nature

Host Family	Valid Scientific Name	Host Origin	Host-Code	Total records	Countries	References
Leguminosae	*Inga edulis* Mart.	Native	*In-edu*	8	Bra, Col, Ecu	Deus et al. (2010), Gonzalez Mendoza (1952), Lemos (2014), Lemos et al. (2017), Malavasi et al. (1980), Tigrero (2009), Zucchi (2000)
Leguminosae	*Inga feuilleei* DC.	Native	*In-feu*	1	Ecu	Tigrero (2009)
Leguminosae	*Inga insignis* Kunth	Native	*In-ins*	1	Ecu	Tigrero (2009)
Leguminosae	*Inga micheliana* Harms	Native	*In-mic*	1	Gua	Eskafi and Cunningham (1987)
Leguminosae	*Inga paterna* Harms	Native	*In-pat*	1	Gua	Eskafi and Cunningham (1987)
Leguminosae	*Inga sellowiana* Benth.	Native	*In-sel*	1	Bra	Garcia and Norrbom (2011)
Leguminosae	*Inga semialata* (Vell.) C. Mart.	Native	*In-sem*	1	Arg	Oroño et al. (2005)
Leguminosae	*Inga* sp.	Native	*In-sp*	4	Bra, Ecu, Ven	Briceño Vergara (1975), Campos (1960), Molineros et al. (1992), Raga et al. (2011)
Leguminosae	*Inga spectabilis* (Vahl) Willd.	Native	*In-spe*	2	Ecu	Tigrero (2009)
Leguminosae	*Inga vera* subsp. *spuria* (Willd.) J. Leon	Native	*In-ver*	1	Col	Gonzalez Mendoza (1952)
Lythraceae	*Punica granatum* L.	Exotic	*Pu-gra*	5	Arg, Bra, Ecu	Molineros et al. (1992), Nasca et al. (1996), Putruele et al. (1996), Raga et al. (2011), Tigrero (2009)
Malpighiaceae	*Malpighia emarginata* DC.	Exotic	*Ml-ema*	4	Col, Ecu	Campos (1960), Pirovani (2011), Raga et al. (2011), Sá et al. (2008)
Malpighiaceae	*Malpighia glabra* L.	Native	*Ml-gla*	13	Bra, Col, Ecu	Aguiar-Menezes et al. (2001), Araujo (2015), Gonzalez Mendoza (1952), Leal et al. (2009), Malavasi et al. (1980), Marinho (2004), Souza-Filho et al. (2000), Uramoto (2002), Uramoto et al. (2004), Zucchi (2000)
Malvaceae	*Quararibea cordata* (Bonpl.) Vischer	Native	*Qu-cor*	3	Ecu	Molineros et al. (1992), Tigrero (2009)
Malvaceae	*Theobroma cacao* L.	Native	*Th-cac*	2	Ven	Caraballo (1981), Hernández-Ortiz and Morales-Valles (2004)
Melastomataceae	*Bellucia grossularioides* (L.) Triana	Native	*Be-gro*	1	FrGuy	Vayssières et al. (2013)

(*Continued*)

TABLE 6.1 (*Continued*)
List of Host Plants for the Nominal Species *Anastrepha fraterculus* Reported in the Literature as Food Plants Infested in Nature

Host Family	Valid Scientific Name	Host Origin	Host-Code	Total records	Countries	References
Melastomataceae	*Mouriri acutiflora* Naudin	Native	*Mo-acu*	1	Bra	Jesus-Barros et al. (2012)
Melastomataceae	*Mouriri glazioviana* Cogn.	Native	*Mo-gla*	2	Bra	Uramoto (2007), Uramoto et al. (2008)
Moraceae	*Ficus carica* L.	Exotic	*Fi-car*	16	Arg, Bra, Ecu	Garcia and Norrbom (2011), Molineros et al. (1992), Nasca et al. (1996), Ovruski et al. (2003), Putruele et al. (1996), Rust (1918), Tigrero (2009)
Moraceae	*Helicostylis* sp.	Native	*He-sp*	3	Bra	Bondar (1950), Kovaleski et al. (1999), Zucchi (2000)
Moraceae	*Helicostylis tomentosa* (Poepp. & Endl.) J. F. Macbr.	Native	*He-tom*	1	Bra	Bondar (1950)
Myrtaceae	*Acca sellowiana* (O. Berg) Burret	Native	*Ac-sel*	31	Arg, Bra, Col, Ecu	Alberti et al. (2002), Cruz et al. (2017), Custódio et al. (2017), Garcia and Norrbom (2011), Hernández-Ortiz et al. (2015), Hickel and Ducroquet (1994), Kovaleski et al. (1999, 2000), Marsaro Júnior (2014), Molineros et al. (1992), Nasca et al. (1996), Nunes et al. (2012), Ovruski et al. (2008), Pereira-Rêgo et al. (2011), Rust (1918), Salles (1995), Schliserman et al. (2010), Segura et al. (2004, 2006), Tigrero (2009), Yepes and Vélez (1989), Zucchi (2000)
Myrtaceae	*Calycolpus moritzianus* (O. Berg) Burret	Native	*Ca-mor*	3	Ven	Hernández-Ortiz and Morales-Valles (2004), Hernández-Ortiz et al. (2012), Katiyar et al. (2000)
Myrtaceae	*Campomanesia adamantium* (Cambess.) O. Berg	Native	*Cm-ada*	8	Bra	Malavasi et al. (1980), Selivon (2000), Souza-Filho et al. (2000), Veloso et al. (2000), Zucchi (2000)
Myrtaceae	*Campomanesia espiritosantensis* Landrum	Native	*Cm-esp*	2	Bra	Uramoto (2007), Uramoto et al. (2008)
Myrtaceae	*Campomanesia guaviroba* (DC.) Kiaersk.	Native	*Cm-gua*	1	Bra	Zucchi (2000)

(*Continued*)

TABLE 6.1 (*Continued*)
List of Host Plants for the Nominal Species *Anastrepha fraterculus* Reported in the Literature as Food Plants Infested in Nature

Host Family	Valid Scientific Name	Host Origin	Host-Code	Total records	Countries	References
Myrtaceae	*Campomanesia guazumifolia* (Cambess.) O. Berg	Native	*Cm-guz*	2	Bra	Garcia and Norrbom (2011), Marsaro Júnior (2014)
Myrtaceae	*Campomanesia lineatifolia* Ruiz & Pav.	Native	*Cm-lin*	2	Bra	Uramoto (2007), Uramoto et al. (2008)
Myrtaceae	*Campomanesia pubescens* (Mart. ex DC.) O. Berg	Native	*Cm-pub*	2	Bra	Guimarães et al. (1999), Raga et al. (2005)
Myrtaceae	*Campomanesia xanthocarpa* (Mart.) O. Berg	Native	*Cm-xan*	8	Arg, Bra	Garcia and Norrbom (2011), Kovaleski et al. (1999, 2000), Malavasi et al. (1980), Marsaro Júnior (2014), Salles (1995), Schliserman et al. (2010), Zucchi (2000)
Myrtaceae	*Eugenia brasiliensis* Lam.	Native	*Eu-bra*	8	Bra	Aguiar-Menezes et al. (2007), Raga et al. (2005), Steck and Malavasi (1988), Steck et al. (1990), Stone (1942b), Uramoto (2007), Uramoto et al. (2008), Wille (1941)
Myrtaceae	*Eugenia dodonaeifolia* Cambess.	Native	*Eu-dod*	1	Bra	Zucchi (2000)
Myrtaceae	*Eugenia dysenterica* DC.	Native	*Eu-dys*	3	Bra	Silva et al. (2010), Veloso et al. (2000), Zucchi (2000)
Myrtaceae	*Eugenia florida* DC.	Native	*Eu-flo*	1	Pan	Stone (1942a)
Myrtaceae	*Eugenia gemmiflora* O. Berg	Native	*Eu-gem*	2	Bra	Uramoto (2007), Uramoto et al. (2008)
Myrtaceae	*Eugenia involucrata* DC.	Native	*Eu-inv*	15	Bra	Araujo (2015), Garcia and Norrbom (2011), Haji and Da Gama Miranda (2000), Kovaleski et al. (1999, 2000), Marsaro Júnior (2014), Nunes et al. (2012), Raga et al. (2005), Salles (1995), Uramoto (2002, 2007), Uramoto et al. (2004, 2008), Zucchi (2000)
Myrtaceae	*Eugenia lambertiana* DC.	Native	*Eu-lam*	2	Bra	Raga et al. (2005), Zucchi (2000)
Myrtaceae	*Eugenia leitonii* D. Legrand	Native	*Eu-lei*	2	Bra	Raga et al. (2005), Zucchi (2000)

(*Continued*)

TABLE 6.1 (*Continued*)
List of Host Plants for the Nominal Species *Anastrepha fraterculus* Reported in the Literature as Food Plants Infested in Nature

Host Family	Valid Scientific Name	Host Origin	Host-Code	Total records	Countries	References
Myrtaceae	*Eugenia myrcianthes* Nied.	Native	*Eu-myr*	2	Arg	Putruele et al. (1996), Segura et al. (2006)
Myrtaceae	*Eugenia platyphylla* O. Berg	Native	*Eu-pla*	3	Bra	Uramoto (2007)
Myrtaceae	*Eugenia platysema* O. Berg	Native	*Eu-plt*	1	Bra	Uramoto et al. (2008)
Myrtaceae	*Eugenia punicifolia* (Kunth) DC.	Native	*Eu-pun*	2	Ven	Fernandez Yepez (1953), Guagliumi (1966)
Myrtaceae	*Eugenia pyriformis* Cambess.	Native	*Eu-pyr*	19	Bra	Araujo (2015), Garcia and Norrbom (2011), Guimarães et al. (1999), Kovaleski et al. (1999, 2000), Malavasi et al. (1980), Marinho (2004), Marsaro Júnior (2014), Nunes et al. (2012), Perre (2016), Raga et al. (2005), Souza-Filho et al. (2000), Uramoto (2002), Uramoto et al. (2004), Zucchi (2000)
Myrtaceae	*Eugenia stipitata* McVaugh	Native	*Eu-sti*	8	Bra, Ecu	Lemos (2014), Lemos et al. (2017), Molineros et al. (1992), Silva et al. (2010), Tigrero (2009), Uramoto (2007), Uramoto et al. (2008)
Myrtaceae	*Eugenia uniflora* L.	Native	*Eu-uni*	48	Arg, Bra, Gua, Mex, Pan	Aguiar-Menezes and Menezes (2000), Aguiar-Menezes et al. (2003), Aguiar-Menezes et al. (2007), Aluja et al. (1987), Alvarenga et al. (2009), Alvarenga et al. (2000), Araujo (2015), Eskafi (1990), Garcia and Norrbom (2011), Guimarães et al. (1999), Hernández-Ortiz et al. (2012), Malavasi et al. (1980), Marchiori et al. (2000), Marinho (2004), Marsaro Júnior (2014), Nunes et al. (2012), Ovruski et al. (2003, 2004), Ovruski et al. (2008), Pirovani (2011), Pirovani et al. (2010), Raga et al. (2005), Salles (1995), Silva et al. (2010, 2011), Souza-Filho et al. (2000), Stone (1942b), Uramoto (2002), Uramoto et al. (2004), Veloso et al. (2000), Zucchi (2000)
Myrtaceae	*Myrceugenia euosma* (O. Berg) D. Legrand	Native	*Me-euo*	3	Bra	Kovaleski et al. (1999, 2000), Zucchi (2000)
Myrtaceae	*Myrcia clausseniana* Berg	Native	*Mc-cla*	2	Bra	Uramoto (2007), Uramoto et al. (2008)
Myrtaceae	*Myrcia popayanensis* Hieron.	Native	*Mc-pop*	1	Col	Olarte Espinosa (1980)

(*Continued*)

TABLE 6.1 (*Continued*)
List of Host Plants for the Nominal Species *Anastrepha fraterculus* Reported in the Literature as Food Plants Infested in Nature

Host Family	Valid Scientific Name	Host Origin	Host-Code	Total records	Countries	References
Myrtaceae	*Myrcianthes fragrans* (Sw.) McVaugh	Exotic	*My-fra*	1	Col	Gonzalez Mendoza (1952)
Myrtaceae	*Myrcianthes pungens* (O. Berg) D. Legrand	Native	*My-pun*	11	Arg, Bra	Garcia and Norrbom (2011), Marsaro Júnior (2014), Ovruski et al. (2003, 2004, 2008), Putruele et al. (1996)
Myrtaceae	*Myrciaria dubia* (Kunth) McVaugh	Native	*Mr-dub*	1	Bra	Custódio et al. (2017)
Myrtaceae	*Myrciaria floribunda* (H.West ex Willd.) O. Berg	Native	*Mr-flo*	1	Mex	Aluja et al. (2000)
Myrtaceae	*Myrciaria glazioviana* (Kiaersk.) G. M. Barroso ex Sobral	Native	*Mr-gla*	1	Bra	Pirovani (2011)
Myrtaceae	*Myrciaria glomerata* O. Berg	Native	*Mr-glo*	5	Bra	Araujo (2015), Marinho (2004), Raga et al. (2005), Uramoto (2002), Uramoto et al. (2004)
Myrtaceae	*Myrciaria strigipes* O. Berg	Native	*Mr-str*	2	Bra	Uramoto (2007), Uramoto et al. (2008)
Myrtaceae	*Plinia cauliflora* (Mart.) Kausel	Native	*Pl-cau*	15	Bra	Aguiar-Menezes and Menezes (2000), Aguiar-Menezes et al. (2007), Araujo (2015), Garcia and Norrbom (2011), Marinho (2004), Pirovani (2011), Pirovani et al. (2010), Raga et al. (2005), Salles (1995), Silva et al. (2011), Souza-Filho et al. (2000), Uramoto (2002), Uramoto et al. (2004), Zucchi (2000)
Myrtaceae	*Plinia edulis* (Vell.) Sobral	Native	*Pl-edu*	2	Bra	Raga et al. (2005), Zucchi (2000)
Myrtaceae	*Psidium acutangulum* Mart. ex DC.	Native	*Ps-acu*	1	Col	Hernández-Ortiz et al. (2015)
Myrtaceae	*Psidium cattleianum* Afzel. ex Sabine	Native	*Ps-cat*	22	Arg, Bra, Gua	Aguiar-Menezes et al. (2001), Eskafi and Cunningham (1987), Garcia and Norrbom (2011), Kovaleski et al. (1999, 2000), Leal et al. (2009), Manni et al. (2015), Marinho (2004), Marsaro Júnior (2014), Nunes et al. (2012), Pereira-Rêgo et al. (2011), Raga et al. (2005), Rust (1918), Zucchi (2000)

(*Continued*)

TABLE 6.1 (*Continued*)
List of Host Plants for the Nominal Species *Anastrepha fraterculus* Reported in the Literature as Food Plants Infested in Nature

Host Family	Valid Scientific Name	Host Origin	Host-Code	Total records	Countries	References
Myrtaceae	*Psidium guajava* L.	Native	*Ps-gua*	298	Arg, Bol, Bra, Col, CR, Ecu, FrG, Gua, Mex, Pan, Per, Ven	Adaime et al. (2017), Aguiar-Menezes and Menezes (1997, 2000), Aguiar-Menezes et al. (2001, 2007), Alberti et al. (2002), Aluja (1984), Aluja et al. (1987, 2000), Alvarenga et al. (2000. 2009), Araujo et al. (2000), Araujo (2012, 2015), Baker et al. (1944), Barros (2008), Bomfim et al. (2007), Boscán de Martínez and Godoy (1996), Campos (1960), Caraballo (1981), Castañeda et al. (2010), Corsato (2004), Cruz et al. (2017), Deus et al. (2010), Eskafi (1990), Eskafi and Cunningham (1987), Fernandez Yepez (1953), Fischer (1934), Garcia and Norrbom (2011), Gonzalez Mendoza (1952), Guagliumi (1966), Haji and Da Gama Miranda (2000), Hedström (1987), Hernández-Ortiz and Morales-Valles (2004), Hernández-Ortiz and Pérez-Alonso (1993), Hernández-Ortiz et al. (1994, 2004, 2012, 2015), Isaac (1905), Jesus-Barros et al. (2012), Jirón and Hedström (1988), Katiyar et al. (2000), Korytkowski et al. (2001), Korytkowski and Ojeda Peña (1968, 1969),Leal et al. (2009), Lemos (2014), Lemos et al. (2017), Malavasi et al. (1980), Manni et al. (2015), Marchiori et al. (2000), Marsaro Júnior (2014), Mascarenhas (2007), McPhail and Bliss (1933), Molineros et al. (1992), Nunes et al. (2012), Núñez Bueno (1981), Nuñez Bueno et al. (2004), Ovruski (1995), Ovruski et al. (2003, 2004, 2005, 2008, 2009), Pereira-Rêgo et al. (2011), Perre (2016), Pirovani (2011), Pirovani et al. (2010), Putruele et al. (1996), Querino et al. (2014), Raga et al. (2005), Sá et al. (2008), Salles (1995), Sarmiento et al. (2012), Schliserman et al. (2010), Segura et al. (2006), Selivon (2000), Silva et al. (2010, 2011a, 2011b), Silva and Silva (2007), Souza-Filho (2005, 2006), Souza-Filho et al. (2009), Steck (1991), Steck et al. (1990), Stone (1942b), Taira (2012), Taira et al. (2013), Tigrero (1998, 2009), Uchôa-Fernandes and Zucchi (2000), Uchôa-Fernandes et al. (2002), Uramoto (2002), Uramoto et al. (2004), Vayssières et al. (2013), Veloso et al. (2000), Wille (1937), Zucchi (2000)
Myrtaceae	*Psidium guineense* Sw.	Native	*Ps-gui*	17	Bra, Col, Mex, Ven	Aluja et al. (2000), Araujo et al. (2000), Bomfim et al. (2007), Castañeda et al. (2010), Fernandez Yepez (1953), Gonzalez Mendoza (1952), Guagliumi (1966), Hernández-Ortiz et al. (2004), Katiyar et al. (2000), Pirovani (2011), Pirovani et al. (2010), Silva et al. (2010, 2011), Uramoto et al. (2008), Veloso et al. (2000), Zucchi (2000)

(Continued)

TABLE 6.1 (*Continued*)
List of Host Plants for the Nominal Species *Anastrepha fraterculus* Reported in the Literature as Food Plants Infested in Nature

Host Family	Valid Scientific Name	Host Origin	Host-Code	Total records	Countries	References
Myrtaceae	*Psidium guyanense* Pers.	Native	*Ps-guy*	1	Bra	Uramoto (2007)
Myrtaceae	*Psidium kennedyanum* Morong	Native	*Ps-ken*	1	Bra	Uchôa-Fernandes and Nicácio (2010)
Myrtaceae	*Psidium myrtoides* O. Berg	Native	*Ps-myr*	2	Bra	Uramoto (2007), Uramoto et al. (2008)
Myrtaceae	*Psidium sartorianum* (O. Berg) Nied.	Native	*Ps-sar*	4	Mex	Aluja et al. (2000), Aluja et al. (2003b)
Myrtaceae	*Psidium* sp.	Native	*Ps-sp*	8	Arg, Bra, Ecu	Molineros et al. (1992), Rust (1918), Salles (1995), Tigrero (2009), Uchôa-Fernandes and Zucchi (2000), Uchôa-Fernandes et al. (2002)
Myrtaceae	*Syzygium aqueum* (Burm. f.) Alston	Exotic	*Sy-aqu*	1	Bra	Zucchi (2000)
Myrtaceae	*Syzygium jambos* (L.) Alston	Exotic	*Sy-jam*	35	Bra, Ecu, Mex, Per, Ven	Aluja (1984), Aluja et al. (1987, 2000), Araujo (2015), Baker et al. (1944), Baker (1945), Bomfim et al. (2007), Bush (1962), Campos (1960), Hernández-Ortiz et al. (2004), Korytkowski and Ojeda Peña (1968, 1969), Marinho (2004), Molineros et al. (1992), Perre (2016), Raga et al. (2005), Souza-Filho et al. (2000), Steck (1991), Steck et al. (1990), Tigrero (2009), Uchôa-Fernandes and Nicácio (2010), Uramoto (2002), Uramoto et al. (2004)
Myrtaceae	*Syzygium malaccense* (L.) Merr. & L. M. Perry	Exotic	*Sy-mal*	4	Bra, Pan	Silva et al. (2010, 2011), Stone (1942b)
Oleaceae	*Olea europaea* L.	Exotic	*Ol-eur*	1	Arg	Nasca et al. (1996)
Oxalidaceae	*Averrhoa carambola* L.	Exotic	*Av-car*	18	Bra	Aguiar-Menezes et al. (2001, 2007), Araujo (2015), Feitosa et al. (2007), Garcia and Norrbom (2011), Leal et al. (2009), Malavasi et al. (1980), Marchiori et al. (2000), Raga et al. (2011), Silva et al. (2010), Souza-Filho et al. (2000), Uramoto (2002), Uramoto et al. (2004), Veloso et al. (2000), Zucchi (2000)
Passifloraceae	*Passiflora alata* Curtis	Native	*Pa-ala*	3	Bra	Pirovani (2011), Pirovani et al. (2010), Souza-Filho et al. (2000)
Passifloraceae	*Passiflora caerulea* L.	Native	*Pa-cae*	3	Arg	Ovruski et al. (2003), Putruele et al. (1996)

(*Continued*)

TABLE 6.1 (*Continued*)
List of Host Plants for the Nominal Species *Anastrepha fraterculus* Reported in the Literature as Food Plants Infested in Nature

Host Family	Valid Scientific Name	Host Origin	Host-Code	Total records	Countries	References
Passifloraceae	*Passiflora edulis* Sims	Native	*Pa-edu*	4	Bra	Araujo (2015), Souza-Filho et al. (2000), Uramoto (2002), Uramoto et al. (2004)
Passifloraceae	*Passiflora* sp.	Native	*Pa-sp*	3	Bra	Kovaleski et al. (1999, 2000), Zucchi (2000)
Passifloraceae	*Passiflora tripartita* (Juss.) Poir.	Native	*Pa-tri*	1	Col	Castro (2009)
Picramniaceae	*Picramnia* sp.	Native	*Pi-sp*	1	Bra	Raga et al. (2011)
Rhamnaceae	*Ziziphus joazeiro* Mart.	Native	*Zi-joa*	1	Bra	Sá et al. (2008)
Rhamnaceae	*Ziziphus jujuba* Mill.	Native	*Zi-juj*	1	Col	Gonzalez Mendoza (1952)
Rosaceae	*Cydonia oblonga* Mill.	Exotic	*Cy-obl*	6	Arg, Per, Ven	Guagliumi (1966), Korytkowski and Ojeda Peña (1968, 1969), Nasca et al. (1996), Ovruski et al. (2010), Putruele et al. (1996)
Rosaceae	*Eriobotrya japonica* (Thunb.) Lindl.	Exotic	*Er-jap*	75	Arg, Bra, Col, Ecu, Mex, Per, Ven	Aguiar-Menezes and Menezes (2000), Aguiar-Menezes et al. (2007), Alberti et al. (2002), Aluja (1984), Araujo (2015), Briceño Vergara (1975, 1979), Caraballo (1981), Cruz et al. (2017), Fernandez Yepez (1953), Garcia and Norrbom (2011), Gonzalez Mendoza (1952), Guagliumi (1966), Hernández-Ortiz and Morales-Valles (2004), Hernández-Ortiz et al. (2012), Katiyar et al. (1995), Kovaleski et al. (1999, 2000), Malavasi et al. (1980), Marinho (2004), Mascarenhas (2007), Molineros et al. (1992), Nunes et al. (2012), Ovruski et al. (2003), Perre (2016), Pirovani (2011), Pirovani et al. (2010), Putruele et al. (1996), Raga et al. (2011), Salles (1995), Schliserman et al. (2010), Segura et al. (2006), Souza-Filho (2006), Souza-Filho et al. (2000, 2009), Steck (1991), Stone (1942b), Tigrero (2009), Uramoto (2002), Uramoto et al. (2004), Yepes and Vélez (1989), Zucchi (2000)
Rosaceae	*Fragaria* × *ananassa* (Duchesne ex Weston) Duchesne ex Rozier	Exotic	*Fg-ana*	1	Bra	Raga et al. (2011)
Rosaceae	*Fragaria vesca* L.	Exotic	*Fg-ves*	2	Bra	Salles (1995), Zucchi (2000)
Rosaceae	*Malus domestica* Borkh.	Exotic	*Ma-dom*	16	Bra, Col, Ecu, Ven	Ballou (1945), Boscán de Martínez et al. (1980), Campos (1960), Caraballo (1981), Gonzalez Mendoza (1952), Guagliumi (1966), Kovaleski et al. (1999, 2000), Malavasi et al. (1980), Molineros et al. (1992), Perre (2016), Raga et al. (2011), Salles (1995), Tigrero (2009), Zucchi (2000)

(*Continued*)

TABLE 6.1 (*Continued*)
List of Host Plants for the Nominal Species *Anastrepha fraterculus* Reported in the Literature as Food Plants Infested in Nature

Host Family	Valid Scientific Name	Host Origin	Host-Code	Total records	Countries	References
Rosaceae	*Prunus armeniaca* L.	Exotic	*Pr-arm*	6	Arg	Nasca et al. (1996), Ovruski et al. (2003, 2004), Putruele et al. (1996), Rust (1918)
Rosaceae	*Prunus avium* (L.) L.	Exotic	*Pr-avi*	1	Bra	Garcia and Norrbom (2011)
Rosaceae	*Prunus domestica* L.	Exotic	*Pr-dom*	17	Arg, Bra, Col, Ecu	Garcia and Norrbom (2011), Guimarães et al. (1999), Kovaleski et al. (1999, 2000), Marin Patiño (2002), Molineros et al. (1992), Nasca et al. (1996), Ovruski et al. (2003, 2004, 2010), Raga et al. (2011), Salles (1995), Tigrero (2009), Zucchi (2000)
Rosaceae	*Prunus domestica* subsp. *insititia* (L.) Bonnier & Layens	Exotic	*Pr-dom*	2	Arg	Putruele et al. (1996), Segura et al. (2006)
Rosaceae	*Prunus dulcis* (Mill.) D. A.Webb	Exotic	*Pr-dul*	1	Arg	Nasca et al. (1996)
Rosaceae	*Prunus mume* (Siebold) Siebold & Zucc.	Exotic	*Pr-mum*	2	Bra	Perre (2016), Raga et al. (2011)
Rosaceae	*Prunus persica* (L.) Batsch	Exotic	*Pr-per*	177	Arg, Bol, Bra, Col, Ecu, Gua, Mex, Per, Ven	Aguiar-Menezes and Menezes (1997), Alberti et al. (2002), Aluja et al. (2000), Araujo (2015), Baker et al. (1944), Boscán de Martínez and Godoy (1996), Boscán de Martínez et al. (1980), Briceño Vergara (1975, 1979), Campos (1960), Caraballo (1981), Cruz et al. (2017), Eskafi and Cunningham (1987), Garcia and Norrbom (2011), Gonzalez Mendoza (1952), Greene (1934), Guagliumi (1966), Hernández-Ortiz and Morales-Valles (2004), Kovaleski et al. (1999, 2000), Malavasi et al. (1980), Marsaro Júnior (2014), Molineros et al. (1992), Nasca et al. (1996), Nunes et al. (2012), Ovruski (1995), Ovruski et al. (2003, 2004, 2009, 2010), Perre (2016), Putruele et al. (1996), Raga et al. (2011), Salles (1995), Schliserman et al. (2010), Segura et al. (2004, 2006), Shaw (1947), Souza-Filho (2006), Souza-Filho et al. (2000, 2009), Steck (1991), Tigrero (2009), Uramoto (2002), Uramoto et al. (2004), Zucchi (2000)
Rosaceae	*Prunus salicina* Lindl.	Exotic	*Pr-sal*	7	Arg, Bra	Alberti et al. (2002), Marinho (2004), Marinho et al. (2009), Perre (2016)
Rosaceae	*Prunus serotina* Ehrh. ssp. *capuli* (Cav.) McVaugh	Native	*Pr-ser*	1	Gua	Eskafi and Cunningham (1987)

(*Continued*)

TABLE 6.1 (*Continued*)
List of Host Plants for the Nominal Species *Anastrepha fraterculus* Reported in the Literature as Food Plants Infested in Nature

Host Family	Valid Scientific Name	Host Origin	Host-Code	Total records	Countries	References
Rosaceae	*Prunus* sp.	Exotic	*Pr-sp*	5	Bra, Col	Gonzalez Mendoza (1952), Malavasi et al. (1980), Salles (1995), Zucchi (2000)
Rosaceae	*Pyrus communis* L.	Exotic	*Py-com*	18	Arg, Bra, Ecu, Per	Aguiar-Menezes and Menezes (2000), Campos (1960), Garcia and Norrbom (2011), Korytkowski and Ojeda Peña (1968, 1969), Kovaleski et al. (2000), Molineros et al. (1992), Nasca et al. (1996), Nunes et al. (2015), Raga et al. (2011), Salles (1995), Souza-Filho et al. (2000), Stone (1942b), Tigrero (2009), Zucchi (2000)
Rosaceae	*Rubus eriocarpus* Liebm.	Native	*Ru-eri*	3	Ecu, Ven	Hernández-Ortiz and Morales-Valles (2004), Tigrero (2009)
Rosaceae	*Rubus idaeus* L.	Exotic	*Ru-ida*	1	Arg	Funes et al. (2017)
Rosaceae	*Rubus* sp.	?	*Ru-sp*	26	Arg, Bra, Col, Ecu, Ven	Araujo (2015), Briceño Vergara (1975, 1979), Castañeda et al. (2010), Funes et al. (2017), Hernández-Ortiz et al. (2012), Katiyar et al. (1995), Kovaleski et al. (2000), Manni et al. (2015), Molineros et al. (1992), Núñez Bueno (1981), Salles (1995), Steck (1991), Steck et al. (1990), Uramoto (2002), Uramoto et al. (2004), Zucchi (2000)
Rosaceae	*Rubus ulmifolius* Schott	Exotic	*Ru-ulm*	1	Bra	Raga et al. (2011)
Rubiaceae	*Coffea arabica* L.	Exotic	*Co-ara*	51	Bra, Col, Mex, Ven	Aluja (1984), Aluja et al. (1987), Araujo (2015), Boscán de Martínez et al. (1980), Caraballo (1981), Castañeda et al. (2010), Cruz et al. (2017), Gonzalez Mendoza (1952), Hernández-Ortiz and Morales-Valles (2004), Hernández-Ortiz et al. (2012), Hernández-Ortiz et al. (2015), Katiyar et al. (1995), Malavasi et al. (1980), Núñez Bueno (1981), Nuñez Bueno et al. (2004), Sarmiento et al. (2012), Souza et al. (2005), Steck (1991), Steck et al. (1990), Stone (1942b), Torres (2004), Uramoto (2002), Uramoto et al. (2004), Yepes and Vélez (1989), Zucchi (2000)
Rubiaceae	*Coffea liberica* Hiern	Exotic	*Co-lib*	1	Ven	Guagliumi (1966)
Rutaceae	*Citrus aurantiifolia* (Christm.) Swingle	Exotic	*Ct-arn*	4	Bra	Salles (1995), Zucchi (2000)
Rutaceae	*Citrus japonica* Thunb.	Exotic	*Ct-jap*	5	Arg, Bra, Col	Gonzalez Mendoza (1952), Raga et al. (2004), Rust (1918), Salles (1995), Zucchi (2000)
Rutaceae	*Citrus limon* (L.) Osbeck	Exotic	*Ct-lim*	3	Bra	Salles (1995). Souza-Filho et al. (2000), Zucchi (2000)

(*Continued*)

TABLE 6.1 (*Continued*)
List of Host Plants for the Nominal Species *Anastrepha fraterculus* Reported in the Literature as Food Plants Infested in Nature

Host Family	Valid Scientific Name	Host Origin	Host-Code	Total records	Countries	References
Rutaceae	*Citrus maxima* (Burm.) Merr.	Exotic	*Ct-max*	8	Arg, Bra, Ecu, Per	Korytkowski and Ojeda Peña (1968, 1969), Malavasi et al. (1980), Raga et al. (2004), Rust (1918), Souza-Filho et al. (2000), Tigrero (2009), Zucchi (2000)
Rutaceae	*Citrus medica* L.	Exotic	*Ct-med*	2	Bra, Gua	Eskafi and Cunningham (1987), Raga et al. (2004)
Rutaceae	*Citrus nobilis* Lour.	Exotic	*Ct-nob*	1	Arg	Rust (1918)
Rutaceae	*Citrus paradisi* Macfad.	Exotic	*Ct-par*	32	Arg, Bra, Ecu, Gua, Ven	Caraballo (1981), Eskafi and Cunningham (1987), Molineros et al. (1992), Nasca et al. (1996), Ovruski et al. (2003), Putruele et al. (1996), Schliserman et al. (2010), Segura et al. (2006), Zucchi (2000)
Rutaceae	*Citrus reticulata* Blanco	Exotic	*Ct-ret*	30	Arg, Bra, Col, Ecu	Araujo (2015), Castañeda et al. (2010), Garcia and Norrbom (2011), Molineros et al. (1992), Ovruski et al. (2003), Putruele et al. (1996), Raga et al. (2004), Schliserman et al. (2010), Segura et al. (2004, 2006), Souza-Filho et al. (2000), Tigrero (2009), Uramoto (2002), Uramoto et al. (2004)
Rutaceae	*Citrus sinensis* (L.) Osbeck	Exotic	*Ct-sin*	44	Arg, Bra, Ecu, Gua, Per	Aguiar-Menezes and Menezes (2000), Ballou (1945), Boscán de Martínez et al. (1980), Eskafi (1990), Eskafi and Cunningham (1987), Guagliumi (1966), Guimarães et al. (1999), Korytkowski and Ojeda Peña (1968, 1969), Malavasi et al. (1980), Molineros et al. (1992), Ovruski et al. (2003), Pirovani (2011), Pirovani et al. (2010), Putruele et al. (1996), Raga et al. (2004), Salles (1995), Selivon (2000), Silva et al. (2006), Stone (1942b), Tigrero (2009), Zucchi (2000)
Rutaceae	*Citrus sinensis x C reticulata*	Exotic	*Ct-sxr*	2	Bra	Silva et al. (2006)
Rutaceae	*Citrus* sp.	Exotic	*Ct-sp*	34	Bra, Col, Ecu, Ven	Araujo (2012, 2015), Briceño Vergara (1979), Caraballo (1981), Gonzalez Mendoza (1952), Molineros et al. (1992), Núñez Bueno (1981), Raga et al. (2004), Salles (1995), Uramoto (2002), Uramoto et al. (2004), Zucchi (2000)
Rutaceae	*Citrus trifoliata* L.	Exotic	*Ct-tri*	2	Ecu	Molineros et al. (1992), Tigrero (2009)
Rutaceae	*Citrus x aurantium* L.	Exotic	*Ct-aur*	39	Arg, Bra, Col, Ecu, Mex, Ven	Briceño Vergara (1975), Campos (1960), Gonzalez Mendoza (1952), Hernández-Ortiz et al. (1994, 2006), Malavasi et al. (1980), Marchiori et al. (2000), Ovruski et al. (2003), Putruele et al. (1996), Raga et al. (2004), Schliserman and Ovruski (2004), Schliserman et al. (2010), Segura et al. (2006), Tigrero (1998, 2009), Yepes and Vélez (1989), Zucchi (2000)

(*Continued*)

TABLE 6.1 (*Continued*)
List of Host Plants for the Nominal Species *Anastrepha fraterculus* Reported in the Literature as Food Plants Infested in Nature

Host Family	Valid Scientific Name	Host Origin	Host-Code	Total records	Countries	References
Rutaceae	*Ravenia wampi* Oliv.	?	*unres. name*	1	Bra	Aguiar-Menezes et al. (2001)
Salicaceae	*Dovyalis abyssinica* (A. Rich.) Warb.	Exotic	*Do-aby*	2	Ecu	Molineros et al. (1992), Tigrero (2009)
Salicaceae	*Dovyalis hebecarpa* (Gardner) Warb.	Exotic	*Do-heb*	1	Pan	Stone (1942b)
Salicaceae	*Zuelania guidonia* (Sw.) Britton & Millsp.	Native	*Zu-gui*	1	Mex	García-Ramírez et al. (2010)
Sapindaceae	*Diatenopteryx sorbifolia* Radlk.	Native	*Di-sor*	2	Bra	Salles (1995), Zucchi (2000)
Sapindaceae	*Melicoccus bijugatus* Jacq.	Native	*Me-bij*	1	Bra	Gonzalez Mendoza (1952)
Sapindaceae	*Melicoccus oliviformis* Kunth	Native	*Me-oli*	1	Mex	García-Ramírez et al. (2010)
Sapotaceae	*Chrysophyllum cainito* L.	Native	*Cs-cai*	3	Ecu, Ven	Campos (1960), Hernández-Ortiz and Morales-Valles (2004)
Sapotaceae	*Chrysophyllum gonocarpum* (Mart. & Eichler ex Miq.) Engl.	Native	*Cs-gon*	3	Arg, Bra	Oroño et al. (2005), Salles (1995), Zucchi (2000)
Sapotaceae	*Chrysophyllum mexicanum* Brandegee	Native	*Cs-mex*	1	Bra	Souza-Filho et al. (2000)
Sapotaceae	*Manilkara zapota* (L.) P. Royen	Native	*Ma-zap*	8	Bra, Col, Ecu, Pan	Araujo (2015), Campos (1960), Gonzalez Mendoza (1952), Núñez Bueno (1981), Raga et al. (2011), Stone (1942b), Uramoto (2002), Uramoto et al. (2004)
Sapotaceae	*Planchonella obovata* (R. Br.) Pierre	Native	*Pl-obo*	1	Per	Korytkowski and Ojeda Peña. (1969)
Sapotaceae	*Pouteria caimito* (Ruiz & Pav.) Radlk.	Native	*Po-cai*	10	Bra, Ecu	Aguiar-Menezes et al. (2007), Araujo (2015), Raga et al. (2011), Souza-Filho et al. (2000), Tigrero (2009), Uramoto (2002), Uramoto et al. (2004), Veloso et al. (2000), Zucchi (2000)

(*Continued*)

TABLE 6.1 (*Continued*)
List of Host Plants for the Nominal Species *Anastrepha fraterculus* Reported in the Literature as Food Plants Infested in Nature

Host Family	Valid Scientific Name	Host Origin	Host-Code	Total records	Countries	References
Sapotaceae	*Pouteria gardneriana* (A. DC.) Radlk.	Native	*Po-gar*	3	Bra	Veloso et al. (1996, 2000), Zucchi (2000)
Sapotaceae	*Pouteria ramiflora* (Mart.) Radlk.	Native	*Po-ram*	3	Bra	Veloso et al. (1996, 2000), Zucchi (2000)
Sapotaceae	*Pouteria torta* (Mart.) Radlk.	Native	*Po-tor*	1	Bra	Taira (2012)
Sapotaceae	*Sideroxylon capiri* (A. DC.) Pittier	Native	*Si-cap*	1	Mex	Aluja et al. (1987)
Simaroubaceae	*Simaba guianensis* Aubl.	Native	*Sm-gui*	1	Bra	Deus et al. (2013)
Siparunaceae	*Siparuna guianensis* Aubl.	Native	*Sp-gui*	1	Bra	Pirovani (2011)
Solanaceae	*Capsicum annuum* L.	Native	*Ca-ann*	1	Col	Castañeda et al. (2010)
Solanaceae	*Cyphomandra betacea* (Cav.) Sendtn.	?	*unres. name*	1	Col	Gonzalez Mendoza (1952)
Solanaceae	*Solanum decompositiflorum* Sendtn.	Native	*So-dec*	1	Bra	Pirovani (2011)
Solanaceae	*Solanum nudum* Dunal	Native	*So-nud*	1	Col	Castañeda et al. (2010)
Solanaceae	*Solanum quitoense* Lam.	Native	*So-qui*	1	Col	Núñez Bueno (1981)
Solanaceae	*Solanum triste* Jacq.	Native	*So-tri*	1	Col	Gonzalez Mendoza (1952)
Staphyleaceae	*Turpinia occidentalis* (Sw.) G. Don	Native	*Tu-occ*	1	Pan	Stone (1942b)

(*Continued*)

TABLE 6.1 (*Continued*)

List of Host Plants for the Nominal Species *Anastrepha fraterculus* Reported in the Literature as Food Plants Infested in Nature

Host Family	Valid Scientific Name	Host Origin	Host-Code	Total records	Countries	References
Ulmaceae	*Ampelocera hottlei* (Standl.) Standl.	Native	*Am-hot*	1	Mex	Aluja et al. (2003b)
Urticaceae	*Pourouma* sp.	Native	*Po-sp*	1	Bra	Bondar (1950)
Vitaceae	*Vitis vinifera* L.	Exotic	*Vi-vin*	3	Arg, Col	Gonzalez Mendoza (1952), Nasca et al. (1996), Stone (1942b)

Note: Nomenclatures of scientific names are currently valid according to *The Plant List* (2013).

Abbreviations: Arg, Argentina; Bra, Brazil; Bol, Bolivia; Col, Colombia; CR, Costa Rica; Ecu, Ecuador; FrG, French Guyana; Gua, Guatemala; Mex, Mexico; Pan, Panama; Per, Peru; Ven, Venezuela.

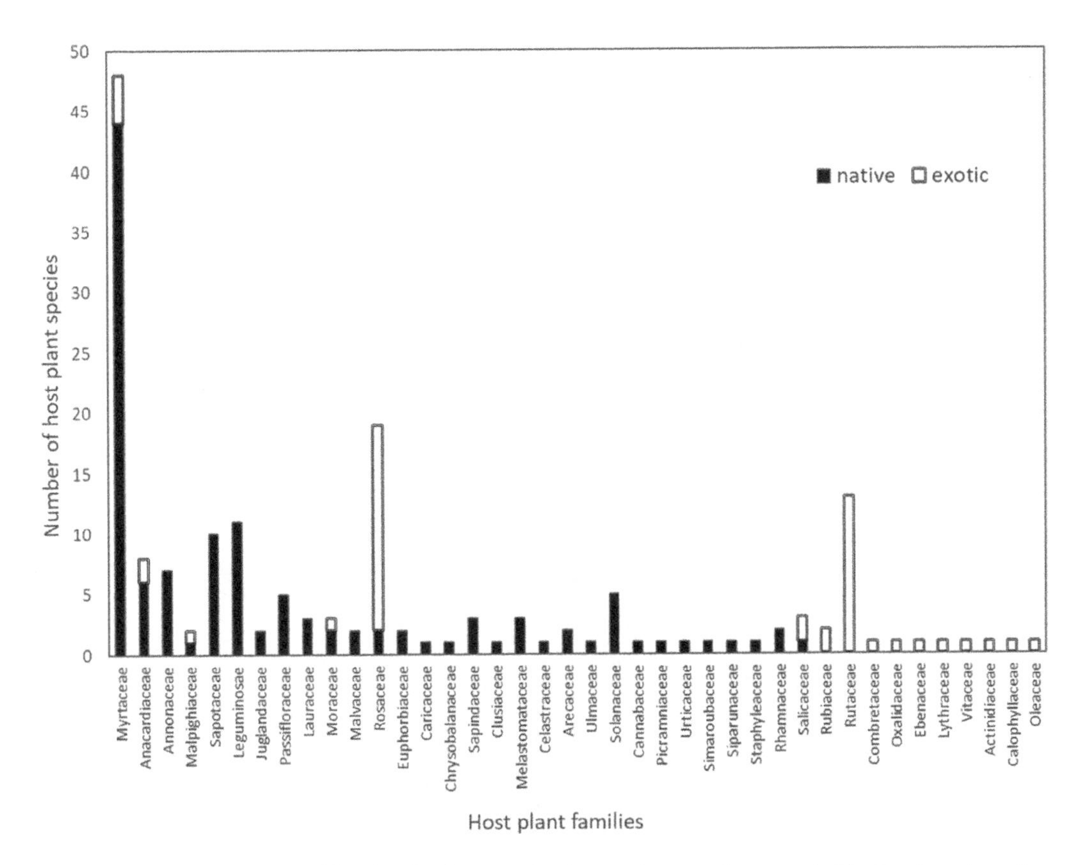

FIGURE 6.1 Host plant families for the *Anastrepha fraterculus* complex recorded in the Americas. Bars indicate the number of plant species; black bars, native species; white bars, exotic species.

In the family Rosaceae, 20 species were found, with most reports featuring *Prunus* spp., *Rubus* spp., and *Eriobotrya japonica*. The family Rutaceae included 14 host species, almost all of which were *Citrus* spp. In the case of the family Combretaceae, all reports were of a single host, the tropical almond *Terminalia cattapa*. A large proportion of at least 18 plant families presented only one to three hosts and a similarly low number of reports; thus, these could be considered rare hosts. These included *Actinidia deliciosa* (Actinidaceae), *Celtis iguanaea* (Cannabaceae), *Endlicheria paniculata* (Lauraceae), *Butia eriospatha* (Arecaceae), and *Simaba guianensis* (Simaroubaceae), among others (Figure 6.1 and Table 6.1).

6.3.2 Host Origin

Examination of the origin of hosts revealed that the SAFF feeds on 50 exotic plant species in the Americas, representing 28.3% of the total number of their host plant species. Five of these, all belonging to different families, accounted for the highest number of records: *Prunus persica* (Rosaceae), *Eriobotrya japonica* (Myrtaceae), *Coffea arabica* (Rubiaceae), *Mangifera indica* (Anacardiaceae), and *Citrus* spp. (Rutaceae). The frequency of records for these species accounted for slightly more than 30% of the total number of reports. The peach is recognized as the exotic host that is most frequently reported throughout the distributional range of the SAFF. However, at a regional level, there is a higher proportion of records for other species, such as coffee, mango, and citrus because each of these have economic significance in specific regions or countries (Figure 6.2). For instance, infestation of *Coffea arabica* coffee cherries was found mainly in

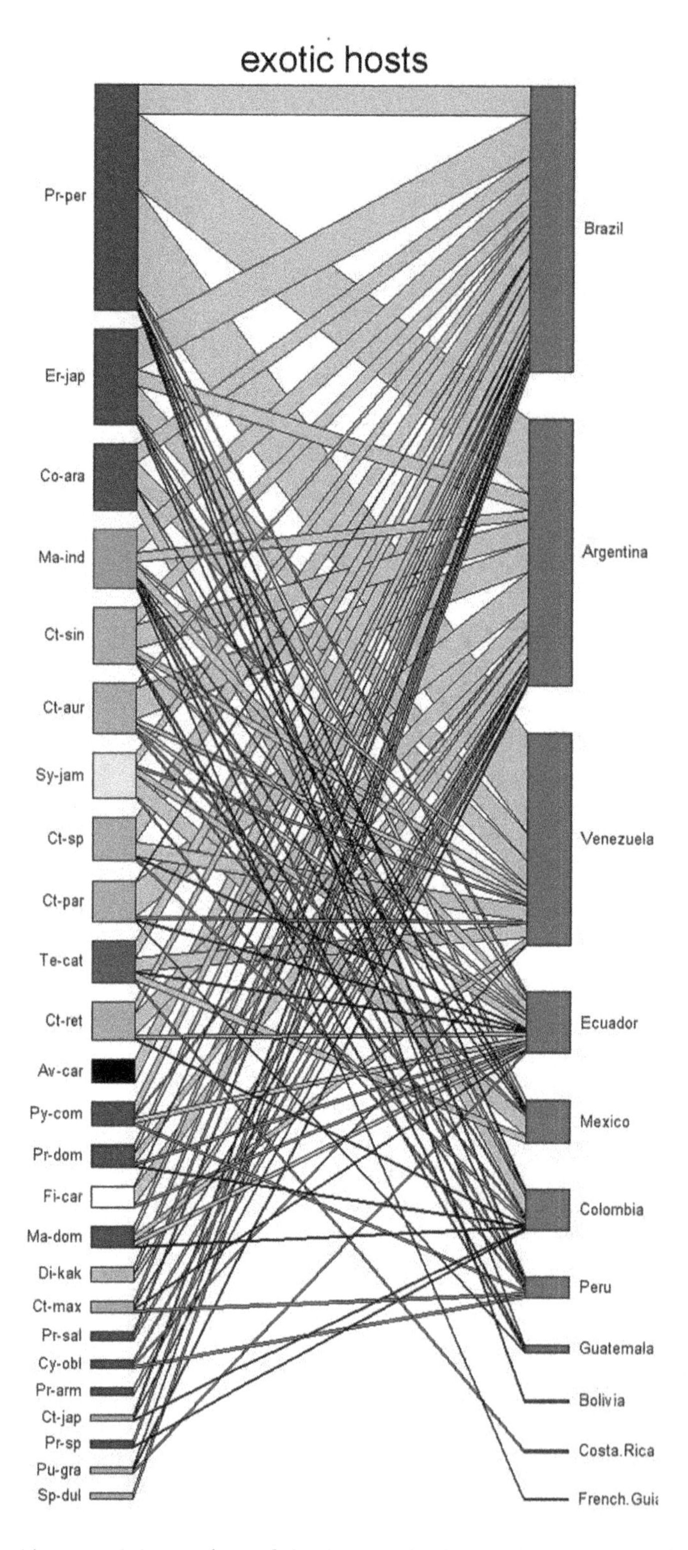

FIGURE 6.2 Trophic network interactions of the *Anastrepha fraterculus* complex with exotic host plants recorded in various countries across the Americas. Plant species with less than five records are omitted. The thickness of the linking gray lines represents the frequency of records. Same color nodes (left) denote that the hosts belong to the same family; the red nodes (right) depict countries of distribution.

the northern region of the Andes in Venezuela and Colombia but also in some Brazilian locations. In *Citrus* spp., almost all of the records were from South America, especially Brazil and Argentina, whereas fruits of *Mangifera indica* have been most commonly reported as hosts in Brazil, Argentina, and Ecuador. These results revealed the occurrence of certain preferences in the use of hosts at a regional level, suggesting the relatively recent adaptation of some fly populations in order to exploit those exotic hosts.

On the other hand, the list of native hosts comprised a total of 124 plant species throughout the American tropics, accounting for 70% of all of the known hosts listed. The highest frequencies of records were found for *Psidium guajava* (298), *Eugenia uniflora* (48), *Acca sellowiana* (31), *Spondias purpurea* (23), *Psidium cattleianum* (22), and *Annona cherimola* (21), among others. It should be noted that 8 of the 10 native species with the highest number of records belonged to the family Myrtaceae, which represented nearly 29% of the total. It is also noteworthy that guava fruits are the only common host shared by different populations throughout the tropical and subtropical regions of the Americas. In contrast, other common host species (e.g., Surinam cherry, feijoa, seriguela, araçá, and chirimoya) are only found infested in certain biogeographical provinces of the neotropical region (Figure 6.3).

6.3.3 Geographical Distribution

Examination of the geographical distribution of host plants used by the SAFF produced records for 12 Latin American countries. The largest numbers of host plants were documented for South America, including Brazil (121), Argentina (40), Ecuador (40), Colombia (38), and Venezuela (24). In Mexico, one of the most well-studied countries in the north of the Americas, the infestation of 19 host plants has been recorded.

The data on host plants recorded for each country allowed us to explore shared resources through a cluster analysis. In this context, Brazil exhibited the highest dissimilarities of hosts shared with all other countries. This is supported by previous studies that highlight the presence of three morphotypes in this territory (Selivon 2004, 2005a, 2005b), which would explain the existence of 120 host species, including many native species. In contrast, Colombia, Ecuador, and Argentina added another subgroup with closer affinities in terms of resource use. These countries shared exotic plants such as *Prunus domestica*, *Punica granatum*, and *Ficus carica*, but also native plants such as *Acca sellowiana*, *Malphigia glabra*, and *Annona cherimola*, among others. Mexico and Venezuela formed a third cluster, sharing hosts such as *Terminalia catappa*, *Syzygium jambos*, and *Psidium guineense* (Figure 6.4).

With respect to taxonomic richness, 30 plant families were found in Brazil, representing nearly twice the number of families found in other countries. This contrasts with the situation in Argentina, where only 16 families were recognized. The Brazilian territory is larger, and three distinct morphotypes have been characterized within the SAFF species complex, whereas just one of them has been reported in Argentina (Hernández-Ortiz et al. 2012). This has led to the hypothesis that the widespread host range found in Brazilian populations could be explained by the presence of different taxa, which requires further investigation.

The richness in the Andean region comprised 10–17 families and 24–40 plant species. Those countries share the common infestation of coffee berries, pineapple guava, custard apple, and berry of the Andes, among others. These are hosts that are rare or absent elsewhere in the Americas. In the entire northern region, from Mesoamerica to Central America, including Mexico, Costa Rica, Guatemala, and Panama, 33 host plant species are recognized. In the northern and central Andean countries, such as Venezuela, Colombia, Ecuador, Peru, and Bolivia, a total of 68 host plants are known. In contrast, in the Amazonian, Atlantic Forest and Chacoan subregions of Brazil and Argentina, a total of 135 plant species are recorded.

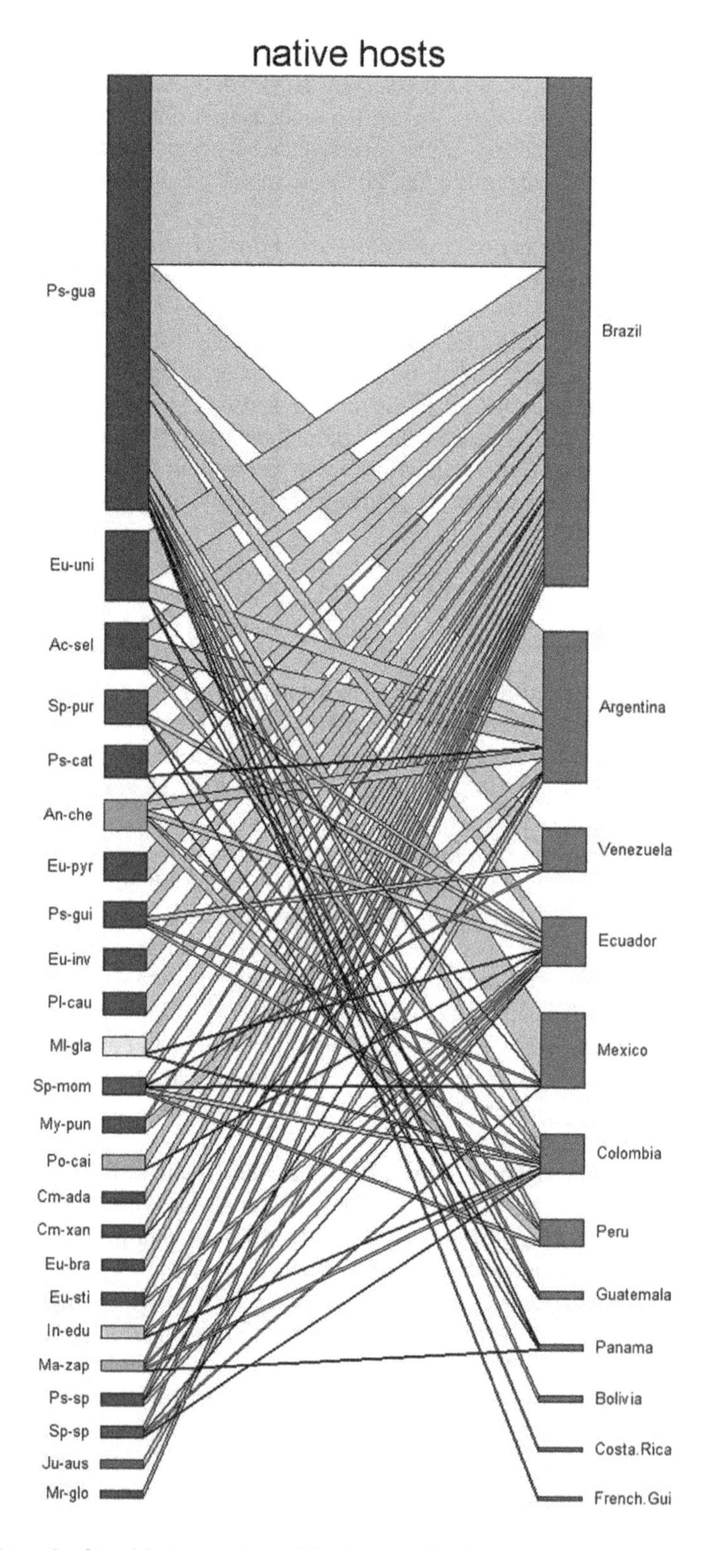

FIGURE 6.3 Network of trophic interactions of the *Anastrepha fraterculus* complex with native host plants recorded in various countries across the Americas. Plant species with less than five records are omitted. The thickness of the linking gray lines represents the frequency of records. Same color nodes (left) denote that the hosts belong to the same family; the red nodes (right) depict countries of distribution.

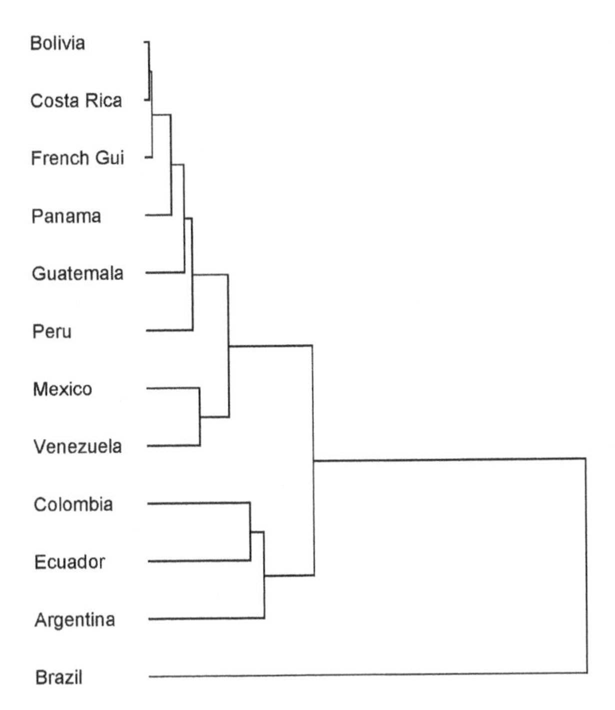

FIGURE 6.4 Dendrogram of dissimilarities (Ward distance: 0–2.33) by countries across the Americas, based on qualitative records (presence–absence) of the host plant species.

6.4 REMARKS AND CONCLUSIONS

The current knowledge of host plants occupied by populations of the nominal species *A. fraterculus* throughout the Americas shows the persistence of gaps in the information. This was evident in nearly all of the countries of Central America and some others from South America (i.e., Bolivia, Peru, Surinam, Paraguay, and Uruguay).

Plant richness showed a great diversification of hosts under natural conditions constituted by 70.1% of native species. Prominent among these are the species of the Myrtaceae family, which accounted for 25% of the total number of host species. These plants could therefore have been the most important influence in the original diversification of the *A. fraterculus* species complex. However, other families, such as the Sapotaceae, Annonaceae, Leguminosae, Anacardiaceae, Passifloraceae, and Solanaceae, might have been used secondarily. The exotic species belonged to 17 botanical families, although most were represented only by a few species. The notable exceptions to this were the two richest families, Rosaceae and Rutaceae, which together accounted for two-thirds of all the exotic species recorded, particularly of plants within the respective genera *Prunus* and *Citrus*.

The high richness of native host plants for the *A. fraterculus* complex denotes an explicit sign of host diversification. However, it should be noted that exotic plants have played a significant role in the divergence of the species complex because some have become almost exclusive hosts of populations in certain biogeographical zones. For example, in the central region of the Andes, *Coffea arabica* is widely used as a preferred host by the SAFF but is virtually an unknown host in other neotropical locations. This suggests that such plants could have played a critical role in the current dispersion of the SAFF following their introduction to the Americas.

The areas occupied by certain populations of this nominal species reflect some preferences in the use of resources. This would reinforce the hypothesis regarding the existence of distinct taxonomic entities because, although these resources exist throughout the range, they are not common hosts for all fly populations. The fact that there are at least eight morphotypes, presumably corresponding to different species within the SAFF complex, leads us to conclude that knowledge on host use at a regional level is critical. The data analyzed here indicated strong preferences at the regional level for both native and introduced plants, and the latter have served as a scattering factor, thus converting some of these populations into distinct species or hosting races of economic importance.

In this context, the potential application of the SIT in certain geographical areas requires knowledge in terms of the host plants consumed by the target species. A holistic view of the trophic interactions among populations of this complex allows the use of much of the published host information. This could help to outline a confidence range of preferred hosts exploited by each biological species and ultimately determine their pest status in different regions of the Americas.

ACKNOWLEDGMENTS

This research received financial support from the International Atomic Energy Agency (Vienna, Austria) through the Research Contract No: 22457/R0–MEX. We thank José F. Dzul-Cauich for technical assistance with capturing of part of the databases. We appreciate the feedback and suggestions from two anonymous reviewers that improved the final version of the manuscript.

REFERENCES

Adaime, R., M. S. M. Sousa, C. R. Jesus-Barros et al. 2017. Frugivorous flies (Diptera: Tephritidae, Lonchaeidae), their host plants, and associated parasitoids in the extreme north of Amapá State, Brazil. *Florida Entomologist* 100:316–324.

Aguiar-Menezes, E. L., & E. B. Menezes. 1997. Natural occurrence of parasitoids of *Anastrepha* spp. Schiner, 1868 (Diptera: Tephritidae) in different host plants, in Itaguai (RJ), Brazil. *Biological Control* 8:1–6.

Aguiar-Menezes, E.L. & E.B. Menezes. 2000. Moscas-das-frutas nos estados brasileiros: Rio de Janeiro. In *Moscas-das-frutas de importância econômica no Brasil. Conhecimento básico e aplicado*, A. Malavasi & R. A. Zucchi (Eds.), pp. 259–263. Holos Ed, Riberão Preto, Brazil.

Aguiar-Menezes, E. L., E. B. Menezes, & M. S. Loiácono. 2003. First record of *Coptera haywardi* Loiácono (Hymenoptera: Diapriidae) as a parasitoid of fruit-infesting Tephritidae (Diptera) in Brazil. *Neotropical Entomology* 32:355–358.

Aguiar-Menezes, E. L., E. B. Menezes, P. S. Silva, A. C. Bittar, & P. C. R. Cassino. 2001. Native Hymenopteran Parasitoids Associated with *Anastrepha* spp. (Diptera: Tephritidae) in Seropedica City, Rio de Janeiro, Brazil. *Florida Entomologist* 84:706–711.

Aguiar-Menezes, E. L., M. Lima-Filho, F. A. A. Ferrara et al. 2007. Estrutura e flutuação das populações de moscas-das-frutas no polo de fruticultura das regiões norte e noroeste Fluminense. *Boletim de Pesquisa e Desenvolvimient* 16:4–40.

Alberti, A.C., M. S. Rodriguero, P. C. Gomez, B. O. Saidman, & J. C. Vilardi. 2002. Evidence indicating that Argentine populations of *Anastrepha fraterculus* (Diptera: Tephritidae) belong to a single biological species. *Annals of the Entomological Society of America* 95:505–512.

Almeida, R. R. 2016. Dípteros (Tephritidae e Lonchaeidae) associados à produção de frutas na ilha de Santana, Amazônia brasileira. MSc Dissertation, Fundação Universidade Federal do Amapá, 93 p.

Aluja, M. 1984. *Manejo integrado de las moscas de la fruta (Diptera: Tephritidae).* Dirección General Sanidad Vegetal SARH, Mexico DF, 241 p.

Aluja, M., A. J. Guillén, G. De la Rosa et al. 1987. Natural host plant survey of the economically important fruit flies (Diptera: Tephritidae) of Chiapas, Mexico. *Florida Entomologist* 70:329–338.

Aluja, M., D. Pérez-Staples, R. Macías-Ordoñez et al. 2003a. Nonhost status of *Citrus sinensis* Cultivar Valencia and *C. paradisi* Cultivar Ruby Red to Mexican *Anastrepha fraterculus* (Diptera: Tephritidae). *Journal of Economic Entomology* 96:1693–1703.

Aluja, M., J. Pinero, M. Lopez et al. 2000. New host plant and distribution records in Mexico for *Anastrepha* spp., *Toxotrypana curvicauda* Gerstacker, *Rhagoletis zoqui* Bush, *Rhagoletis* sp., and *Hexachaeta* sp. (Diptera: Tephritidae). *Proceedings Entomological Society of Washington* 102:802–815.

Aluja, M., J. Rull, J. Sivinski et al. 2003b. Fruit flies of the genus *Anastrepha* (Diptera: Tephritidae) and associated native parasitoids (Hymenoptera) in the tropical rainforest biosphere reserve of Montes Azules, Chiapas, Mexico. *Environmental Entomology* 32:1377–1385.

Alvarenga, C. D., C. A. R. Matrangolo, G. N. Lopes et al. 2009. Moscas-das-frutas (Diptera: Tephritidae) e seus parasitóides em plantas hospedeiras de três municípios do norte do estado de Minas Gerais. *Arquivos do Instituto Biológico* 76:195–204.

Alvarenga, C. D., N. A. Canal, & R. A. Zucchi. 2000. Moscas-das-frutas nos estados brasileiros: Minas Gerais. In *Moscas-das-frutas de importância econômica no Brasil. Conhecimento básico e aplicado* A. Malavasi & R. A. Zucchi (Eds.), 265–270. Holos Ed, Riberão Preto, Brazil.

Araujo, E. L., J. L. Batista, & R. A. Zucchi. 2000. Moscas-das-frutas nos estados brasileiros: Paraíba. In *Moscas-das-frutas de importância econômica no Brasil. Conhecimento básico e aplicado*, A. Malavasi & R. A. Zucchi (Eds.), 227–228. Holos Ed, Riberão Preto, Brazil.

Araujo, M. R. 2015. Análise comparativa e modelagem espacial de espécies de *Anastrepha* (Diptera, Tephritidae) coletadas em armadilhas e em plantas hospedeiras. MSc thesis. São Paulo, Brazil: Escola Superior de Agricultura Luiz de Queiroz, Universidade de São Paulo, 98 p.

Araujo, N. S. 2012. Análise de espécies crípticas do complexo *Anastrepha fraterculus* (Diptera: Tephritidae) no Brasil através de sequências do gene mitocondrial Cytochrome Oxidase I. MSc thesis. São Paulo, Brazil: Instituto de Biociências, Universidade de São Paulo, 79 p.

Baker, A. C., W. E. Stone, C. C. Plummer, & M. McPhail. 1944. A review of studies on the Mexican fruit-fly and related Mexican species. *United States Department of Agriculture Miscellaneous Publication* 531:1–155.

Baker, E. W. 1945. Studies on the Mexican fruitfly known as *Anastrepha fraterculus*. *Journal of Economic Entomology* 38:95–100.

Ballou, C. H. 1945. Notas sobre insectos dañinos observados en Venezuela 1938–1943. Comisión Organizadora. Tercera Conferencia Interamericana de Agricultura, Cuaderno Verde 34:125.

Barros, N. E. L. 2008. Ecologia de moscas-das-frutas (Diptera: Tephritidae) em goiaba (*Psidium guajava* L.; Myrtaceae) cultivada em sistema agroflorestal, em Santana, Amapá. MSc thesis. Amapá, Brazil: Universidade Federal do Amapá, 84 p.

Bomfim, D. A., M. A. Uchôa-Fernandes, & M. A. Bragança. 2007. Hosts and parasitoids of fruit flies (Diptera: Tephritoidea) in the State of Tocantins, Brazil. *Neotropical Entomology* 36:984–986.

Bondar, G. 1950. Moscas de frutas na Bahia. *Boletin Campo* 6:13–15.

Boscán de Martínez, N., & F. Godoy. 1996. Nuevos parasitoides de moscas de las frutas de los generos *Anastrepha* y *Ceratitis* en Venezuela. *Agronomía Tropical* 46:465–471.

Boscán de Martínez, N., J. R. Dedordy, & J. R. Requena. 1980. Estado actual de la distribucion geografica y hospederas de *Anastrepha* spp. (Diptera-Trypetidae) en Venezuela. *Agronomía Tropical* 30:55–63.

Bressan, S., & M. Da Costa Teles. 1991. Host range and infestation by species of the genus *Anastrepha* (Diptera: Tephritidae) in the region of Ribeirão Preto-SP, Brazil. *Anais Sociedade Entomologica Brasileira* 20:5–15.

Briceño Vergara, A. J. 1975. Distribución de las moscas de las frutas (*Anastrepha* spp., Diptera: Tephritidae) y sus plantas hospederas en los Andes venezolanos. *Revista Facultad de Agronomía* (Maracaibo) 3:45–49.

Briceño Vergara, A. J. 1979. Las moscas de las frutas *Anastrepha* spp. (Diptera: Tephritidae), en los Andes venezolanos. *Revista Facultad de Agronomía* (Maracaibo) 5:449–457.

Bush, G. L.1962. The cytotaxonomy of the larvae of some Mexican fruit flies in the genus *Anastrepha* (Tephritidae, Diptera). *Psyche* 68:87–101.

Cáceres, C., D. F. Segura, M. T. Vera et al. 2009. Incipient speciation revealed in *Anastrepha fraterculus* (Diptera; Tephritidae) by studies on mating compatibility, sex pheromones, hybridization, and cytology. *Biological Journal of Linnean Society* 97:152–165.

Campos, R. F. 1960. Las moscas (Brachycera) del Ecuador. *Revista Ecuatoriana Higiene y Medicina Tropical* 17:1–66.

Caraballo, C. J. 1981. Las moscas de frutas del género *Anastrepha* Schiner, 1868 (Diptera: Tephritidae) de Venezuela. MSc Dissertation, Universidad Central de Venezuela, Maracay, 210 p.

Castañeda, M. R., F. A. Osorio, N. A. Canal, & P. E. Galeano. 2010. Especies, distribución y hospederos del género *Anastrepha* Schiner en el departamento del Tolima, Colombia. *Agronomía Colombiana* 28:265–271.

Castro, P. E. 2009. Dinâmica populacional das Moscas-das-frutas (Diptera: Tephritidae) em agroecosistemas nos Estados de Guayas e Santa Elena, Equador. Dissertation PhD, Universidade Federal do Recôncavo da Bahia, 107 p.

Chiaradia, L. A., J. M. Milanez, & R. Dittrich. 2004. Flutuação populacional de moscas-das-frutas em pomares de citros no oeste de Santa Catarina, Brasil. *Ciência Rural* 34:337–343.

Corsato, C. D. A. 2004. Moscas-das-frutas (Diptera: Tephritidae) em pomares de goiaba no norte de Minas Gerais: biodiversidade, parasitóides e controle biológico. PhD diss. São Paulo, Brazil: Escola Superior de Agricultura "Luiz de Queiroz," Universidade de São Paulo, 83 p.

Cruz, M. I., T. Bacca, & N. A. Canal. 2017. Diversidad de las moscas de las frutas (Diptera: Tephritidae) y sus parasitoides en siete municipios del departamento de Nariño. *Boletín Científico Museo de Historia Natural Universidad de Caldas* 21:81–98.

Custódio, A. C., C. L. Lago, M. F. Souza-Filho, L. R. F. Louzeiro, & A. Raga. 2017. Moscas-das-frutas do gênero *Anastrepha* (Diptera: Tephritidae): Novas associações hospedeiras. *Biológico* 79:1–2.

Deus, E. G., L. D. S. Pinheiro, C. R. Lima et al. 2013. Wild hosts of frugivorous dipterans (Tephritidae and Lonchaeidae) and associated parasitoids in the Brazilian Amazon. *Florida Entomologist* 96:1621–1625.

Deus, E. G., R. A. Silva, D. B. Nascimiento, C. F. Marinho, & R. A. Zucchi. 2010. Hospedeiros e parasitóides de especies de *Anastrepha* (Diptera, Tephritidae) em dois municípios do estado do Amapá. *Revista de Agricultura* 84:194–203.

Dias, S. V., J. G. Silva, K. M. Lima et al. 2016. An integrative multidisciplinary approach to understanding cryptic divergence in Brazilian species of the *Anastrepha fraterculus* complex (Diptera: Tephritidae). *Biological Journal of Linnean Society* 117:725–746.

Dormann, C. F., & B. Gruber. 2009. Package "Bipartite": visualizing bipartite networks and calculating some ecological indices. R statistical software. "R group". Available at: https://github.com/biometry/bipartite.

Duarte, P. A. S., F. R. M. Garcia, & V. Andaló. 2016. Faunal analysis and population density of fruit flies (Diptera: Tephritidae) in an orchard located in the central western region of Minas Gerais, Brazil. *Bioscience Journal* 32:960–968.

Eskafi, F. M. 1990. Parasitism of fruit flies *Ceratitis capitata* and *Anastrepha*-spp. (Diptera: Tephritidae) in Guatemala. *Entomophaga* 35:355–362.

Eskafi, F. M., & R.T. Cunningham. 1987. Host plants of fruit flies (Diptera: Tephritidae) of economic importance in Guatemala. *Florida Entomologist* 70:116–123.

Feitosa, S. S., P. P. R. Silva, L. E. M. Pádua, M. P. S. Sousa, E. P. Passos, & A. A. R. Soares. 2007. Primeiro registro de moscas-das-frutas (Diptera: Tephritidae) em carambola nos municípios de Teresina, Altos e Parnaíba no estado do Piauí. *Ciências Agrárias* 28:629–634.

Fernandez Yepez, F. 1953. Contribución al estudio de las moscas de las frutas del género *Anastrepha* Schiner (Diptera: Trypetidae) de Venezuela. II Congreso de Ciencias Naturales y Afines, Caracas. *Cuaderno* 77:1–42.

Ferreira, H. J., V. R. S. Veloso, R. V. Naves, & J. R. Braga Filho. 2003. Infestação de moscas-das-frutas em variedades de manga (*Mangifera indica* L.) no estado de Goiás. *Pesquisa Agropecuaria Tropical* 33:43–48.

Fischer, C. R. 1934. Variação das cerdas frontaes e outras notas sobre duas especies de *Anastrepha* (Dipt. Trypetidae). *Revista de Entomologia* 4:17–22.

Funes, C. F., L. I. Escobar, N. G. Meneguzzi, S. M. Ovruski, & D. S. Kirschbaum. 2017. Occurrence of *Anastrepha fraterculus* and *Ceratitis capitata* (Diptera: Tephritidae) in organically grown *Rubus* (Rosales: Rosaceae), in two contrasting environments of northwestern Argentina. *Florida Entomologist* 100:672–674.

Garcia, F. R. M., & A. L. Norrbom. 2011. Tephritoid flies (Diptera, Tephritoidea) and their plant hosts from the state of Santa Catarina in southern Brazil. *Florida Entomologist* 94:151–157.

García-Ramírez, M. J., H. R. E. Medina, V. López-Martínez, L. M. Vázquez, U. I. E. Duarte, & H. Delfín-González. 2010. *Talisia olivaeformis* (Sapindaceae) and *Zuelania guidonia* (Flacourtiaceae): New host records for *Anastrepha* spp. (Diptera: Tephritidae) in México. *Florida Entomologist* 93:633–634.

Gonzalez Mendoza, R. 1952. Contribucion al estudio de las moscas *Anastrepha* en Colombia. *Revista Facultad Nacional de Agronomía* 12:423–549.

Greene, C. T. 1934. A revision of the genus *Anastrepha* based on a study of the wings and on the length of the ovipositor sheath (Diptera: Trypetidae). *Proceedings Entomological Society of Washington* 36:127–179.

Guagliumi, P. 1966. Insetti e aracnidi delle piante comuni del Venezuela segnalati nel periodo 1938–1963. *Relazione Monografica Agricola Subtropical y Tropical (ns)* 86:1–392.

Guimarães, J. A., R. A. Zucchi, N. B. Díaz, M. F. Souza Filho, & M. A. Uchôa. 1999. Espécies de Eucoilinae (Hymenoptera: Cynipoidea: Figitidae) parasitóides de larvas frugívoras (Diptera: Tephritidae e Lonchaeidae) no Brasil. *Anais Sociedade Entomologica Brasileira* 28:263–273.

Haji, F. N. P., & I. Da Gama Miranda. 2000. Moscas-das-frutas nos estados brasileiros: Pernambuco. In *Moscas-das-frutas de importância econômica no Brasil. Conhecimento básico e aplicado*, A. Malavasi & R. A. Zucchi (Eds.), 229–233. Holos Ed, Riberão Preto, Brazil.

Hedström, I. 1987. Fruit flies (Diptera: Tephritidae) infesting common guava (*Psidium guajava* L.) (Myrtaceae) in Ecuador. *Revista de Biología Tropical* 35:373–374.

Hernández-Ortiz, V., & M. Aluja. 1993. Lista preliminar del género neotropical *Anastrepha* Schiner (Diptera: Tephritidae) con notas sobre su distribución y plantas hospederas. *Folia Entomológica Mexicana* 88:89–105.

Hernández-Ortiz, V., A. F. Bartolucci, P. Morales-Valles, D. Frías, & D. Selivon. 2012. Cryptic species of the *Anastrepha fraterculus* complex (Diptera: Tephritidae): A multivariate approach for the recognition of South American morphotypes. *Annals of the Entomological Society of America* 105:305–318.

Hernández-Ortiz, V., & P. Morales-Valles. 2004. Distribución geográfica y plantas hospederas de *Ananstrepha fraterculus* (Diptera: Tephritidae) en Venezuela. *Folia Entomológica Mexicana* 43:181–189.

Hernández-Ortiz, V., N. A. Canal, J. O. Tigrero Salas, F. M. Ruíz-Hurtado, & J. F. Dzul-Cauich. 2015. Taxonomy and phenotypic relationships of the *Anastrepha fraterculus* complex in the Mesoamerican and Pacific Neotropical dominions (Diptera, Tephritidae). *Zookeys* 540:95–124.

Hernández-Ortiz, V., H. Delfín-González, A. Escalante-Tió, & P. Manrique-Saide. 2006. Hymenopteran parasitoids of *Anastrepha* fruit flies (Diptera: Tephritidae) reared from different hosts in Yucatan, Mexico. *Florida Entomologist* 89:508–515.

Hernández-Ortiz, V., J. A. Gómez-Anaya, A. Sánchez, B. A. McPheron, & M. Aluja. 2004. Morphometric analysis of Mexican and South American populations of the *Anastrepha fraterculus* complex (Diptera: Tephritidae) and recognition of a distinct Mexican morphotype. *Bulletin of Entomological Research* 94:487–499.

Hernández-Ortiz, V., & R. Pérez-Alonso. 1993. The natural host plants of *Anastrepha* (Diptera: Tephritidae) in a tropical rain forest of Mexico. *Florida Entomologist* 76:447–460.

Hernández-Ortiz, V., R. Pérez-Alonso, & R. A. Wharton. 1994. Native parasitoids associated with the genus *Anastrepha* (Dipt.: Tephritidae) in Los Tuxtlas, Veracruz, Mexico. *Entomophaga* 39:171–178.

Hickel, E. R., & J-P. H. J. Ducroquet. 1994. Ocorrência de moscas-das-frutas *Anastrepha fraterculus* (Wied.) em frutas de goiabeira serrana. *Anais Sociedade Entomologica Brasileira* 23:311–315.

Hickel, E. R., & E. Schuck. 1993. Ocorrência da moscas-das frutas, *Anastrepha fraterculus* (Diptera: Tephritidae) em frutos de quivi. *Pesquisa Agropecuaria Brasileira* 28:1345–1347.

Isaac, J. 1905. Report on the Mexican orange worm (*Trypeta ludens*) in Mexico. *California State Horticulture Communications* 428:7–48.

Jesus-Barros, C. R., R. Adaime, M. N. Oliveira, W. R. Silva, S. V. Costa-Neto, & M. F. Souza-Filho. 2012. *Anastrepha* (Diptera: Tephritidae) species, their hosts and parasitoids (Hymenoptera: Braconidae) in five municipalities of the state of Amapá, Brazil. *Florida Entomologist* 95:694–705.

Jirón, L. F., & I. Hedström. 1988. Occurrence of fruit flies of the genera *Anastrepha* and *Ceratitis* (Diptera: Tephritidae), and their host plant availability in Costa Rica. *Florida Entomologist* 71:62–73.

Katiyar, K. P., J. M. Camacho, & R. Matheus. 2000. Fruit flies (Diptera: Tephritidae) infesting fruits of the genus *Psidium* (Myrtaceae) and their altitudinal distribution in Western Venezuela. *Florida Entomologist* 83:480–486.

Katiyar, K. P., J. M. Camacho, F. Geraud, & R. Matheus. 1995. Parasitoides hymenópteros de moscas de las frutas (Diptera: Tephritidae) en la región occidental de Venezuela. *Revista Facultad de Agronomía* 12:303–312.

Korytkowski, C. A. 2001. Situación actual del género *Anastrepha* Schiner, 1868 (Diptera: Tephritidae) en el Perú. *Revista Peruana de Entomología* 42:97–158.

Korytkowski, C. A., & D. Ojeda Pena. 1968. Especies del género *Anastrepha* Schiner 1868 en el nor-oeste peruano. *Revista Peruana de Entomología* 11:32–70.

Korytkowski, C.A., & D. Ojeda Peña. 1969. Distribution ecológica de especies del género *Anastrepha* Schiner en el nor-oeste peruano. *Revista Peruana de Entomología* 12:71–95.

Kovaleski, A., R. L. Sugayama, K. Uramoto, & A. Malavasi. 2000. Moscas-das-frutas nos estados brasileiros: Rio Grande do Sul. In *Moscas-das-frutas de importância econômica no Brasil. Conhecimento básico e aplicado*, A. Malavasi & R. A. Zucchi (Eds.), 285–290. Holos Ed, Riberão Preto, Brazil.

Kovaleski, A., K. Uramoto, R. L. Sugayama, N. A. Canal, & A. Malavasi. 1999. A survey of *Anastrepha* Schiner (Diptera, Tephritidae) species in the apple growing area of the state of Rio Grande do Sul, Brazil. *Revista Brasileira de Entomologia* 43:229–234.

Lemos, L. N. 2014. Moscas-das-frutas (Diptera: Tephritidae e Lonchaeidae) em sistemas de cultivo e entorno no Estado do Amapá, Brasil. PhD diss. Amapá, Brasil: Fundação Universidade Federal do Amapá, 78 p.

Lemos, L. N., E. G. Deus, D. B. Nascimento, C. R. Jesus-Barros, S. V. Costa-Neto, & R. Adaime. 2017. Species of *Anastrepha* (Diptera: Tephritidae), their host plants, and parasitoids in small fruit production areas in the state of Amapá, Brazil. *Florida Entomologist* 100:403–410.

Malavasi, A., J. S. Morgante, & R. A. Zucchi. 1980. Biologia de "moscas-das-frutas" (Diptera, Tephritidae). I: lista de hospedeiros e ocorrência. *Revista Brasileira de Biologia* 40:9–16.

Manni, M., K. M. Lima, C. R. Guglielmino et al. 2015. Relevant genetic differentiation among Brazilian populations of *Anastrepha fraterculus* (Diptera, Tephritidae). *ZooKeys* 540:157–173.

Marchiori, C. H., A. M. S. Oliveira, F. F. Martins, F. S. Bossi, & A. T. Oliveira. 2000. Espécies de moscas-da-fruta (Diptera: Tephritidae) e seus parasitóides em Itumbiara-GO. *Pesquisa Agropecuaria Tropical* 30:73–76.

Marin Patiño, M. L. 2002. Identificación y caracterización de moscas de las frutas en los Departamentos del Valle del Cauca, Tolima y Quindío. Dissertation, Facultad de Ciencias Agropecuarias, Universidad de Caldas, 27 p.

Marinho, C. F. 2004. Espécies de parasitóides (Hymenoptera: Braconidae) de moscas-das-frutas (Diptera: Tephritidae) no estado de São Paulo: caracterização taxonômica, distribuição geográfica e percentagem de parasitismo. MSc Dissertation, Escola Superior de Agricultura "Luiz de Queiroz", Universidade de São Paulo, 88 p.

Marinho, C. F., M. F. Souza-Filho, A. Raga, & R. A. Zucchi. 2009. Parasitóides (Hymenoptera: Braconidae) de Moscas-das-Frutas (Diptera: Tephritidae) no Estado de São Paulo: Plantas Associadas e Parasitismo). *Neotropical Entomology* 38:321–326.

Marsaro Júnior, A. L. 2014. Novos registros de hospedeiros de moscas-das-frutas (Diptera: Tephritidae) para o Rio Grande do Sul. *Revista de Agricultura* 89:65–71.

Mascarenhas, R. O. 2007. Endossimbionte *Wolbachia* em moscas-das-frutas do gênero *Anastrepha* (Tephritidae) e em vespas parasitóides (Braconidae) associadas. MSc thesis. São Paulo, Brazil: Instituto de Biociências, Universidade de São Paulo, 54 p.

McPhail, M., & C. I. Bliss. 1933. Observations on the Mexican fruitfly and some related species in Cuernavaca, Mexico, in 1928 and 1929. *Unites States Department of Agriculture Circular* 255:1–24.

Molineros, J., J. O. Tigrero, & D. Sandoval. 1992. Diagnóstico de la situación actual del problema de las moscas de la fruta en el Ecuador. Comisión Ecuatoriana de Energía Atómica, Dirección de Investigaciones, 53 p.

Nasca, A. J., J. A. Zamora, L. E. Vergara, & H. E. Jaldo. 1996. Hospederos de moscas de los frutos en el Valle de Antinaco-Los Colorados, Provincia de La Rioja, República Argentina. *Revista de Investigación* 10:19–24.

Norrbom, A. L. 2004. Fruit fly (Diptera, Tephritidae) host plant database. Available at http://www.sel.barc.usda.gov:591/diptera/Tephritidae/TephHosts/search.html [last accessed, January 10, 2015].

Nunes, A. M., F. Appel Müller, R. S. Gonçalves, M. S. Garcia, V. A. Costa, & D. E. Nava. 2012. Moscas frugívoras e seus parasitoides nos municípios de Pelotas e Capão do Leão, Rio Grande do Sul, Brasil. *Ciência Rural* 42:6–12.

Nunes, M. Z., I. C. Boff Mari, S. S. Dos Santos Régis et al. 2015. Damage and development of *Anastrepha fraterculus* (Diptera: Tephritidae) in fruits of two pear cultivars. *Agrociencia Uruguay* 19:42–48.

Núñez Bueno, L. 1981. Contribución al reconocimiento de las moscas de las frutas (Diptera: Tephriridae) en Colombia. *Revista del Instituto Colombiano Agropecuario* 16:173–179.

Nuñez Bueno L., R. Gómez Santos, G. Guarín, & G. León. 2004. Moscas de las frutas (Díptera: Tephritidae) y sus parasitoides asociados con *Psidium gujava* L. y *Coffea arabica* L. en tres municipios de la Provincia de Vélez (Santarder, Colombia). *Revista Corpoica* 5:5–12.

Olarte Espinosa, W. 1980. Dinámica poblacional de complejo constituido por las moscas de las frutas *Anastrepha striata* Schiner y *Anastrepha fraterculus* Wiedemann en el medio ecológico del sur de Santander. Dissertation, Universidad Industrial de Santander, Colombia, 69 p.

Oroño, L. E., S. Ovruski, A. L. Norrbom, P. Schliserman, C. Colin, & C.B. Martin. 2005. Two new native host plant records for *Anastrepha fraterculus* (Diptera: Tephritidae) in Argentina. *Florida Entomologist* 88:228–232.

Ovruski, S. 1995. Pupal and larval-pupal parasitoids (Hymenoptera) obtained from *Anastrepha* spp. and *Ceratitis capitata* (Dipt.: Tephritidae) pupae collected in four localities of Tucuman Province, Argentina. *Entomophaga* 40:367–370.

Ovruski, S., P. Schliserman, & M. Aluja. 2003. Native and introduced host plants of *Anastrepha fraterculus* and *Ceratitis capitata* (Diptera: Tephritidae) in northwestern Argentina. *Journal of Economic Entomology* 96:1108–1118.

Ovruski, S., P. Schliserman, & M. Aluja. 2004. Indigenous parasitoids (Hymenoptera) attacking *Anastrepha fraterculus* and *Ceratitis capitata* (Diptera: Tephritidae) in native and exotic host plants in Northwestern Argentina. *Biological Control* 29:43–57.

Ovruski, S., P. Schliserman, G. A. van Nieuwenhove, L. P. Bezdjian, S. Núñez-Campero, & P. Albornoz-Medina. 2010. Occurrence of *Ceratitis capitata* and *Anastrepha fraterculus* (Diptera: Tephritidae) on cultivated, exotic fruit species in the highland valleys of Tucuman in Northwest Argentina. *Florida Entomologist* 93:277–282.

Ovruski, S., P. Schliserman, L. E. Oroño et al. 2008. Natural occurrence of hymenopterous parasitoids associated with *Anastrepha fraterculus* (Diptera: Tephritidae) in Myrtaceae species in Entre Rios, northeastern Argentina. *Florida Entomologist* 91:220–227.

Ovruski, S., P. Schliserman, S. R. Nuñez-Campero et al. 2009. A survey of hymenopterous larval-pupal parasitoids associated with *Anastrepha fraterculus* and *Ceratitis capitata* (Diptera: Tephritidae) infesting wild guava (*Psidium guajava*) and peach (*Prunus persica*) in the southernmost section of the Bolivian Yungas forest. *Florida Entomologist* 92:269–275.

Ovruski, S., R. A. Wharton, P. Schliserman, & M. Aluja. 2005. Abundance of *Anastrepha fraterculus* (Diptera: Tephritidae) and its associated native parasitoids (Hymenoptera) in "feral" guavas growing in the endangered Northernmost Yungas forests of Argentina with an update on the taxonomic status of Opiine parasitoids previously reported in this country. *Environmental Entomology* 34:807–818.

Pereira-Rêgo, D. R. G., S. M. Jahnke, L. R. Redaelli, & N. Schaffer. 2011. Morfometria de *Anastrepha fraterculus* (Wied.) (Diptera: Tephritidae) relacionada a hospedeiros nativos, Myrtaceae. *Arquivos do Instituto Biológico* 78:37–43.

Perre, P. 2016. Utilização diferencial de frutos hospedeiros por *Anastrepha* sp.1 affinis *fraterculus* (Diptera, Tephritidae): Aspectos morfológicos e reprodutivos. Dissertation PhD, Instituto de Biociências, Universidade de São Paulo, 124 p.

Pirovani, V. D. 2011. Moscas-das-frutas (Diptera: Tephritidae): Diversidade, hospedeiros e parasitóides em áreas nativa e cultivadas na região de Viçosa, Minas Gerais, Brasil. MSc Dissertation, Universidade Federal de Viçosa, 66 p.

Pirovani, V. D., D. S. Martins, S. A. S. Souza, K. Uramoto, & P. S. F. Ferreira. 2010. Moscas-das-frutas (Diptera: Tephritidae), seus parasitoides e hospedeiros em Viçosa, Zona da mata mineira. *Arquivos do Instituto Biológico* 77:727–733.

Putruele, M. T. G. 1996. Hosts for *Ceratitis capitata* and *Anastrepha fraterculus* in the northeastern Province of Entre Ríos, Argentina. In *Fruit Fly Pests: A World Assessment of Their Biology and Management*, B. A. McPheron, & G. J. Steck (Eds.), pp. 343–345. St. Lucie Press, Delray Beach, FL.

Querino, R. B., J. B. Maia, G. N. Lopes, C. D. Alvarenga, & R. A. Zucchi. 2014. Fruit fly (Diptera: Tephritidae) community in guava orchards and adjacent fragments of native vegetation in Brazil. *Florida Entomologist* 97:778–786.

R Core Team. 2014. R: A language and environment for statistical computing. R Foundation for Statistical Computing, Vienna, Austria. Available at: http://www. R-project.org/.

Raga, A., D. A. O. Prestes, M. F. Souza Filho et al. 2004. Fruit fly (Diptera: Tephritoidea) infestation in *Citrus* in the State of São Paulo, Brazil. *Neotropical Entomology* 33:85–89.

Raga, A., M. F. D. Souza-Filho, R. A. Machado, M. E. Sato, & R. C. Siloto. 2011. Host ranges and infestation indices of fruit flies (Tephritidae) and lance flies (Lonchaeidae) in São Paulo State, Brazil. *Florida Entomologist* 94:787–794.

Raga, A., R. A. Machado, M. F. Souza Filho, M. E. Sato, & R. C. Siloto. 2005. Tephritoidea (Diptera) species from Myrtaceae fruits in the State of São Paulo, Brazil. *Entomotropica* 20:11–14.

Rust, E. W. 1918. *Anastrepha fraterculus* Wied. (Trypetidae)—A severe menace to the southern United States. *Journal of Economic Entomology* 11:457–467.

Sá, R. F., M. A. Castellani, A. S. do Nacimiento, M. H. T. Silva, A. N. Silva, & R. Pérez-Maluf. 2008. Índice de infestação e diversidade de moscas-das-frutas em hospedeiros exóticos e nativos no pólo de fruticultura de anagé, BA. *Bragantia* 67:401–411.

Salles, L. A. 1995. Bioecologia e controle da mosca-das-frutas sul-americana EMBRAPA, Centro de Pesquisa Agropecuária de Clima Temperado, Pelotas, 58 p.

Santos, W. S. 2003. Moscas frugívoras (Diptera: Tephritoidea) associadas ao umbu-cajá (*Spondias* sp.) no Recôncavo Baiano. MSc Dissertation, Escola de Agronomia. Universidade Federal da Bahia, 50 p.

Sarmiento, E. C., H. Aguirre, & J. Martínez. 2012. *Anastrepha* (Diptera: Tephritidae) y sus asociados: dinámica de emergencia de sus parasitoides en frutos de tres especies de plantas. *Boletín Museo de Entomología de la Universidad del Valle* 13:25–32.

Savaris, M., S. Lampert, A. L. Marsaro-Júnior, R. Adaime, & M. F. Souza-Filho. 2013. First record of *Anastrepha fraterculus* and *Ceratitis capitata* (Diptera, Tephritidae) on Arecaceae in Brazil. *Florida Entomologist* 96:1597–1599.

Schliserman, P., & S. Ovruski. 2004. Incidencia de moscas de la fruta de importancia económica sobre *Citrus aurantium* (Rutaceae) en Tucumán, Argentina. *Manejo Integrado de Plagas y Agroecología* 72:44–53.

Schliserman, P., S. Ovruski, O. R. De Coll, & R. A. Wharton. 2010. Diversity and abundance of hymenopterous parasitoids associated with *Anastrepha fraterculus* (Diptera: Tephritidae) in native and exotic host plants in Misiones, Northeastern Argentina. *Florida Entomologist* 93:175–182.

Segura, D. F., M. T. Vera, & J. L. Cladera. 2004. Fluctuación estacional en la infestación de diversos hospedadores por la mosca del Mediterráneo, *Ceratitis capitata* (Diptera: Tephritidae), en la provincia de Buenos Aires. *Ecología Austral* 14:3–17.

Segura, D. F., M. T. Vera, C. L. Cagnotti et al. 2006. Relative abundance of *Ceratitis capitata* and *Anastrepha fraterculus* (Diptera: Tephritidae) in diverse host species and localities of Argentina. *Annals of the Entomological Society of America* 99:70–83.

Selivon, D. 2000. Relaçoes com as plantas hospedeiras. In *Moscas-das-frutas de importância econômica no Brasil. Conhecimento básico e aplicado*, A. Malavasi & R.A. Zucchi (Eds.), 87–91. Holos Ed, Riberão Preto, Brazil.

Selivon, D., A. L. P. Perondini, & J. S. Morgante. 2005a. A genetic-morphological characterization of two cryptic species of the *Anastrepha fraterculus* complex (Diptera: Tephritidae). *Annals of the Entomological Society of America* 98:367–381.

Selivon, D., A. L. P. Perondini, & L. S. Rocha. 2005b. Karyotype characterization of *Anastrepha* fruit flies (Diptera: Tephritidae). *Neotropical Entomology* 34:273–279.

Selivon, D., C. Vretos, L. Fontes, & A. L. P. Perondini. 2004. New variant forms in the *Anastrepha fraterculus* complex (Diptera, Tephritidae). In *Proceedings 6th International Symposium on Fruit Flies of Economic Importance*, B.N. Barnes (Ed.), 253–258. Isteg Scientific Publications, Stellenbosch.

Shaw, J. G. 1947. Hosts and distribution of *Anastrepha serpentina* in northeastern Mexico. *Journal of Economic Entomology* 40:34–40.

Silva, F. F. D., R. N. Meirelles, L. R. Redaelli, & F. K. Dal Soglio. 2006. Diversity of flies (Diptera: Tephritidae and Lonchaeidae) in organic citrus orchards in the Vale do Rio Caí, Rio Grande do Sul, southern Brazil. *Neotropical Entomology* 35:666–670.

Silva, G. J., V. S. Dutra, M. S. Santos et al. 2010. Diversity of *Anastrepha* spp. (Diptera: Tephritidae) and associated braconid parasitoids from native and exotic hosts in Southeastern Bahia, Brazil. *Environmental Entomology* 39:1457–1465.

Silva, L. N., M. S. Santos, V. S. Dutra, E. L. Araujo, M. A. Costa, & J. G. Silva. 2011a. First survey of fruit fly (Diptera: Tephritidae) and parasitoid diversity among Myrtaceae fruit across the state of Bahia, Brazil. *Revista Brasileira de Fruticultura* 33:757–764.

Silva, N. M. 1993. Levantamento e análise faunística de moscas-das-frutas (Diptera: Tephritidae) em quatro locais do Estado do Amazonas. Dissertation, Escola Superior de Agricultura "Luiz de Queiroz," Universidade do São Paulo, Piracicaba. 155 p.

Silva, N. M., S. Silveira Neto, & R. A. Zucchi. 1996. The natural host plants of *Anastrepha* in the state of Amazonas, Brazil. In *Fruit Fly Pests: A World Assessment of Their Biology and Management*, B. A. McPheron & G. J. Steck (Eds.), 352–357. St. Lucie Press, Delray Beach, FL.

Silva, R. A., A. L. Lima, S. L. O. Xavier, W. R. Silva, C. F. Marinho, & R. A. Zucchi. 2011b. *Anastrepha* species (Diptera: Tephritidae), their hosts and parasitoids in southern Amapá State, Brazil. *Biota Neotropica* 11:429–434.

Silva, W. R., & R. A. Silva. 2007. Levantamento de moscas-das-frutas e de seus parasitóides no município de Ferreira Gomes, Estado do Amapá. *Ciência Rural* 37:265–268.

Souza, S. A. S., A. L. S. Resende, P. C. Strikis, J. R. Costa, M. S. F. Ricci, & E. L. Aguiar-Menezes. 2005. Infestação Natural de Moscas Frugívoras (Diptera: Tephritoidea) em Café Arábica, sob Cultivo Orgânico Arborizado e a Pleno Sol, em Valença, RJ. *Neotropical Entomology* 34:639–648.

Souza-Filho, M. F. 2006. Infestação de moscas-das-frutas (Diptera: Tephritidae e Lonchaeidae) relacionada à fenologia da goiabeira (*Psidium guajava* L.), nespereira (*Eriobotrya japonica* Lindl.) e do pessegueiro (*Prunus persica* Batsch). PhD diss. São Paulo, Brazil: Escola Superior de Agricultura Luiz de Queiroz, Universidade de São Paulo. 125 p.

Souza-Filho, M. F., A. Raga, & R. A. Zucchi. 2000. Moscas-das-frutas nos estados brasileiros: São Paulo. In *Moscas-das-frutas de importância econômica no Brasil. Conhecimento básico e aplicado*, A. Malavasi & R. A. Zucchi (Eds.), 277–283. Holos Ed, Riberão Preto, Brazil.

Souza-Filho, M. F., A. Raga, J. A. Azevedo-Filho, P. C. Strikis, J. A. Guimarães, & R. A. Zucchi. 2009. Diversity and seasonality of fruit flies (Diptera: Tephritidae and Lonchaeidae) and their parasitoids (Hymenoptera: Braconidae and Figitidae) in orchards of guava, loquat and peach. *Brazilian Journal of Biology* 69:31–40.

Souza-Filho, Z. A. 2005. Estudos populacionais de moscas-das-frutas (Diptera: Tephritidae) em um pomar de goiaba (*Psidium guajava* L.) em Una–Bahia. MSc Dissertation, Universidade Estadual de Santa Cruz, 60 p.

Steck, G. J. 1991. Biochemical systematics and population genetic structure of *Anastrepha fraterculus* and related species (Diptera: Tephritidae). *Annals Entomological Society of America* 84:10–28.

Steck, G. J. & A. Malavasi. 1988. Description of immature stages of *Anastrepha bistrigata* (Diptera: Tephritidae). *Annals of the Entomological Society of America* 81:1004–1006.

Steck, G. J., L. E. Carroll, H. Celedonio-Hurtado & J. Guillen-Aguilar. 1990. Methods for identification of *Anastrepha* larvae (Diptera: Tephritidae), and key to 13 species. *Proceedings Entomological Society of Washington* 92:333–346.

Stone, A. 1942a. New species of *Anastrepha* and notes on others. *Journal of Washington Academy of Sciences* 32:298–304.

Stone, A. 1942b. The fruitflies of the genus *Anastrepha*. *United States Department of Agriculture Miscellaneous Publication* 439:1–112.

Taira, T. L. 2012. Moscas-das-frutas (Diptera: Tephritidae) e seus parasitóides em hospedeiros cultivados e silvestres no ecótono Cerrado-Pantanal sul-mato-grossense, Brasil. MSc Dissertation, Universidade Estadual de Mato Grosso do Sul, 59 p.

Taira, T. L., A. R. Abot, J. Nicácio, M. A. Uchôa, S. R. Rodrigues & J. A. Guimarães. 2013. Fruit flies (Diptera, Tephritidae) and their parasitoids on cultivated and wild hosts in the Cerrado-Pantanal ecotone in Mato Grosso do Sul, Brazil. *Revista Brasileira de Entomologia* 57:300–308.

The Plant List. 2013. Version 1.1. Published on the Internet; http://www.theplantlist.org/ (last accessed July 10, 2018).

Tigrero, J. O. 1998. Revisión de especies de moscas de la fruta presentes en el Ecuador. Published by the author, Sangolquí, Ecuador. 55 p.

Tigrero, J. O. 2009. Lista anotada de hospederos de moscas de la fruta presentes en Ecuador. *ESPE Serie Zoológica Boletín Técnico* 8:107–116.

Torres, C. A. S. 2004. Diversidade de espécies de moscas-das-frutas (Diptera: Tephritidae) e de seus parasitóides em cafeeiro (*Coffea arabica* L.). MSc Dissertation, Universidade Estadual do Sudoeste da Bahia–UESB/Campus de Vitória da Conquista-BA, 71 p.

Uchôa-Fernandes, M.A. & J. Nicácio. 2010. New records of Neotropical fruit flies (Tephritidae), lance flies (Lonchaeidae) (Diptera: Tephritoidea), and their host plants in the South Pantanal and adjacent areas, Brazil. *Annals of the Entomological Society of America* 103:723–733.

Uchôa-Fernandes, M. A., & R. A. Zucchi. 2000. Moscas-das-frutas nos estados brasileiros: Mato Grosso e Mato Grosso do Sul. In *Moscas-das-frutas de importância econômica no Brasil. Conhecimento básico e aplicado*, A. Malavasi and R. A. Zucchi (Eds.), 241–245. Holos Ed, Riberão Preto, Brazil.

Uchôa-Fernandes, M. A., I. Oliveira, R. M. S Molina, & R. A. Zucchi. 2002. Species diversity of frugivorous flies (Diptera: Tephritoidea) from hosts in the Cerrado of the State of Mato Grosso do Sul, Brazil. *Neotropical Entomology* 31:515–524.

Uramoto, K. 2002. Biodiversidade de moscas-das-frutas do gênero *Anastrepha* (Diptera, Tephritidae) no campus Luiz de Queiroz. MSc Dissertation, Escola Superior de Agricultura "Luiz de Queiroz", Universidade de São Paulo. Piracicaba, Brasil. 85 p.

Uramoto, K. 2007. Diversidade de moscas-das- frutas (Diptera, Tephritidae) em pomares comerciais de papaia e em áreas remanescentes da Mata Atlântica e suas plantas hospedeiras nativas, no município de Linhares, Espírito Santo. Dissertation PhD, Escola Superior de Agricultura Luiz de Queiroz, Universidade de São Paulo. 105 p.

Uramoto, K., D. S. Martins, & R. A. Zucchi. 2008. Fruit flies (Diptera: Tephritidae) and their asscociations with native host plants in a remnant area of the highly endangered Atlantic Rain Forest in the State of Espirito Santo, Brazil. *Bulletin of Entomological Research* 98:457–466.

Uramoto, K., J. M. M. Walder, & R. A. Zucchi. 2004. Biodiversidade de moscas-das-frutas do gênero *Anastrepha* (Diptera, Tephritidae) no campus da ESALQ-USP, Piracicaba, São Paulo. *Revista Brasileira de Entomologia* 48:409–414.

Vaníčková, L., V. Hernández-Ortiz, I. S. J. Bravo et al. 2015. Current knowledge of the species complex *Anastrepha fraterculus* (Diptera, Tephritidae) in Brazil. *ZooKeys* 540:211–237.

Vayssières, J. F., J. P. Cayol, P. Caplong et al. 2013. Diversity of fruit fly (Diptera: Tephritidae) species in French Guiana: their main host plants and associated parasitoids during the period 1994–2003 and prospects for management. *Fruits International Journal of Tropical and Subtropical Horticulture* 68:219–243.

Veloso, V. R. S., P. M. Fernandes, & R. A. Zucchi. 2000. Moscas-das-frutas nos estados brasileiros: Goiás. In *Moscas-das-frutas de importância econômica no Brasil. Conhecimento básico e aplicado*, A. Malavasi and R.A. Zucchi (Eds.), 247–252. Holos Ed, Riberão Preto, Brazil.

Veloso, V. R. S., G. A. Ferreira, P. M. Fernandes, N. A. Canal Daza, & R. A. Zucchi. 1996. Ocorrência e índice de infestação de *Anastrepha* spp. (Dip., Tephritidae) em *Pouteria gardneriana* Radlk. e *Pouteria ramiflora* (Mart.) Radlk. (Sapotaceae), nos cerrados de Goiás. *Anais Escola Agron. Vet. Universidade Fed. Goias* 26:109–120.

Vera, M. T., C. Cáceres, V. Wornoayporn et al. 2006. Mating incompatibility among populations of the South American fruit fly *Anastrepha fraterculus* (Diptera: Tephritidae). *Annals of the Entomological Society of America* 99:387–397.

White, I. M., & M. M. Elson-Harris. 1992. *Fruit Flies of Economic Significance: Their Identification and Bionomics*. CAB International, Wallingford, UK.

Wiedemann, C. R. W. 1830. Aussereuropaische zweiflugelige Inseckten. Vol. 2. Schulz, Hamm.

Wille, J. E. 1937. Fruitflies in the Republic of Ecuador. *United States Department of Agriculture Bureau of Entomology and Plant Quarantine Services* 130:25–26.

Wille, J. E. 1941. Tres informes de observaciones entomólogicas en la costa en 1940. *Informe Estación Experimental Agrícola La Molina* 53:1–26.

Yepes, R. F., & A. R. Vélez. 1989. Contribución al conocimiento de las moscas de las frutas (Tephritidae) y sus parasitoides en el departamento de Antioquia. *Revista Facultad Nacional de Agronomía Medellin* 42:73–98.

Zucchi, R. A. 2000. Espécies de *Anastrepha*, sinonímias, plantas hospedeiras e parasitóides. In *Moscas-das-frutas de importância econômica no Brasil. Conhecimento básico e aplicado*, A. Malavasi & R.A. Zucchi (Eds.), 41–48. Holos Ed, Riberão Preto, Brazil.

Zucchi, R. A. 2007. Diversidad, distribución y hospederos del género *Anastrepha* en Brasil. In *Moscas de la fruta en Latinoamérica (Diptera: Tephritidae): Diversidad, Biología y Manejo*, V. Hernández-Ortiz (Ed.), 77–100. SYG Editores, Mexico City.

7 Preliminary Report of *Anastrepha* Species Associated with "Kaniste" Fruits (*Pouteria campechiana*) (Sapotaceae) in the State of Campeche, Mexico

María de Jesús García Ramírez, Enrique Antonio Hernández, Juan José Vargas Magaña, Marvel del Carmen Valencia Gutiérrez, Ivonne Esmeralda Duarte Ubaldo, Enrique A. González Durán, and Lisandro Encalada Mena*

CONTENTS

Abstract Campeche is a state located in the southeast of the Gulf of Mexico and has a great variety of microclimates and native vegetation. Backyard orchards in rural communities in this state provide both introduced and native host fruits for *Anastrepha* spp. fruit flies, such as the "Kaniste" fruit (*Pouteria campechiana*) (Sapotacea). In this study, we sampled Kaniste fruits from backyards located in the municipality of Escarcega, Campeche, Mexico. Three species of the genus *Anastrepha* were recorded feeding on Kaniste fruits: *Anastrepha fraterculus*, *Anastrepha serpentina*, and *Anastrepha hamata*. These findings contribute to our knowledge of the hosts and diversity of the genus *Anastrepha* in Campeche, Mexico.

7.1 INTRODUCTION

Fruit flies (Diptera: Tephritidae) are one of the most important agricultural pests in the world because they represent an economic impact on fruit production worldwide (Aluja 1994). Moreover, national and international quarantines are imposed on the mobilization of commercial hosts infected by

* Corresponding author.

several species of fruit flies. Campeche is a state located in the southeast of Mexico, and it is a region with a high diversity of plants, which provides various habitats for tephritid species such as those of the genus *Anastrepha* (García-Ramírez et al. 2018). *Anastrepha* includes species of economic importance, as well as species with ecological roles that do not require phytosanitary management (Aluja 1994). "Kaniste," *Pouteria campechiana* (HBK) Baehni (Sapotaceae), is a native tree from southern Mexico with a distribution that extends to South America (Martin and Malo 1999). Kaniste fruits are orange-yellow with an ovoid-round shape; they measure around 4 cm in diameter and weigh around 7 g. The fruiting period extends from July to January. This study aimed to determine the occurrence of *Anastrepha* species in fruits of *P. campechiana* in Campeche, Mexico.

7.2 MATERIALS AND METHODS

Fruit sampling was conducted in familiar backyard orchards in the municipality of Escarcega, Campeche, Mexico (18°29′25.1″N and 90°55′30.9″W, altitude of 42 m. a. s. l.). The region is surrounded by evergreen medium elevation forests and is characterized by a subhumid climate with rainfall in the summer. Fruit sampling was carried out in two periods: from August to September 2017 and from November to December 2017. Ripe fruits of *P. campechiana* were collected directly from the ground, they were inspected for the presence of tephritid larvae, and infested fruits were placed in plastic containers (20 × 20 cm). Samples were transported to the Entomology Laboratory of the Escuela Superior de Ciencias Agropecuarias, Universidad Autónoma de Campeche (ESCA-UAC) in Escarcega, Campeche, Mexico.

Infested fruits were placed in individual transparent cylindrical plastic chambers (10 cm in diameter × 20 cm high). In the second period of sampling, larvae were found in the seeds of the fruits; thus, they were placed in different containers. Moist vermiculite was used as a pupation substrate in the chambers described. Pupae were recovered and placed in small plastic containers of 5 × 5 cm and were inspected daily. Emerged adults were identified by Enrique Antonio Hernández and María de Jesús García-Ramírez using the taxonomic keys provided by Korytkowski (2008). The entomological material was deposited in the Entomology Laboratory of the ESCA-UAC.

7.3 RESULTS

Fruit sampling data are summarized in Table 7.1. Overall, three species of the genus *Anastrepha* were found to be associated with fruits of *P. campechiana*. Specimens of *Anastrepha serpentina* larvae (Wiedemann, 1830) were observed feeding on the pulp of fruits collected in both the

TABLE 7.1
***Anastrepha* spp. Recorded in Fruits of *Pouteria campechiana* Collected in Familiar Backyard Orchards in the Municipality of Escarcega, Campeche, Mexico**

				Adults			
Sampling Period	Sample Weight	Infested Fruit Part	No. of Pupae	♀	♂	Date of Emergence	Species
August–September 17	340 g	Mesocarp	77	24	26	August 18, 2017	*Anastrepha serpentina*
November–December 17	400 g	Mesocarp	87	31	28	August 12, 2017	*Anastrepha serpentina*
				—	2	December 2, 2017	*Anastrepha fraterculus*
		Seed	32	8	12	December 25, 2017	*Anastrepha hamata*

first and the second sampling periods. In the second sampling period, *Anastrepha fraterculus* specimens (Wiedemann, 1830) were also observed feeding on the pulp of the fruits, whereas *Anastrepha hamata* specimens (Loew, 1983) were observed feeding on the seeds. It is important to note that in the second sampling period, there were fruit fly larvae feeding on both the pulp and the seeds of the same Kaniste fruit. In some cases, all three *Anastrepha* species emerged from the same fruit.

7.4 DISCUSSION AND CONCLUSION

We report for the first time the occurrence of three species of the genus *Anastrepha* (*A. fraterculus, A. serpentina,* and *A. hamata*) simultaneously infesting fruits of *P. campechiana* in Campeche, Mexico. This is also the first record of *A. fraterculus* infesting fruits of *P. campechiana.* Even though we only obtained two male adults of this species from the infested fruits (Table 7.1), its occurrence is not considered to be accidental because *A. fraterculus* has been reported with complete development in *Pouteria caimito* fruits in the province of Guayas, Ecuador (Tigrero 2009). However, in this study, most of the infestation in fruits of *P. campechiana* was by *A. serpentina*, a multivoltine species commonly associated with plants of the family Sapotaceae in Mexico (Sosa-Armenta et al. 2015).

The simultaneous infestation of *P. campechiana* by three species of *Anastrepha* is an important finding. This could be explained by the inability of these species to recognize the fruit-marking pheromone used by heterospecifics. This recognition has been reported by Aluja et al. (2000) for *Anastrepha bahiensis*, *A. fraterculus*, and *Anastrepha obliqua* infesting *Myrciaria floribunda* (Myrtaceae) in Apazapan, Veracruz, and by Antonio-Hernández (2006) for *A. fraterculus* and *Anastrepha striata* in *Psidium guajava* in Tehuantepec, Oaxaca. Another possible explanation is the low availability of host fruits in the region of Escarcega at the time of the samplings, which could have resulted in more than one species of *Anastrepha* infesting these fruits (Hernandez-Ortiz et al. 2002).

Compared to *A. serpentina* and *A. fraterculus*, *A. hamata* showed a relatively long pupation period, with a time of emergence of 28–34 days. This physiological condition is likely because of the atypical temperatures recorded at the time of the year (22°C–25°C) in the sampling site, which are lower than the temperatures that normally occur in this location (35°C–40°C). It is possible that some *Anastrepha* species in an immature state are able to synchronize their physiological mechanisms according to abiotic factors such as climate and temperature (Aluja et al. 1998), which would allow them to regulate the time of adult emergence. This could be the case in *A. hamata* based on the observations made in this study.

REFERENCES

Aluja, M. 1994. Biomomics and management of *Anastrepha*. *Annual Review of Entomology* 39: 155–178.

Aluja, M., Piñero, J., López, M., Ruiz, C., Zúñiga, A., Piedra, E., Díaz-Fleischer, F., and Sivinski J. 2000. New host plant and distribution records in México for *Anastrepha* spp., *Toxotrypana curvicauda* Gerstaecker, *Rhagoletis zoqui* Bush, *Rhagoletis* sp. and *Hexachaeta* sp. (Diptera: Tephritidae). *Proceedings of the Entomological Society of Washington* 102: 802–815.

Aluja, M., Lopez, M., and Sivinski, J. 1998. Ecological evidence for diapause in four native and one exotic species of larval–pupal fruit fly (Diptera: Tephritidae) parasitoids in tropical environments. *Annals of the Entomological Society of America* 91: 821–833.

Antonio-Hernández, E. 2006. Determinación de especies de moscas de la fruta del género *Anastrepha* Schiner (Diptera: Tephritidae) mediante trampeo y muestreo en cuatro áreas del Istmo de Tehuantepec, Oaxaca. Oaxac, Mexico, Tesis Ing. Agrónomo, Instituto Tecnológico de Comitancillo. 123 p.

García-Ramírez. M. J., Antonio-Hernández, E., and Valencia-Gutiérrez, M. del C. 2018. Diversidad de Moscas de la fruta del género *Anastrepha* spp., como parte importante del patrimonio natural presente en el estado de Campeche, México. In *Estado del Patrimonio en América Latina (Estudios de caso)*, ed. R. M. Velázquez-Sánchez, N. del J. Bolivar-Fernández, and S. B. Rangel-Rojas, 16–21. TECCIS A. C. Campeche, México.

Hernández-Ortiz, V., Manrique-Saide, P., Delfin-Gonzalez, H., and Novelo-Rincon, L. 2002. First report of *Anastrepha compresa* in Mexico and New Record for other *Anastrepha* species in the Yucatan Peninsula (Diptera: Tephritidae). *Florida Entomologist* 85: 389–391.

Korytkowski, Ch. A. 2008. Manual Para la Identificación de Moscas de la fruta del genero *Anastrepha* SCHINER, 1868. Universidad de Panamá, Vicerrectoría de Investigación y Posgrado.

Martin, F. W., and Malo, S. E. 1999. Canistel and its close relatives. *Revista de Geografia Agrícola*. Chapingo, Mexico.

Sosa-Armenta, J. M., López-Martínez, V., Alia-Tejacal, I., Jiménez-García, D., Guillen-Sánchez, D., and Delfín-González, H. 2015. Hosts of five *Anastrepha* species (Diptera: Tephritidae) in the state of Quintana Roo, Mexico. *Florida Entomologist* 98: 1000–1002.

Tigrero, S. J. O. 2009. Lista anotada de hospederos de moscas de la fruta presentes en Ecuador. *Boletín Técnico 8, Serie Zoológica* 4–5: 107–116.

Section III

Chemical Ecology and Attractants

8 Bait Stations for Control of Mexican Fruit Flies (*Anastrepha ludens*), First Year

Hugh Conway, Guadalupe Gracia, Pedro Rendón, and Christopher Vitek*

CONTENTS

Abstract The Mexican fruit fly, *Anastrepha ludens* (Loew) (Mexfly), is a pest of economic importance with the potential to cause millions of dollars in damage to citrus and other fruits. This chapter presents the first-year results from a three-year field study (2014–2017) conducted to evaluate the effectiveness of bait stations with Spinosad embedded in a wax matrix to control Mexfly. Spinosad is produced by a naturally occurring bacteria, *Saccharopolyspora spinosa*, and is considered an organic insecticide acceptable for use by organic growers. The flies feed on the wax matrix, and the Spinosad acts as a stomach poison killing the flies. Each bait station contains a two-component lure consisting of the attractants, putrescine, and ammonium acetate. The study used 500 bait stations strategically placed based on historic wild fly capture data at 12 locations in the Lower Rio Grande Valley in south Texas, United States. Results indicate that a hat or protective covering over the bait station extends the residual killing effect with fly mortality of up to 3 months from aged bait stations taken from the field. In addition, a reduction in wild Mexfly capture was observed in the areas where the bait stations were used. Potential uses would be around wild fly finds, in abandoned or poorly maintained groves, in organic groves, or with permission, in residential citrus plantings. This study indicates that bait stations are another valuable integrated pest management (IPM) tool for program managers in their effort to control Mexfly.

* Corresponding author.

8.1 INTRODUCTION

Fruit fly managers are investigating unique methods of controlling invasive fruit fly species including the use of attract-and-kill devices. A bait station is a type of attract-and-kill device described as a discreet type of container (with or without a visual component) with attractants and toxins that target specific pests (Piñero et al. 2014). Bait stations may require servicing to remain active during the season, and insects attracted and killed, if retained, should be discarded and not counted (Piñero et al. 2014). Mangan et al. (2006) tested a mixture of GF 120 fruit fly bait with a killing component of Spinosad that had been purified from soil bacteria actinomycete, *Saccharopolyspora spinosa* Mertz. Mangan and Moreno (2007) developed and tested a form of bait station using protein-based baits with a photoactive dye toxicant in small discrete containers resulting in a 70%–90% sterile Mexican fruit fly (Mexfly), *Anastrepha ludens* (Loew), reduction compared to the control but only 22% as effective as spot insecticidal sprays. In 2008, Food and Agricultural Organization/International Atomic Energy Association (FAO/IAEA) initiated the development of bait stations for fruit fly suppression in support of sterile insect technique (SIT) (IAEA 2009). The bait station initiative investigated the type and category of different bait stations, including the attributes and components needed for a successful attract-and-kill device, and the potential application of bait stations (IAEA 2009). Epsky et al. (2012) demonstrated that wax matrix bait stations embedded with Spinosad were effective in controlling the Mediterranean fruit fly, *Ceratitis capitata* (Wiedemann). Since 2010, Animal and Plant Health Inspection Service (APHIS) Plant Protection and Quarantine (PPQ) Guatemala has been working on developing a bait station incorporating the wax matrix reported on by Epsky et al. (2012). This bait station would be able to accept a variety of lures allowing its use against a number of different fruit fly species. Important characteristics in the designed bait station, include: easy to deploy, biodegradable, and ability to retrieve (if necessary). The bait station should satisfy organic production concerns and concerns with use in urban settings. Such a bait station was eventually developed, and a US Department of Agriculture patent claim presented and granted in early 2018. The bait station has been tested in several locations including Guatemala, Texas, Florida, and Hawaii to validate the technology. We report here on the findings of its use for Mexfly control in the Lower Rio Grande Valley, Texas.

The fruit fly trapping and surveillance program in the Lower Rio Grande Valley of South Texas includes 13 Multilure traps per km^2 across the 109.3 km^2 citrus growing area employing 2,905 traps. Each trap contains a two-component lure (putrescine and ammonium acetate) and 300 mL of a 10% propylene glycol mixture as a drowning solution and fly preservative. The trapping design provides a means for conducting tests to compare the Multilure traps with the effect of bait stations in controlling Mexfly.

A 3-year study was conducted in the Lower Rio Grande Valley of south Texas evaluating the effectiveness in the field of bait stations. First-year data analysis was used to compare capture of feral and sterile Mexfly at (1) sites with bait stations versus (2) sites without bait stations. Laboratory bioassays on bait stations aged in the field was used to compare fly mortality between (1) bait stations, (2) bait station with protective hat, and (3) no bait station control. Efficacy of aged bait stations from servicing intervals across the year were compared in laboratory bioassays using bait stations with or without hats to a no bait station control. Results from the first year of the study are presented in this chapter.

8.2 METHODS AND MATERIALS

8.2.1 INSECTS

Anastrepha ludens used in Hidalgo County comparisons were obtained as pupae from colonies maintained at the Mexican fruit fly rearing facility in San Miguel Petapa, Guatemala. The arriving Tapachula-7 strain, also referred to as Black Pupae Strain (BPS), were presorted by brown pupae color, resulting in ~90% male flies. The aerial release rate for BPS flies was 61,775 flies per km^2 (250 flies per acre). Mexican fruit flies used in the majority of Cameron Country (Texas), Willacy

County (Texas), and for laboratory bioassays comparisons were from a colony maintained by the Department of Agriculture (APHIS) at the Moore Air Base Mexican fruit fly mass-rearing facility near Edinburg, Texas. The colony is an isofemale line originating from an outbreak of wild flies captured in Willacy County, Texas, in April 2008. The aerial release rate for Texas reared flies was 123,550 flies per km^2 (500 flies per acre).

8.2.2 Wax-Based Bait Stations

The bait station used in this study was made of a cardboard box approximately 7.6 × 10.2 × 1.3 cm with an opening to slide a cardboard strip with the attractant fruit fly lures using two-component patches (putrescine and ammonium acetate) inside the box (Figure 8.1). The box has 24 holes in the front and back, allowing movement for the odor from the lures. The box is dipped into melted wax that contains Spinosad (kill-component active ingredient) and phagostimulants, including corn syrup and sugar (Epsky et al. 2012). The bait station containing the Spinosad toxicant must attract and then stimulate the flies to feed on the waxy matrix. Spinosad must be ingested to obtain maximum effect (Prokopy et al. 2003). The wax matrix serves to extend the insecticide killing effect or longevity over a longer period of time. Fruit flies attracted by the lure land on the wax bait station and begin feeding. Within 24–48 hours after ingesting the waxy matrix, the fly will die. The second bait station configuration involved the addition of a conical shaped covering (hat) dipped in the wax matrix with Spinosad positioned over and covering the bait station (Figure 8.2). The conical hat was made of a cardboard circle with a diameter of 17.8 cm (7 in), having one cut to the middle of the circle, which allowed the circle to be formed into a cone shape and stapled in place above the bait station. The conical hat protected the bait station from sun and weather (rain) with the aim of extending the efficacy of the bait station.

A total of 500 bait stations (half with and half without protective hats) were placed in the field at each placement period. Often, field placement of bait stations would require up to 1 week in duration to place all of the bait stations into a test comparison site. Bait stations were placed in the field at approximately 4-week intervals on August 25, September 22, and again on October 25, 2014, when the two-component lures were replaced in the test traps. For the rest of the year, bait stations were added to the field when the lures in the traps were replaced at 8-week intervals with bait stations

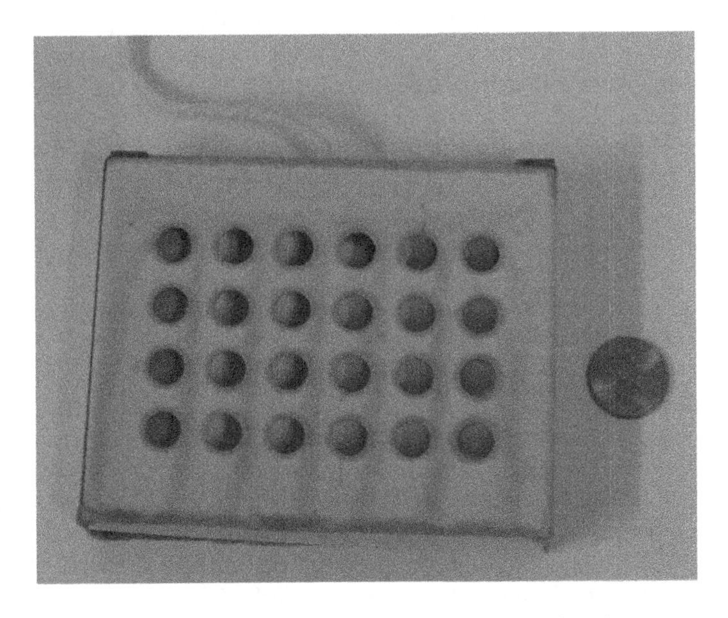

FIGURE 8.1 Bait station covered with waxy matrix mixed with Spinosad (toxin).

FIGURE 8.2 Bait station with conical hat protection.

placement occurring on December 25, 2014, and on February 25, April 25, and June 25, 2015. For each placement period, bait stations were marked with different colored plastic ties with the date the bait station was placed into the field.

8.2.3 Field Tests

Texas was declared eradicated of Mexican fruit fly in the beginning of 2012. Since then, only sporadic and localized wild Mexfly finds have occurred: 38 adults and six larvae were captured in the 2013 citrus-growing season. In this study, sterile Mexfly captures were used for field comparisons because of the low number of wild flies. Previous research by Mangan and Moreno (1995) showed that sterile mass-reared Mexflies are equivalent to fertile flies in attraction and feeding behavior.

Two-piece plastic Multilure traps with two-component lures (putrescine and ammonium acetate) and 300 mL of 10% propylene glycol capture solution were used for fly capture. The field comparison tests used sterile Mexican fruit flies released by the program at 123,550 flies per km^2 (500 flies per acre) weekly in Cameron and Willacy Counties, and 61,775 flies per km^2 (250 flies per acre) in Hidalgo County where the Guatemala male-sorted Tapachula-7 BPS flies were released.

Test comparison sites were based on previous hot-spot locations with historic recurring feral *A. ludens* capture and larva finds (Figure 8.3). Specific comparison metrics were identified at different locations as indicated in Table 8.1. Grove comparisons were made by using two small groves and placing 40 bait stations (20 with hats and 20 without hats) in one orchard and no bait stations in the control orchard. For both grove tests, four Multilure traps were placed in the bait station and control orchards. The grove site comparison in La Feria, Texas, received aerial-released sterile Mexican fruit flies and the south orchard received 40 bait stations (Table 8.1). The grove site comparison in Bayview, Texas, in Cameron County did not receive sterile flies. The northwest grove in Bayview received the 40 bait stations and the southeast grove was the control (Table 8.1). The city of Lyford, Texas, in Willacy County, was used to compare urban areas with primarily residential properties by dividing the town in half using Business 77 and placing 40 bait stations on the east side and no bait stations on the west side. Both sides of the Lyford test received six Multilure traps (Table 8.1). Two similar-sized trailer parks near Donna, Texas, were used for fly capture comparison in a residential park setting with Country Sunshine receiving 50 bait stations and Southern Comfort no bait stations. Both trailer parks received four Multilure traps each (Table 8.1). The next comparison site was the trailer park Ranchero Village with 50 bait stations compared to the Estero Llano Grande

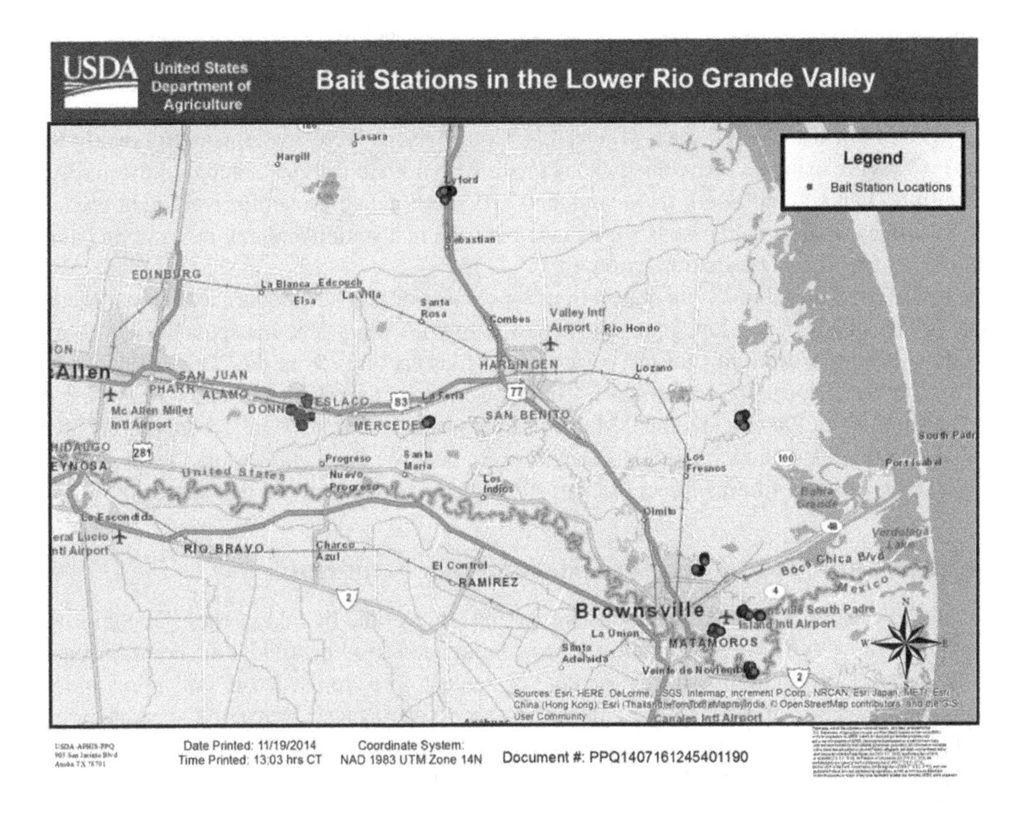

FIGURE 8.3 Site locations for bait stations test marked by dots.

TABLE 8.1
Site Locations and Number of Traps and Bait Stations per Site

Comparison	Site Name	Size m²	Location Center	County	BS	Traps
Field Test Fly Releases	La Feria S	22,674	26°08′17″N 97°49′02″W	Cameron	40	4
	La Feria N	20,738	26°08′25″N 97°48′53″W	Cameron	0	4
Field Test No Fly Release	Bayview N	24,402	26°08′07″N 97°24′12″W	Cameron	40	10
	Bayview S	23,776	26°07′59″N 97°23′46″W	Cameron	0	10
Abandoned Grove	Abandoned	24,965	25°58′32″N 97°27′15″W	Cameron	40	8
	Sol grove	25,900	25°58′45″N 97°27′08″W	Cameron	0	8
Urban Lyford, TX	Bus 77 E	73,996	26°24′42″N 97°47′19″W	Willacy	40	6
	Bus 77 W	80,935	26°24′42″N 97°47′31″W	Willacy	0	6
RV Parks	Country Sunshine	139,211	26°08′35″N 97°58′22″W	Hidalgo	50	4
	Southern Comfort	138,402	26°08′43″N 97°58′17″W	Hidalgo	0	4
RV Park Bird Center	Ranchero	63,523	26°08′15″N 97°58′56″W	Hidalgo	50	4
	Estero Llano	45,052	26°08′28″N 97°57′30″W	Hidalgo	0	4
Test Spots Without Comparison Sites Based on Larvae or Wild Fly Captures in 2013						
Larvae	South Point Brownsville, TX		25°50′23″N 97°23′44″W	Cameron	60	12
Fly Find	Monica Brownsville, TX		25°53′22″N 97°26′02″W	Cameron	40	8
Fly Find	Toronja Brownsville, TX		25°54′39″N 97°24′14″W	Cameron	40	10
Larvae	Nevada St. Weslaco		26°10′06″N 97°58′54″W	Hidalgo	50	6
Fly Find	Illinois St. Weslaco		26°09′11″N 97°59′95″W	Hidalgo	25	4
Fly Find	7th St. Donna		26°09′13″N 97°59′06″W	Hidalgo	25	2

State Park (Weslaco Birding Center) with no bait stations. The Weslaco Birding Center contains the remnants of an old recreational vehicle (RV) Park started in the 1950s with a variety of fruit fly host trees. Both locations received four Multilure traps each with traps in the Birding Center placed on trees in the old RV Park (Table 8.1). Comparisons were made near Robindale Road/ Sol in Brownsville using an abandoned citrus grove with 40 bait stations compared to small citrus grove with no bait stations. Both the abandoned and small active grove received eight traps each (Table 8.1). In each site, comparisons were made between bait station and non-bait station based on the numbers of feral and sterile flies captured.

The rest of the bait stations were placed in hot spot locations. A hot spot is a location of recurring wild fly or larva finds in previous years and across numerous years. The hot spots used in Cameron and Hidalgo Counties were in sites of larvae or wild flies captures from the 2013–2014 citrus season. Three hot-spot locations in Cameron County were near the Mexican border in Brownsville, Texas, at South Point (60 bait stations), Monica Street (40 bait stations), and Toronja Street (40 bait stations). The three locations in Hidalgo County were near Weslaco and Donna, Texas, at Nevada Street (50 bait stations), Illinois Street (25 bait stations), and 7th Street (25 bait stations).

8.2.4 Laboratory Bioassay of Bait Station Efficacy in the Field

The length of effectiveness of bait stations was tested across time by taking eight random samples (four bait station with hats and four bait station without hats) from the field at no comparison locations. Bait stations from the initial release date of August 25 were used with collection and comparisons occurring at 2- to 4-week intervals. Laboratory bioassay tests were conducted on mortality inside Plexiglas observation cages (30 cm × 30 cm × 40 cm). Each clear Plexiglas observation cage contained 50 Mexican fruit flies, food, and water ad libitum and one bait station, either with or without hat. Each bait station was suspended downward from the top in the middle of the observation cage. Mortality comparisons were made at 24-hour intervals across 72 hours against two control cages with 50 adult Mexican fruit flies, food, and water ad libitum. Because each cage started with 50 live flies, percentage of mortality per cage was calculated by taking the number of dead flies in an observation cage and multiplying by two.

An additional test was conducted at the end of the year on bait stations with and without a protective hat that had been weathered in the field for 8–52 weeks (Figure 8.4). Twelve random samples (six bait station with hats and six bait station without hats) were taken from the field at comparison locations by time of placement. Half of the tested bait stations were obtained from Hidalgo County and half from Cameron and Willacy Counties. The same procedures described previously were used for the mortality test. Because each cage started with 50 live flies, percentage of mortality per cage was calculated by taking the number of dead flies in an observation cage and multiplying by two.

8.2.5 Data Analysis

In this study, feral Mexican fruit fly capture was small and presented by location and the number of flies captured. For field tests, Mexican fruit fly capture comparisons between bait station and without bait stations for each site were conducted for sterile fly capture using t-tests and means comparisons using ALL-Pairs Tukey's honestly significant difference (HSD) tests ($P = 0.05$) (SAS JMP 13 2017). Data results were presented in table form.

For both the laboratory bioassay tests, fly mortality was analyzed using analyses of variance (ANOVA) with means separation using ALL-PairsTukey's HSD test ($P = 0.05$) by date, comparing percentage of mortality in observation cage from bait station with hat, bait stations without hat, and control (SAS JMP 13 2017). The mean ± standard error (SE) values by treatment were graphed for comparisons of fly mortality using bait station, bait stations with hat, and no bait station control. Only the mortality at 72 hours is presented because Spinosad is a stomach poison that may take up

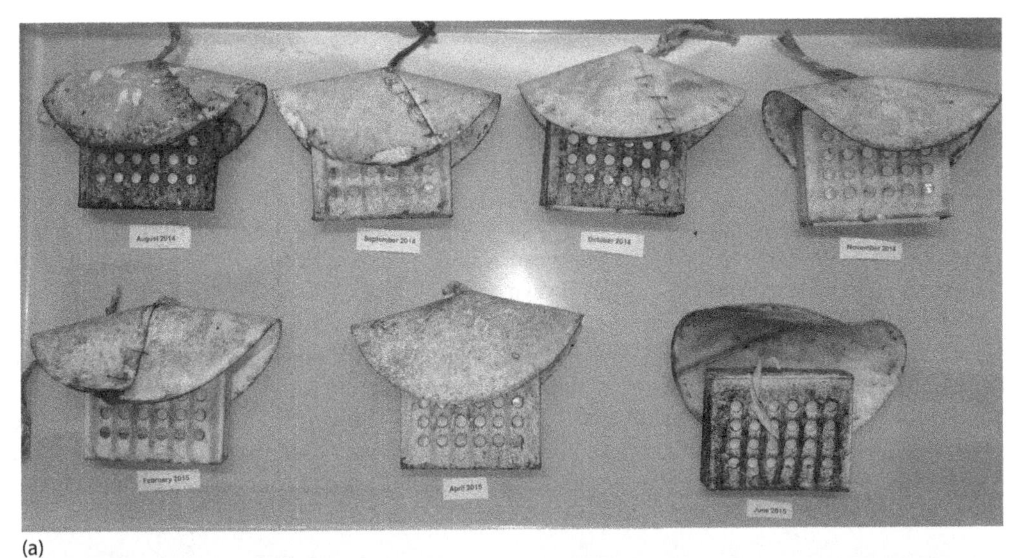

(a)

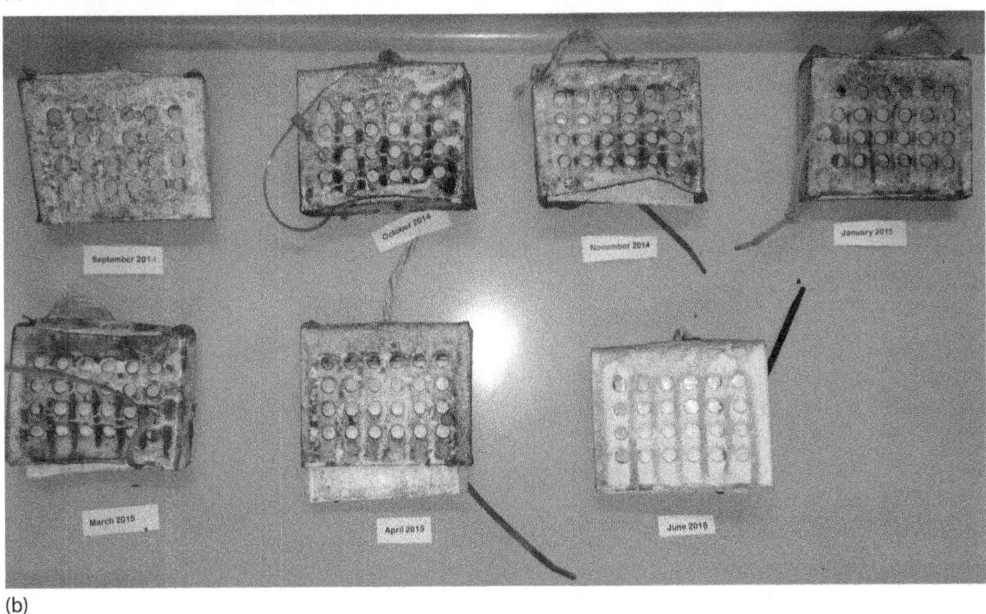

(b)

FIGURE 8.4 Photos of weathered bait stations (a) with hats and (b) aged for 8 weeks (bottom right) to 52 weeks (upper left).

to 48 hours to kill (Cisneros et al. 2002); thus, the mean mortality reading at 72 hours provides the most accurate measure of total mortality obtained from the treatments. Mexican fruit fly mortality comparisons were made by length of time bait stations were aged in the field by number of weeks; 8–9, 17–18, 25–26, 33–34, 41–42, 45–46, and 51–52 weeks. Placing bait stations in each test location required more than 1 week, which is indicated by the values spanning 2 weeks. The same procedures as listed previously for laboratory bioassays were used for conducting mortality testing. The mean ± SE values by treatment were graphed for comparisons of fly mortality using bait station, bait stations with hat, and no bait station control by the number of weeks the bait stations had aged in the field.

8.3 RESULTS

8.3.1 Field Tests

In the 2014 growing season, there were 75 feral (wild) adult Mexflies captured in traps and seven larvae found inside fruit in the south of Texas. There were no feral flies captured in traps on the bait stations side in the field in any of the test comparison sites compared to three feral flies captured at two control locations. One feral fly was captured in a trap on the control side in Lyford and two were captured in the control side in the small grove at Sol in Brownsville, Texas. There was one feral fly in a no comparison site at South Point near Brownsville (Table 8.2). Sterile fly capture was variable in the test based on site location. Significantly more sterile Mexflies were captured in the bait stations at Ranchero Way compared to Birding Center, the Brownsville abandoned orchard, and overall comparison of all fly captures in the test (Table 8.2). Similar capture results occurred at the La Feria field test, the Bayview field test, and the Lyford urban test. The control side (non-bait station side) captured significantly more Mexflies at Southern Comfort RV park compared to Country Sunshine RV park. Across the whole year, only one sterile fly was captured on the control side in the Bayview grove comparison where no sterile fly release occurred (Table 8.2).

TABLE 8.2
Wild and Sterile Mexican Fruit Fly Capture ± SE by Location

				Fly Capture					
Comparison	**Site Name**	**BS**	**Trap**	**Wild**	**SIT**	**Mean ± SE**	**DF**	**T-ratio**	**Prob > |t|**
Field test fly releases	**La Feria S**	40	4	0	4420	49.9 ± 7.4	150	−1.747	0.08
	La Feria N	0	4	0	2838	38.2 ± 6.4			
Field test no fly release	**Bayview N**	40	10	0	0	One SIT fly captured			
	Bayview S	0	10	0	1				
Abandoned grove	**Abandoned**	40	8	0	24282	111.5 ± 12.7	340	−2.781	0.006
	Sol grove	0	8	2	11597	66.3 ± 7.4			
Urban Lyford, TX	**Bus 77 E**	40	6	0	9281	75.8 ± 18.9	240	−1.082	0.28
	Bus 77 W	0	6	1	5504	55.1 ± 7.7			
RV Parks	**Country Sunshine**	50	4	0	836	8.7 ± 1.7	190	3.078	0.002
	Southern Comfort	0	4	0	3545	36.4 ± 8.6			
RV Park Bird Center	**Ranchero Way**	50	4	0	1934	18.0 ± 3.7	206	3.870	0.0001
	Estero Llano	0	4	0	368	3.4 ± 0.6			
Capture comparison across all		**Bait Station**		0	40753	61.3 ± 6.0	1130	−3.075	0.002
Sites with comparison tests		**Control**		3	23853	43.9 ± 3.4			
Test spots without comparison sites based on larvae or wild fly captures in 2013									
Site name		**BS**	**Trap**	**Wild**	**SIT flies captured**		**Mean capture ± SE**		
South Point Brownsville, TX		60	12	1	9835		32.5 ± 4.0		
Monica Brownsville, TX		40	8	0	12073		56.2 ± 6.4		
Toronja Brownsville, TX		40	10	0	10406		53.5 ± 7.7		
Nevada St. Weslaco		50	6	0	1768		17.4 ± 4.0		
Illinois St. Weslaco		25	4	0	1528		14.4 ± 3.5		
7th St. Donna		25	2	0	479		5.8 ± 1.9		

DF, degree of freedom; BS, bait station; SE, standard error; SIT, sterile insect technique.

In the noncomparison sites in the areas of Brownsville, Texas, and the Weslaco/Donna, Texas areas, only 1 feral fly was captured compared to 10 Mexfly adults and larvae in the previous citrus season. The feral fly was captured in a trap near bait stations at South Point located close to Brownsville, Texas.

8.3.2 Laboratory Bioassay of Bait Station Efficacy in the Field

Some of the bait stations were partially covered with mold from rains, and many of the older bait stations lacked much of the original waxy coating (Figure 8.4). The bait stations were effective in controlling and killing adult Mexican fruit flies across 12 weeks under field conditions. After 14 weeks, bait station mortality decreased to ~35% at 72 hours in the laboratory bioassay tests as compared to controls at ~9% (Figure 8.5).

The bait stations with hats were effective for approximately 30 weeks with mortality at or above 80% compared to control Mexfly mortality of 8%–11% (Figure 8.5). At 32 weeks, Mexfly mortality dropped to 60% on the bait station with hats and gradually decreased in each subsequent testing period up to week 52 (Figure 8.5). Mortality in the control cages ranged from 2% to 11% with a mean across time of 6.1% mortality at 72 hours in the cage.

After 8–9 weeks of exposure in the field, the tested bait stations without hats only obtained 38% mortality in the laboratory efficacy trial (Figure 8.6). The bait stations with hats did maintain efficacy having more than 60% mortality for over 41–42 weeks of exposure in the field (Figure 8.6). There was variability between bait stations received from Hidalgo County compared to Cameron County. Cameron County is closer to the Gulf of Mexico and tends to receive much more precipitation than Hidalgo County. The hat protected the bait station and maintained the waxy coating for a longer period of time, especially on the bottom (unexposed) portion of the hat.

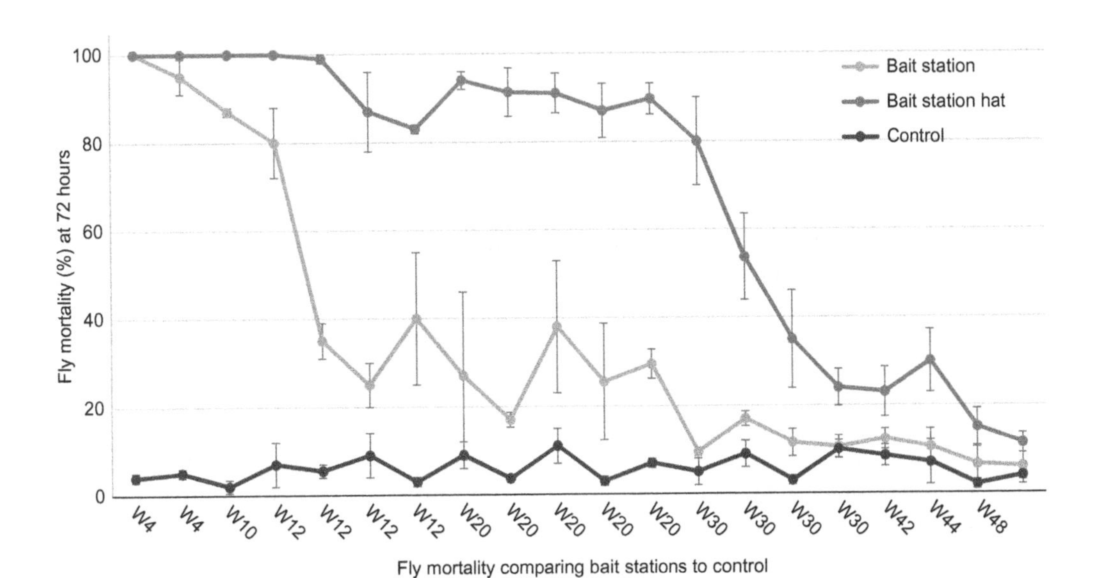

FIGURE 8.5 Mexican fruit fly percentage of mortality at 72 hours in the laboratory bioassay comparing bait stations with and without hat to control at distinct periods across the year.

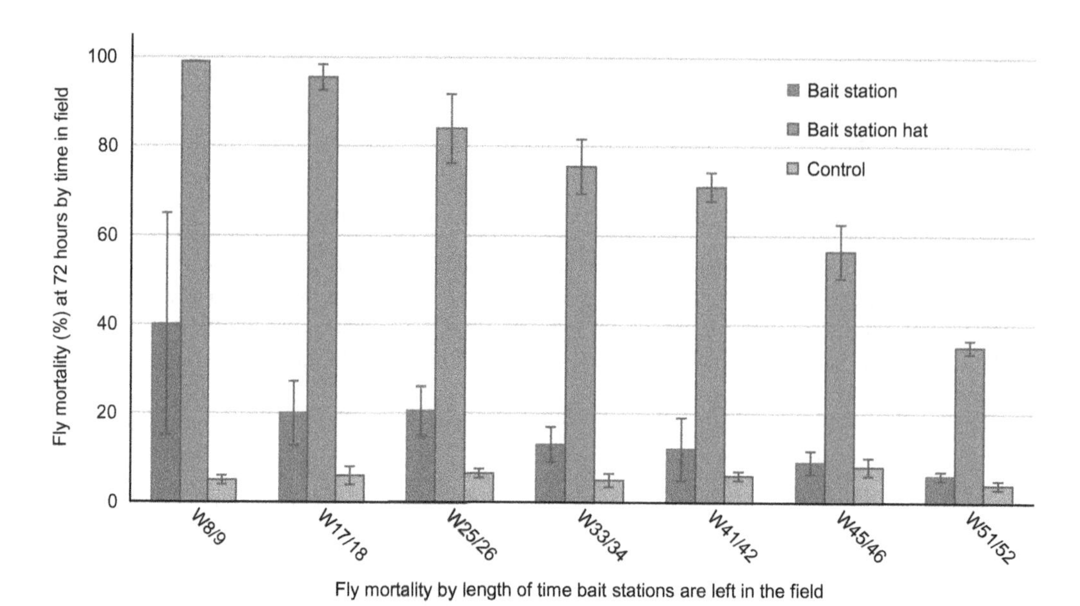

FIGURE 8.6 Mexican fruit fly percentage of mortality at 72 hours based on length of time the bait station was left in the field comparing bait stations with and without hat to control.

8.4 DISCUSSION

In the field study, more flies were captured in traps located in the sites with bait stations except for the RV Park comparison site. There were only three feral Mexflies captured in the comparison tests and all three wild flies were captured in traps on the sites without bait stations. Bait stations with protective hat coverings were significantly more effective in attracting and killing adult Mexican fruit flies in laboratory bioassay tests based on bait station samples taken from fields across time and from tested bait station samples brought in from the field by release dates.

The type of trap and most effective attractant are important factors to consider. Two means of attracting Mexican fruit flies are aqueous slurries often made of torula yeast and two component biolures (putrescine and ammonium acetate) (Conway and Forrester 2007). Thomas et al. (2001) found that open-bottomed, plastic traps baited with a two-component synthetic lure (putrescine and ammonium acetate) captured as many and sometimes more Mexflies than the standard glass McPhail trap baited with aqueous solution of torula yeast. In a citrus orchard in Mexico, more feral Mexflies were captured in traps using two-component biolures (putrescine and ammonium acetate) in Multilure traps with 10% Prestone Low Tox antifreeze than in similar type traps baited with *Anastrepha* fruit fly lure (Robacker and Thomas 2007). Thomas (2008) found that a capture solution of 10% propylene glycol based antifreeze captured significantly more flies than the more toxic automotive antifreeze.

The bait stations function as attract-and-kill stations using the two-component patch (putrescine and ammonium acetate) as the attractant. Flies landing on the surface of the bait station detect the phagostimulant with the fly's taste sensors on the feet (Reinhard 2010), activating a feeding response. The flies feed on the waxy outer covering and also ingest the insecticide Spinosad imbedded in the waxy matrix, killing the flies within 72 hours. Spinosad acts as a stomach poison (Mangan et al. 2006), requiring a period of time to kill. During the time to kill, the flies with poison in their stomach seek a source of water, which often is the capture solution in the Multilure traps. The bait stations side of the test sites have the lures inside the Multilure traps as well as 40–50 additional sets of the attractant lures containing putrescine and ammonium acetate inside the bait station compared to the 4–8 putrescine and ammonium acetate lures inside the Multilure traps in the control site. Yet, the bait station sites captured more sterile Mexflies in all but one comparison site at an RV Park. Home owners in RV parks are protective of their fruit trees and may have applied pesticides to control other insect pests, which could have negative effects on Mexfly.

Bait stations are a form of attract-and-kill device that have been effective on a number of pestiferous fruit fly species. Díaz-Fleischer et al. (2017) and Flores et al. (2017) used bait stations for control of *Anastrepha* fruit flies in mango orchards. Piñero et al. (2009) helped develop bait stations to control *Bactrocera* species. Epsky et al. (2012) used bait stations to help control Mediterranean fruit flies in Guatemala, and Rahman and Broughton (2016) used them for control of Mediterranean fruit flies in Australia.

Laboratory bioassay tests with observation cages indicate significantly higher mortality from bait stations with hats taken from the field over both control and regular bait stations. The regular bait stations provide control of Mexican fruit flies for up to 12 weeks in the field. The bait stations with hats provided additional protection, resulting in good fly mortality for up to 30 weeks. The addition of the hats increases the efficacy of the bait stations by delaying degradation of the active ingredient (Spinosad) due to protection from exposure to the sun and rain. The hat provides a larger quantity (nearly double the amount found in a standard bait station) of waxy coating with Spinosad for the flies to feed.

This study indicates that bait stations can be another valuable tool for integrated pest management (IPM) for program managers in their effort to control invasive fruit flies. Potential bait station uses would include: around wild fly finds, in specific locations with historic recurrent fly captures, in abandoned or poorly maintained groves, or with permission, in residential backyard plantings of citrus. Discussions are ongoing with Dow Chemical Company and the US Environmental Protection Agency to register commercial use of these bait stations in US fruit fly programs.

DISCLAIMER

The findings and conclusions in this preliminary publication have not been formally disseminated by the US Department of Agriculture and should not be construed to represent any agency determination or policy.

ACKNOWLEDGMENTS

This research was supported by the intramural research program of the US Department of Agriculture, APHIS, PPQ, Science and Technology, CPHST Mission Lab.

We thank the workers in Guatemala for producing the bait stations, field operations employees for placing the bait stations and servicing the traps, Laurie Morales and Heather Quintanilla for data entry of comparison sites, and Elma (Josie) Salinas, Matt Ciomperlik, and Ken Bloem for reviewing the manuscript. Thanks to Kari Pina and numerous University of Texas–Rio Grande Valley (UTRGV) cooperators for technical support.

REFERENCES

Cisneros, J., D. Goulson, L. C. Derwent, D. I. Penagos, O. Hernández, and T. Williams. 2002. Toxic effects of Spinosad on predatory insects. *Biological Control* 23: 156–163.

Conway, H. E., and O. T. Forrester. 2007. Comparison of Mexican fruit fly (Diptera: Tephritidae) capture between McPhail traps with torula and Multilure traps with Biolures in South Texas. *Florida Entomologist* 90: 579–580.

Díaz-Fleischer, F., D. Pérez-Staples, H. Cabrere-Mireles, P. Montoya, and P. Liedo. 2017. Novel insecticides and bait stations for the control of *Anastrepha* fruit flies in mango orchards. *Journal of Pest Science* 90: 865–872.

Epsky, N. D., D. Midgarden, R. Rendón, D. Villatoro, and R. R. Heath. 2012. Efficacy of wax matrix bait stations for Mediterranean fruit flies (Diptera: Tephritidae). *Journal of Economic Entomology* 105: 471–479.

Flores, S., E. Gómez, S. Campos, F. Gálvez, J. Toledo, P. Liedo, R. Pereira, and P. Montoya. 2017. Evaluation of mass trapping and bait stations to control *Anastrepha* (Diptera: Tephritidae) fruit flies in mango orchards of Chiapas, Mexico. *Florida Entomologist* 100: 358–365.

IAEA-314.D408CT11588. 2009. Development of bait stations for fruit fly suppression in support of SIT. Vienna, Austria: IAEA.

Mangan, R. L., and D. S. Moreno. 1995. Development of phloxine B and uranine bait for control of Mexican fruit fly. In *Light Activated Pest Control, ACS Symposium Series 616*, eds. J. R. Heitz and K. Downum, pp. 116–126. Washington, DC: American Chemical Society.

Mangan, R. L., and D. S. Moreno. 2007. Development of bait stations for fruit fly population suppression. *Journal of Economic Entomology* 100: 440–450.

Mangan, R. L., D. S. Moreno, and G. D. Thompson. 2006. Bait dilution, Spinosad concentration, and efficacy of GF-120 based fruit fly sprays. *Crop Protection* 25: 125–133.

Piñero, J. C., W. Enkerlin, and N. D. Epsky. 2014. Recent developments and applications of bait stations for integrated pest management of Tephritid fruit flies. In *Trapping and the Detection, Control, and Regulation of Tephritid Fruit Flies*, eds. T. Shelly, N. Epsky, E. Jang, J. Reyes-Flores, and R. I. Vargas, pp. 457–492. Dordrecht, the Netherlands: Springer.

Piñero, J. C., F. F. L. Mau, G. T. McQuate, and R. I. Vargas. 2009. Novel bait stations for attract-and-kill of pestiferous fruit flies. *Entomologia Experimentalis et Applicata* 133: 208–216.

Prokopy, R. J., N. W. Miller, J. C. Piñero, J. D. Barry, L. C. Tran, L. Oride, and R. I. Vargas. 2003. Effectiveness of GF-120 fruit fly bait spray applied to border area plants for control of melon flies (Diptera: Tephritidae). *Journal of Economic Entomology* 96: 1485–1493.

Rahman, T., and S. Broughton. 2016. Suppressing Mediterranean fruit fly (Diptera: Tephritidae) with and attract-and-kill device in pome and stone fruit orchards in Western Australia. *Crop Protection* 80: 108–117.

Reinhard, J. 2010. Taste: Invertebrates. In *Encyclopedia of Animal Behavior*, volume 3, eds. M. D. Breed, and J. Moore, pp. 379–385. Oxford, UK: Academic Press.

Robacker, D. C., and D. B. Thomas. 2007. Comparison of two synthetic food-odor lures for captures of feral Mexican fruit flies (Diptera: Tephritidae) in Mexico and implications regarding use of irradiated flies to assess lure efficacy. *Journal of Economic Entomology* 100: 1147–1152.

SAS JMP 13. 2017. *Windows Version*. Cary, NC: SAS Institute.

Thomas, D. B. 2008. A safe and effective propylene glycol based capture liquid for fruit fly (Diptera: Tephritidae) traps baited with synthetic lures. *Florida Entomologist* 91: 210–213.

Thomas, D. B., T. C. Holler, R. R. Heath, E. J. Salinas, and A. L. Moses. 2001. Trap-lure combinations for surveillance of *Anastrepha* flies (Diptera: Tephritidae). *Florida Entomologist* 84: 344–351.

9 Assessment of Modified Waste Brewery Yeast as an Attractant for Fruit Flies of Economic Importance in Mauritius

*Nausheen A. Patel**, *Sunita Facknath, and Preeaduth Sookar*

CONTENTS

Abstract Yeast and yeast products are widely considered alternative sources of protein for baits used in fruit fly suppression. The objective of this study was to develop new protein bait formulations from locally available materials in Mauritius, thereby making baits more affordable and reducing the cost of the fruit fly monitoring and control programs. Locally available waste brewery yeast (WBY) was modified in a digester. The WBY was exposed to different boiling and proteolysis conditions. A two-choice bioassay was conducted and each of the 64

* Corresponding author.

resulting baits were tested against water in a noncompetitive situation with two fruit fly species of economic importance: the peach fruit fly, *Bactrocera zonata* (Saunders), and the melon fly, *Zeugodacus cucurbitae* (Coquillett). Three baits, F1, F2, and F3, showed significantly more fly attraction ranging from 0.7 to 1.1 mean fly catches. These baits were used for further testing for optimal concentrations (7.5, 10, 12.5, and 15% v/v) in field cages. With *B. zonata*, bait attractiveness increased significantly with increasing bait concentrations for both male and female flies. With *Z. cucurbitae*, an increase in attraction was observed but attractiveness was not significantly different. Bait concentration (10% v/v) was selected for open field trials using the three preselected baits (F1, F2, and F3). Two cucurbit plantations were chosen to test the baits against commercial protein hydrolysate in Tephri Traps®, targeting *Z. cucurbitae*. A similar trial was conducted in a fruit orchard to test selected baits in attracting *B. zonata*. The results of traps baited with modified WBY at 10% v/v were comparable to commercial protein hydrolysate in attracting flies. A 5-year cost-benefit analysis indicated that a net benefit of US$283,558.60 is possible if modified WBY is used instead of imported commercial protein hydrolysate. Thus, modified WBY is a promising cost-effective alternative to the imported costly protein hydrolysate in fruit fly suppression programs for Mauritius.

9.1 INTRODUCTION

Fruit flies (Diptera: Tephritidae) are some of the most destructive and important pests of fruits and vegetables worldwide. Fruit fly problems in Mauritius date back to the beginning of this century (Sookar et al. 2006). Fruit flies can be controlled by regular insecticide applications of cover sprays; however, there are known negative side effects, including residues of insecticides in crops, health problems for farmers, contamination of water and soil, development of insecticide resistance, and a decrease in natural enemy populations (Guaman Sarango 2009).

The sterile insect technique (SIT) is one of the major corner stones for fruit fly control programs (Barnes 2008). However, SIT is not a stand-alone technique and should always be integrated with other control methods including baiting (Anon. 2018b).

Flies require sugars and proteinaceous food to survive and mature and are attracted to high-quality protein and sugar baits (Vargas et al. 2002; Bharati et al. 2004). Studies conducted by Prokopy and Roitberg (1984) on searching behavior for food, water, mating, and egg-laying have led to new methodologies by using food baits for monitoring and control of several important fruit fly species. Ammonia-releasing compounds in lures such as protein hydrolysate play an important role in fruit fly attraction (Thomas et al. 2008). The success of any suppression program relies on the ability of protein-based bait formulations to induce good levels of attraction (Mazor et al. 1987).

Protein hydrolysate is the highest-cost component of bait in Mauritius, comprising 15% of the total cost. The import cost of commercial protein amounts to US$15/L. The annual requirement is estimated to be 9000 L, to be used mainly for spraying with respect to the local fruit fly suppression program, for monitoring purposes, and for free distribution to planters and backyard owners. Moreover, because this protein bait needs to be imported, there is always a risk of shortage or delay in procurement as a result of shipping problems (Gopaul et al. 2001). Hence, there is an imperative need to develop new bait formulations, preferably from locally available materials, to bring down the cost of protein used in SIT programs and make bait more affordable. The formulation must be effective, economically feasible, and lack environmental and health hazards.

Lloyd and Drew (1997) reported that waste brewery yeast (WBY) could be developed into suitable locally produced baits because WBY is a rich source of B-complex vitamins, protein, and minerals. Extensive field trials have been conducted in Tonga using Royal Tongalure, which is a protein bait derived from brewer's yeast. Royal Tongalure bait is as effective as the expensive imported protein bait. Preliminary studies have been carried out on the modification of WBY as a protein source for the control of *Zeugodacus cucurbitae* (Coquillett) in Mauritius (Sookar et al. 2002). There is presently no protocol for developing fruit fly bait from free local WBY, which is available from the Phoenix beer factory.

This study aims to (i) develop and test various formulations using local brewery yeast, (ii) develop a protocol for producing an optimum formulation, and (iii) assess its attractancy and efficacy against *Bactrocera zonata* (Saunders) and *Zeugodacus cucurbitae.*

9.2 MATERIALS AND METHODS

The study was divided into three main parts. The first section included laboratory studies to narrow down the number of prospective baits. The second part involved using field cages to determine the most effective bait's concentration for fly capture. Field studies were then conducted using identified hosts for *Z. cucurbitae* in cucurbit fields and for *B. zonata* in an orchard setting.

9.2.1 Laboratory Bioassay

The WBY was boiled and cooled before addition of papain enzyme. Different conditions were tested by altering:

1. Boiling time (24, 48, 72, and 96 hrs) at 95°C;
2. Proteolysis time (24, 48, 72, and 96 hrs) at 60°C;
3. Papain concentration (0.1%, 0.2%, 0.3%, and 0.4% w/v).

B. zonata flies were reared under standard rearing conditions for 210 generations and *Z. cucurbitae* flies were reared for 30 generations. A two-choice bioassay was used with each of the resulting 64 baits tested independently against water in a non-competitive situation using flies of 10–16 days of age.

From eclosion, adult *B. zonata* and *Z. cucurbitae* flies were fed sugar, water, and protein. Two days prior to testing, they were deprived of the protein source. On the testing day, 20 female flies were released in a small fiberglass cage (30 cm × 30 cm × 30 cm) without sugar or water. Four replicates for each tested bait were set under control conditions of 28 ± 1°C and relative humidity varying from 70% to 75%.

Two dry square sponges (4 cm^2) were placed on top of the gauze cages in diagonally opposite corners. This step enabled flies to get acquainted with the dry sponge before introducing the test bait. At the start of each test in each test cage, 1 mL of water was applied in one sponge (control) and 1 mL of diluted bait in the second sponge.

At the start of the experiment, the impregnated sponge was inverted on the top of the cage to enable flies to have direct access to both water and the test bait. Five measurements were taken; that is, every 2 minutes the number of flies on each sponge were counted across a 10-minute duration. After 5 minutes, the test cages were rotated 180° to ensure optimal light conditions and reduce potential position and light biases.

The total number of flies attracted to each sponge during the observation time was recorded as the attraction of the bait versus the water control. Flies were used only once and then were discarded. A clean cage and new sponges were used for each individual test to reduce any potential interference from possible bait residue.

Tests were run four times daily: from 08:30 to 09:30 am, from 09:30 to 10:30 am, from 10:30 to 11:30 am, and from 11:30 am to 12:30 pm. Each bait was tested 16 times. The attraction of each bait relative to the standard was expressed as the ratio of the mean maximum number of flies attracted to the standard control.

Data were analyzed using an analysis of variance (ANOVA) in Minitab, and means were separated by Tukey's test ($P < 0.05$). Interactions among the three factors (boiling time, proteolysis time, and papain concentration) were tested using the factorial plot of the general linear models.

9.2.2 Comparison of Attraction of Different Concentrations of Modified WBY in Field Cages for *B. zonata* and *Z. cucurbitae*

Three baits, namely F1 (boiled for 72 hrs, hydrolyzed for 72 hrs with 0.2% w/v papain), F2 (boiled for 72 hrs, hydrolyzed for 72 hrs with 0.3% w/v papain), and F3 (boiled for 72 hrs, hydrolyzed for 72 hrs with 0.4% w/v papain), showing significant relative attraction ranging from 0.7 to 1.1 mean fly catches, were selected for further trials.

The experiment was conducted in field cages (2.9 m × 2.9 m × 2 m) resembling semi-natural conditions using six potted fruit trees consisting of guava, mango, and Annona species for *B. zonata* flies. For *Z. cucurbitae*, cucurbit plants (*Cucurbita pepo*) were placed in the field cages. Tephri Traps® with 300 mL of bait solution were used for trappings. There were five treatments with five replicates and each experiment was repeated independently four times with male and female flies using different concentrations of modified WBY (Table 9.1).

Traps were hung 50 cm above the plant canopy on an H-shaped base and arranged along a 1-m-radius imaginary circle around the center of the field cage. Four traps with treatments F1, F2, F3, and protein hydrolysate were placed randomly at the beginning of each experiment and then moved 90° clockwise along the circle every 2 hours to minimize position bias. The fifth trap was placed at the center as a control, where it received exposure comparable to the other traps. Once the traps were in place, 100 flies (males or females) were released in each cage. The traps were serviced after 24 hours. Flies were collected in vials at the end of the experiment and counted. The attraction of the different baits, at varying concentrations for FI, F2, and F3, was analyzed with an analysis of variance (ANOVA) in Minitab for the determination of the differences between the means, and when significant differences were observed, a Tukey's test was used for mean separation. A P value of 0.05 or less was considered for statistical significance level. Analysis of treatment values across rows and down columns was done using a one-way ANOVA.

9.2.3 Comparison of Fly Attraction of Selected Baits in Tephri Traps® in Cucurbit Fields Targeting *Z. cucurbitae*

The experiment was carried out in two local regions, namely Saint Pierre and Albion. Saint Pierre is situated at 20.22° S latitude and 57.52° E longitude, with minimum and maximum temperatures of 10°C–31°C, respectively. Albion is situated at 20.21° S latitude and 57.4° E longitude, with minimum and maximum temperatures of 12°C–35°C, respectively (Anon. 2017a).

TABLE 9.1
Comparison of Different Concentrations of Selected Baits in Field Cage Trials

SN	Treatments	Amount of Protein Bait Used (v/v)			
1	Control	Water only			
2	FI	7.5%	10%	12.5%	15%
3	F2	7.5%	10%	12.5%	15%
4	F3	7.5%	10%	12.5%	15%
5	Protein Hydrolysate	2%	2%	2%	2%

TABLE 9.2
Concentration of Baits Used as Treatments in St Pierre, Albion, and Labourdonnais

SN	Treatments	Amount of Bait Used (v/v)	Borax (w/v)
1	Control	Water only	2%
2	Protein Hydrolysate	2%	
3	F1	10.0%	
4	F2	10.0%	
5	F3	10.0%	

The trial in St. Pierre started on February 16, 2017, and were conducted in a pumpkin plantation when crops were in the fruiting stage. The second trial started on May 17, 2017, and was conducted in Albion in a bitter gourd plantation in the fruiting stage. The five treatments are shown in Table 9.2.

Borax (Sodium Tetraborate) purchased from Loba Chemie PVT. Ltd (India) was added to all the treatments (food baits) at 2% w/v to help prevent disintegration of the fruit flies (Sookar et al. 2002). Tephri Traps® with 300 mL of bait solution were placed at 10-m intervals and hung just above the crop canopy using bamboo stands in a randomized complete block design. Traps were serviced weekly with trap washing and addition of fresh solution, and collected insects were placed in vials with ethanol (70%) for subsequent identification in the laboratory. Trials were conducted across a 5-week period with traps rotated sequentially within a block at each service.

Trapping data recorded from the five treatments (Table 9.2) were converted to flies per trap per day and then log transformed to meet the assumption of homogeneity of variance. The transformed data from the Randomized block design (RBD) experiment were subjected to an analysis of variance (ANOVA) using Minitab and a Tukey's means separation test was used if F values were significant ($P < 0.05$).

9.2.4 Comparison of Fly Attraction of Selected Baits in Tephri Traps® in a Fruit Orchard Targeting *B. zonata*

The experiment was conducted in a 45-hectare area located at Mapou in the Labourdonnais Orchard, having an annual fruit production of around 500 tons/year (Anon. 2018a). Mapou is situated at 20.07° S latitude and 57.62° E longitude, with Minimum and maximum temperatures of 24°C–30°C, respectively (Anon. 2017b).

Trials were conducted in a guava plantation in the fruiting stage. The trial started in August 2017 and lasted 5 weeks. The same treatments shown in Table 9.2 were used. The same methods and procedures for locating traps, setting traps, servicing traps, identifying trap catches, and analyzing data used in the previous test with *Z. cucurbitae* were used in this test.

9.2.5 Cost–Benefit Analysis

A cost–benefit analysis template provides a simple tool for calculating financial futures. A free Excel online version was used for the purpose of this study. This program facilitated financial forecasting over a 5-year period using an estimated yearly increase rate of 2%. All the costs associated with the production of the novel bait were inserted as costs versus the benefit of not importing commercial protein hydrolysate.

9.3 RESULTS

9.3.1 Laboratory Bioassay

A total of 64 baits were obtained following the treatments applied to the crude WBY (i.e., after boiling and proteolysis). The laboratory bioassay on *Z. cucurbitae* indicated that increasing proteolysis time increased fly attraction to the modified baits (F statistic (F) = 1.60, degrees of freedom (df) = 15, p-value (P) = 0.242). Additionally, increasing the papain concentration significantly increased fly attraction (F = 3.62, df = 15, P = 0.046). Increasing boiling time from 24 to 96 hrs significantly increased fly attraction (F = 6.01, df = 15, P = 0.01) (Figures 9.1 through 9.4).

Time of test influenced the results, with fly attraction being significantly higher when the baits were tested between 08:30 and 09:30 am (F = 6.82, df = 15, P = 0.006), compared to other time periods (Figure 9.5).

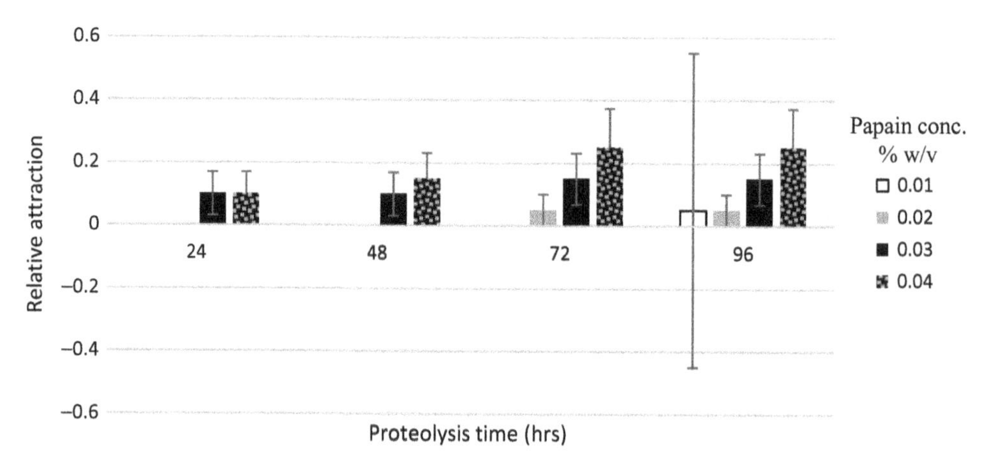

FIGURE 9.1 *Zeugodacus cucurbitae* relative attraction to baits (mean ± SE) boiled for 24 hrs, modified with papain enzyme powder at different proteolysis times.

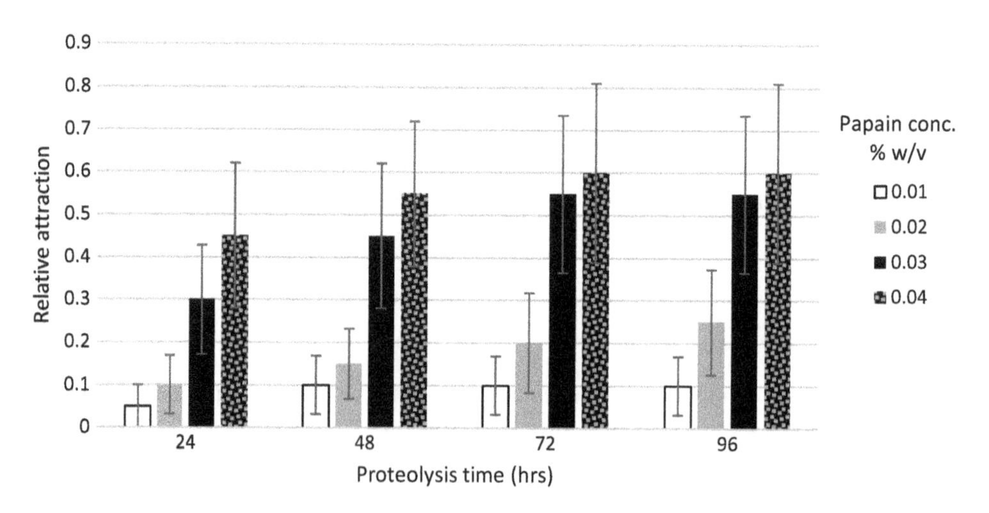

FIGURE 9.2 *Zeugodacus cucurbitae* relative attraction to baits (mean ± SE) boiled for 48 hrs, modified with papain enzyme powder at different proteolysis times.

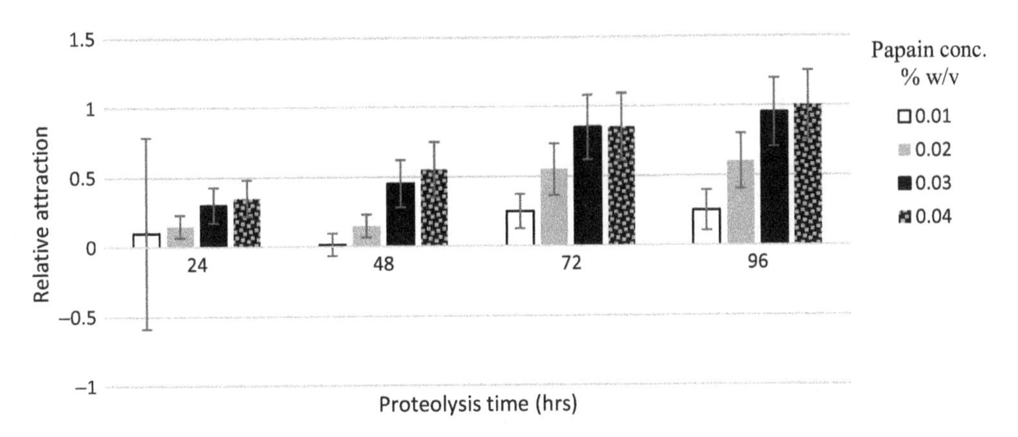

FIGURE 9.3 *Zeugodacus cucurbitae* relative attraction to baits (mean ± SE) boiled at 72 hrs, modified with papain enzyme powder at different proteolysis times.

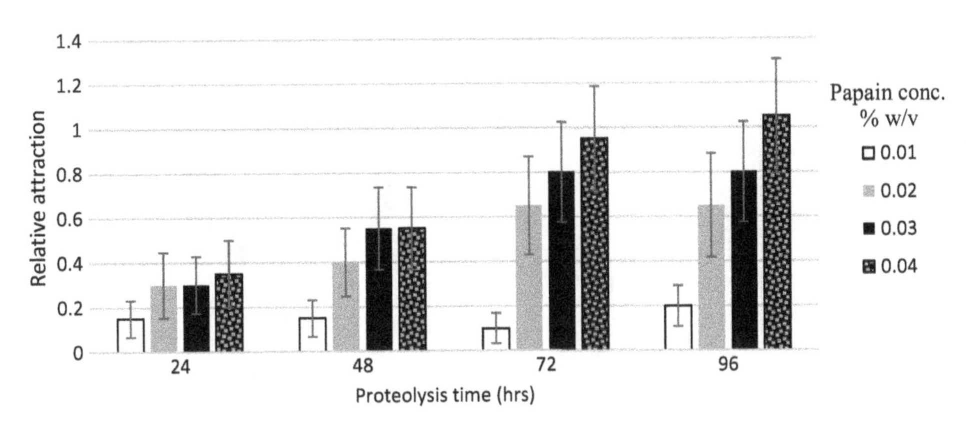

FIGURE 9.4 *Zeugodacus cucurbitae* relative attraction to baits (mean ± SE) boiled at 96 hrs, modified with papain enzyme powder at different proteolysis times.

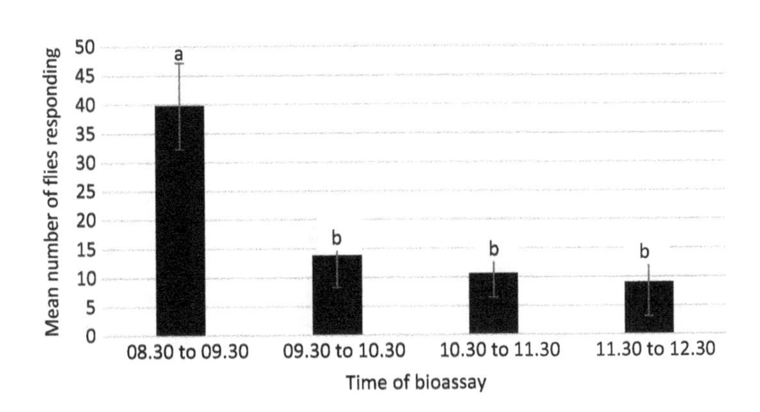

FIGURE 9.5 Number of *Zeugodacus cucurbitae* flies responding (Mean ± SE) by collection time. Means followed by the same letter were not significantly different at $P > 0.05$ level.

Further analyses to determine the effect of the interactions revealed the following: (i) boiling time and proteolysis time was significant (F = 2.46, df = 9, P = 0.023), (ii) boiling time and papain concentration was significant (F = 2.61, df = 9, P = 0.016), and (iii) proteolysis time and papain concentration was not significant (F = 1.00, df = 9, P = 0.445).

In the case of *B. zonata*, increased proteolysis time had a positive effect on fly attraction (F = 0.98, df = 15, P = 0.288). Increasing the concentration of papain during proteolysis increased fly attraction (F = 1.4, df = 15, P = 0.433). Fly attraction increased significantly with increasing boiling time of waste brewery yeast from 24 to 96 hrs (F = 7.93, df = 15, P = 0.004) (Figures 9.6 through 9.9).

Fly attraction was significantly higher when tested between 08:30 and 09:30 am and between 09:30 and 10:30 am, compared to other time periods (F = 6.67, df = 15, P = 0.007) (Figure 9.10).

An interaction effect occurs when the effect of one variable depends on the value of another variable, and the data analysis revealed the following interaction effects: (i) boiling time and proteolysis time was significant (F = 2.60, df = 9, P = 0.017), (ii) boiling time and papain concentration was significant (F = 2.31, df = 9, P = 0.031), and (iii) proteolysis time and papain concentration was not significant (F = 1.34, df = 9, P = 0.245).

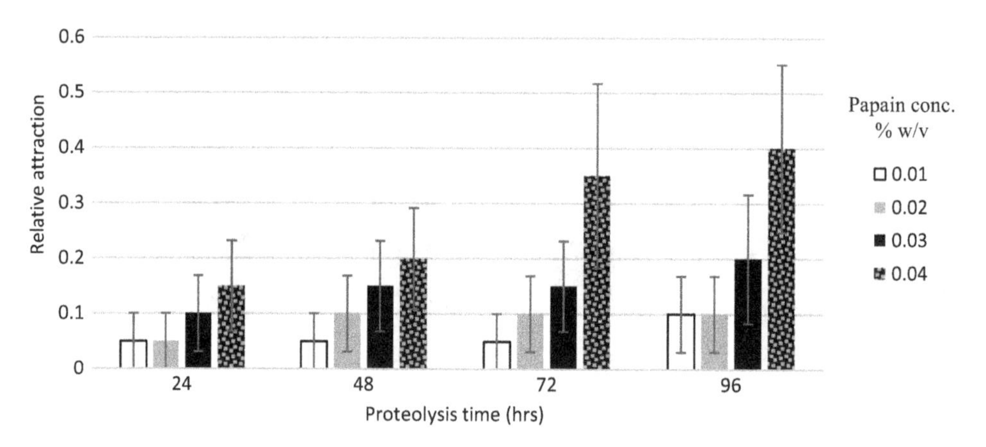

FIGURE 9.6 *Bactrocera zonata* relative attraction to baits (mean ± SE) boiled at 24 hrs, modified with papain enzyme powder at different proteolysis times.

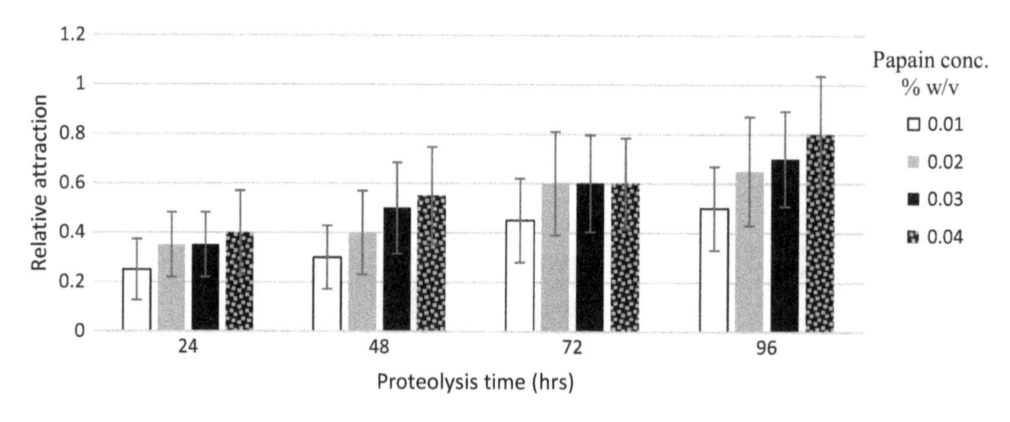

FIGURE 9.7 *Bactrocera zonata* relative attraction to baits (mean ± SE) boiled at 48 hrs, modified with papain enzyme powder at different proteolysis times.

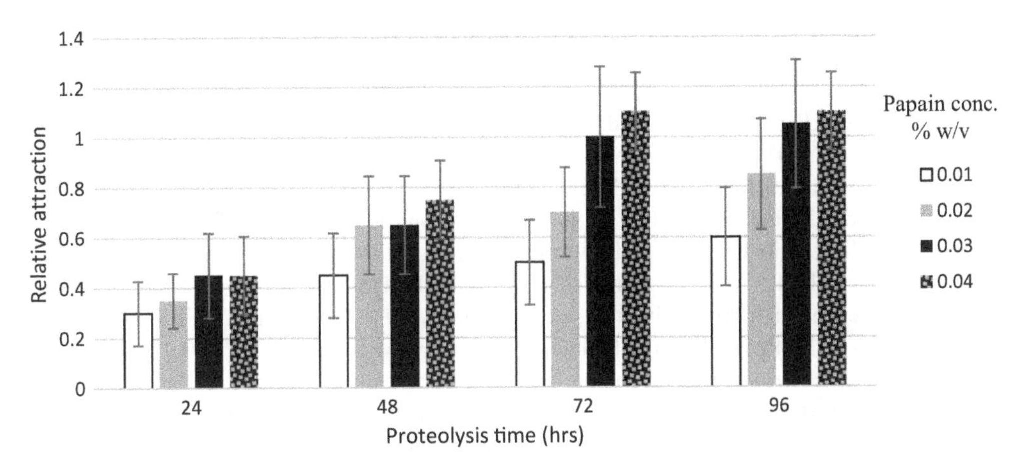

FIGURE 9.8 *Bactrocera zonata* relative attraction to baits (mean ± SE) boiled at 72 hrs, modified with papain enzyme powder at different proteolysis times.

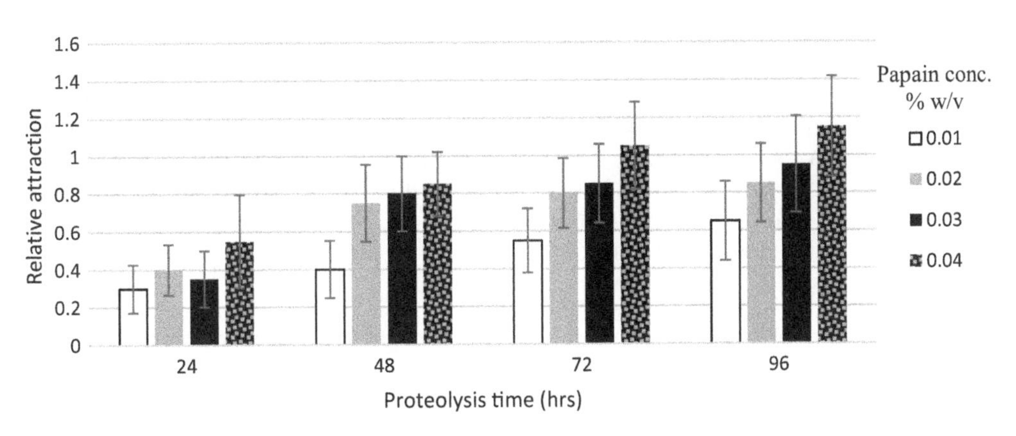

FIGURE 9.9 *Bactrocera zonata* relative attraction to baits (mean ± SE) boiled at 96 hrs, modified with papain enzyme powder at different proteolysis times.

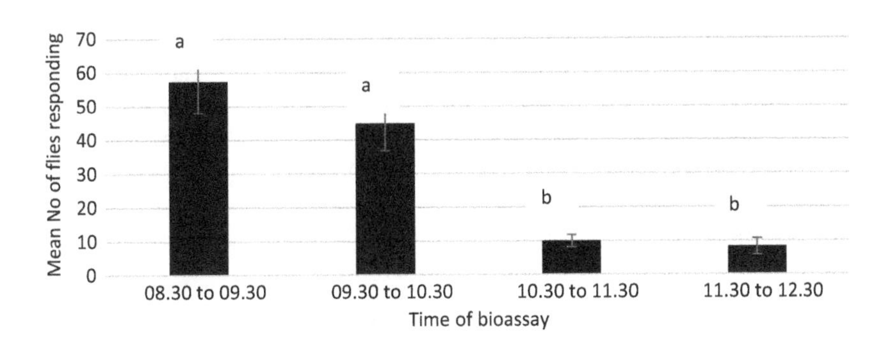

FIGURE 9.10 Number of *B. zonata* flies responding (Mean ± SE) by collection time. Means followed by the same letter were not significantly different at $P > 0.05$ level.

9.3.2 Comparison of Attraction of Different Concentrations of Modified WBY in Field Cages for *B. zonata* and *Z. cucurbitae*

The mean number of *B. zonata* (male and female) collected in Tephri Traps® increased with increasing concentration of WBY from 7.5% v/v to 15% v/v. Trap catches from the different bait concentrations were significantly different for both female flies (F = 7.08, df = 11, P = 0.012) (Table 9.3) and male flies (F = 6.65, df = 11, P = 0.014) (Table 9.4). There was no significant difference in attractiveness among the three baits F1, F2, and F3 when they were compared to each other for both female flies (F = 1.52, df = 11, P = 0.269) and male flies (F = 1.60, df = 11, P = 0.254). There was also no significant difference when protein hydrolysate was compared to the modified WBY for both male flies (F = 1.82, df = 15, P = 0.197) and female fruit flies (F = 1.52, df = 15, P = 0.259).

In the case of *Z. cucurbitae*, the mean number of male and female flies collected in Tephri Traps® increased with increasing concentration of WBY from 7.5% v/v to 15% v/v, as shown in Tables 9.5 and 9.6. Trap catches for different bait concentrations were not significantly different for either female (F = 2.24, df = 11, P = 0.161) or male flies (F = 2.59, df = 11, P = 0.126).

TABLE 9.3
Fly Attraction of Bait Treatments at Different Concentrations for Adult *Bactrocera zonata*

	No. of Female Flies Captured/Trap/Day				
WBY	**7.5% v/v**	**10.0% v/v**	**12.5% v/v**	**15.0% v/v**	**Pooled Mean**
Water	0.1 + 0.1 cA	0.2 + 0.1 cA	0.4 + 0.1 bA	0.4 + 0.13c A	0.275 b
Protein (20% v/v)	13.4 + 1.1 aA	12.7 + 0.9 aA	13.9 + 1.2 aA	11.8 + 1.1b A	13.0 a
WBY F1	5.7 + 0.8 bA	7.9 + 0.5 bA	11.4 + 1.4 aA	12.0 + 1.3 bA	9.3 a
WBY F2	7.4 + 0.6 bB	10.6 + 0.7 abAB	13.1 + 1.6 aAB	15.5 + 1.1 abA	11.7 a
WBY F3	8.8 + 1.0 bB	11.7 + 0.6 aAB	15.9 + 1.8 aAB	18.5 + 1.0 aA	13.7 a
Mean (WBY)	7.3 b	10.1 a, b	13.5 a	15.3 a	

In a column or row, means followed by the same letter (capital letters within a row and small letters within a column) are not significantly different at *P* >0.05 level.
WBY, waste brewery yeast.

TABLE 9.4
Fly Attraction of Bait Treatments at Different Concentrations for Adult *Bacterocera zonata*

	No. of Male Flies Captured/Trap/Day				
WBY	**7.5 % v/v**	**10.0 % v/v**	**12.5 % v/v**	**15.0 % v/v**	**Pooled Mean**
Water	0.1 + 0.7 d A	0.1 + 0.1 cA	0.2 + 0.1 bA	0.2 + 0.09 cA	0.15 b
Protein (20% v/v)	8.9 + 0.6aA	9.5 + 0.9 aA	9.1 + 0.9 aA	9.9 + 0.8 abA	9.35 a
WBY F1	3.9 + 0.4 c B	4.9 + 0.4bB	7.8 + 0.8 aA	8.7 + 0.8bA	6.3 a
WBY F2	4.9 + 0.4 bcC	5.9 + 0.5bBC	8.5 + 0.8 aAB	10.4 + 1.1abA	7.4 a
WBY F3	5.8 + 0.6bC	8.2 + 0.6 aBC	11.9 + 1.1 aAB	13.0 + 1.2 aA	9.73 a
Mean (WBY)	4.9 b	6.3 a, b	9.4 a, b	10.7 a	

In a column or row, means followed by the same letter (capital letters within a row and small letter within a column) are not significantly different at *P* >0.05 level.
WBY, waste brewery yeast.

TABLE 9.5
Fly Attraction of Bait Treatments at Different Concentrations for Adult *Zeugodacus cucurbitae*

	No. of Male Flies Captured/Trap/Day				
WBY	**7.5% v/v**	**10.0% v/v**	**12.5% v/v**	**15.0% v/v**	**Pooled Mean**
Water	0.35 + 0.2 dA	0.35 + 0.2 eA	0.3 + 0.1 cA	0.35 + 0.2 cA	0.338 c
Protein (20% v/v)	15.2 + 0.6 cA	15.65 + 0.7 aA	14.85 + 1.0 aA	16.0 + 1.3 aA	15.43 a
WBY F1	4.8 + 0.4cB	5.3 + 0.4 dB	6.4 + 0.5 bB	9.3 + 0.9 bA	6.45 b
WBY F2	6.75 + 0.7b cC	7.9 + 0.5 cC	11.2 + 0.7 aB	14.15 + 0.4 aA	10.0 ab
WBY F3	8.5 + 0.6 bC	10.5 + 0.6 bBC	14.85 + 0.8 aAB	17.7 + 1.2 aA	12.89 a
Mean (WBY)	6.683 a	7.9 a	10.817 a	13.72 a	

In a column or row, means followed by the same letter (capital letters within a row and small letter within a column) are not significantly different at $P > 0.05$ level.
WBY, waste brewery yeast.

TABLE 9.6
Fly Attraction of Bait Treatments at Different Concentrations for Adult *Zeugodacus cucurbitae*

	No. of Female Flies Captured/Trap/Day				
WBY	**7.5% v/v**	**10.0% v/v**	**12.5% v/v**	**15.0% v/v**	**Pooled Mean**
Water	0.4 + 0.1 dA	1.25 + 0.3 dA	0.55 + 0.2 cA	0.5 + 0.2 c A	0.68 d
Protein (20% v/v)	21.8 + 1.3aA	19.05 + 1.3 aA	19.85 + 0.9 a A	22.6 + 0.7 a A	16.40 a
WBY F1	5.95 + 0.7 cA	6.75 + 0.8 cA	9.1 + 1.1bA	10.35 + 1.0 bA	8.04 c
WBY F2	8.3 + 0.9bcB	10.9 + 1.0 bc AB	17.1 + 1.3 aA	18.25 + 1.5 aA	13.64 bc
WBY F3	10.6 + 1.0 bC	14.45 + 1.0 abBC	18.95 + 1.3 aAB	21.6 + 1.6 aA	16.40 ab
Mean (WBY)	8.28 a	10.7 a	15.05 a	16.73 a	

In a column or row, means followed by the same letter (capital letters within row and small letter within a column) are not significantly different at $P > 0.05$ level.
WBY, waste brewery yeast.

When protein baits were compared to each other, there was a significant difference in attractiveness for both female flies (F = 8.54, df = 15, $P = 0.003$) and male flies (F = 7.26, df = 15, $P = 0.005$).

9.3.3 Comparison of Attraction of Selected Baits in Tephri Traps® in Cucurbit Fields (St. Pierre) Targeting *Z. cucurbitae*

The mean number ± standard error (SE) of melon flies (male and female) trapped per day by each bait treatment is shown in Table 9.7. An analysis of variance was done on transformed data using log (x + 1). A significantly higher number of melon flies were caught in the protein baits compared to the water control (F = 7.50, df = 29, $P = 0.0001$). There was no significant difference (F = 0.39, df = 23, $P = 0.765$) between the tested protein hydrolysate and modified protein baits F1, F2, and F3. Similar results were obtained when the analysis was done for captured females only (F = 8.29, df = 29, $P = 0.000$). Results were not significant when the analysis was done for males only (F = 2.43, df = 29, $P = 0.074$).

The sex ratio was significantly different (F = 8.99, df = 9, $P = 0.017$), with a higher number of female melon fly catches compared to male melon fly catches.

TABLE 9.7
Mean Number ± SE of Melon Flies Collected per Day in St. Pierre Using Different Attractants

Catch	Protein	Water	F1	F2	F3	Mean
Females	0.41 + 0.1 aA	0.04 + 0.03 aA	0.51 + 0.2 aA	0.60 + 0.2 aA	0.68 + 0.3 aA	0.45 a
Males	0.05 + 0.01 bA	0.01 + 0.01 aA	0.11 + 0.04 aA	0.13 + 0.05 bA	0.19 + 0.1 aA	0.10 b
Mean	0.23 a	0.025 b	0.31 a	0.37 a	0.44 a	

In a column or row, means followed by the same letter (capital letters within a row and small letter within a column) are not significantly different at $P > 0.05$ level.
SE, standard error.

9.3.4 Comparison of Fly Attraction of Selected Baits in Tephri Traps® in Cucurbit Fields (Albion) Targeting *Z. cucurbitae*

The mean number of melon flies (male and female) trapped per day is shown in Table 9.8. Significantly higher catches of melon flies were noted for the protein baits when compared to water (F = 23.79, df = 24, $P = 0.001$). Similar results were observed when captured males (F = 15.32, df = 24, $P = 0.001$) and captured females were analyzed separately (F = 21.93, df = 24, $P = 0.0001$). However, the ratio of male-to-female catches was not significantly different (F = 1.43, df = 9, $P = 0.266$).

TABLE 9.8
Mean Number ± SE of Melon Flies Collected per Day in Albion Using Different Attractants

Catch	Protein	Water	F1	F2	F3	Mean
Females	1.15 + 0.2 aA	0.06 + 0.04 aB	0.70 + 0.2 aAB	0.90 + 0.2 aA	0.97 + 0.2 a A	1.16 a
Males	1.55 + 0.4 aA	0.08 + 0.04 aB	1.18 + 0.3 aAB	1.44 + 0.3aA	1.54 + 0.3aA	0.76 a
Mean	1.35 a	0.07 b	0.94 ab	1.17 ab	1.26 ab	

In a column or row, means followed by the same letter (capital letters within a row and small letter within a column) are not significantly different at $P > 0.05$ level.
SE, standard error.

9.3.5 Comparison of Fly Attraction of Selected Baits in Tephri Traps® in a Fruit Orchard Targeting *B. zonata*

Total trap catches for the three selected baits (Fl, F2, and F3) were not significantly different (F = 0.35, df = 14, $P = 0.709$). Trap catches in protein hydrolysate were significantly lower (F = 29.97, df = 24, $P = 0.0001$) compared to modified WBY (Table 9.9).

TABLE 9.9
Mean Capture ± SE of Peach Fruit Flies per Day in Labourdonnais Using Different Attractants

Catch	Protein	Water	F1	F2	F3	Total
Females	0.19 + 0.02 aA	0 aB	0.39 + 0.1 aA	0.26 + 0.00 aA	0.25 + 0.02 a A	1.09 a
Males	0.17 + 0.03 aB	0 aC	0.19 + 0.02 aB	0.34 + 0.04 aAB	0.39 + 0.02 a A	1.09 a
Total	0.36 a	0	0.58 a	0.60 a	0.64 a	

In a column or row, means followed by the same letter (capital letters within a row and small letter within a column) are not significantly different at $P > 0.05$ level.
SE, standard error.

The ratio of male-to-female trap catches was not significantly different between protein hydrolysate ($F = 0.29$, $df = 9$, $P = 0.608$) and bait F2 ($F = 3.09$, $df = 9$, $P = 0.117$), whereas trap catches were significantly different between F1 ($F = 5.62$, $df = 9$, $P = 0.045$) and F3 ($F = 22.53$, $df = 9$, $P = 0.01$).

9.3.6 Cost–Benefit Analysis

Total benefits over 5 years were US$ 811,010.62, and the total costs amounted to US$ 527,452.02. The resulting net benefits were US$ 283,558.60 (Table 9.10) when using modified WBY instead of imported protein hydrolysate.

9.4 DISCUSSION

Ekesi et al. (2016) stated that WBY has promising chemical properties. However, to be effective, there is a need to release the ammonium compound and its derivatives efficiently, which serve as volatile cues to locate protein rich food (Piñero et al. 2017). In the laboratory bioassay, boiling of the WBY served to remove as much alcohol as possible (Ekesi et al. 2016), increasing solid content (Lloyd and Drew 1997) and causing lysis of the yeast cell wall to release protein compounds that are very attractive to tephritids (Vargas and Prokopy 2006). Papain, known to digest most protein substrates (Anon. 2015), was added to accelerate this process and an increase in bait attraction was observed by increasing papain concentration from 0.1% to 0.4% w/v. The higher the concentration of the enzyme, the higher the proteolysis of the yeast cells; thus, higher amounts of free amino acids were available in the protein bait. A direct correlation between dry matter content of a protein bait, which is the result of proteolysis of the yeast cell content, and relative attractiveness has been reported by Sookar et al. in 2003 (Aggrey-Korsay 2014). Results of the laboratory bioassay demonstrated that combining boiling for more than 72 hrs with proteolysis for more than 72 hrs did not have a significant effect. Therefore, it is not required to boil for 96 hrs followed by proteolysis for 96 hrs.

The response of the flies to the time at which the tests were carried out was significantly different for both species. This agrees with a study carried out by Prokopy and Roitberg (1984) on the foraging behavior and daily activity of different fruit fly species which varied based on time of the day.

TABLE 9.10
5-Year Cost–Benefit Analysis Using an Estimated Increase Rate of 2%

	Current Year (CY) ($)	CY + 1 ($)	CY + 2 ($)	CY + 3 ($)	CY + 4 ($)	CY + 5 ($)	
Costs							
Cost of Digestor	73, 529.00						
Electrical Cost	1, 233.00	1, 257.66	1, 282.81	1, 308.47	1, 334.64	1, 361.33	
Water Cost	2, 136.00	2, 178.72	2, 222.29	2, 226.74	2, 312.08	2, 358.32	
Cost of Papain	4, 764.00	4, 859.28	4, 956.47	5, 055.59	5, 156.71	5, 259.84	
Cost of Potassium Sorbate	36, 529.00	37,259.58	38, 004.77	38, 764.87	39, 540.16	40, 330.97	
Cost of Diesel	6, 195.00	6, 318.90	6, 445.28	6, 574.18	6, 705.67	6,839.78	
Labor	11, 847.00	12, 083.94	12,325.62	12, 572.13	12, 823.57	13, 080.05	
Transport Cost	466.00	475.32	484.83	494.52	504.41	514.50	
Training Cost	150.00	153.00	156.06	159.18	162.36	165.61	
Infrastructure	73, 529.00						
Total Costs (Future Value)	210, 378.00	64, 586.40	65,878.13	67, 195.69	68, 539.60	69, 910.40	
Total Costs (Present Value)	210, 378.00	63, 320.00	63, 344.35	63, 392.16	63, 462.60	63, 554.91	527, 452.02
Benefits	135, 000.00	137,700.00	140, 454.00	143, 263.08	146, 128.34	149, 050.91	
No Importation of Protein Hydrolysate							
Total Benefits (Future Value)	135, 000.00	137,700.00	140, 454.00	143, 263.08	146, 128.34	149, 050.91	
Total Benefits (Present Value)	135, 000.00	135,000.00	135, 051.92	135, 153.85	135, 304.02	135, 500.83	811,010.62
	Cost–Benefit Analysis (US dollars)						
	Total PV Benefits		$811,010.62				
	Total PV Costs		$527,452.02				
	NET BENEFIT		$ 283,558.60				

Sookar et al. (2002) reported that all their tested baits caught all flies before 11:30 am. The results obtained in this study are consistent with previous findings, as fly attraction for *Z. cucurbitae* was most significant between 08:30 and 09:30 am, compared to *B. zonata* flies, which were most responsive between 08:30 and 10:30 am. Thus, food baited catches reflect the feeding activity of fruit flies (Bharati et al. 2004). This information is important for a more effective planning of fruit fly spraying programs using attract and kill methods targeted toward specific fly species.

To be effective at suppressing fruit fly populations, protein-based bait formulations must induce good levels of attraction. Ammonia is the principal component of the protein bait that attracts fruit flies, mainly females (Mazor et al. 1987; Piñero et al. 2015). Attraction of baits increases with increased concentration or amount of ammonia emitted (Bateman and Morton 1981). The field cage trial with *B. zonata* showed a significantly increased fly attraction with increased bait concentration of the three selected baits (F1, F2, and F3) from 7.5% to 15% v/v. Field cage trials with *Z. cucurbitae* showed that increasing bait concentration did not significantly increase fly attraction. This may be due to the fact that different fly species vary in their response to a particular stimulus (Kotikal and Math 2017). For both species, the number of female fruit flies caught in the protein baits was much higher than the number of male fruit flies. The main purpose of liquid protein baits is to capture female fruit flies. The bait targets the female fruit fly's need for protein for the development and maturation of eggs. With protein attractants, recently emerged female fruit flies enter the trap, get caught, and eventually die by drowning in the capture fluid (Anon. 2016).

Protein sources are an important component of food baits and commercial lures for *Z. cucurbitae* (Steiner 1952; Narayanan and Batra 1960; Vijaysegaran 1985; Satpathy and Rai 2002; Fabre et al. 2003 cited in Nagaraj et al. 2014). In trials conducted in open vegetable fields, protein baits were highly attractive for capturing both male and female *Z. cucurbitae* flies. Results indicated that, for melon flies, there was no significant difference in attractiveness between the selected baits and the commercial protein hydrolysate, which indicates an acceptable quality of the prepared baits. Comparable results were obtained with *Bactrocera invadens* (now *B. dorsalis*) in Nigeria using modified brewery waste and commercial torula (Umeh and Onukwu 2010).

Female fruit flies require a protein source to mature sexually and to develop eggs (Bateman 1972). This is supported by Kotikal and Math (2017), who concluded that baits offer one of the most effective methods of control, especially in the pre-oviposition stage when fruit flies require plenty of water to drink and are easily attracted to protein sources. This helps to explain the significantly higher mean trap catches of female melon flies compared to male melon flies in St. Pierre. In Albion, the ratio of male-to-female captures was not significantly different. One possible reason could be the higher prevailing temperatures (Anon. 2017a), which would increase the attraction of males to a wet trap (Barry et al. 2006).

Climate plays a critical role either directly or indirectly as a control for Tephritids. The results of the open field trials conducted in the north of Mauritius (Labourdonnais orchard) showed promising results on the attraction of *B. zonata* to the low-cost prepared bait. Trap catches from the modified WBY were significantly higher compared to the trap catches from the commercial protein hydrolysate, indicating the quality of the prepared baits. Trap catches of male and female *B. zonata* were not significantly different. The climatic conditions prevailing in the region, which has a hot climate (Anon. 2017b), may explain similar male and female captures. A significant positive correlation has been reported between trap catches and maximum temperature in studies carried out by Gajalakshmi et al. (2011) and Boopathi et al. (2013). Flies were attracted to a wet trap irrespective of whether they were male or female.

The results of the field trial targeting the two fruit fly species shows a marked attraction of *Z. cucurbitae* compared to *B. zonata*. Piñero et al. (2017) documented an inherent stronger response to protein baits by *Z. cucurbitae* compared to *B. dorsalis*. These findings are supported by studies done by Vargas and Prokopy (2006) on attraction and feeding propensity of different fly species relative to different protein baits.

9.5 CONCLUSION

An effective protocol has been developed and tested for the modification of WBY. The study determined the most effective boiling time (72 hrs at 95°C), proteolysis time (72 hrs at 60°C), and concentration of papain (0.4% w/v) required to prepare an effective protein bait. The protocol has been evaluated in both field cage and open field trials against both *B. zonata* and *Z. cucurbitae*.

An important finding from the study was determining the time (early morning) at which the flies are most active. The overall results are encouraging and indicate that the prepared WBY bait may prove to be a useful monitoring tool for both male and female flies of the two species. One drawback of the modified WBY is its non-specificity, which causes non-target flies to be attracted to the bait, resulting in large fly captures. However, this non-specific attractiveness can aid in general surveillance programs to detect other pest species of economic importance.

The cost–benefit analysis supports the feasibility of the project by supporting the objective of this study in developing a cost-effective replacement for the expensive protein hydrolysate imported from Europe or North America. The developed WBY bait is financially promising for smaller growers. The WBY bait will also enhance the sustainability of fruit fly monitoring and fruit fly management programs by making them more economically feasible. The horticulture industry as a whole is assumed to receive the benefits from avoided production loss and gains in export value from improved international market access (Abdalla et al. 2012).

Further research is required on the amount of bait solution needed per trap and the effective servicing intervals when using WBY as a monitoring tool. During hot summer days, the baits dry up faster and may have a negative effect on fruit fly monitoring results. Assessment of the shelf life of the prepared bait requires investigation for production, especially if switching to a large-scale basis.

ACKNOWLEDGMENTS

The authors would like to acknowledge the International Atomic Energy Agency (IAEA) for the financial support for this work. We would like to thank the Ministry of Agro Industry and Food Security (Entomology Division) for providing the required facilities and resources to carry out the study, and Mr. A. Ruggoo, lecturer at the University of Mauritius, for his guidance in the statistical analysis and interpretation of the data. We are also grateful to the Officer in Charge of Belle Vue Experimental Station, Mr. Wiehe, Manager of Labourdonnais orchard, and the farmers who willingly collaborated with us to facilitate our field trials. Lastly but not least, our extended thanks to the Laboratory attendants of the Entomology division for their devotion in assisting us.

REFERENCES

Anon. 2015. Papain from papaya latex. Sigma Aldrich. https://www.sigmaaldrich.com/catalog/substance/papainfrompapayalatex12345900173411?lang=en®ion=MU. Accessed July 9, 2017.

Anon. 2016. Baiting. Plant Health Australia. http://preventfruitfly.com.au Accessed July 7, 2017.

Anon. 2017a. Acuweather, Albion, Mauritius. https://www.accuweather.com/en/mu/albion/1122855/daily-weather-forecast/1122855?day=1. Accessed July 10, 2017.

Anon. 2017b. Mapou current weather report. Riviere Du Rempart, Mauritius. https://worldweatheronline.com. Accessed February 10, 2018.

Anon. 2018a. Domaine de Labourdonnais, Mapou, Mauritius. http://www.domainedelabourdonnais.com. Accessed February 10, 2018.

Anon. 2018b. The concept of Sterile Insect Technique (SIT). Fruitfly Africa. http://www.fruitfly.co.za/how-sit-works/the-concept-of-sit/. Accessed February 5, 2018.

Abdalla, A., N. Milist, B. Buetre, and B. Bowen. 2012. Benefit-cost analysis of the National Fruit Fly strategy action plan. Australian Bureau of Agricultural and Resource Economics and Sciences, Canberra.

Aggrey-Korsay, R. 2014. Efficacy of locally produced papain enzyme for the production of protein bait for *Bactrocera invadens* (Diptera: Tephritidae) control in Ghana. BSc thesis. Accra, Ghana: University of Ghana.

Barnes, B. N. 2008. *The Sterile Insect Technique*. Stellenbosch, South Africa: Arc Infruitec-Nietvoorbij.

Barry, J. D., N. W. Miller, J. C. Piñero, A. Tuttle, R. F. Mau, and R. I. Vargas. 2006. Effectiveness of protein baits on melon fly and oriental fruit fly (Diptera: Tephritidae): attraction and feeding. *Journal of Economic Entomology* 99(4):1161–1167.

Bateman, M. A. 1972. The ecology of fruit flies. *Annual Review of Entomology* 17:493–518.

Bateman, M. A., and T. C. Morton. 1981. The importance of ammonia in proteinaceous attractants for fruit flies (Family: Tephritidae). *Australian Journal of Agricultural Research* 32(6):883–903.

Bharati, T. E., V. K. R. Sathiyanandam, and P. M. M. David. 2004. Attractiveness of some food baits to the melon fruit fly, *Bactrocera cucurbitae* (Coquilett) (Diptera: Tephritidae). *International Journal of Tropical Insect Science* 24(2):125–134.

Boopathi, T., S. B. Singh, S. V. Ngachan, T. Manju, Y. Ramakrishna, and L. Alhruaipuii. 2013. Influence of weather factors on the incidence of fruit flies in chilli (*Capsicum annuum* L.) and their prediction model. *Pest Management in Horticulture ecosystems* 19(2):194–198.

Ekesi, S., S. Mohamed, and M. Meyer. 2016. *Fruit Fly Research and Development in Africa—Towards a Sustainable Management Strategy to Improve Horticulture*. Cham, Switzerland: Springer.

Gajalakshmi, S., K. Revathi, S. Sithanantham, and A. Anbuselvi. 2011. The effects of weather factors on the population dynamics of mango fruit flies, *Bactrocera* spp. (Diptera: Tephritidae). *Hexapoda* 18(2):148–149.

Gopaul, S., N. S. Price, R. Soonoo, J. Stonehouse, and R. Stravens. 2001. Technologies of fruit fly monitoring and control in the Indian Ocean region, In *Indian Ocean Regional Fruit Fly Programme*, pp. 3–17. Réduit, Mauritius: Division of Entomology, Ministry of Agriculture, Food Technology and Natural Resources.

Guaman Sarango, V. M. 2009. Monitoring and pest control of fruit flies in Thailand: New knowledge for integrated pest management. Uppsala, Thailand: Department of Ecology.

Kotikal, Y. K., and M. Math. 2017. Management of fruit flies through traps and attractants—A review. *Journal of Farm Sciences* 30(1):1–11.

Lloyd, A., and R. A. I. Drew. 1997. Modification and testing of brewery waste yeast as a protein source for fruit fly bait, In *Fruit Fly Management in the Pacific*, eds. A. J. Allwood and R. A. I. Drew, pp. 192–198. Nadi, Fiji: ACIAR.

Mazor, M. R., S. Gothilf, and R. Galun. 1987. The role of ammonia in the attraction of females of the Mediterranean fruit fly to Protein hydrolysate baits. *Entomologia Experimentalis et Applicata* 43:25–29.

Nagaraj, K. S., S. Jaganath, and G. S. K. Swamy. 2014. Effect of protein food baits in attracting fruit flies in mango orchard. *The Asian Journal of Horticulture* 9(1):190–192.

Piñero J. C., S. K. Souder, T. R. Smith, A. J. Fox, and R. I. Vargas. 2015. Ammonium Acetate enhances the attractiveness of a variety of protein-based baits to female *Ceratitis capitata* (Diptera: Tephritidae). *Journal of Economic Entomology* 108(2):694–700.

Piñero, J. C., S. K. Souder, T. R. Smith, and R. I. Vargas. 2017. Attraction of *Bactrocera cucurbitae* and *Bactrocera dorsalis* (Diptera: Tephritidae) to beer waste and other protein sources laced with ammonium acetate. *Florida Entomologist* 100(1):70–76.

Prokopy, R. J., and B. D. Roitberg. 1984. Foraging behaviour of true fruit flies. *American Scientist* 72(1):41–49.

Sookar, P., S. Facknath, S. Permalloo, and S. I. Seewooruthun. 2002. Evaluation of modified waste brewer's yeast as a protein source for the control of the melon fly, *Bactrocera cucurbitae* (Coquilett). In *Proceedings of the 6th International Fruit Fly Symposium*, ed. B. N. Barnes, pp. 295–299. Stellenbosch, South Africa.

Sookar, P., Permalloo, S., Gungah, B., Alleck, M., Seewooruthun, S. E., and R. A. Soonoo. 2006. An Area wide control of fruit flies in Mauritius. In *Proceedings of the 7th International Symposium on Fruit Flies of Economic Importance*, pp. 261–269.

Thomas, D. B., N. D. Epsky, P. Kendra, R. Serra, and D. Hall. 2008. Ammonia formulations and capture of *Anastrepha* fruit flies (Diptera: Tephritidae). *Journal of Entomological Science* 43(1):76–85.

Umeh, V., and D. Onukwu. 2010. Effectiveness of foliar protein bait sprays in controlling *Bactrocera invadens* (Diptera: Tephritidae) on sweet oranges. *National Horticulture Institution* 66(5):307–314.

Vargas, R., E. Jang, R. Mau, and L. Wong. 2002. Fruit Fly Bait. Hawaii Area-Wide Fruit Fly Integrated Pest Management. https://www.extento.hawaii.edu/fruitfly

Vargas. R. I., and R. Prokopy. 2006. Attraction and feeding responses of melon flies and Oriental fruit flies (Diptera: Tephritidae) to various protein baits with and without toxicants. *Hawaiian Entomological Society Proceedings* 38:49–60.

Section IV

Risk Assessment, Quarantine, and Post-Harvest

10 International Database on Commodity Tolerance (IDCT)

Emilia Bustos-Griffin[*], *Guy J. Hallman, Abdeljelil Bakri, and Walther Enkerlin*

CONTENTS

Abstract An important factor for increasing the commercialization of phytosanitary irradiation (PI) is the adoption of generic doses in international and national regulatory frameworks. A limiting factor to accelerating the use of PI is the availability of information on commodity tolerance for the wide range of horticultural products that might be eligible for treatment with PI. The International Database on Insect Disinfestation and Sterilization (IDIDAS) created by the Joint Food and Agriculture Organization/International Atomic Energy Agency (FAO/IAEA) Program for Nuclear Techniques in Food and Agriculture contains an extensive collection of international research on PI. The International Database on Commodity Tolerance (IDCT; https://nucleus.iaea.org/sites/naipc/IDCT/Pages/default.aspx), also created under the auspices of the Joint Program, contains information on the reaction of fresh horticultural commodities including fruit, vegetables, flowers, roots, and tubers, to radiation. Data were extracted from scientific publications from 1950 to the present. The procedure was to collect defined data elements for reporting in a consistent manner. The information used for the database focused on the parameters of specific treatments and conditions. The concepts of "market acceptance" and "market rejection" were determined based on factors associated with how the radiation dose (or range of doses) affected acceptance or rejection of the commodity by taking into account the damage from the treatment and handling conditions in each research scenario. Approximately 415 articles were reviewed: 336 articles corresponded to 48 different fruit species; 47 articles corresponded to tubers and vegetables; and 35 articles covered 21 species of flowers. The database can be searched by commodity using the common name, cultivar, or Latin name as well as by genus and family. Each study lists the respective reference and listings are illustrated with the Google photo gallery. The availability of this information in the IDCT database greatly facilitates the process of identifying potential trade opportunities using PI and helps highlight where commodity tolerance research is sufficient or is still needed.

[*] Corresponding author.

10.1 BACKGROUND

One of the primary uses of food irradiation is as a phytosanitary treatment for fresh commodities (IPPC 2003). Another benefit of this treatment can be to increase the shelf life of commodities in some instances (Arvanitoyannis et al. 2009). Doses in the range of 0.05–2.5 kGy are useful to achieve both purposes in some fresh horticultural products (Bustos-Griffin et al. 2012).

Unlike other phytosanitary treatments (e.g., fumigation and heat or cold treatment) the efficacy of phytosanitary irradiation (PI) is not measured on the basis of acute mortality of the target pests but on preventing the development of the life stages (e.g., non-emergence of adults when larvae are irradiated) or in affecting the ability of the pest to reproduce (e.g., reproductive sterility of irradiated adults) as when irradiated females lay eggs that hatch, but the F_1 neonates die (Hallman et al. 2016). Radiation at the doses applied for PI alters certain physiological processes in the commodities while controlling the pests that may be associated with the fresh products. PI offers an alternative to many traditional quarantine treatments such as fumigation and creates new treatment opportunities where commodity treatments were not previously available.

PI was first used in 1986 when a commercial shipment of irradiated mango from Puerto Rico was shipped to markets in Florida, USA (Phillips 1986). Commercialization followed slowly because of challenges associated with the lack of regulations, concerns about consumer acceptance, and the absence of adequate facilities, among others (Bustos-Griffin et al. 2015). As many other quarantine treatments were banned, such as ethylene Dibromide (EDB) (Ruckelshaus 1984), PI has steadily grown as a mainstream quarantine treatment.

Over time, a number of countries including the United States, Australia, New Zealand, India, Thailand, Vietnam, Mexico, South Africa, and Malaysia have become involved in the import and export of PI-treated fresh horticultural products (Bustos-Griffin et al. 2015). This growth can be attributed in large part to changing perceptions, market forces, and the adoption of international phytosanitary standards that provide the basis for national programs that incorporate PI as a viable treatment option. These conditions have helped to better position PI for large-scale commercialization.

One important characteristic of PI that distinguishes it from other quarantine treatments is the potential to adopt generic treatments for similar pest species across a broad range of commodities (APHIS 2006). For example, studies done on different species of the Tephritidae family demonstrate that a 150 Gy treatment dose can control their normal development (Hallman 2012). Once the dose for a pest is established by research, it is the same no matter what commodity is being treated. Thus, the dose is "generic" for the pest in all commodities that can tolerate the dose. When this dose is then expanded to whole groups of pest organisms, it opens the door for many new commodity treatments with generic doses. The key to taking advantage of this opportunity is having available PI host tolerance information. This information explains which commodities are potential candidates for PI using generic doses.

The quality of most fresh horticultural commodities is not affected by the radiation doses used for PI. However, a few commodities (such as avocado) cannot tolerate the doses required to control most common pest species (Balock et al. 1966). Before investing in PI programs and treatment facilities, marketers need to understand which aspects of quality are most important for the commodities in question and the effects of the radiation dose under specific commercial conditions.

The scientific literature on commodity tolerance to PI is highly variable and can be confusing with vague or incomplete conclusions and conflicting reports. For instance, the quality of several avocado cultivars is negatively affected at doses irradiated below or equal to 250 Gy (Balock et al. 1966; Arevalo et al. 2002), but Simon and Vietes (2014) report the quality of avocado is not affected at 1000 Gy. Interpretation and synthesis of the information in a consistent and easily searchable format is important to make it useful for the business decisions needed to commercialize PI.

The International Database on Insect Disinfestation and Sterilization (IDIDAS) was created by the Joint Food Agriculture Organization/International Atomic Energy Agency (FAO/IAEA) Program for Nuclear Techniques in Food and Agriculture as a searchable international repository for irradiation treatment research relevant to agricultural applications, including PI. The International Database on Commodity Tolerance (IDCT) has been created to complement IDIDAS with consolidated commodity tolerance information based on critical review and analysis of the available literature. The objective was to create a database that substantially supports the decision-making processes associated with the expansion and commercialization of PI.

10.1.1 Phytosanitary Irradiation

Extensive studies have demonstrated that irradiation used for phytosanitary purposes offers significant benefits and has many characteristics that make it unique as a phytosanitary treatment (Hallman 2011). Irradiation offers a range of possible responses other than only mortality. The treatment response can also be sterility, limited fertility, limited development, non-emergence of adult from pupae (Bustos et al. 2004), devitalization in the case of seed (Wage and Kwon 2007), inactivation in the case of microorganisms (Dickson 2001), and sprout inhibition (ICGFI 1991). The integrity of the treatment is assured by research that identifies the appropriate dose, and dosimetry to assure that treatment achieves the dose under specific conditions.

A key difference between PI and all other phytosanitary treatments is that the presence of live pests post-treatment is acceptable by plant protection organizations. Quarantine security is not compromised by the presence of live target pests if research has shown that they are unable to grow or produce viable offspring after treatment (Bakri et al. 2005; Hallman 2012).

The source of radiation for a PI treatment can be a radioactive gamma source such as Co-60 or Cs-137, or machine-generated radiation in the form of X-rays or electron beams (FAO 1984). Dosimetry assures that the minimum dose is absorbed at every point in the treatment load. The dose is the same for every commodity treated for the same pest. There should be no need for additional commodity dose research or regulatory approvals for each commodity (IPPC 2003).

In the case of generic doses, research has demonstrated the effectiveness of one dose for groups of pests. For example, 150 Gy is currently recognized by the International Plant Protection Convention as the generic dose for all tephritid fruit flies, (IPPC 2008), and 400 Gy is recognized by the Animal and Plant Health Inspection Service (APHIS) and New Zealand for all insects except lepidopteran pupae and adults. These doses create an enormous opportunity to treat a wide range of commodities affected by the pests of concern if the commodity can tolerate the dose, hence the importance of commodity tolerance for accelerating the commercialization of PI.

10.1.2 Tolerance of Irradiated Fresh Commodities

The concept of tolerance is related to the desired characteristics of the final product following treatment. Tolerance is not a regulatory parameter, but it is crucial to determining if a treatment is practical and the treated product is acceptable for marketing purposes. PI that effectively treats a pest but renders the product unusable for the desired market is not commercially viable.

Consumers expect safety and quality of their food. The safety of irradiated food has been demonstrated with extensive research carried out in many countries over many decades. One of the principal studies was The International Project on Food Irradiation that was active from 1970 to 1982. Its work included feeding studies contracted to cover a range of commodities irradiated at very high doses. None of the studies gave any indication of the presence of radiation induced carcinogens or other toxic substances (Elias and Cohen 1983). The data generated by this project and other related investigations were reviewed by the Expert Committee on the Wholesomeness of Irradiated Food at World Health Organization (WHO) Headquarters. After several meetings and extensive review, this committee, which was formed by members of the WHO, IAEA, and FAO concluded that the irradiation of any commodity up to an overall average dose of 10 kGy (10,000 Gy) presented no toxicological hazard and no special nutritional or microbiological problems. Hence, toxicological testing of foods so treated was no longer required (FAO 1984).

The quality characteristics in fresh horticultural products for consumers are shelf life, appearance, odor, flavor, firmness, and texture. Some of these characteristics can be subjectively judged by simple observation (e.g., shape, appearance, defects, color, odor) and others can be precisely measured (texture, color, size) or another characteristic may be more important depending on the commodity and end use. This means that acceptable quality is variable. For example, appearance will be more important for apples to be sold as fresh fruit than apples that will be used for applesauce.

The ability to extend commodity shelf life is generally a desirable characteristic, and some fresh horticultural products benefit from this characteristic of PI. Treatment of radio-phylic commodities such as mango, papaya, and rambutan either delay the ripening or the senescence process or tolerate high doses (1 kGy) with little or no negative effects (Kader 1986). At the opposite end of the scale are radio-phobic commodities such as avocado, soursop, and some leafy vegetables that are very susceptible to damage by radiation even at low doses (Kader 1986). Because of their sensitivity to irradiation treatment, they cannot be considered viable commodities for PI. Between these extremes is a range of commodities that tolerate some level of radiation under specific conditions. PI may be a viable treatment for this large category of commodities, but the successful use of PI requires a clear understanding of the tolerance of the commodity and the optimal conditions for treatment at the dose required for the pest(s) of concern.

Fresh horticultural commodities contain a high percentage of water, and this compound is an excellent medium for chemical and biochemical reactions in living tissues. Exposure to irradiation excites molecules in the product, creating free radicals and ions, which almost instantaneously initiate reactions that affect the metabolism of the commodity. The resulting physiological changes will be reflected in characteristics that may affect the quality of an irradiated fresh product, depending on the dose (Arvanitoyannis et al. 2009).

10.2 METHODS

10.2.1 The IDCT Project

As interest in the commercialization of irradiation for phytosanitary purposes has grown, IAEA through its Joint FAO/IAEA Program recognized the need for information that supports business decisions as well as research. Commodity tolerance is a central point to trust in the viability of PI on a commercial scale.

Some commodity tolerance information is available from specific studies in peer-reviewed literature. Other information is found in related research, proceedings, and industry studies. The IDCT aims at identifying, collecting, reviewing, analyzing, and summarizing the information in a form that makes it easily searchable and useful. Researchers, regulators, marketers, and investors benefit from being able to quickly determine the viability of a particular commodity treatment and whether additional research is needed or not for a commodity tolerance for a particular dose and under specific storage or shipment conditions.

10.2.2 Collecting and Selecting Information

The review includes available research from 1950 to 2018. Information regarding commodity tolerance was extracted, interpreted, compiled, and submitted in a defined format for inclusion in an IAEA database created for the purpose. Information on the quality of different irradiated fresh commodities was provided. The information was adjusted to a format that met the objectives of the database (i.e., to be both useful and user friendly). The scope of the database was limited to specific information on commodity tolerance for irradiated fresh plant products. Live plants, seeds, wood, and other nonhorticultural products were not included. The procedure was to collect defined data elements for reporting in a consistent manner (see Table 10.1). The information used for the database focused on the conditions of pre-treatment, treatment, and post-treatment for each commodity. The concepts of "Market acceptance" and "Market rejection" listed in the summaries refer to the parameters used or insinuated by the authors of each study to determine if an irradiated commodity was marketable.

TABLE 10.1
IDCT Data Collection Framework

Conditions						
		Post-Treatment				
Pre-Treatment	**Treatment**	**Bioassay**	**Market Acceptance**	**Market Rejection**	**Dose (Gy)**	**References**
Commodity origin (country, city), maturity etc.	Radiation source, dose rate, DUR.	Tests done on tolerance and quality of commodities irradiated (T, RH, air, packaging, time tested parameters, quality assessment), etc.	Basis for acceptance	Basis for rejection	Minimum to maximum range yielding acceptance; marketability of the doses tested	Published results of each study.

Abbreviations: DUR, dose uniformity ratio; IDCT, International Database on Commodity Tolerance; RH, relative humidity; T, temperature.

10.3 RESULTS

Tolerance information was obtained from scientific articles published in peer-reviewed journals, proceedings of symposia, reports, and any sources that could be cited. Approximately 415 sources were initially reviewed: 336 corresponded to fruit and covered 48 different fruit; 47 articles corresponded to tubers and vegetables, and 35 articles were reviewed for flowers, covering 21 different flowers for each commodity including cultivars or varieties. Table 10.2 identifies the commodities reviewed. This list is continually being expanded as new information comes to our attention.

The analysis of the results shows that the tolerance of a product is not only a function of the radiation dose, but also other parameters such as the species and cultivar or variety, the stage of maturity at the time of treatment, and the physical conditions before, during, and after the treatment (e.g., temperature, relative humidity, type of package, and type/time of storage and transportation).

TABLE 10.2
Fresh Horticultural Products in the IDCT as Well as Those Planned to Be Added in the Future

Fruit	Apple, apricot, avocado, banana, blackberry, blueberry, cantaloupe, cherry, clementine, curuba, custard apple, Dragon fruit, durian, feijoa, fig, granadilla, grapes, grapefruit, guava, gulupa, hazelnut, kiwifruit, lemon, lime, longan, lychee, lulo, mandarin, mango, mangosteen, nectarine, orange, papaya, passion fruit, peach, persimmon, pear, pineapple, plantain, plum, pomelo, rambutan, starfruit or carambola, soursop, strawberry, uchuva
Vegetables, bulbs, and tubers	Asparagus, capsicum, cucumber, eggplant, mushroom, onion, potato, tomato, spinach, sweet potato, zucchini
Flowers	Bellflower, bird of paradise, Bouvardia, carnation, chrysanthemum, foliage, freesia, gentian, gerbera, ginger, gladiolus, gloriosa, hoary tock, iris, lily, orchids, palm, Rosa, statice, summer tulip, sweet pea

Abbreviation: IDCT, International Database on Commodity Tolerance.

TABLE 10.3
Dose Tolerated in Different Cherry Cultivars

Cultivar	Dose Tolerated (Gy)	References[a]
Bing	≤600 or 2000	Salunkhe (1961), Drake (1997), Drake (1998), Neven and Drake (2000)
Lambert	1000 or <2000	Salunkhe (1961), Eaton (1970)
Napoleon	<2000	Salunkhe (1961)
Rainer	300 or ≤600 or 1000	Drake (1994), Drake (1997), Drake (1998), Neven and Drake (2000)
Van	1000	Eaton (1970)
Windsor	<2000	Salunkhe (1961)
0900Zirat	300	Akbudak (2008)

[a] See the complete citation on the International Database on Commodity Tolerance (IDCT) database.

The range of variables affecting tolerance may be manifest in differences in research results. Studies done at different times in different countries with different conditions can have different results for the same variety. As can be seen in Table 10.3, the doses tolerated by different cultivars of cherry are highly variable. Additional cases can be found by exploring IDCT. For this reason, it is important to consider the range of results and conditions affecting commodity tolerance.

It is important to note that the search was principally for fresh commodities. The idea is to later complement the IDCT with other plant products that need PI, such as nuts. Some of these, like grapes, apples, and strawberries, combine irradiation with low temperature or modified atmosphere, but the commodities are always fresh. It is also important to mention that there are some other very particular commodities that belong to specific regions or countries. All dosage information was normalized to gray units (Gy) and temperatures to degrees Celsius (°C).

10.3.1 Database Structure

The IDCT (2018) was launched by the IAEA Insect Pest Control Section of the Joint FAO/IAEA Division on March 9, 2017. The database can be searched by commodity using the common name, the cultivar, or Latin name as well as by genus and family. Each study lists the respective reference, and all commodity listings are illustrated with links to a Google photo gallery. The database tracks user traffic metrics because the number of hits and visits are generated in a monthly report by Google analytics.

10.4 CONCLUSIONS

The phytosanitary applications for irradiation have become mainstream tools for safe trade in horticultural commodities. To realize the full potential of PI, efforts that have previously emphasized research and regulatory frameworks must also support commercialization. This means that the experience and information developed over the past decades needs to be made available in a way that facilitates decision making for regulators, marketers, and investors to clearly understand the opportunities and limitations of the technology. A central strategy in this transformation is providing easily accessed and digestible information on commodity tolerance.

IDCT is an important source of collective scientific knowledge to learn and know about the quality of irradiated commodities. It clearly demonstrates that the response of a commodity to irradiation is not only a function of the radiation dose but also conditions of handling and storage. The data contained in this system also indicates that the commodity cultivar type and the country or region of origin are also important parameters to consider in the application of the treatment.

ACKNOWLEDGMENTS

The authors are grateful to IAEA and the Joint FAO/IAEA Program for supporting the IDCT project and for the assistance of USDA, APHIS, Plant Protection and Quarantine, especially Lucy Reid and Kathryn Mordecai with the Plant Epidemiology and Risk Analysis Laboratory in Raleigh, North Carolina, for their excellent assistance with the references.

REFERENCES

Akbudak, B., H. Tezcan and A. Eris. 2008. Effect of low-dose gamma irradiation on the quality of sweet cherry during storage. *Italian Journal of Food Science* 20(3): 381–390.

Animal and Plant Health Inspection Service (APHIS). 2006. Treatments for fruits and vegetables. *Federal Register* 71(18): 4451–4464.

Arevalo, L., M. E. Bustos, and C. Saucedo. 2002. Changes in the vascular tissue of fresh Hass avocados treated with Cobalt 60. *Radiation Physics and Chemistry* 63: 375–377.

Arvanitoyannis, I. S., A. C. Stratakos, and P. Tsarouhas. 2009. Irradiation applications in vegetables and fruits: A review. *Critical Reviews in Food Science and Nutrition* 49(5): 427–462.

Bakri, A., K. Metha, and E. R. Lance. 2005. Sterilizing insects with ionizing radiation. In *Sterile Insect Technique, Principles, and Practice in Area-Wide Integrated Pest Management*, eds. V. A. Dyck, J. Hendrichs, and A. S. Robinson. Dordrecht, the Netherlands: Springer Press.

Balock, J. W., A. K. Burditt, E. K. Stanley et al. 1966. Gamma radiation as a quarantine treatment for Hawaiian fruit flies. *Journal of Economic Entomology* 59(1): 202–204.

Bustos M. E., W. Enkerlin, J. Reyes, and J. Toledo. 2004. Irradiation of mangoes as a postharvest quarantine treatment for fruit flies (Diptera: Tephritidae). *Journal of Economic Entomology* 97: 286–292.

Bustos-Griffin, E., G. J. Hallman, and R. Griffin. 2012. Current and potential trade in horticultural products irradiated for phytosanitary purposes. *Radiation Physics and Chemistry* 81: 1203–1207.

Bustos-Griffin, E., G. J. Hallman, and R. Griffin. 2015. Phytosanitary irradiation in ports of entry a practical solution for developing countries. *International Journal of Food Science and Technology* 50: 249–255.

Dickson, J. S. 2001. Radiation inactivation of microorganism. *In Irradiation Principles and Application*, ed. R. A. Mollins. New York: John Wiley & Sons.

Drake, S. R., H. R. Mottiff, and D. E. Eakin. 1994. Low dose irradiation of a 'Rainer' sweet cherries as a quarantine treatment. *Journal of Food Processing and Preservation* 18: 473–481.

Drake, S. R., and L. G. Neven. 1997. Quality response of 'bing' and 'rainer' sweet cherries to low dose electron beam irradiation. *Journal of Food Processing and Preservation* 21(4): 345–351.

Drake, S. R., and L. G. Neven. 1998. Irradiation as an alternative to methyl bromide for quarantine treatment of stone fruits. *Journal of Food Quality* 22: 529–538.

Eaton, G. W., C. Mechan, and N. Turner. 1970. Some physical effects of postharvest gamma radiation on the fruit of sweet cherry, blueberry and cranberry. *Canadian Institute of Food Technology Journal* 394: 152–156.

Elias P. S., and A. J. Cohen. 1983. *Resent Advances in Food Irradiation*. Amsterdam, the Netherlands: Elsevier Biomedical Press.

FAO. 1984. Codex General Standard for Irradiated Food and Recommended International Code of Practice for the Operation of Irradiation Facilities Used for the Treatment of Foods. Food and Agriculture Organization of the United Nations. Vol XV FAO WHO Rome Rev.1-2003.

Hallman, G. J. 2011. Phytosanitary application of irradiation. Comprehensive reviews. *Food Science and Food Safety* 10: 143–151.

Hallman, G. J. 2012. Generic phytosanitary irradiation treatments. *Radiation Physics and Chemistry* 81: 861–866.

Hallman, G. J., Y. Henon, G. Andrew et al. 2016. Phytosanitary irradiation: An overview. *Florida Entomologist* 99(2): 1–13.

International Consultative Group of Food Irradiation (ICGFI). 1991. Code of good irradiation practice for sprout inhibition of bulb and tuber crops. Document No. 8. FAO/IAEA/WHO.

International Database on Commodity Tolerance (IDCT). 2018. https://nucleus.iaea.org/sites/naipc/IDCT/Pages/Browse-IDCT.aspx

International Plant Protection Convention (IPPC). 2003. International Standards for Phytosanitary Measures #18. Guidelines for the use of irradiation as a phytosanitary treatment, FAO, Rome, Italy.

International Plant Protection Convention (IPPC). 2008. International Standards for Phytosanitary Measures. Annex to ISPM No. 28. Phytosanitary treatments for regulated pests. FAO, Rome, Italy.

Kader, A. 1986. Potential applications of ionizing radiation in postharvest handling of fresh fruits and vegetables. *Food Technology* 40: 117–121.

Neven, L. G., and S. R. Drake. 2000. Comparison of alternative post-harvest quarantine treatments of sweet cherries. *Postharvest Biology and Technology* 20(2): 107–114.

Phillips, D. 1986. Irradiated fruit gets Miami okay. *The Packer* 93: 1A and 6A.

Ruckelshaus, W. D. 1984. Ethylene dibromide, amendment of notice of intent to cancel registration of pesticide products containing ethylene. *Federal Register* 49: 14182–14185.

Salunkhe, D. K. 1961. Gamma radiation effects on fruits and vegetables. *Economic Botany* 15(1): 28–56.

Simon, W. J., and L. R. Vieites. 2014. Postharvest quality of avocados "hass" held in ambient temperature and treated with different radiation. *Energia na Agricultura* 29(3): 213–219.

Wage, C., and J. H. Kwon. 2007. Improving the food safety of seed sprouts through irradiation treatment. *Food Science and Biotechnology* 16(2): 171–175.

11 Gamma-H2AX
A Promising Biomarker for Fruit Fly Phytosanitary Irradiation Exposure

*Mohammad Sabbir Siddiqui**, *Phillip Taylor, and Peter Crisp*

CONTENTS

Abstract DNA double-strand breaks (DSBs) are one of the most biologically significant DNA damage lesions. Exposure to ionizing radiation (IR) causes DSBs in living organisms, which trigger intrinsic DNA repair mechanisms. Phosphorylation of the C-terminal of the core histone protein H2AX (termed γH2AX when phosphorylated) is an early known response to DNA DSBs. Quantification of the γH2AX response offers a highly sensitive and specific assay for detecting DSB formation and repair. Postharvest exposure to IR of 150–400 Gy is an increasingly prominent phytosanitary measure in a variety of Australian (and imported) fruit. The radiation-induced γH2AX response has been shown to be highly persistent in the Queensland fruit fly ("Q-fly"; *Bactrocera tryoni*), Australia's most economically damaging insect pest of horticultural crops, lasting at least 17 days after exposure to IR. The presence of persistent γH2AX, indicating ongoing repair of impaired DNA, can be used to assess irradiation exposure in fruit flies. A direct and reliable assay using γH2AX as a marker of prior IR exposure in fruit flies has the potential to facilitate domestic and international trade in commodities that have been irradiated for disinfestation.

11.1 INTRODUCTION

Fruit flies are the most economically damaging insect pest of Australian horticulture. Between 2006 and 2009, the average value of fruit fly susceptible production in Australia was approximately AU$5.3 billion and exports of fruit fly susceptible horticulture products were around AU$406.9 million (Abdalla et al. 2012; Hyam 2007; Plant Health Australia 2018). The risk of exotic fruit flies—in the form of eggs, larvae, pupae, or adult—entering and establishing in Australia is increasing (Abdalla et al. 2012; Hallman 2011; Hallman et al. 2011). Phytosanitary treatments, such as fumigation and other chemical and physical (e.g., heat, cold) treatments, are commonly used to disinfest imported and exported commodities of quarantine pests (Hallman 2011; Hallman et al. 2011, 2018). Over the past 40 years, the standard postharvest insect disinfestation chemicals dimethoate and fenthion have provided phytosanitary assurance, but the use of these insecticides has been

* Corresponding author.

greatly restricted (Richard et al. 2003). Finding alternatives to chemical treatments is necessary to prevent introduction and establishment of exotic pests in new areas (Hallman 2011).

Ionizing radiation (IR) is a safer alternative than fumigation and other chemical and physical (heat/cold) disinfestation methods (Follett 2009, Follett et al. 2011; Hallman 2011; IAEA-TECDOC-1427 2004). Numerous countries use IR to disinfest fruit and vegetables from a multitude of quarantine pests (Richard et al. 2003; Hallman 2011; Hallman et al. 2011), including approximately 30,000 metric tons (and increasing by ~10% each year) of sweet potatoes. Increasing quantities of irradiated tropical fruit, such as mangoes, papayas, litchis, capsicums, and tomatoes, are now successfully being exported from Australia to New Zealand consumer markets (Lynch 2010; Lynch and Nalder 2015).

For biosecurity treatments, fresh produce in finished pallet loads is exposed to a minimum generic dose of 150–400 Gy of IR (e.g., electron beam, X-ray, or gamma ray from cobalt-60) (Follett 2009; Hallman et al. 2011). When IR comes into contact with a cell of a pest insect, it breaks chemical bonds in DNA and other molecules, rendering the insect unable to complete development and to reproduce, and thus preventing the establishment of viable pest populations. Verifying irradiation treatment is difficult because quarantine pests are often found alive during inspection in exported and imported commodities. Currently, the only means of assessing quality of imported and exported fruit is through quarantine audits and treatment facility certification. For commercial disinfestation, a regulatory framework exists with the use of generic irradiation doses for a wide range of pest groups. However, the lack of a reliable test to retrospectively confirm radiation exposure can reduce market confidence in a situation where live pests are detected in exported and imported fruit and costs must be incurred to destroy or export the infested consignment.

11.1.1 Can Gamma-H2AX Be Used as a Biomarker of Phytosanitary Radiation?

On exposure to IR, DNA double-strand breaks (DSBs) are induced in the nuclei of all living cells, inducing a DNA repair mechanism characterized by the phosphorylation of the histone protein H2AX (producing the active form gamma-H2AX [γH2AX]) (Rogakou et al. 1998, 1999). Gamma-H2AX is highly conserved across a wide taxonomic range of organisms and is a well-characterized histone protein known to be responsive to IR-induced DNA DSBs (Downs et al. 2000; Foster and Downs 2005; Redon et al. 2002). Gamma-H2AX assay is a standard and well-established method for biological dosimetry of IR exposure. Quantification of the γH2AX response has been used widely as a highly sensitive and specific assay in radiation biodosimetry and cellular radiosensitivity responses during chemotherapy and radiotherapy and to identify regions of the genome where DSBs fail to repair (Bhogal et al. 2010; Ivashkevich et al. 2012; Redon et al. 2012). However, the γH2AX test has not yet been exploited as a retrospective test for identifying the irradiation status of live insects found in exported or imported consignments of fruit and vegetables.

In the γH2AX assay, the DSB level and corresponding IR dose exposure in the nuclei of cells are measured either by measuring the overall γH2AX protein level or by counting discrete "foci" in individual nuclei, which can be visualized and quantified using numerous methods, including fluorescence microscopy and flow cytometry (Figure 11.1) (Hamasaki et al. 2007; Nakamura et al. 2006; Pilch et al. 2004). Two types of γH2AX foci have been found in cells: the first is transient γH2AX foci that are associated with rapid DSB repair and dephosphorylation of γH2AX to H2AX, usually in minutes to hours (Markova et al. 2007, 2011). The second type of γH2AX foci is residual and tends to persist for days to months (Figure 11.1). The measurement of persistent γH2AX signals has been widely used in many applications in recent years, such as for monitoring cancer patients' response to chemotherapy and radiotherapy, radiation biodosimetry, drug biodosimeters, environmental genotoxicity, and in disease (Siddiqui et al. 2015). A study on mini-pig skin cells showed that γH2AX was significantly elevated in irradiated cells after 70 days post-IR exposure (Ahmed et al. 2012). Another study on mouse skin found γH2AX signals up to 7 days post exposure

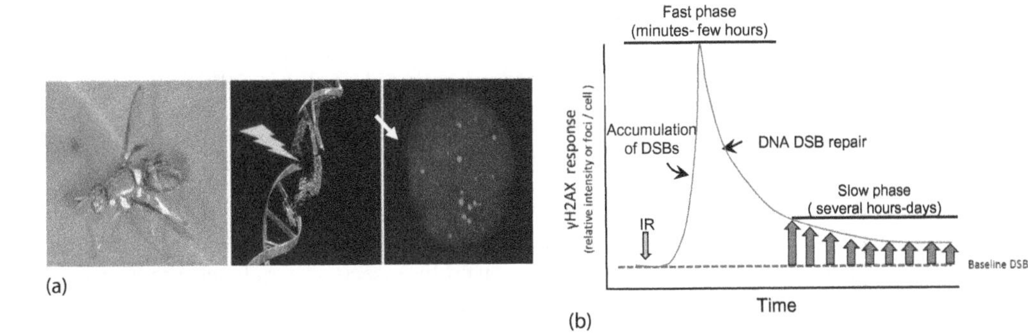

FIGURE 11.1 (a) Ionizing radiation (IR) causes DNA double-strand breaks (DSBs) in Q-fly (arrow indicates γH2AX foci). The number of γH2AX foci represents the number of DSBs. Representative fluorescence image of human buccal cell nuclei containing discrete or diffuse γH2AX foci. Human buccal cell nuclei were visualized (stained with DAPI) with a fluorescence microscope. (b) Schematic representation of the short-term kinetics and persistent γH2AX response in relation to DSB repair. The kinetics of DNA DSB repair follows two phases, a fast phase lasting up to a few hours, which is followed by a slower phase that may persist for several hours to days. On exposure to DNA damaging agents, such as IR, the γH2AX foci appear in the fast phase within minutes after the DSBs are formed and reach a maximum level after about 30 min. This level then declines rapidly and corresponds to repair of DNA DSBs. A small portion of γH2AX (above baseline, as indicated by the dashed line) may persist for up to several months (slower phase) after the initial DSB-induction event and is known as the persistent γH2AX response (as indicated by the bold red arrows). Persistent γH2AX may represent unrepaired DSBs, which are either in the process of slow ongoing repair or too complex to repair. (Adapted from Siddiqui, M.S. et al., *Mutat. Res-Rev. Mutat. Res.*, 766, 1–19, 2015.)

and proposed that they may be used as a biodosimeter in accident scenarios (Bhogal et al. 2010). Linking radiation-induced DNA damage and persistent γH2AX signals is of fundamental importance in establishing a molecular tests capable of detecting and quantifying a prior radiation dose and the resulting DNA damage. The objective of this short communication is to propose the use of the γH2AX test for confirming whether fruit fly species of quarantine concern found in irradiated exported and imported consignment have actually been irradiated and to quantify the dose absorbed.

The proof-of-concept study uses the commercially important pest, Queensland fruit fly (Q-fly; *Bactrocera tryoni* [Froggatt]), as a model to test whether irradiation exposure can be measured retrospectively. In Q-fly, IR exposure leads to a persistent γH2AvB response (a fruit fly variant of γH2AX) for up to 17 days after exposure and can be detected using the γH2AvB test (Siddiqui et al. 2013). Because H2AvB is conserved for all fruit flies of major quarantine concern in which the histone has been sequenced (including *Ceratitis capitata* [Wiedemann], *Bactrocera dorsalis* [Hendel], *Rhagoletis zephyria* Snow, *Bactrocera latifrons* [Hendel], *Bactrocera oleae* [Rossi], *Zeugodacus cucurbitae* [Coquillet]), the γH2AvB test may offer promise in providing rapid, sensitive, and specific detection of prior irradiation exposure to a wide range a fruit flies of market access and biosecurity concern.

11.1.2 Potential Limitations of Gamma-H2AvB as a DNA Damage Biomarker in Fruit Flies

Specimen processing challenges: Once fruit fly specimens are acquired, they must be handled during transport and in the laboratory according to rigorously defined and controlled processes to avoid γH2AvB protein degradation (Valdiglesias et al. 2013).

γH2AvB kinetics differs among species: Persistent γH2AX levels vary in different tissues and cell types and may be affected by genomic status as well as by the type of DNA-damaging agent.

The kinetics (e.g., persistent response) of γH2AvB response in different fruit flies is still unknown. It would be interesting to test whether γH2AvB can be used to assess the kinetics of persistent γH2AvB responses in diverse fruit flies of quarantine concern (Siddiqui et al. 2015).

11.2 CONCLUSION

Currently, for recipients of shipments, certification by the treatment facility is the only available assurance of prior irradiation treatment of live pests discovered in imported and exported fruits and vegetables. A test that directly assesses the dose received by insects in irradiated produce would improve confidence in commercial irradiation treatments, thus offering potential production and market access advantages. Because persistent γH2AX responses have been reported in different cell and tissue types, an assay based on measuring these responses should be investigated for its potential as a method to detect and quantify prior phytosanitary irradiation exposure. This γH2AX test may provide producers an advantage by increasing market acceptance of irradiation as a phytosanitation treatment. A key advantage of the test focusing on measuring the persistence of γH2AvB is that the biomarker has been identified in many insect species and could form the basis of a similar test in diverse pest insects of quarantine concern. The next steps involve broadening the range of insects in which γH2AvB can be detected and validating or modifying the γH2AvB test for operational conditions so that it can be incorporated in commercial and quarantine facilities.

ACKNOWLEDGMENTS

The project Raising Q-fly Sterile Insect Technique to World Standards (HG14033) is funded by the Hort Frontiers Fruit Fly Fund, part of the Hort Frontiers strategic partnership initiative developed by Hort Innovation, with co-investment from Macquarie University and contributions from the Australian Government.

REFERENCES

Abdalla, A., N. Millist, B. Buetre, and B. Bowen. 2012. Benefit–cost Analysis of the National Fruit Fly Strategy Action Plan. ABARES report to client prepared for Plant Health Australia, Canberra, (December): 1–41. http://www.planthealthaustralia.com.au/wp-content/uploads/2012/12/BCA-Fruit-Fly-Strategy-Action-Plan.pdf.

Ahmed, E. A., D. Agay, G. Schrock, M. Drouet, V. Meineke, and H. Scherthan. 2012. Persistent DNA damage after high dose *in vivo* gamma exposure of minipig skin. *PLoS One* 7(6) e39521.

Bhogal, N., P. Kaspler, F. Jalali, O. Hyrien, R. Chen, R. P. Hill et al. 2010. Late residual Gamma-H2AX foci in murine skin are dose responsive and predict radiosensitivity in vivo. *Radiation Research* 173(1): 1–9.

Downs, J. A., N. F. Lowndes and S. P. Jackson. 2000. A role for *Saccharomyces cerevisiae* histone H2A in DNA repair. *Nature* 408(6815): 1001–1004.

Follett, P. A. 2009. Generic radiation quarantine treatments: The next steps. *Journal of Economic Entomology* 102(4): 1399–1406.

Follett, P. A., T. W. Phillips, J. W. Armstrong, and J. H. Moy. 2011. Generic phytosanitary radiation treatment for Tephritid fruit flies provides quarantine security for *Bactrocera latifrons* (Diptera: Tephritidae). *Journal of Economic Entomology* 104(5): 1509–1513.

Foster, E. R., and J. A. Downs. 2005. Histone H2A phosphorylation in DNA double-strand break repair. *The FEBS Journal* 272(13): 3231–3240.

Hallman, G. J. 2011. Phytosanitary applications of irradiation. *Comprehensive Review in Food Science and Food Safety* 2: 143–151.

Hallman, G. J., K. Guo, and Liu T.X. 2011. Phytosanitary irradiation of *Liriomyza Trifolii* (Diptera: Agromyzidae). *Journal of Economic Entomology* 104(6): 1851–1855.

Hallman, G. J., L. Wang, G. Demirbas Uzel, E. Cancio-Martinez, C. E. Caceres-Barrios, S. W. Myers, and M. J. B. Vreysen. 2018. Comparison of populations of *Ceratitis capitata* (Diptera: Tephritidae) from three continents for susceptibility to cold phytosanitary treatment and implications for generic cold treatments. *Journal of Economic Entomology* 112(1): 127–133.

Hamasaki, K., K. Imai, K. Nakachi, N. Takahashi, Y. Kodama, and Y. Kusunoki. 2007. Short-term culture and gammaH2AX flow cytometry determine differences in individual radiosensitivity in human peripheral T lymphocytes. *Environmental and Molecular Mutagenesis* 48(1): 38–47.

Hyam, L. 2007. National Program to Coordinate Fruit Fly Fight, Plant Health Australia, Canberra. http://www.planthealthaustralia.com.au/wp-content/uploads/2012/12/Draft-National-Fruit-Fly-Strategy-Mar-2008.pdf

IAEA-TECDOC-1427. 2004. Irradiation as a Phytosanitary Treatment of Food and Agricultural Commodities. Proceedings of a Final Research Coordination Meeting Organized by the Joint FAO/IAEA Division of Nuclear Techniques in Food and Agriculture 2002. *International Atomic Energy Agency (IAEA)*: 181 pp.

Ivashkevich, A., C. E. Redon, A. J. Nakamura, R. F. Martin, and O. A. Martin. 2012. Use of the Gamma-H2AX assay to monitor DNA damage and repair in translational cancer research. *Cancer Letters* 327(1–2): 123–133.

Lynch, M. 2010. Irradiated Tropical Fruit Exports to New Zealand. https://steritech.com.au/wpcontent/uploads/downloads/Mango%20Trade%20Fact%20Sheet.pdf.

Lynch, M., and K. Nalder. 2015. Australia export programmes for irradiated fresh produce to New Zealand. *Stewart Postharvest Review* 3: 8.

Markova, E., N. Schultz, and I. Y. Belyaev. 2007. Kinetics and dose-response of residual 53BP1/gamma-H2AX Foci: Co-Localization, relationship with DSB repair and clonogenic survival. *International Journal of Radiation Biology* 83(5): 319–329.

Markova, E., J. Torudd, and I. Belyaev. 2011. Long time persistence of residual 53BP1/gamma-H2AX Foci in human lymphocytes in relationship to apoptosis, chromatin condensation and biological dosimetry. *International Journal of Radiation Biology* 87(7): 736–745.

Nakamura, A., O. A. Sedelnikova, C. Redon, D. R. Pilch, N. I. Sinogeeva, R. Shroff et al. 2006. Techniques for Gamma-H2AX detection. *Methods in Enzymology* 409: 236–250.

Pilch, D. R., C. Redon, O. A. Sedelnikova, and W. M. Bonner. 2004. Two-dimensional gel analysis of histones and other H2AX-related methods. *Methods in Enzymology* 375: 76–88.

Plant Health Australia. 2018 Http://www. Planthealthaustralia. Com.au/national-programs/fruit-Fly/

Redon, C., D. Pilch, E. Rogakou, O. Sedelnikova, K. Newrock, and W. Bonner. 2002. Histone H2A variants H2AX and H2AZ. *Current Opinion in Genetics & Development* 12(2): 162–169.

Redon, C., U. Weyemi, P. R. Parekh, D. Huang, A. S. Burrell, and W. M. Bonner. 2012. Gamma-H2AX and other histone post-translational modifications in the clinic. *Biochimica Et Biophysica Acta* 1819(7): 743–756.

Richard, D., S. Winter, D. Michael, and J. Oakeshott. 2003. Irradiation in Horticulture-An Australian Perspective. Australian horticulture industry prepared for Horticulture Australia Limited. http://www.hortaccess.com.au/files/irradiation_in_horticulture_-_april_2003.pdf. Accessed April 2003.

Rogakou, E. P., C. Boon, C. Redon, and W. M. Bonner. 1999. Megabase chromatin domains involved in DNA double-strand breaks in vivo. *The Journal of Cell Biology* 146(5): 905–916.

Rogakou, E. P., D. R. Pilch, A. H. Orr, V. S. Ivanova, and W. M. Bonner. 1998. DNA double-stranded breaks induce histone H2AX phosphorylation on serine 139. *The Journal of Biological Chemistry* 273(10): 5858–5868.

Siddiqui, M. S., E. Filomeni, M. Francois, S. R. Collins, T. Cooper, R. V. Glatz et al. 2013. Exposure of insect cells to ionising radiation in vivo induces persistent phosphorylation of a H2AX homologue (H2AvB). *Mutagenesis* 28(5): 531–541.

Siddiqui, M. S., M. François, M. F. Fenech, and W. R. Leifert. 2015. Persistent γH2AX: A promising molecular marker of DNA damage and aging. *Mutation Research/Reviews in Mutation Research* 766: 1–19.

Valdiglesias, V., S. Giunta, M. Fenech, M. Neri, and S. Bonassi. 2013. γH2AX as a marker of DNA double strand breaks and genomic instability in human population studies. *Mutation Research* 753(1): 24–40.

Section V

Sterile Insect Technique

12 Performance of the Tap-7 Genetic Sexing Strain Used to Control *Anastrepha ludens* Populations in the Citrus Region of Tamaulipas, Mexico

Salvador Flores, Sergio Campos, Enoc Gómez, Rubén Leal Mubarqui, Jorge Luis Morales-Marin, Jorge Vélez, Arturo Bello-Rivera, and Pablo Montoya*

CONTENTS

Abstract In fruit fly pest programs applying the sterile insect technique (SIT), it is of crucial importance to evaluate in situ the performance of sterile males competing for females against wild insects. Consequently, we performed a comprehensive evaluation of the genetic sexing strain (GSS) Tap-7 of *Anastrepha ludens* at the end of emergence, chilling, and release procedures under field conditions in the citrus growing region of Tamaulipas, Mexico. We evaluated survival, sterility induction, the effect of male age on attractiveness of different attractants, and the effect of sterile male releases of this strain on wild populations of *A. ludens*. Our results revealed that survival in the field did not differ between the Tap-7 GSS and the standard A. ludens bisexual strain. Although slightly higher, the sexual competitiveness

* Corresponding author.

of the standard strain was not different from that of Tap-7. In addition, the age of *A. ludens* males of both strains did not influence the response to the attractants used for their monitoring. The performance of both strains in citrus areas of Tamaulipas reduced population peaks and decreased the levels of fruit infestation. Based on the response to attractants, survival, sexual competitiveness, and performance in the field, we consider that the use of the Tap-7 strain of *A. ludens* is a viable option in SIT.

12.1 INTRODUCTION

Area-wide integrated pest management (AW-IPM) is defined as the systematic reduction of population size of one or more target pests through the application of mitigation measures on geographical areas clearly defined by biological criteria (Klassen 2005, Faust 2008). For the reduction of pest populations, chemical control is the most commonly used method because of its effectiveness, but it is questioned for the undesirable effects on the environment and beneficial insects (Mangan 2005, Urbaneja et al. 2009). Since its development, the sterile insect technique (SIT) has represented a viable alternative and environmentally friendly approach to suppress pest populations of many insect species (Knipling 1955, Enkerlin 2005). In Mexico, this technique has been successfully applied in the eradication of the screwworm, achieved in 1991 (Klassen and Curtis 2005), to avoid the establishment and spread of the Mediterranean fly in the southern border of Mexico (Villaseñor et al. 2000) and to obtain areas free of fruit flies of economic importance of the genus *Anastrepha* in the north of Mexico (Gutiérrez 2013, Orozco-Davila et al. 2017).

The objective of the SIT is to induce sterility in wild populations (Hendrichs et al. 2002) by releasing sterile males with the necessary attributes, such as the ability to survive and to locate and copulate with wild females (Calkins and Parker 2005, FAO/IAEA/USDA 2014). However, the mass-rearing process and packing, shipping, and handling procedures of the sterile pupae prior to the release may affect the quality of sterile adults (Rull et al. 2005, 2012) and, consequently, reduce the suppressive effect on pest populations. For *Ceratitis capitata* Wiedemann, it has been reported that male-only releases greatly increase the efficiency of the SIT (Rendón et al. 2004). In this sense, the Moscafrut program initiated a project in coordination with the International Atomic Energy Agency (IAEA) to develop a genetic sexing strain (GSS) of *Anastrepha ludens* Loew, the Mexican fruit fly, which culminated with the obtention of the Tapachula 7 (Tap -7) strain (Zepeda-Cisneros et al. 2014), which is characterized by black pupae that give rise to females and males that emerge from brown pupae. Prior to irradiation, the black pupae of this strain are mechanically separated with an optical sensor, leaving only brown pupae for release. Several studies have compared the quality of the Tap-7 males to that of the standard bisexual strain of *A. ludens*, suggesting that both strains are compatible and competitive for their use in the SIT program of Mexico (Orozco-Dávila et al. 2007, 2015, Hernández et al. 2007, Flores et al. 2014).

Releases of both Tap-7 and bisexual strains were carried out in Chiapas in 2014. During monitoring activities, a reduction in the recapture of Tap-7 sterile flies compared to recapture numbers of the standard strain was observed. This difference may be attributed to a lower survival of the Tap-7 strain in the field or a differential response to the attractants used for monitoring. So far, comparative studies between the Tap-7 and standard strains have focused on evaluating the quality obtained in the mass-rearing facility (Orozco-Dávila et al. 2007, 2015), the effect of the packing and chilling processes (Arredondo et al. 2016), and longevity and dispersion in the field (Flores et al. 2015). Considering that Tap-7 is the newest strain to be released against *A. ludens* pest populations, it is necessary to generate information about its performance in the field, with special emphasis on those regions where it is being released so that it serves as a reference for the comparison and monitoring of the quality of sterile adults. In the field, the capture of adults is taken as a reference to estimate the sterile-to-fertile ratio at the time of applying the SIT (Rendón et al. 2004). In *A. ludens*, the age of sterile males of the bisexual strain shows a bimodal effect on the response to synthetic attractants such as Biolure (Déctor et al. 2016). For the Tap-7 strain, the response to different attractants as a function of age has not been determined. This is necessary so that estimation of the sterile-to-fertile ratio in the field can be interpreted more accurately.

As a part of a comprehensive evaluation of the Tap-7 *A. ludens* GSS for programs applying the SIT, we aimed to: (1) determine the effect of the age of sterile males on their response to the attractants used for monitoring, (2) determine male-induced sterility by the Tap-7 strain in field cage tests after chilled adult collection, (3) determine the effects of the release of Tap-7 and standard bisexual strain males on wild populations of *A. ludens* and the levels of citrus fruit infestation in Tamaulipas, and (4) compare the performance of Tap-7 and standard bisexual strain males in the citrus growing zone of Tamaulipas, Mexico.

12.2 MATERIALS AND METHODS

12.2.1 Biological Material

For preliminary tests, the Moscafrut mass-rearing facility, located in Metapa de Dominguez, Chiapas, provided irradiated pupae to the Mediterranean fruit fly Packing Center (CEMM) adjacent to the Tapachula city airport. The irradiated *A. ludens* Tap-7 strain pupae were marked with signal green dye (Day-Glo Color Corp., Cleveland, OH) and the pupae of the standard bisexual strain were marked with aurora pink dye (Day-Glo Color Corp., Cleveland, OH). For adult emergence, pupae were distributed on shelves of Mexican-type towers. Each tower is composed of 18 shelves consisting of an aluminum frame of 80 × 70 cm and 10 cm high. At each level, 18,460 pupae were placed with a density of 1.3 adults per cm^2 and kept at 22°C in dark conditions. A mixture of 120 g of sugar and protein in a ratio of 24:1 (Liedo et al. 2013) was placed as a food source, and two sponges of 20 × 10 cm saturated with water were also provided. On the fifth day after adult emergence, the towers were moved to rooms at 0°C, where they remained for approximately 45 min until the flies became immobilized (Zavala et al. 2010). For each strain, adults were collected and placed in boxes or cages to perform the different tests described.

12.2.2 Evaluation of Strains Under Controlled Conditions

12.2.2.1 Effect of Male Age on the Response to Different Attractants

The sensitivity of males to different attractants as a function of age was compared between Tap-7 and standard strain adults in field cage tests in a mixed mango (*Mangifera indica* L.) and guava (*Psidium guajava* L.) orchard in Metapa, Chiapas (105 masl, 14.825825°N, −92.196471°W). The pairwise response of males of both strains to each of three attractants was contrasted for males of 5, 9, and 12 d of age. The evaluated attractants were: Captor 300® (Hydrolyzed Protein, Promotora Agropecuaria Universal SA de CV Mexico, DF), CeraTrap® (Hydrolyzed Protein, Bioibérica, SA Barcelona), and the two component lures (putrecine and ammonium acetate) of Biolure® (Suterra LCC, Columbia, Bend, Oregon, USA). Each attractant was exposed in a Multilure trap containing 250 mL of Captor, prepared at a concentration of 4% along with 2% of borax in tap water, or 250 mL of Ceratrap at a concentration of 100%. For the Biolure treatments, 250 mL of a 20% glycol propylene solution in tap water were used to retain the flies. Each repetition consisted of nine field cages (2 m high, 3 m diameter), one for each combination of attractant and fly age. Inside each field cage, we placed six small mango trees (approximately 90 cm high) to simulate natural conditions. In each age, we hung from the field cage ceiling one trap baited with one of the attractants described previously and released 50 males of each strain. After 24 h of exposure, the trap was inspected. Recovered males were examined under an ultraviolet light and the number of adults of each strain was recorded. For each combination of age and attractant, eight replicates were made, each from a different batch of fly strain. The average temperature during the tests was of 24.7 ± 0.9°C and the accumulated precipitation was of 161 mm.

12.2.2.2 Determination of Induced Sterility

Wild *A. ludens* larvae were obtained from "matasano" fruits (*Casimiroa sapota* Oerst.) collected in La Independencia, Chiapas. The recovered larvae were placed in containers with moist vermiculite and maintained for 14 d at 25°C until pupal maturity. Emerged flies were separated by sex in 30 × 30 × 30 cm wooden cages supplied with water and a mixture of sugar: hydrolyzed protein at a ratio of 3:1 as food. The cages were maintained for 14 d at 24°C until adult flies reached sexual maturity. Additionally, a sample of sterile males from the Tap-7 strain and the standard bisexual strain was taken after the packing, emergence, and chilling processes as described previously. Males were kept for 3 days at 25°C to reach sexual maturity prior to mating competition tests.

Seven 3-m diameter and 2-m high field cages were set up with 10 mango plants inside within a mixed mango and guava orchard in Metapa de Dominguez, Chiapas. At 12:00 h on the test day, in each of three cages, 90 sterile males of the Tap-7 strain were released, and in each of the other three cages, 90 sterile males of the standard bisexual strain were released. Thirty wild males were subsequently released, and 30 min later, 30 wild females were added. One field cage with 30 pairs of wild flies was left as a control. One day later (24 h), females were collected and transported to the laboratory. Recovered flies were placed inside seven 30 × 30 × 40 cm Plexiglas cages with water and food and kept at 25°C. This was conducted to evaluate induction of sterility. In each Plexiglas cage, five 5-cm-diameter Fucellerone (Tic Gums, Inc., Belcamp, MD) spheres (as described in Boller 1968) stained with food-grade green dye (McCormick®) were placed as an oviposition substrate. These artificial hosts were replaced every 24 h and the recovered eggs were quantified and placed over a black satin cloth in a humid chamber and maintained at 25°C following Orozco et al. (2013a). After 5 days, the number of hatched and unhatched eggs was quantified. For each strain, 27 replicates were made with nine batches of flies. Recorded climatic conditions were an average temperature of 24.4°C ± 0.7°C and 577.1 mm of precipitation.

12.2.3 Large-Scale Evaluation of the Tap-7 Strain in Citrus Orchards in Tamaulipas

This study was carried out in the citrus growing area of the Municipalities of Hidalgo and Padilla (24°02′56″ N, −98°54′ 20″ W) and Corona (23°58′02″ N, −99°03′24″ W), at an elevation of 175 masl, in the state of Tamaulipas in Northeastern Mexico from January 2016 to June 2017. In this region, four 494 ha (1.3 × 3.8 km) plots were delimited, where the following treatments were distributed (Figure 12.1): (1) aerial release of males of the Tap-7 GSS at a rate of 2,000 males per Ha per week, (2) aerial release of adults of the standard bisexual strain at a rate of 4,000 adults per Ha (2,000 males and 2,000 females per Ha) per week, (3) chemical control at the beginning of the 2017 season, and (4) untreated control.

In each plot, 16 Multilure® traps (Better World Manufacturing Inc., Fresno, California) were installed uniformly; approximately 1 trap per 25 ha (SAGARPA/SENASICA 2012a). Traps were serviced weekly during the study period.

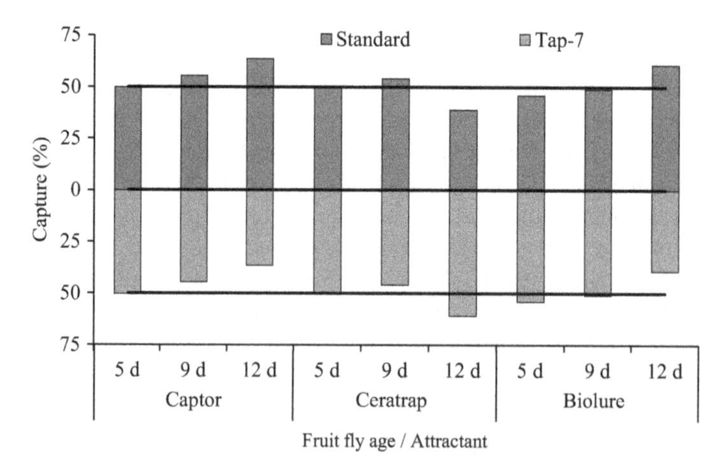

FIGURE 12.1 Percentage of males of the standard strain and the Tap-7 strain of *Anastrepha ludens* at different ages captured in Multilure traps baited with different attractants. For each bar, there is no significant difference in percentage of capture between strains for each day or attractant (P >0.05).

During the 2016 season, only trapping activities were performed in the plots. For the 2017 season, with the aim to suppress the inicial wild populations in all treatments (except the untreated control), in November and December 2016, we applied four aerial bait sprays over 4,400 ha at a rate of approximately one application every 10 days (depending on the weather conditions). The treated surface covered all the above treatment plots, except the control. Toxic bait was prepared as a 1:4 mixture of ULV Malathion (Fyfanon ULV, FMC, Philadelphia): hydrolyzed protein (Atralat®, Agrodesarrollos Nutricionales y Especialidades S de RL, Mexico) and sprayed at a rate of one liter per hectare in alternate bands (SAGARPA/SENASICA 2012b). From February to May 2017, 18 weekly separate releases were made in SIT plots.

Sterile flies (the two strains) were provided by the Moscafrut facility in Metapa de Dominguez, Chiapas. Before the irradiation procedure, every pupae batch was painted, on a weekly basis, with pink, green, or yellow fluorescent dye (Aurora Pink®, Signal Green®, and Saturn Yellow® DyeGlo Color Corp., Cleveland, OH) to differentiate adults according to release date. Sterile flies were shipped as pupae batches of 2,000,000 and 4,000,000 irradiated pupae of the Tap-7 strain and the standard bisexual strain, respectively. The irradiation dose for both strains under hypoxia conditions was of 80 Gy using a cobalt-60 source (dry storage gamma irradiator, Model GB-127, Nordion International Inc., Ottawa, Ontario, Canada). After arrival to the release areas, the pupae were confined in Mexico-type towers and chilled as described previously (Zavala et al. 2010). Chilled adult flies from each strain were released by air following SAGARPA/SENASICA (2012c).

To assess fruit infestation, six fruits still attached to the trees and on the ground (when available) were collected weekly from each plot. Number of fruits and total weight were recorded for each sample. Fruits were opened in the laboratory and present larvae were identified and quantified, providing estimates of levels of larval infestation per kilogram of fruit, number of larvae per fruit, and percentage of infested fruit per treatment.

12.2.4 Estimation of Survival of Sterile Flies

To estimate the survival of released adults inside of each aereal release block of the Tap-7 and standard bisexual strains in Tamaulipas, five Multilure traps were inspected three times per week. The traps were baited with 250 mL of Strepha-Trap® protein (Productos Biologicos, S.A., Barcelona) and were replaced every 28 days. Captured insects were examined under an ultraviolet (UV) light.

12.2.5 Analysis of Data

The effect of age and attractant combination on the response of males of each strain was analyzed applying a generalized linear model with a binomial response on the recapture data. Sterility induced by males of each strain for each replicate was estimated using the "*C*" competitiveness index (Fried 1971, FAO/IAEA/USDA 2014). Differences in competitiveness between strains were inferred using a student's *t*-test.

The capture of sterile and wild adults in Tamaulipas is presented as a fly per trap per day (FTD) index. FTDs were compared using a two-factor design with four treatments and two seasons considering the weekly trap services (from January to June in both seasons) as repeated measures. Mean separation was performed using a Tukey's test. Fruit infestation levels were compared after the third week of release. Number of larvae per kilogram of fruit was converted into a ranks (Conover and Iman 1981, Potvin and Roff 1993, Díaz-Fleischer and Aluja 2003) and compared using an analysis of variance followed by a Tukey's test. Percentage of infested fruits was compared among treatments with a generalized linear model and orthogonal contrasts for pairwise comparisons (Zar 1996). Recapture of adults of each strain released in the aereal release blocks for the Tap-7 and standard bisexual strains was compared between treatment plots across days of trap service. Survival trends derived from the recapture data were contrasted using the log-rank test as in Hernandez et al. (2007). This method is based on the following assumptions: all ages are equally trappable and flies do not leave the trapping grid (Utges et al. 2013). All analyses were performed using JMP 7.0 at a significance level of 95% (SAS Institute 2007).

12.3 RESULTS

12.3.1 Effect of Age on Attraction to Traps

The percentage of capture of males of the Tap-7 strain of *A. ludens* was not different from that of the standard strain (Figure 12.1). Age ($\chi^2 = 4.0$, d.f. = 2, $P = 0.133$) and attractant ($\chi^2 = 5.8$, d.f. = 2, $P = 0.054$) were not factors that affected the capture of males of each strain. However, there was a significant interaction between male age and attractant ($\chi^2 = 18.4$, d.f. = 4, $P = 0.001$). For Captor and Biolure, there was a slight tendency of a higher number of captures of the standard strain as male age increased; whereas for CeraTrap, captures were reduced for males of 12 d of age.

12.3.2 Competitiveness in Field Cages

Egg fertility differed significantly among treatments (F = 36.5, d.f. = 2, 60, P <0.001). The highest percentage of egg hatching was recorded for the wild control, whereas egg hatching between the two laboratory strains did not differ significantly (Figure 12.2). The male competitiveness index was slightly higher for the standard bisexual strain (0.38 ± 0.07) compared to that of the Tap-7 strain (0.28 ± 0.02), although the difference was no statistically significant (F = 1.3, d.f. = 1, 52, $P = 0.259$) (Figure 12.3).

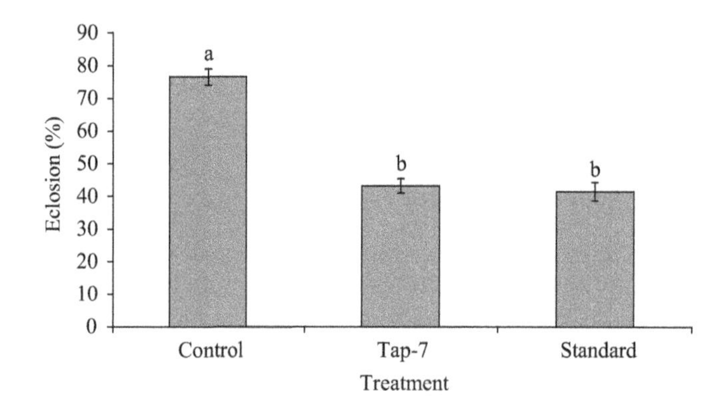

FIGURE 12.2 Percentage of egg-hatching (Mean ± SE) of wild females in competitiveness tests using sterile males of the Tap-7 and standard strains of *Anastrepha ludens*. Columns with the same letter are not statistically different (P >0.05).

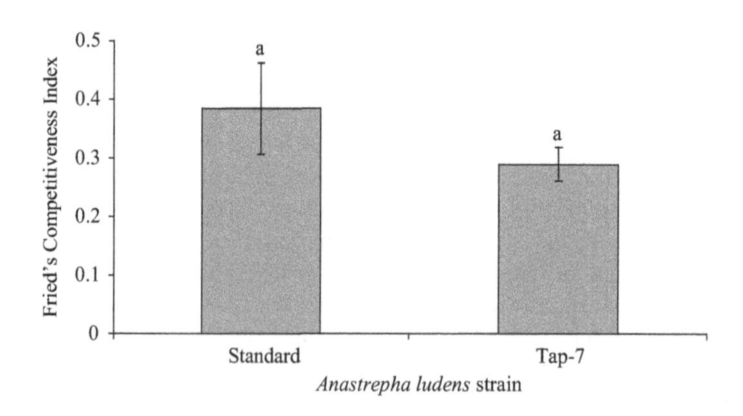

FIGURE 12.3 Index of competitivity of sterile males (Mean ± SE) of the Tap-7 and standard strains of *Anastrepha ludens*. Columns with the same letter are not statistically different (P >0.05).

12.3.3 Survival and Longevity in Citrus Orchards

The survival curve did not differ significantly between strains (χ^2 = 1.6, d.f. = 1, P = 0.200). Longevity was significantly greater for the Tap-7 strain (6.22 days) than for the standard bisexual strain (4.85 days) (t = 3.8, d.f. = 16, P = 0.067) (Figure 12.4).

12.3.4 Evaluation of the Tap-7 Strain in Citrus Orchards

Anastrepha ludens population fluctuations are shown in Figure 12.5, where it can be observed that in plots chosen for the release of sterile adults, population levels were higher than for plots chosen for chemical control and the untreated control. During the 2016 season, for all plots, there was an increase in FTDs of March and no captures from October to December. For plots with chemical control and the untreated control, populations in the 2017 season displayed very similar levels and dynamics to those observed in the 2016 season; however, in the plots under release treatments during the 2017 season, population levels were lower than those observed in the 2016 season.

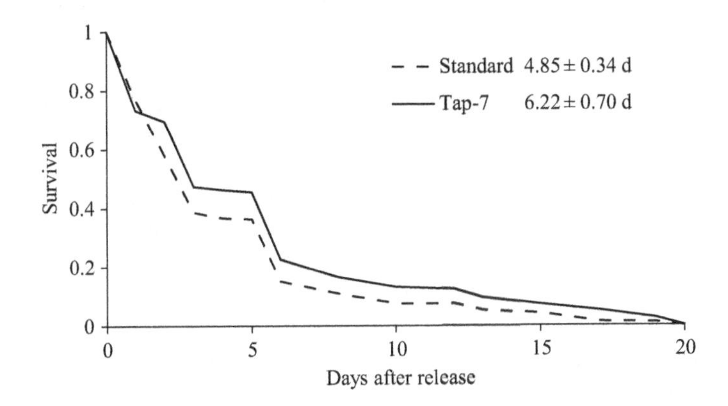

FIGURE 12.4 Survival curve of sterile males of *Anastrepha ludens* of the Tap-7 and standard strains after the emergence, adult chilling, and air release processes in the citrus zone of Tamaulipas, Mexico.

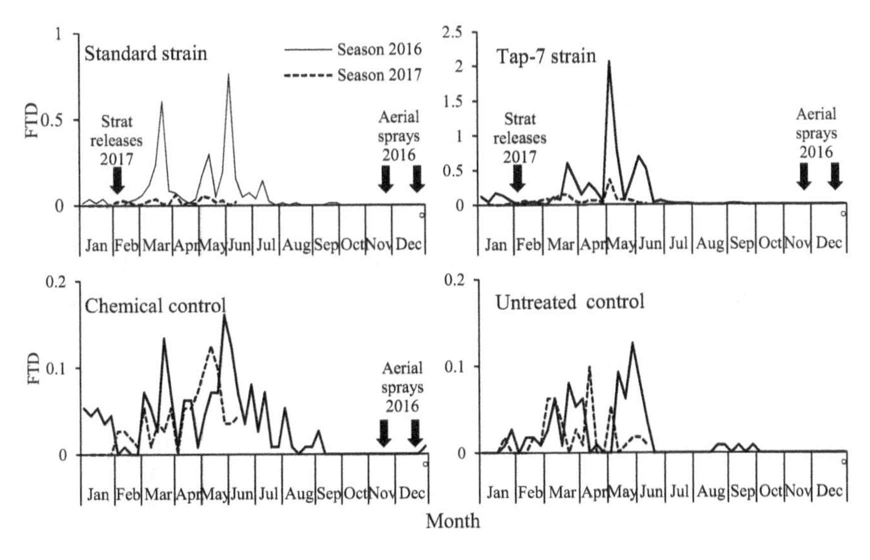

FIGURE 12.5 Population fluctuations of *Anastrepha ludens* in the different treatments applied in the citrus zone of Tamaulipas, Mexico.

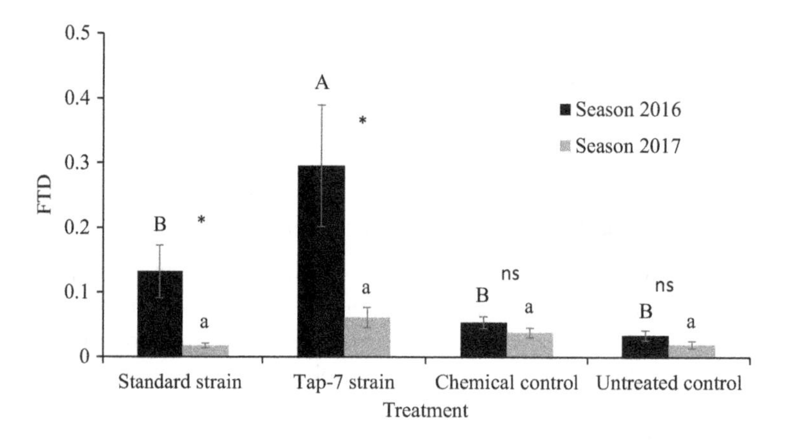

FIGURE 12.6 Capture (Mean ± SE) of wild *Anastrepha ludens* in the different treatments. Same letters, capital letters for year 2016 and lowercase letters for year 2017, indicate nonsignificant differences between treatments. For each treatment, "ns" indicates nonsignificant differences between years and * indicates significant differences.

The FTD index was significantly higher in the 2016 season than in 2017 ($F = 17.4$, d.f. = 1, 109, P <0.001). Plots assigned to different treatments showed significant differences in population densities in the 2016 season prior to treatment application ($F = 8.2$, d.f. = 3, 109, P <0.001), but not in the 2017 season when releases were made ($F = 0.4$, d.f. = 3, 109, $P = 0.766$). In the 2016 season, the highest pest density was recorded in the plot assigned for the release of the Tap-7 strain. The comparison between both seasons indicated that the application of the SIT by releasing the standard bisexual strain ($F = 5.0$, d.f. = 1, 109, $P = 0.028$) or the Tap-7 strain ($F = 19.8$, d.f. = 1, 109, P <0.001) resulted in a significant reduction of the FTD index. Pest density did not differ significantly between the 2016 and 2017 seasons for the plots under chemical control ($F = 0.1$, d.f. = 1, 109, $P = 0.745$) and the untreated control ($F = 1.2$, d.f. = 1, 109, $P = 0.268$) (Figure 12.6).

According to the fruit varieties found in the work area, pink grapefruit presented the greatest infestation both in larvae per kilogram and in percentage of infested fruit. Red grapefruit, sour orange, and navel orange were all infested with *A. ludens*. In mandarin, early orange, white grapefruit, and tangerine, there was no infestation in the samples collected. There were no significant differences in larvae per kilogram among treatments ($F = 1.4$, d.f. = 3, 20, $P = 0.263$), but differences in percentage of larvae were significant ($\chi^2 = 46.6$, d.f. = 3, P <0.001). Percentage of infested fruit was significantly different among treatments, except between the plot under chemical control and the plot with Tap-7 strain releases (Table 12.1).

TABLE 12.1
Total Sampled Fruits and *Anastrepha ludens* Larval Infestation Levels for Plots under Different Pest Management Treatments in the Citrus-Growing Region of Tamaulipas, Mexico

Treatments	Total (kg)	Number of Fruits	Infested Fruits	Total Larvae	Larvae/kg	% Infested Fruit
Standard strain	280.85	1066	12	63	0.22 a	1.13 a
Tap-7 strain	173.95	620	17	105	0.60 a	2.74 b
Chemical control	82.6	301	12	83	1.00 a	3.99 b
Untreated control	110.3	321	30	254	2.30 a	9.35 c

Different letters indicate significant differences.

12.4 DISCUSSION

The type of survivorship, sterility induction, and mating competitiveness studies performed here represent the quality control measures that should be adopted to evaluate the performance of released sterile insects (Calkins and Parker 2005). Our studies allowed us to validate the capacity of the Tap-7 *A. ludens* strain compared to the standard bisexual strain after the process of holding, emergence, chilling, and release of sterile adults. These data are important to support the planning of strategies in programs that apply the SIT, as well as for the follow-up of the effect and efficiency of the use of this technique for the suppression of pest populations. Our data also determined that the application of the SIT by releasing either the standard strain or the GSS resulted in significant reductions of seasonal pest population peaks.

The Tap-7 strain and the standard strain were subjected to the same packing and release procedures, and both exhibited a longevity that ranged between 2.6 and 2.9 days. These values were close to 2.03 and 2.39 days for males and females of the standard strain, respectively, reported by Hernández et al. (2007). However, they are lower than those recently reported by Flores et al. (2015), who obtained 4.2 and 3.6 d for the standard and the Tap-7 *A. ludens* strains, respectively. Nevertheless, survival is also affected by factors such as host availability, weather, environmental stress (Vreysen 2005), and food availability (Calkins and Parker 2005, Gómez-Cendra et al. 2007, Utgés et al. 2013).

Sterility induction by and competitiveness of sterile males are basic parameters to estimate the performance of a strain and to establish an adequate sterile-to-fertile ratio (Barry et al. 2003, Shelly et al. 2005, 2007, Flores et al. 2014). In field cages, we obtained a sterility induction of 45% for both strains using a sterile-to-fertile ratio of 3: 1. Similar values (48%–55%) were reported by Flores et al. (2014) for *A. ludens*. Therefore, the Tap-7 strain subjected to the adult chilling process maintains values of sterility induction rates and competitiveness at levels established in the FAO/IAEA/USDA (2014) quality control manual and are similar to those reported for the standard strain in competition with different wild populations of Mexico (Orozco-Davila et al. 2007, Quintero-Fong et al. 2018).

Attraction of sterile *A. ludens* males to different baits neither differed between strains nor was affected by male age. These results are important for the interpretation of monitoring data for sterile insects because preliminary observations in action programs indicated that males of the Tap-7 strain were captured less frequently in the trapping network, which led to the assumption that males of this strain either lived less or were less attracted to baits. However, our results indicate that neither assumption was correct. Studies that relate the response of different strains to attractants are scarce. Shelly and Edu (2009) reported a lower attraction to trimedlure for the *C. capitata* TSL strain compared to that of a standard bisexual strain, whereas Barry et al. (2003) found that irradiation increases the response of flies to trimedlure. Some authors such as Kendra et al. (2005) reported that immature young females showed a better electroantennographic response than mature females of *A. suspensa* when exposed to ammonia-based attractants, which was rooted on the biological basis of the need of young females to obtain protein for oocyte formation. Robacker (1991) pointed out that feeding history at an early age would influence the need to feed in adult flies. In our case, all insects were fed with a mixture of sugar and protein at a 24:1 ratio, thus protein hunger was not a factor that affected captures by an attractant. Díaz-Fleischer et al. (2009) reported that *A. ludens* adults were significantly less sensitive to Nulure® and Biolure® when they were fed with a mixture of protein and sugar compared to adults fed with other diets. In addition, Arredondo et al. (2104) noted that the preference for attractants occurs only in a small segment of the fly population, which are those that have a specific need for aminoacids (immature flies) and those that require proteins. According to Díaz-Fleischer et al. (2009), the response to baits may vary according to species, sex, physiological state, and age, as well as strain and sterilization and handling history (Shelly and Edu 2009, Weldon and Meats 2010), all of which show the complex nature of the response to chemical stimuli in the environment. Our results indicate that the quality of the adults released in the field was similar between strains; thus, the use of different trapping approaches for different strains released by operational areas is not justified. However, because the number of released flies is different for GSS and bisexual strains (Arredondo et al. 2016), the lower capture of Tap-7 strain males in traps could be the product of a lower (half) fly density.

The release of sterile adults of *A. ludens* in the citrus orchard of Tamaulipas, whether of the standard bisexual or the Tap-7 strain, resulted in a significant decrease in population peaks compared to the previous year without application of the SIT. Additionally, there was a reduction in the levels of fruit infestation compared to an area without application of control methods. Vanoye-Eligio et al. (2015a, 2015b) reported that populations in the region become detectable from October to April, with the highest population peaks in February and March. Such seasonal population peaks did not occur in areas under sterile fly release, which is an indication that the decrease was a result of the application of the sterile insect technique. On the other hand, fly population densities in the control and the chemical control plots did not show a significant difference in pest density between 2016 and 2017. In the region, there are different host species that have different degrees of susceptibility to infestation by *A. ludens* and whose availability varies throughout the year. This exploitation of resources by the pest plays an important role in its population dynamics and in the efficacy of control methods (Clarke et al. 2011). The reduction of populations and damage levels when applying the SIT has been documented by Stainer et al. (1979) and Koyama et al. (2004). The release of the standard strain of *A. ludens* for the control of wild populations has been effective (Orozco-Davila et al. 2007, 2017, Orozco et al. 2013a). However, studies by McInnis et al. (1994), Rendón et al. (2004), Vreysen et al. (2006), and Orozco et al. (2013b) report that males are more competitive when released alone. Moreover, the release of only males represents an advantage because of a reduction of transportation, packing, maintenance, and release costs (FAO/IAEA 2016).

Based on the results of our study, we can conclude that after packing, emergence, and chilling procedures prior to the release in the field, males of the Tap-7 GSS of *A. ludens* display a similar behavior in response to attractants at different ages, similar survival, and equal sexual competitiveness to the standard bisexual *A. ludens* strain. The release of sterile adults of the Tap-7 strain or the standard bisexual strain managed to reduce pest population levels of host infestation of *A. ludens* in Tamaulipas. Both strains showed similar mortality curves, although longevity was slightly greater for the Tap-7 strain, which indicates a good performance of sterile males of both strains released in citrus-growing areas.

ACKNOWLEDGMENTS

We are grateful to David Nestel for the critical review of this document. To the staff of the Tamaulipas State Plant Health Committee for their technical support. Mrs. María Eugenia Álvarez, owner of the Galicia garden, for the use of facilities during the field tests. To Emigdio Espinoza, Willy Mac Wilson, Carmen Garcia, and Orlando Rivera for their technical support. This research was supported by the Moscafrut Program, Direccion General de Sanidad Vegetal-Servicio Nacional de Sanidad Inocuidad y Calidad Agroalimentaria (DGSV-SENASICA).

REFERENCES

Arredondo, J., S. Flores, P. Montoya, and F. Díaz-Fleischer. 2014. Effect of multiple endogenous biological factors on the response of the tephritids *Anastrepha ludens* and *Anastrepha obliqua* (Diptera: Tephritidae) to MultiLure traps baited with BioLure or NuLure in mango orchards. *Journal of Economic Entomology* 107:1022–1031.

Arredondo, J., L. Ruiz, E. Hernández, P. Montoya, and F. Díaz-Fleischer. 2016. Comparison of *Anastrepha ludens* (Diptera: Tephritidae) bisexual and G (Tapachula-7) strains: Effect of hypoxia, fly density, chilling period, and food type on fly quality. *Journal of Economic Entomology* 109:572–579.

Barry, J. D., D. O. McInnis, D. Gates, and J. G. Morse. 2003. Effects of irradiation on Mediterranean fruit flies (Diptera: Tephritidae): Emergence, survivorship, lure attraction, and mating competition. *Journal of Economic Entomology* 96:615–622.

Boller, E. F. 1968. An artificial egging device for the European cherry fruit fly *Rhagoletis cerasi. Journal of Economic Entomology* 61:850–852.

Calkins, C. O., and A. G. Parker. 2005. Sterile insect quality. In *Sterile Insect Technique. Principles and Practice in Area-Wide Integrated Pest Management*, ed. V. A. Dyck, J. Hendrichs, and A. S. Robinson, pp. 269–296. Dordrecht, the Netherlands: Springer.

Clarke, A. R., K. S. Powell, C. W. Weldon, and P. W. Taylor. 2011. The ecology of *Bactrocera tryoni* (Froggatt) (Diptera: Tephritdae): What do we know to assist pest management? *Annals of Applied Biology* 158:26–55.

Conover, W. J., and R. L. Iman. 1981. Rank transformations as a bridge between parametric and nonparametric statistics. *The American Statistician* 35:124–133.

Déctor, N., E. A. Malo, J. C. Rojas, and P. Liedo. 2016. Comparative responses of *Anastrepha ludens* and *Anastrepha obliqua* (Diptera: Tephritidae) to the synthetic attractant BioLure. *Journal of Economic Entomology* 109:2054–2060.

Díaz-Fleischer, F., and M. Aluja. 2003. Behavioural plasticity in relation to egg and time limitation: The case of two fly species in the genus *Anastrepha* (Diptera: Tephritidae). *Oikos* 100:125–133.

Díaz-Fleischer, F., J. Arredondo, S. Flores, P. Montoya, and M. Aluja. 2009. There is no magic fruit fly trap: Multiple biological factors influence the response of adult *Anastrepha ludens* and *Anastrepha obliqua* (Diptera: Tephritidae) individuals to MultiLure traps baited with BioLure or NuLure. *Journal of Economic Entomology* 102:86–94.

Enkerlin, W. R. 2005. Impact of fruit fly control programmes using the sterile insect technique. In *Sterile Insect Technique. Principles and Practice in Area-Wide Integrated Pest Management*, ed. V. A. Dyck, J. Hendrichs, and A. S. Robinson, pp. 651–676. Dordrecht, the Netherlands: Springer.

FAO/IAEA. 2016. *Guidelines for the Use of Mathematics in Operational Area-Wide Integrated Pest Management Programmes Using the Sterile Insect Technique with a Special Focus on Tephritid Fruit Flies*, eds. H. L. Barclay, W. R. Enkerlin, N. C. Manoukis, and J. Reyes-Flores. Rome, Italy: Food and Agriculture Organization of the United Nations.

FAO/IAEA/USDA. 2014. *Product Quality Control for Sterile Mass-Reared and Released Tephritid Fruit Flies, Version 6.0.* Vienna, Austria: International Atomic Energy Agency.

Faust, R. M. 2008. General introduction to areawide pest management. Publications from USDA-ARS/UNL Faculty. 645. http://digitalcommons.unl.edu/usdaarsfacpub/645. Accessed November 25, 2017.

Flores, S., P. Montoya, J. Toledo, W. Enkerlin, and P. Liedo. 2014. Estimation of populations and sterility induction in *Anastrepha ludens* (Diptera: Tephritidae) fruit flies. *Journal of Economic Entomology* 107:1502–1507.

Flores, S., S. Campos, E. Gómez, E. Espinoza, W. Wilson, and P. Montoya. 2015. Evaluation of field dispersal and survival capacity of the genetic sexing strain Tapachula-7 of *Anastrepha ludens* (Diptera: Tephritidae). *Florida Entomologist* 98:209–214.

Fried, M. 1971. Determination of sterile-insect competitiveness. *Journal of Economic Entomology* 64:869–872.

Gómez-Cendra, P., D. Segura, A. Allinghi, J. Cladera, and J. Vilardi. 2007. Comparison of longevity between a laboratory strain and a natural population of *Anastrepha fraterculus* (Diptera: Tephritidae) under field cage conditions. *Florida Entomologist* 90: 147–153.

Gutiérrez, J. M. 2013. *Los Programas de moscas de la fruta en México, su Historia Reciente*. Mexico City, Mexico: IICA.

Hendrichs, J., A. S. Robinson, J. P. Cayol, and W. Enkerlin. 2002. Medfly areawide sterile insect technique Programmes for prevention, suppression or eradication: The Importance of mating behavior studies. *Florida Entomologist* 85:1–13.

Hernández, E., D. Orozco, S. Flores, and J. Domínguez. 2007. Dispersal and longevity of wild and mass-reared *Anastrepha ludens* and *Anastrepha obliqua* (Diptera: Tephritidae). *Florida Entomologist* 90:125–135.

Kendra, P. E., W. S. Montgomery, D. M. Mateo, H. Puche, N. E. Epsky, and R. R. Heath. 2005. Effect of age on EAG response and attraction of female *Anastrepha suspensa* (Diptera: Tephritidae) to ammonia and carbon dioxide. *Environmental Entomology* 34:584–590.

Klassen, W. 2005. Area-wide integrated pest management and the sterile insect technique. In *Sterile Insect Technique. Principles and Practice in Area-Wide Integrated Pest Management*, ed. V. A. Dyck, J. Hendrichs, and A. S. Robinson, pp. 40–68. Dordrecht, the Netherlands: Springer.

Klassen, W., and C. F. Curtis. 2005. History of the sterile insect technique. In *Sterile Insect Technique: Principles and Practice in Area-Wide Integrated Pest Management*, ed. V. A. Dyck, J. Hendrichs, and A. S. Robinson, pp. 3–36. Dordrecht, the Netherlands: Springer.

Knipling, E. F. 1955. Possibilities of insect control or eradication through the use of sexually sterile males. *Journal of Economic Entomology* 48:459–462.

Koyama, J., H. Kakinohana, and T. Miyatake. 2004. Eradication of the melon fly, *Bactrocera cucurbitae*, in Japan: Importance of behavior, ecology, genetics, and evolution. *Annual Review of Entomology* 49:331–349.

Liedo, P., D. Orozco, L. Cruz-López, J. L. Quintero, C. Becerra-Pérez, M. del Refugio Hernández, A. Oropeza, and J. Toledo. 2013. Effect of post-teneral diets on the performance of sterile *Anastrepha ludens* and *Anastrepha obliqua* fruit flies. *Journal of Applied Entomology* 137(1):49–60.

Mangan, R. L. 2005. Population suppression in support of the sterile insect technique. In *Sterile Insect Technique: Principles and Practice in Area-Wide Integrated Pest Management*, eds. V. A. Dyck, J. Hendrichs, and A. S. Robinson, pp. 407–425. Dordrecht, the Netherlands: Springer.

McInnis, D. O., S. Tam, C. Grace, and D. Miyashita. 1994. Population suppression and sterility rates induced by variable sex-ratio sterile insect releases of *Ceratitis capitata* (Diptera: Tephritidae) in Hawaii. *Annals of the Entomological Society of America* 87:231–240.

Orozco, D., M. R. Hernández, J. S. Meza, and J. L. Quintero. 2013b. Do sterile females affect the sexual performance of sterile males of *Anastrepha ludens* (Diptera: Tephritidae)? *Journal of Applied Entomology* 137:321–326.

Orozco, D., J. S. Meza, S. Zepeda, E. Solís, and J. L. Quintero-Fong. 2013a. Tapachula-7, a new genetic sexing strain of the Mexican fruit fly (Diptera: Tephritidae): Sexual compatibility and competitiveness. *Journal of Economic Entomology* 106:735–741.

Orozco-Dávila, D., M. L. Adriano-Anaya, L. Quintero-Fong, and M. Salvador-Figueroa. 2015. Sterility and sexual competitiveness of Tapachula-7 *Anastrepha ludens* males irradiated at different doses. *PLoS ONE* 10(8): e0135759.

Orozco-Davila, D., R. Hernández, S. Meza, and J. Domínguez. 2007. Sexual competitiveness and compatibility between mass-reared sterile flies and wild populations of *Anastrepha ludens* (Diptera: Tephritidae) from different regions in Mexico. *Florida Entomologist* 90:19–26.

Orozco-Dávila, D., L. Quintero, E. Hernández, E. Solís, T. Artiaga, R. Hernández, and P. Montoya. 2017. Mass rearing and sterile insect releases for the control of *Anastrepha* spp. (Diptera: Tephritidae) pests in Mexico, *Entomologia Experimentalis et Applicata* 164:176–187.

Potvin, C., and Roff, D. A. 1993. Distribution-free and robust statistical methods: Viable alternatives to parametric statistics? *Ecology* 74:1617–1628.

Quintero-Fong, L., J. H. Luis, P. Montoya, and D. Orozco-Dávila. 2018. *In situ* sexual competition between sterile males of the genetic sexing Tapachula-7 strain and wild *Anastrepha ludens* flies of San Luis Potosí, Mexico. *Crop Protection* 106:1–5.

Rendón, P., D. Mcinnis, D. Lance, and J. Stewart. 2004. Medfly (Diptera: Tephritidae) genetic sexing: Large-scale field comparison of males-only and bisexual sterile fly releases in Guatemala. *Journal of Economic Entomology* 97:1547–1553.

Robacker, D. C. 1991. Specific hunger in *Anastrepha ludens* (Diptera: Tephritidae): Effects on attractiveness of proteinaceous and fruit-derived lures. *Environmental Entomology* 20:1680–1686.

Rull, J., A. Birke, R. Ortega, P. Montoya, and L. López. 2012. Quantity and safety vs. quality and performance: Conflicting interests during mass rearing and transport affect the efficiency of sterile insect technique programs. *Entomologia Experaimentalis et Applicata* 142:78–86.

Rull, J., O. Brunel, and M. E. Mendez. 2005. Mass rearing history negatively affects mating success of male *Anastrepha ludens* (Diptera: Tephritidae) reared for sterile insect technique programs. *Journal of Economic Entomology* 98:1510–1516.

SAGARPA/SENASICA. 2017a. Manual técnico para las operaciones de campo de la campaña nacional contra moscas de la fruta sección I: trampeo y muestreo de frutos. 36 pp. https://www.gob.mx/cms/uploads/attachment/file/262603/MT_Operaciones_de_campo_CNMF_Secci_n_I_TyM.pdf. Accessed November 11, 2017.

SAGARPA/SENASICA. 2017b. Manual técnico para las operaciones de campo de la campaña nacional contra moscas de la fruta sección II: control químico. 39 pp. https://www.gob.mx/cms/uploads/attachment/file/239235/MT_Operaciones_de_campo_CNMF_Secci_n_II_CQ.pdf. Accessed September 17, 2017.

SAGARPA/SENASICA. 2017c. Manual técnico del control autocida de moscas de la fruta. 26 pp. https://www.gob.mx/cms/uploads/attachment/file/262619/MT_Operaciones_de_campo_CNMF_Secci_n_V_CA.pdf. Accessed November 5, 2017.

SAS Institute. 2007. JMP version 7.0.1. Cary, NC: SAS Institute.

Shelly, T. E., and J. Edu. 2009. Capture of mass-reared vs. wild-like males of Ceratitis capitata (Diptera, Tephritidae) in trimedlure-baited traps. *Journal of Applied Entomology* 133:640–646.

Shelly, T. E., D. O. McInnis, and Pedro Rendón. 2005. The sterile insect technique and the Mediterranean fruit fly: Assessing the utility of aromatherapy in large field enclosures. *Entomologia Experimentalis et Applicata* 116:199–208.

Shelly, T. E., D. O. McInnis, C. Rodd, J. Edu, and E. Pahio. 2007. Sterile insect technique and Mediterranean fruit fly (Diptera: Tephritidae): Assessing the utility of aromatherapy in a Hawaiian coffee field. *Journal of Economic Entomology* 100:273–282.

Steiner, L. F., W. G. Hart, E. J. Harris, R. T. Cunningham, K. Ohinata, and D. C. Kamakahi. 1970. Eradication of the oriental fruit fly from the Mariana Islands by the methods of male annihilation and sterile insect release. *Journal of Economic Entomology* 63:131–135.

Urbaneja, A., P. Chueca, H. Montón, S. Pascual-Ruiz, O. Dembilio, P. Vanaclocha, R.l Abad-Moyano, T. Pina, and P. Castañera. 2009. Chemical alternatives to Malathion for controlling *Ceratitis capitata* (Diptera: Tephritidae), and their side effects on natural enemies in Spanish citrus orchards. *Journal of Economic Entomology* 102:144–151.

Utgés, M. E., J. C. Vilardi, A. Oropeza, J. Toledo, and P. Liedo. 2013. Pre-release diet effect on field survival and dispersal of *Anastrepha ludens* and *Anastrepha obliqua* (Diptera: Tephritidae). *Journal of Applied Entomology* 137 (Suppl. 1):163–177.

Vanoye-Eligio, V., L. Barrientos-Lozano, R. Pérez-Castañeda, G. Gaona-García, and M. Lara-Villalon. 2015a. Regional-scale spatio-temporal analysis of *Anastrepha ludens* (Diptera: Tephritidae) populations in the citrus region of Santa Engracia, Tamaulipas, Mexico. *Journal of Economic Entomology* 108:1655–1664.

Vanoye-Eligio, V., L. Barrientos-Lozano, R. Pérez-Castañeda, G. Gaona-García, and M. Lara-Villalon. 2015b. Population dynamics of *Anastrepha ludens* (Loew) (Diptera: Tephritidae) on citrus areas in Southern Tamaulipas, Mexico. *Neotropical Entomology* 44:565–573.

Villaseñor, A., J. Carrillo, J. Zavala, J. Stewart, C. Lira, and J. Reyes. 2000. Current progress in the medfly program Mexico-Guatemala. In *Proceedings of Area-Wide Control Insect Pest, 5th International Symposium of Fruit Flies of Economic Importance*, ed. K. H. Tan, pp. 361–368. Pulau Pinang, Malaysia.

Vreysen, M. J. B. 2005. Monitoring sterile and wild insects in area-wide integrated pest management programmes. In *Sterile Insect Technique. Principles and Practice in Area-wide Integrated Pest Management*, eds. V. A. Dyck, J. Hendrichs, and A. S. Robinson, pp. 325–361. Dordrecht, the Netherlands: Springer.

Vreysen, M. J. B., H. J. Barclay, and J. Hendrichs. 2006. Modeling of preferential mating in areawide control programs that integrate the release of strains of sterile males only or both sexes. *Annals of the Entomological Society of America* 99:607–616.

Weldon, C., and A. Meats. 2010. Dispersal of mass-reared sterile, laboratory-domesticated and wild male Queensland fruit flies. *Journal of Applied Entomology* 134:16–25.

Zar, J. H. 1996. *Biostatistical Analysis*. Upper Saddle River, NJ: Prentice Hall.

Zavala, J.L., E. Hernandez, and P. Montoya. 2010. Empaque y liberacion de Moscas Estériles. In *Moscas de la Fruta: Fundamentos y Procedimientos para su Manejo,* eds. P. Montoya, J. Toledo, and E. Hernandez, pp. 319–330. México City, México: S y G editores.

Zepeda-Cisneros, C. S., J. S. Meza Hernández, V. García-Martínez, J. Ibañez-Palacios, A. Zacharopoulou, and G. Franz. 2014. Development, genetic and cytogenetic analyses of genetic sexing strains of the Mexican fruit fly, *Anastrepha ludens* Loew (Diptera: Tephritidae). *BMC Genetics* 15:S1.

13 Toxicological Evaluation of Corncob Fractions on the Larval Performance of *Anastrepha obliqua*

Marysol Aceituno-Medina, Rita Teresa Martínez-Salgado, Arseny Escobar, Carmen Ventura, and Emilio Hernández**

CONTENTS

Abstract Assuring the quality of artificial diet ingredients is the first step to ensure the mass production of quality insects for the sterile insect technique (SIT). Ingredients used as a bulking agent for larval diets, such as corncob fractions, are vulnerable to chemical or microbiological contamination that can make the diet toxic to larvae. Traditional methods of physicochemical and microbiological testing of diet ingredients have been inefficient at guaranteeing the safety of ingredients, making quality assurance of the material a challenge. Thus, we developed a toxicological test to detect the presence of pathogenic bacteria and fungi and their mycotoxins. Sixteen batches were evaluated. Each batch was composed of 26 pallets containing 40 packages of corncob fractions of 20 kg each. The bacteria and fungi were isolated from a sample of 2.6 kg of corncob fractions per batch, which was obtained by sampling three packages per pallet to complete 100 g for each of the 26 pallets of a batch. The toxicological tests involved four steps: (1) identification of bacteria and fungi, (2) toxicological test for larval survival, (3) mycotoxin determination, and (4) effect of the incorporation of mycotoxin sequestrants on larval yield. The toxicological test of the corncob fractions allowed to classify a batch as

* Corresponding authors.

acceptable or non-acceptable. A nonacceptable batch presented at least a colony of any these species: *Morganella morganii, Serratia marcescens, Fusarium oxysporum, F. graminearum,* or *Aspergillus* spp., or any of these mycotoxins: aflatoxin, fumonisin, ochratoxin, or T2-HT2. Fungi contamination by *F. oxysporum, F. graminearum*, and *Aspergillus* spp. caused high mortality either in single contaminations/alone or in association with the bacteria *M. morganii* or *S. marcescens*. The addition of Aflaban™ to a contaminated diet reduced the effects of mycotoxins and increased larval yield. A toxicological test should be included during quality control evaluations of diet ingredients to classify diets as acceptable or non-acceptable.

13.1 INTRODUCTION

The sterile insect technique (SIT) requires the production of large numbers of sterile insects to be released in a given area to control a target pest species population. An important aspect of rearing fruit flies (Diptera: Tephritidae) is the rearing of larvae. This involves the adoption of low-cost suitable diets by using raw ingredients that allow the mass production of insects with high-quality life history traits that ensure the sexual competitiveness for the success of the SIT. Currently in mass-rearing facilities, quality parameters of the insects produced are well established (e.g., larval weight, pupal weight, percentage of fliers; FAO/IAEA/USDA 2014). However, parameters related to the quality of ingredients destined to the preparation of artificial larval diets are defined according to the average of the previous 5-year period record.

Corncob fractions are a key ingredient for the mass rearing of larvae of fruit flies of the genus *Anastrepha* at the Moscafrut mass rearing facility SADER-SENASICA located in Metapa de Domínguez, Chiapas, Mexico (Orozco-Dávila et al. 2017). However, as a derivative product of corn, it has significant disadvantages for rearing fruit flies, including quality variability, which depends on season of the year and cultivar, and susceptibility to contamination by microorganisms that produce mycotoxins, which can inhibit larval development or cause high larval mortality (Aceituno-Medina et al. 2016). Mycotoxins are toxic secondary metabolites produced by fungi. Contamination may occur in the field during crop growth or during the storage of food-derivative products (Ji et al. 2016). It is estimated that about 25% of world crops contain mycotoxins (Iheshiulor et al. 2011).

Corn derivatives are usually infested by fungi from the genera *Aspergillus* and *Fusarium*, and these organisms produce mycotoxins (Hernández et al. 2007, Montes et al. 2009, Perrone et al. 2014, Kara et al. 2015, Majid et al. 2015). *Aspergillus* grows on food products and contaminates them with aflatoxins, which are the result of the interaction among fungi, the host, and the environment. The interaction of such factors determines the infection and colonization of the substrate, as well as the type and quantity of aflatoxins produced (García and Heredia 2006). *Fusarium* causes high insect mortality and shows high specificity for certain insect species. For example, *Fusarium oxysporum* Schltdl. (Hypocreales: Nectraciales) causes 100% of mortality in the guava shield scale, *Pulvinaria* (*Chloropulvinaria*) *psidii* Maskel (Hemiptera: Coccidae), at 5 days after treatment with 4.8×10^8 conidia/mL (Gopalakrishnan and Narayan 1989) and causes 100% of mortality in *Nilaparvata lugens* Stål (Hemiptera: Delphacidae) at 3 days after treatment (Kuruvilla and Jacob 1979).

Another major concern for the mass rearing and production of any insect is the risk of bacterial contamination of the larval diet (Sikorowski et al. 2001, Cohen 2004). For example, Sikorowski et al. (2001) found that *Stenotrophomonas* (*Pseudomonas*) *maltophilia* Palleroni and Bradbury (Xanthomonadales: Xanthomonadaceae) caused high mortality in the parasitoid wasp *Microplitis croceipes* (Cresson) (Hymenoptera: Braconidae). Bacterial species included in the genera *Enterobacter, Proteus*, and *Serratia* can also become facultative pathogens of insects (Tanada and Kaya 1993, Sikorowski et al. 2001). A concentration of 10^5 colony-forming units (CFU)/mL or higher levels of inoculum of *Morganella* spp. resulted in 100% of mortality in larvae of *Anastrepha ludens* (Loew) (Diptera: Tephritidae) (Salas et al. 2017).

Assuring that larval diet ingredients meet all established quality parameters is the first step in the mass-rearing production process. Nevertheless, evaluation and acceptance protocols currently established at the Moscafrut Facility as quality parameters for ingredients (Hernández-Ibarra et al. 2015) do not include tests for entomopathogens, bacteria, or fungi and their toxins. In 2014, the mass rearing of *A. ludens* and *Anastrepha obliqua* (Macquart) (Diptera: Tephritidae) in the Moscafrut Facility was seriously threatened due to a problem of an undetected fungal and mycotoxin contamination in several batches of the corncob fractions used as a larval diet bulking agent (Aceituno-Medina et al. 2016). Therefore, the objective of this chapter is to report the results and conclusions of an experiment designed to determine the toxicological effect of corncob fractions on the larval performance of *A. obliqua*.

13.2 MATERIALS AND METHODS

Sixteen batches of corncob fractions were sampled. One batch was composed of 26 pallets containing 40 packages of corncob fractions of 20 kg each. The bacteria and fungi were isolated from a sample of 2.6 kg of corncob fractions per batch, which was obtained by sampling 3 packages per pallet to complete 100 g for each of the 26 pallets of a batch. The sample was taken from the entire vertical profile of the package stacked horizontally on the pallet, which was selected randomly.

13.2.1 Identification of Bacteria and Fungi

The identification of *Morganella morganii* Brenner and *Serratia marcescens* Bizio (Enterobacteriales: Enterobacteriaceae) was performed through molecular techniques. Bacteria were isolated from corncob fractions by using specific culture media. Bacterial DNA was extracted using the boiling and freezing method (15 min) and was then subjected to amplification by means of the polymerase chain reaction (PCR) technique using primers ERIC1 (5′-TGAATCCCCAGGAGCTTACAT-3′) and ERIC2 (5′- AAGTAAGTGACTGGGGTGAGCG-3′). A restriction fragment polymorphism analysis was performed after electrophoresis in 2% agarose gel, followed by visualization in an ultraviolet (UV) transilluminator. Samples were identified by using the pyrosequencing technique.

The identification of *Fusarium* spp. was also performed by using molecular techniques. Isolates were grown in a 125 mL Erlenmeyer flask with 50 mL of potato and dextrose culture medium under agitation at 30°C for 4 days. The obtained mycelium was filtered with Whatman # 1 paper under vacuum at −20°C and lyophilized. Approximately 100 mg of mycelium were used to extract the DNA following the modified method of cetyl-trimethyl-ammonium bromide (CTAB) (Murray and Thompson 1980). We added 750 μL of 2X CTAB buffer (100 mM Tris-ClH, pH 8, 100 mM EDTA, 250 mM NaCl, 2% CTAB) and 15 μL of 2-mercaptoethanol to the mycelium contained in an Eppendorf tube and incubated it at 65°C during 30 min. We then added 300 μL of 3M potassium acetate pH 4.8 and centrifuged it at maximum speed for 10 min. The supernatant was transferred to a new tube with 200 μL phenol and 200 μL chloroform-to-isoamyl alcohol (24: 1). It was centrifuged at maximum speed until the upper phase remained clear. The upper phase was transferred to a new tube, 750 μL of cold isopropanol were added, and the sample was centrifuged again. The obtained DNA precipitate was washed twice with 500 μL of 70% ethanol, dried at room temperature, and resuspended in 50 μL sterile bidistilled water. The quality and concentration of the DNA was verified by an electrophoretic run on 1% agarose gel. The concentration was standardized at 10 ng/μL. All DNA samples were amplified in duplicate. Each reaction was carried out in a final volume of 20 μL containing 0.5 mM of each primer, 1X buffer (Invitrogen Carlsbad, CA), 1.5 mM $MgCl_2$, 0.2 mM of each dNTP, 1U of Taq DNA polymerase (Invitrogen Carlsbad, CA), and 10 ng of DNA. PCR reactions were carried out in a Mastercycler gradient thermal cycler (Eppendorf AG, Hamburg Germany). The thermal cycle started at 94°C for 2 minutes, followed by 30 cycles of denaturation for 30 seconds, hybridization at the temperature of each primer for 1 minute and extension at 72°C for 1 minute, and ending with 5 minutes of extension at 72°C. The species-specific primers used were: Fg16NF/R

Fusarium graminearum, FP82F/R for *F. oxysporum*, FACF/R for *Fusarium acuminatum*, and J1Af for *Fusarium avenaceum*. Hybridization temperature was 62°C for *F. graminearum*, 56°C for *F. oxysporum*, and 57°C for *F. acuminatum* and *F. avenaceum*. The resulting PCR products were visualized on 2% agarose gels stained with ethidium bromide. Electrophoresis was run for 1 h at 120V. Molecular markers used to determine the molecular weight of the bands were 100 and 400 bp (Invitrogen). Batches were classified as not-contaminated and contaminated. A batch was classified as contaminated when at least one colony of any of the following species was isolated: *M. morganii*, *S. marcescens F. oxysporum*, *F. graminearum*, and *Aspergillus* spp. The experimental unit consisted of a tray (250 mL) containing 100 g of diet.

13.2.2 Toxicological Test for Larval Survival

A toxicological test was applied to 16 batches and the response variable was larval survival. Larval survival was defined in terms of yield, which was expressed as the number of larvae obtained per gram of diet. In the case of *A. obliqua*, for an artificial diet inoculated with 3.62 eggs per gram of diet, the yield was established at 2.7 larvae per gram of diet, which equals to 75% of larval survival. A yield of 2.7 larvae per gram of diet is the standard production parameter that ensures the goals of production. This value was defined according to the minimum average of the previous 5-year period. Batches were classified as suitable (yield > 2.7 larvae per gram of diet) and unsuitable (yield <2.7 larvae per gram of diet).

The diet was prepared following the recipe and procedures of the Moscafrut Facility: 19% corncob fractions (Mafornu, Cd. Guzman, Jalisco, Mexico), 5.3% corn flour (Maíz Industrializado del Sureste, Arriaga, Chiapas, Mexico), 7% Torula yeast (Lake States, Div. Rhinelander Paper, Rhinelander, WI, USA), 9.2% sugar (Ingenio Huixtla, Chiapas, Mexico), 0.4% sodium benzoate (Cia. Universal de Industrias, S.A. de C.V., Mexico), 0.2% nipagin (Mallinckrodt Specialty, Chemicals Co. St. Louis Mis.), and 0.44% citric acid (Anhidro Acidulantes FNEUM, Mexana S. A. de C.V., Morelos, Mexico). All the dry ingredients were mixed using a CRT mixer (Model CPM-30, 127 V, 60 Hz, 1100 W, CRT Global S.A. de C.V., Santa Catarina, N.L., Mexico) for 5 minutes, and then water (58.46%) was added and they were mixed for an additional 5 min. The experimental unit consisted of a tray (250 mL) containing 100 g of diet. Three experimental units (replications) were independently inoculated with a density of five 4-day-old eggs per gram of diet. Before inoculation, eggs were incubated until they reached 30% of hatching, after which they were disinfected with chlorine (100 ppm) and suspended in 28.5 mL of 0.4% gum agar solution. The eggs used in this study were provided by the Moscafrut Facility. Insects were reared following the procedures described by Stevens (1991) and Artiaga-López et al. (2004).

During the first 2 d after egg seeding, trays were kept at 29 ± 1°C and 90% relative humidity (RH) to allow the completion of egg hatching. Thereafter, trays were kept at 27°C ± 1°C and 85%–90% RH for another 7 d until larvae were mature. The separation of mature larvae from the diet was done by using the Venturi-bubbling system (Kuo and Acharya 2012), which consists in injecting pressurized air and water (approximately 10 liters per 5 kg-diet tray) into the larval diet to dilute the diet. This allows separating third instar larvae using a sieve no. 14 (1.41 mm hole size). The recovered larvae were counted and weighed and the yield was estimated.

13.2.3 Mycotoxin Determination

Batches were also inspected to determine the presence of the following mycotoxins: aflatoxin, fumonisin, ochratoxin, T2-HT2, zearalenone, and DON. The presence of mycotoxins was evaluated both qualitatively and quantitatively. Analyses were carried out with diluted extracts of each corncob powder sample using a lateral flow immunoassay technique with specific test strips-ROSA® for each type of mycotoxin. The 10-g sample of corncob fractions was dissolved and mixed, and the extract was separated, clarified, and then diluted. An aliquot was placed in the corresponding container

TABLE 13.1
Incorporation of Mycotoxin Sequestrants in the Artificial Diet for *Anastrepha obliqua*

Treatment Name	Noncontaminated Corncob Fraction (g/kg diet)	Contaminated Corncob Fraction (g/kg diet)	Carbovet™ (g/kg diet)	Aflaban™ (g/kg diet)
Positive control	182	0	0	0
Negative control	0	182	0	0
Carbovet™(ALLTECH)	0	182	3	0
Aflaban™(ALLTECH)	0	182	0	5

strip, and then it was incubated in the Charm E2-M system following the methods described in the Manual for Complete Test Procedures. The specific test strips used were: (1) ROSA® FAST Aflatoxin Quantitative Test, (2) ROSA® FAST5 DON Quantitative Test, (3) ROSA® FAST5 Fumonisin Quantitative Test, (4) ROSA® Ochratoxin Quantitative Test, (5) ROSA® T2-HT2 Quantitative Test, and (6) ROSA® FAST5 Zearalenone Quantitative Test. Batches were classified as contaminated when at least one of the aforementioned mycotoxins was detected.

13.2.4 Effect of the Incorporation of Mycotoxin Sequestrants

Among the strategies used to reduce the toxic effects of mycotoxins on agricultural livestock, the inclusion of sequestering agents in diets is the most common practice. The test to determine the effect of the addition of sequestrants consisted of a one-way completely randomized design, where treatments were defined by the sequestrant types described in Table 13.1. Two control diets were included, a "positive control" that contained noncontaminated corncob fractions and a "negative control" containing contaminated corncob fractions as a bulking agent. Mycotoxin sequestrants were incorporated by adding the optimum concentration determined in previous experiments. The preparation and management of the diets were as described in the previous experiments. The response variables quantified in this experiment were larval yield (larvae/g of diet) and larval weight (mg). The experimental unit consisted of a tray with 5.5 kg of diet containing a given mycotoxin sequestrant. Each tray was inoculated with 1.2 mL of eggs (~16,600 eggs/mL).

13.2.5 Data Analysis

Yield data from diets prepared using 16 different batches of corncob fractions was analyzed with two approaches. One involved a generalized linear model (GLM) with Poisson distribution and a log function where the fixed factor corresponded to batch category (contaminated or non-contaminated). The second approach involved a multiple regression equation, $Y = a + b_{1 \times 1} + b_{2 \times 2} + b_{3 \times 3} + b_{4 \times 4} + b_{5 \times 5}$, that describes the effect of each microorganism on larval yield (Y) per gram of diet (Lang 2007). The effects of concentration of mycotoxins on larval weight were analyzed using an ANOVA with an arc-sin data transformation. Analyses were carried out with R Statistical Software (R Development Core Team 2014).

13.3 RESULTS

13.3.1 Identification of Bacteria and Fungi

Two species of bacteria (*M. morganii, S. marcescens*) and three species of fungi (*F. oxysporum, F. graminearum, Aspergillus* spp.) were isolated from 16 batches, each with 24–26 pallets of corncob fractions (Table 13.2).

TABLE 13.2
Contamination Levels of Bacteria and Fungi, and Yield of *Anastrepha obliqua* in 16 Corncob-Based Batches of Larval Diet

	Bacteria		Fungi			
Batch	*Morganella morganii*	*Serratia marcescens*	*Fusarium oxysporum*	*Fusarium graminearum*	*Aspergillus* spp.	Yield (larvae/g of diet)
B01	–	–	–	–	–	3.32 ± 0.84
B02	–	–	–	–	–	3.43 ± 0.79
B03	–	–	–	+	+	0.25 ± 0.05
B04	+	–	++	–	+	0.39 ± 0.16
B05	–	–	+	–	+	0.62 ± 0.18
B06	+	–	++	++	+	0.52 ± 0.22
B07	+	–	++	–	–	2.00 ± 0.19
B08	+	+	–	–	–	2.15 ± 0.53
B09	–	–	+	–	–	1.77 ± 0.31
B10	–	–	++	–	–	1.77 ± 0.80
B11	+	–	–	–	+	1.56 ± 0.11
B12	–	–	++	–	–	2.02 ± 0.34
B13	+	–	–	–	–	1.04 ± 0.22
B14	–	+	++	–	++	0.17 ± 0.09
B15	–	+	–	–	–	3.46 ± 0.48
B16	+	–	–	–	–	3.41 ± 0.76

– Absent, + (10^2), ++ (10^3).

13.3.2 Toxicological Test for Larval Survival

Of the 16 batches evaluated, 4 were classified as acceptable (batches 1, 2, 15, and 16) and the rest were rejected (batches 3 to 14) because they did not meet the criterion of 2.7 larvae/g of diet yield (Table 13.2). The GLM assuming a Poisson distribution and a log function showed that yield was significantly affected when at least a colony of *M. morganii*, *S. marcescens*, *F. oxysporum*, *F. graminearum*, or *Aspergillus* spp. was isolated ($\chi^2 = 6.53$, df = 1, $P = 0.0106$). The presence of bacteria did not decrease larval yield (Tables 13.2 and 13.3). In contrast, fungi caused mortality when species were alone (*F. oxysporum*, *F. graminearum*, and *Aspergillus* spp.), together, or in association with the bacteria *M. morganii* or *S. marcescens*. Parameters of the multiple regression analysis are presented in Table 13.3.

TABLE 13.3
Parameters of the Multiple Regression Equation, $Y = a + b_1 \times_1 + b_2 \times_2 + b_3 \times_3 + b_4 \times_4 + b_5 \times_5$, that Describes the Effect of Microorganisms on Larval Yield per Gram of *Anastrepha obliqua*

Parameter	Estimate	F Ratio	Prob > F	R-Square
Intercept	2.6693			
Aspergillus spp.	−0.0176	14.6830	0.0033	0.6070
Fusarium oxysporum	−0.0068	2.2380	0.1655	0.6767
Morganella morganii	−0.0104	0.0580	0.8147	0.6778
Serratia marcescens	0.0105	0.0390	0.8472	0.6786
Fusarium graminearum	0.0032	0.1100	0.7473	0.6821

13.3.3 Mycotoxin Determination

Aflatoxin, fumonisin, ochratoxin, and T2-HT2 toxins were detected only in the contaminated corncob powder batches. Zearalenone and DON were detected in both contaminated and noncontaminated corncob fraction samples (Table 13.4).

13.3.4 Effect of the Incorporation of Mycotoxin Sequestrants

The diet prepared with contaminated corncob fractions (negative control) showed the lowest yield. Diets containing Aflaban™, a mycotoxin sequestrant, presented higher yield in comparison with the results obtained for the negative control and showed no significant difference with the positive control (Table 13.5).

TABLE 13.4
Mycotoxins Isolated from Contaminated Corncob Powder

	Uncontaminated Corncob Fractions			Contaminated Corncob Fractions		
Mycotoxin	Mean	SE	Significance	Mean	SE	Significance
Aflatoxin (ppm)	0.00	0.00	B	9.50	2.00	a
Fumonisin (ppm)	1.00	0.20	Ab	2.40	0.80	a
Ochratoxin (ppb)	0.00	0.00	B	1.00	0.00	a
T2-HT2 toxin (ppb)	0.00	0.00	B	1.00	0.00	a
Zearalenone (ppb)	102.00	32.70	A	106.90	30.50	a
DON (ppm)	1.20	0.60	ab	0.70	0.50	a

Different letters in a column indicate significant differences ($P \leq 0.05$).

TABLE 13.5
Yield and Larval Weight of *Anastrepha obliqua* Obtained from Diets That Were Supplied with Mycotoxin Sequestrants in Contaminated Corncob Powder Used to Prepare the Artificial Diet

	Yield (No. larvae/gram of diet)			Larval Weight (mg)		
Sequestrant	Mean	SE	Significance	Mean	SE	Significance
Carbovet™	3.39	1.21	Ab	19.76	0.92	a
Aflaban™	4.06	0.26	A	19.36	0.42	a
Negative control	1.95	0.37	B	17.85	0.71	a
Positive control	4.33	0.34	A	20.83	0.33	a

Different letters in a column indicate significant differences ($P \leq 0.05$).

13.4 DISCUSSION

The main finding of our research demonstrated that the toxicological test used to evaluate the impact of corncob fraction contamination on the yield of mass-reared *A. obliqua* allowed us to classify a batch as acceptable or non-acceptable. A nonacceptable batch presented at least a colony of *M. morganii*, *S. marcescens*, *F. oxysporum*, *F. graminearum*, or *Aspergillus* spp., or at least detectable amounts of aflatoxin, fumonisin, ochratoxin, and T2-HT2. Fungi caused mortality when species were present alone (*Aspergillus* spp., *F. oxysporum*, *F. graminearum*), together, or in association with the bacteria *M. morganii* or *S. marcescens*. The addition of Aflaban™ to a contaminated diet increased the yield, reducing the effects of the mycotoxins.

Our results are similar to those observed by Gopalakrishnan and Narayan (1989) and Kuruvilla and Jacob (1979), who found that *N. lugens* showed high mortality when infected with *F. oxysporum*, whereas *S. marcescens* was often found in insect infections (Pineda-Castellanos 2015). Here, we observed that *M. morganii* and *S. marcescens*, when present each by itself, showed low concentrations ($<10^2$ CFU/mL) and did not cause mortality. However, when they were present together or with some fungi species, they were a significant mortality factor for the first instar of *A. obliqua*. Salas et al. (2017) observed that an inoculum of *M. morganii* of 10^5 CFU/mL resulted in 100% mortality of larvae of *A. ludens*. The difference between our results and Salas et al. (2017) could be explained by the insecticidal pathogenic effect of fungi and their mycotoxins, which delay larval development (Zeng et al. 2006) and weaken *A. obliqua* larvae, thus compromising their immune system and making *M. morganii* and *S. marcescens* pathogenic opportunistic bacteria. According to Flyg et al. (1980), the immunity response of insects plays an important role in the overall pathogenicity of *S. marcescens*. In addition, the larval developmental medium modifies the pathogenicity of these bacteria, for example, larval diet (Sikorowski et al. 2001).

On the other hand, the presence of mycotoxins in the mass-rearing diet was an event that alerted the need to refocus quality parameters toward the specific detection of microorganisms and their metabolites that are potentially pathogenic for larvae of the genus *Anastrepha*. The mycotoxins produced by *Aspergillus*, such as aflatoxins, ochratoxins, citrinin, and sterigmatocystin, can contaminate food commodities and stored products such as flour (Kara et al. 2015) and maize (Perrone et al. 2014), and in our case, corncob fractions. Our results indicated that the mycotoxins aflatoxin, fumonisin, ochratoxin, and T2-HT2 were highly toxic for the larvae of *A. obliqua*. According to Teetor-Barsch and Roberts (1983), secondary metabolites, like mycotoxins such as trichothecenes (T2), contribute to insect mortality (e.g. termites, mealworms, flour beetles, maize borers, and blowflies). Variability of insect sensitivity to aflatoxin toxicity depends on the developmental stage. Here, we observed that when we inoculated the larval diet with a sample of eggs with 30% of hatching (70% eggs and 30% neonate larvae), even after 5 days, eggs of *A. obliqua* did not hatch and neonate larvae mortality was high. Kirk et al. (1971) determined that aflatoxin-B1 at 10 ppm prolonged larval and pupal stages and increased mortality in all developmental stages in *Drosophila melanogaster* Meigen (Diptera: Drosophilidae). Although, in general, the newly hatched first instar is the most sensitive stage, this mycotoxin also decreased egg-to-adult viability and fertility (Chinnici et al. 1976, Llewellyn and Chinnici 1978), as well as pupal weight and adult body length (Chinnici et al. 1979). Aflatoxins also have been tested for their toxicity to newly hatched larvae of *Heliothis virescens* F. (Lepidoptera: Noctuidae) (Gudauskas et al. 1967). *Aspergillus*-infected diets have also caused growth delay in the silkworm *Bombyx mori* L. (Lepidoptera: Bombycidae) (Ohtomo et al. 1975). Zeng et al. (2006) described the toxicity of AFB1 in *Helicoverpa zea* Boddie (Lepidoptera: Noctuidae) larvae at different larval stages. Tolerance to AFB1 increases with larval stages. Third instar larvae can tolerate 20 ppb AFB1 with only moderate reduction of pupal weight, whereas at 200 ppb, AFB1 significantly affected mortality, pupation rate, pupae development, pupal weight, and resulted in malformation of pupae.

The negative effect of aflatoxin, fumonisin, ochratoxin, and T2-HT2 was counteracted by adding the sequestrant Aflaban™. Zearalenone and DON did not show any effects. Teetor-Barsch and Roberts (1983) observed that zearalenone (F-2) exhibited a beneficial effect on egg production in flour beetles.

In insects, the impact or function of *M. morganii* and *S. marcescens* and their association with fungi of the genera *Fusarium* and *Aspergillus* and their mycotoxins are varied, from transmitters of microorganisms to acting as pathogens to the host. Reports on the impact of these fungi and their mycotoxins on insects are still limited and rarely reported. This is why it is recommended to continue with this type of experiments aimed at explaining the effect of some toxic agents on the survival and reproduction of mass reared fruit flies.

13.5 CONCLUSIONS

The present research is the first evidence demonstrating that, under *A. obliqua* mass-rearing conditions, the presence of *Aspergillus* spp., *F. oxysporum*, and *F. graminearum* and their mycotoxins (Aflatoxin, ochratoxin, T2-HT2) caused mortality when present each by itself, together, or in association with the bacteria *M. morganii* or *S. marcescens*. The addition of Aflaban™ to a contaminated diet increased the yield, reducing the effects of the mycotoxins. We recommend including a toxicological quality control test to detect the presence of bacteria (*M. morganii, S. marcescens*) and fungi (*Aspergillus* spp., *F. oxysporum*, and *F. graminearum*) and their mycotoxins in corncob fractions to classify them as acceptable or non-acceptable (i.e., to be rejected) batches for the mass rearing of *A. obliqua*.

ACKNOWLEDGMENTS

The authors express their gratitude to Margoth García, Julio Lanza, and Pedro Rivera for their technical support. This project was funded by Moscafrut Program SADER-SENASICA Project "Formulación de dietas larvarias y de adulto de *Anastrepha ludens* y *A. obliqua* (Diptera: Tephritidae) con ingredientes alternativos – SDM-004/2013-2018."

REFERENCES

Aceituno-Medina M., C. Ventura, and E. Hernández. 2016. A novel approach for quality assurance of bulking agents for larval mass-rearing of *Anastrepha ludens*: Mycotoxins content. In *9th Meeting of the Tephritid Workers of the Western Hemisphere, Buenos Aires, Argentina*. Abstract Book. p. 119. http://9twwh.senasa.gob.ar/sites/default/files/libro_de_resumenes_9twwh.pdf. Accessed August 26, 2019.

Artiaga-López T., E. Hernández, J. Domínguez-Gordillo, D. S. Moreno, and D. Orozco- Dávila. 2004. Mass-production of *Anastrepha obliqua* at the Moscafrut Fruit Fly Facility, Mexico. In *Proceedings of the 6th International Symposium on Fruit Flies of Economic Importance*. May 6–10, 2002, ed. B. N. Barnes, pp. 389–392. Stellenbosch, South Africa: Isteg Scientific Publications, Irene, South Africa.

Chinnici J. P., M. A. Booker, and G. C. Llewellyn. 1976. Effect of aflatoxin B1 on viability, growth, fertility, and crossing over in *Drosophila melanogaster* (Diptera). *Journal of Invertebrate Pathology* 27:255–258.

Chinnici J. P., L. Erlanger, M. Charnock, M. Jones, and J. Stein. 1979. Sensitivity differences displayed by *Drosophila melanogaster* larvae of different ages to the toxic effects of growth on media containing aflatoxin B1. *Chemico-Biological Interactions* 24:373–380.

Cohen A. C. 2004. *Insect Diets: Science and Technology*. Boca Raton, FL: CRC Press.

FAO/IAEA/USDA (2014) *Product Quality Control and Shipping Procedures for Sterile Mass-Reared Tephritid Fruit Flies*. Joint FAO/IAEA Programme of Nuclear Techniques in Food and Agriculture. Manual, Version 5.0. IAEA, Vienna, Austria. 85 pp. Available at: http://www-naweb.iaea.org/nafa/ipc/public/ipc-mass-reared-tephritid.html.

Flyg C., K. Kenne, and H. G. Boman. 1980. Insect pathogenic properties of *Serratia marcescens*: Phage-resistant mutants with a decreased resistance to *Cecropia* immunity and a decreased virulence to *Drosophila*. *Journal of General Microbiology* 120:173–181.

García S., and N. Heredia. 2006. Mycotoxins in Mexico: Epidemiology, management, and control strategies. *Mycophatologia* 162:255–264.

Gopalakrishnan C., and K. Narayanan. 1989. Occurrence of *Fusarium oxysporum* Schlecht and its pathogenicity on guava scale *Chloropulvinaria psidii* Maskell (Hemiptera: Coccidae). *Current Science* 58:92–93.

Gudauskas R. T., N. D. Davis, and U. L. Diener. 1967. Sensitivity of *Heliothis virescens* larvae to aflatoxin in ad libitum feeding. *Journal of Invertebrate Pathology* 9:132–133.

Hernández D. S., L. A. Reyes, M. C. A. Reyes, O. J. G. García, and P. N. Mayek. 2007. Incidencia de hongos potencialmente toxígenos en maíz (*Zea mays* L.) almacenado y cultivado en el norte de Tamaulipas, México. *Revista Mexicana de Fitopatología* 25:127–133.

Hernández-Ibarra M. D. R., J. J. Bravo-López, C. A. Reyna-Cigarroa, J. C. Coutiño-Molina, R. Villatoro-Telles, and D. D. Meza-Arriaga. 2015. ITCC01—Instructivo de Trabajo del Laboratorio de Control de Calidad del Proceso de *Anastrepha* spp. Dirección General de Sanidad Vegetal/ Dirección del

Programa Nacional de Moscas de la Fruta/ Subdirección de Producción Moscafrut. 2016. Metapa de Domínguez, Chiapas: Secretaría de Agricultura, Ganadería, Desarrollo Rural, Pesca y Alimentación. Servicio Nacional de Sanidad, Inocuidad y Calidad Agroalimentaria.

Iheshiulor O. O. M., B. O. Esonu, O. K. Chuwuka, A. A. Omede, I. C. Okoli, and I. P. Ogbuewu. 2011. Effects of mycotoxins in animal nutrition: A Review. *Asian Journal of Animal Sciences* 5:19–33.

Ji C., Y. Fan, and L. Zhao. 2016. Review on biological degradation of mycotoxins. *Animal Nutrition* 2:127–133.

Kara G. N., F. Ozbey, and B. Kabak. 2015. Co-occurrence of aflatoxins and ochratoxin A in cereal flours commercialised in Turkey. *Food Control* 54:275–281.

Kirk H. D., A. B. Ewen, H. E. Emson, and D. G. Blair. 1971. Effect of aflatoxin B1 on development of *Drosophila melanogaster* (Diptera). *Journal of Invertebrate Pathology* 18:313–315.

Kuo K. K., and R. Acharya. 2012. *Fundamentals of Turbulent and Multiphase Combustion.* Hoboken, NJ: John Wiley & Sons.

Kuruvilla S., and A. Jacob. 1979. Comparative susceptibility of nymphs and adults of *Nilaparvata lugens* to *Fusarium oxysporum* and its use in microbial control. *Agricultural Research Journal Kerala* 17:287–288.

Lang, T. 2007. Documenting research in scientific articles: Guidelines for authors: 3. Reporting multivariate analyses. *Chest Journal* 131:628–632. doi:10.1378/chest.06-2088.

Llewellyn G. C., and J. P. Chinnici.1978. Variation in sensitivity to aflatoxin B1 among several strains of *Drosophila melanogaster* (Diptera). *Journal of Invertebrate Pathology* 31:37–40.

Majid A. H .A, Z. Zahran, A. H. A. Rahim, N. A. Ismail, W. A. Rahman, K. S. M. Zubairi, H. Dieng, and T. Satho. 2015. Morphological and molecular characterization of fungus isolated from tropical bed bugs in Northern Peninsular Malaysia, *Cimex hemipterus* (Hemiptera: Cimicidae). *Asian Pacific Journal of Tropical Biomedicine* 5(9):707–713.

Montes G. N., M. C.A. Reyes, R. N. Montes, and A. M. A. Cantú. 2009. Incidence of potentially toxigenic fungi in maize (*Zea mays* L.) grain used as food and animal feed. *Journal of Food* 7:119–125.

Murray M. G., and W. F. Thompson.1980. Rapid isolation of high molecular weight plant DNA. *Nucleic Acids Research* 8:4321–4325.

Ohtomo T., S. Murakoshi, J. Sugiyama, and H. Kurata. 1975. Detection of aflatoxin B1 in silkworm larvae attacked by an *Aspergillus flavus* isolate from a sericultural farm. *Applied Microbiology* 30:1034–1035.

Orozco-Dávila D., L. Quintero, E. Hernández, E. Solís, T. Artiaga, R. Hernández, C. Ortega, and P. Montoya. 2017. Mass rearing and sterile insect releases for the control of *Anastrepha* spp. pests in Mexico—A review. *Entomologia Experimentalist et Applicata* 164:176–187.

Perrone G., M. Haidukowski, G. Stea, F. Epifani, R. Bandyopadhyay, and J. F. Leslie. 2014. Population structure and aflatoxin production by *Aspergillus* Sect. Flavi from maize in Nigeria and Ghana. *Food Microbiology* 41:52–59.

Pineda-Castellanos M. L., Z. Rodríguez-Segura, F. J. Villalobos, L. Hernández, L. Lina, and M.E. Nuñez-Valdez. 2015. Pathogenicity of isolates of *Serratia marcescens* towards larvae of the scarab *Phyllophaga Blanchardi* (Coleoptera). *Pathogens* 4:210–228.

R Development Core Team. (2014). R: A language and environment for statistical computing. R Foundation for Statistical Computing, Vienna, Austria. http://www. Rproject.org/. Accessed August 26, 2019.

Salas B., H. E. Conway, E. L. Schuenzel, K. Hopperstad, C. Vitek, and D. C. Vacek. 2017. *Morganella morganii* (Enterobacterailes: Enterobacteriaceae) is a lethal pathogen of Mexican fruit fly (Diptera: Tephritidae) larvae. *Florida Entomologist* 100:743–751.

Sikorowski P. P., A. M. Lawrence, and G. D. Inglis. 2001. Effects of *Serratia marcescens* on rearing of the tobacco budworm (Lepidoptera: Noctuidae). *American Entomology* 47:51–60.

Stevens L. 1991. Manual of Standard Operating Procedures (SOP) for the Mass-Rearing and Sterilization of the Mexican Fruit Fly, *Anastrepha ludens* (*Loew*). South Central Region, Mission, TX: USDA-APHIS.

Tanada Y., and H. K. Kaya. 1993. *Insect Pathology*, San Diego, CA: Academic Press.

Teetor-Barsch G. H., and D. W. Roberts.1983. Entomogenous *Fusarium* species. *Mycopathologia* 84:3–16.

Zeng R. S., G. Niu, Z. Wen, M. A. Schuler, and M. R. Berenbaum. 2006. Toxicity of aflatoxin B1 to *Helicoverpa zea* and bioactivation by cytochrome P450 monooxygenases. *Journal of Chemical Ecology* 32:1459–1471.

14 Exploring Cost-Effective SIT

Verification via Simulation of an Approach Integrating Reproductive Interference with Regular Sterile Insect Release

Atsushi Honma[*] *and Yusuke Ikegawa*

CONTENTS

Abstract The sterile insect technique (SIT) is one of the most effective methods to control tephritid pests. However, sustainable production and release of sterilized insects is economically costly. Recently, a new approach incorporating interspecific negative mating interaction, known as reproductive interference, into a pest control program using regular SIT (called "sterile interference") has been proposed. Sterile interference would add value to regular SIT because one could control multiple pest species by releasing only the sterile insect of the main target species. To verify the effectiveness of the combined approach for a pest control program, we conducted a simulation analysis. The result suggests that, even with weak reproductive interference, it is possible to control, with an acceptable level, both the wild-type of the main target species and a closely related pest species just by increasing sterile release. Additionally, when both species can be eradicated, the eradication occurred almost simultaneously. We conclude that this new approach may help to develop a more cost-effective and value-added pest management program.

14.1 INTRODUCTION

The sterile insect technique (SIT) is an environment-friendly pest control method, which is often part of large-scale integrated pest management (IPM) programs and, at present, is widely applied to control Tephritids (Dyck et al. 2005). In this chapter, we explore the possibility of introducing novel technology to improve cost-effectiveness of SIT programs by making use of sterilized insects not only to suppress the wild populations of the same pest species but also that of a closely related pest species through reproductive interference.

Reproductive interference refers to any kind of interspecific mating interaction that reduces the fitness of the species involved (Gröning and Hochkirch 2008). It includes interspecific sexual

[*] Corresponding author.

harassment that can reduce individual reproductive success (e.g., Kishi and Nishida 2009; Friberg et al. 2013; Kitano et al. 2018), as well as interspecific copulation and hybridization. Interspecific copulation and hybridization can cause "direct" fitness reduction (i.e., physical damage on sexual organs and gamete loss), whereas interspecific sexual harassment can cause "indirect" fitness reduction. Females lose opportunities to copulate with conspecific males (Friberg et al. 2013; Kitano et al. 2018) or for oviposition (Kishi and Nishida 2009). Reproductive interference can cause the extinction of a competitor species through a positive feedback mechanism (Kuno 1992), which is also the case with SIT: A reduction in population density amplifies the suppressive effect, leading to an accelerated decline in successive generations. Therefore, reproductive interference can cause species exclusion much easier than resource competition (Kuno 1992). Several recent studies have reported displacement of resident species by invasive species both in plants (Takakura et al. 2009; Takakura 2013) and animals (e.g., Liu et al. 2007; Bargielowski et al. 2013). There is also an example of an attempt to control a pest species using reproductive interference before the development of SIT (Vanderplank 1944; Klassen and Curtis 2005). To suppress the population of a tsetse fly, *Glossina swynnertoni* Austen, hybrid sterility with a closely related species, *Glossina morsitans*, was exploited. Soon after the start of mass releases of the latter species, population density of the former species drastically decreased. Synchronous increase of hybrid individuals indicated that reproductive interference was the cause of the population decline in *G. swynnertoni* (Vanderplank 1944; Klassen and Curtis 2005; see also figure 3 in Honma et al. 2019).

Reproductive interference and SIT operate on the same theoretical principles in reducing the population density of the focal species:

$$\frac{dN}{dt} = b\left(\frac{N}{N + rN_h}\right)N - dN - hN^2$$

This simple differential equation model was proposed by Kuno (1992) to describe competitive interaction through reproductive interference. In this equation, N and N_h are the population densities of the focal species and the heterospecies; b and d are the birth and death rates of the focal species; h is the crowding effect coefficient; and r is the coefficient expressing the intensity of reproductive interference. By replacing the density of the heterospecific species, that is, N_h, with the density of the sterile insects, that is, S, in this equation, we obtain a model describing the effect of sterile-insect release on the wild populations of the pest species targeted in an SIT program:

$$\frac{dN}{dt} = b\left(\frac{N}{N + cS}\right)N - dN - hN^2$$

where c is an index of the sexual competitiveness of the sterilized males relative to that of wild males (Haisch 1970; Fried 1971), which coincides with the coefficient expressing the intensity of reproductive interference, r.

The theoretical similarity between these two mechanisms suggests that by combining SIT and reproductive interference, one can suppress or eradicate a pest species targeted in the program and its closely related species, both through a single operation (called "sterile interference" in Honma et al. 2019). It is to be noted that the suppressive effect of a sterile insect against the related species (i.e., reproductive interference) is, in general, weaker than that against the wild-type of the target species because mating attempts in the former would more often be terminated before "successful" interspecific mating and hybridization occur. Therefore, Honma et al. (2019) conducted a simple simulation analysis to demonstrate that sterile interference can eradicate the latter just by increasing the number of sterile releases within an acceptable level, even when the interference effect from the sterile insects to the related species is weak. The model, however, did not consider the population dynamics of the wild-type of the target species. In practical pest-eradication programs using SIT, not only the availability of eradication but also the time to accomplish the eradication is important. To make SIT more effective by sterile interference, the related species should be eradicated without long periods of sterile release after the eradication of the target species. Thus, we constructed a theoretical model describing population dynamics of two closely related pest species (the target species and the related species) incorporating reproductive interference by sterile target insects on the related species. With the model, we conducted numerical simulations assuming that reproductive interference is so weak that the two species can coexist without sterile release; it verifies if it is possible to eradicate the related species (i) with an acceptable level of increase in the number of sterile insects released and (ii) within an acceptable extended period of sterile releases.

14.2 THE MODEL

Let us assume that we have two pest species competing through reproductive interference and that we must try to suppress both pests simultaneously by releasing sterile insects of one of the two species. For simplicity, we assume that the birth and death rates are identical in the two species, and reproductive interference is unidirectional (i.e., it is directed by the target species toward the related species only). The suppression effect of releasing sterile insects on the target species (i.e., normal SIT) can be described as follows:

$$\frac{dN_1}{dt} = b\left(\frac{N_1}{N_1 + cS}\right)N_1 - dN_1 - hN_1^2$$

where N_1 is density of wild populations of the target pest species. In this condition, the suppression effect of releasing sterile insects on the related species through reproductive interference can be described through the following differential equations:

$$\frac{dN_2}{dt} = b\left[\frac{N_2}{N_2 + r(N_1 + cS)}\right]N_2 - dN_2 - hN_2^2$$

where N_2 is population density of the pest species related to the target species.

To start the numerical simulation, we assume that the initial population densities of the two species, $N_1(0)$ and $N_2(0)$, are their equilibrium densities in the absence of the sterile insects ($S = 0$).

$$N_1(0) = \frac{b-d}{h}$$

$$N_2(0) = \frac{(1-r)(b-d) + \sqrt{\left[(1-r)(b-d)\right]^2 - 4rd(b-d)}}{2h}$$

To verify the effect of reproductive interference by the sterile target insect on the related species, we only consider the scenario where the target species and the related species coexist in the absence of sterile release. Thus, the intensity of reproductive interference, r, should be,

$$r < \frac{b + d - 2\sqrt{bd}}{b-d}$$

The parameter values other than r and S are $b = 2.2$, $d = 1$, $h = 0.00000048$, so that the equilibrium population density of the target species, when no sterilized insects are released (i.e., N1(0)), becomes 2,500,000. We set the index of sexual competitiveness of the mass-reared sterile insect, c, to be 0.7.

Under these circumstances, we conducted numerical simulations to verify the effect of the intensity of reproductive interference, *r*, and the number of released sterile insects, *S*, on the population size of the two species.

14.3 RESULTS AND DISCUSSION

The combined application of normal SIT and reproductive interference generated three types of outcomes, depending on the intensity of reproductive interference, *r*, and the number of released sterile insects, *S* (Figure 14.1): (1) both pests were eradicated (white area), (2) only the target species was eradicated (gray area), or (3) neither pest was eradicated (black area). However, for a broad range of parameter values for *r*, both species can be eradicated by releasing <10 times (= 25,000,000) more sterile insects than the initial population density of the target species at the initial state (i.e., before starting sterile release), $N_1(0)$ = 2,500,000 (Figures 14.1 and 14.2). The results suggest, at least in theory, that eradication of the related species through reproductive interference by the sterile individuals of the target species is possible, just by increasing the number of the released sterile insects (Figures 14.1 and 14.2). Ten times more than the estimated density of the wild population of the target pest is a basic criterion for sterile insect release in SIT programs (see Itô and Kawamoto 1979). Additionally, in the scenario where both species could be eradicated, the density of the two species reached zero almost simultaneously with reasonable levels of increase in sterile releases (Figure 14.3). The amount of time required for eradicating the related species relative to that for the target species is long only when the number of sterile insects released is the minimum level for eradicating both species. It rapidly approaches 1.0 (i.e., simultaneous eradication) as sterile release is slightly increased (Figure 14.3). This indicates that the eradication program targeting multiple species by sterile interference would not necessitate the continued sterile release long after the eradication of the main target species.

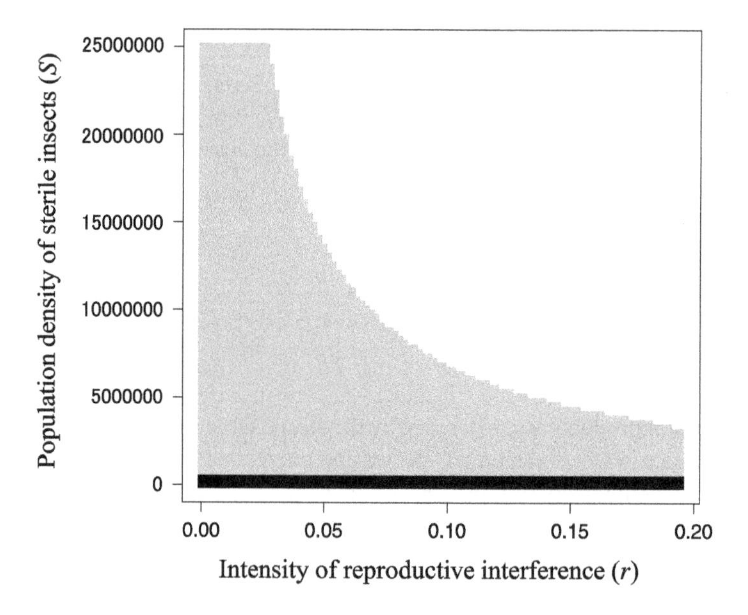

FIGURE 14.1 Outcome of the combined application of regular sterile insect technique (SIT) and sterile interference depending on the intensity of reproductive interference, *r*, and the number of released sterile insects, *S*. Both species are eradicated (white), only the target species is eradicated (gray), and neither of the two species is eradicated (black).

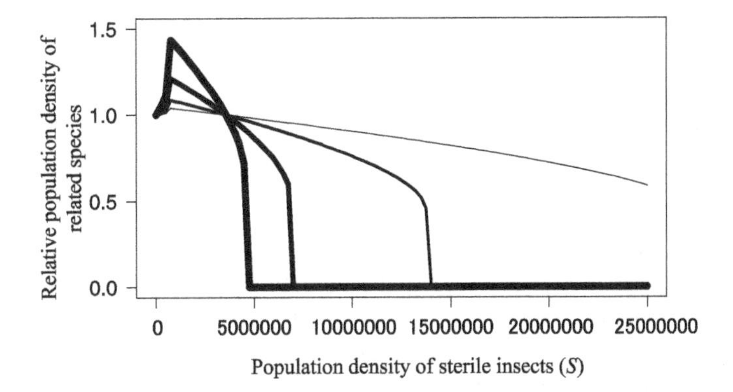

FIGURE 14.2 Effect of the intensity of reproductive interference, r, and the number of released sterile insects, S, on the population size of the pest species related to the target pest used in sterile insect technique (SIT). The vertical axis is the ratio of the equilibrium population density of the related species with sterile interference, N_2^*, divided by that without sterile interference, $N_2(0)$. Each line shows the results when the intensity of reproductive interference, $r = 0.0253, 0.0506, 0.1011, 0.1517$ (from thin to thick), respectively.

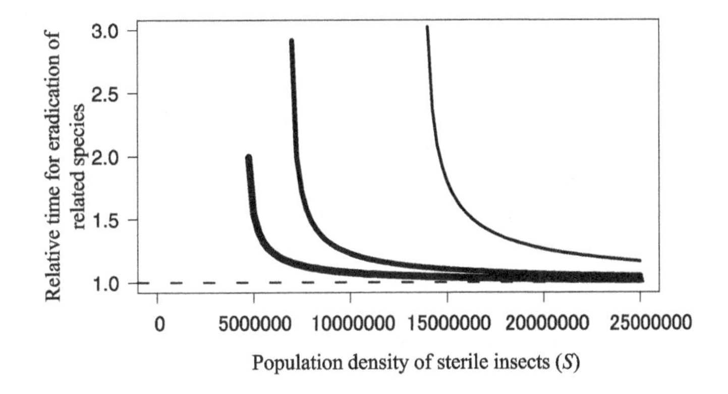

FIGURE 14.3 Amount of time required to eradicate the related species relative to the eradication of the target species. The vertical axis is the value of eradication time for the related species divided by that for the target species. When both species are eradicated simultaneously, the value becomes 1.0. Each line shows the results when the intensity of reproductive interference, $r = 0.0506, 0.1011, 0.1517$ (from thin to thick), respectively.

Although we assumed fixed reproductive characters in our model, it is important to notice that evolution of female mate preferences to reject mass-reared sterile males has been reported in some conventional SIT programs (Hibino and Iwahashi 1991; McInnis et al. 1996), which may result in a reduction of the suppressive effect of sterile interference. Similarly, reproductive character displacement (Konuma and Chiba 2007; Bargielowski et al. 2013) between sterile males and the females of the related species could be caused with this novel approach. A simulation model study, however, revealed that even if this type of behavioral resistance evolves in wild females, the reduction of the suppressive effect can be overcome by increasing the number of sterile-males released in regular SIT programs (Tsubaki and Bunroongsook 1990). Additionally, Takakura et al. (2015) modeled the evolutionary dynamics of signal traits and mate recognition in both sexes in two closely related species diverged through allopatric speciation. They examined whether interspecific sexual interactions diminish as a result of reproductive character displacement in the subsequent secondary contact of the two species. The model predicted that even after reproductive character displacement occurred (mainly in females), reproductive interference between the species would persist because males remain promiscuous. Similar modeling analyses should be conducted in SIT incorporating reproductive interference to verify the population outcome of both the target and the related species.

There are only limited studies reporting reproductive interference in Tephritid species. *Bactrocera carambolae* Drew & Hancock and *Bactrocera dorsalis* (Hendel) are a sibling species pair which are partly sympatric but exhibit resource partition. Kitano et al. (2018) conducted interspecific mating experiments with different frequencies in the proportion of each species and observed both intra- and interspecific mating interactions, and they also measured the reproductive success of each species (i.e., the number of offspring that pupate). In both species, males were promiscuous, whereas females nonselectively refused males' mating attempts. With an increase in frequency of the opponent species, mating attempts by males of the opponent species (i.e., interspecific sexual harassment) increased in both species, whereas frequency of interspecific mating did not change. In contrast, with the increase in frequency of the opponent species, successful mating with conspecific individuals decreased only in *B. carambolae*, leading to offspring reduction. The results indicate that interspecific sexual harassment was the main cause of the unidirectional reproductive interference in these fruit fly species and, more importantly, would deprive of mating opportunities, resulting in population decline of the inferior species, *B. carambolae.*

In Tephritid species, there are several cases of species displacement, indicating strong negative competitive interactions among species (Duyck et al. 2004). Contrarily, candidate mechanisms explaining the phenomenon, such as competition for resources among larvae and interference competition among adult females, do not seem to have major roles except in the case of heterogeneric competitive displacement (Ekesi et al. 2009); as such, it would be beneficial to verify hitherto unexplored reproductive interference as a candidate mechanism in these systems, not only by small scale lab experiments but in large-scale field experiments.

The novel approach can be adopted via relatively simple steps: first, choosing the target species which happens to be the "strongest" among all closely related species in a species group in terms of reproductive interference; although verifying the strength of reproductive interference of all possible pairs of pest species would be laborious, one can start with the species which has often displaced related species as a likely candidate. In the case of Tephritid fruit flies, for example, it would be *B. dorsalis* (Duyck et al. 2004; Honma et al. 2019). Secondly, according to the estimated strength of reproductive interference, one would choose strategies of sterile interference in the program (suppression, eradication, containment, prevention, or host limitation, see Honma et al. 2019), which is similar to regular SIT programs, and conduct sterile release of the target species. When reproductive interference is not too weak, one can accomplish eradication or containment with fewer sterile releases in a shorter period. In contrast, it would be better to choose the other options (suppression or host limitation) to save costs and time. In the present study, we only considered the situation of one pair of closely related pest species that coexist in natural conditions. However, the combined approach would be most effective for preventive release of sterile insects where the invasion risk of several related pests is high; since, to predict when and which species invades is impossible, managing all possible pests by SIT would be costly. The novel approach, however, may make it possible to control the invasion risk of all the members of a species group by releasing only the strongest species in terms of reproductive interference. Moreover, because the densities of the invading pests are usually low, they can be eradicated even if the effect of reproductive interference from heterospecific sterile insects is relatively weak.

Although the approach requires to increase the number of sterile males released, the cost would be largely reduced compared to the cost of constructing and running separate systems for multiple species. Regular SIT requires extensive infrastructure for each target species to rear a large number of insects. Additionally, even after the target species has been eradicated, preventive release of the sterile insects should be continued against its possible reinvasion. In an SIT program incorporating reproductive interference, it would be possible to control multiple pests with one operation.

In conclusion, the present study verified that the approach integrating reproductive interference with a sterile-insect–release program would be feasible and cost effective because the program makes it possible to suppress multiple pest species almost simultaneously by releasing sterilized males of a single species. We believe that this approach will add value to future pest management programs that involve the SIT.

ACKNOWLEDGMENTS

This study was supported by Grants-in-Aid for Scientific Research (No. 16K14864) from the Japan Society for the Promotion of Science. We thank Diana Pérez-Staples for editing the MS and two anonymous reviewers for their valuable comments on the earlier versions of the MS.

REFERENCES

Bargielowski I. E., L. P. Lounibos, and M. C. Carrasquilla. 2013. Evolution of resistance to satyrization through reproductive character displacement in populations of invasive dengue vectors. *Proceedings of the National Academy of Science* 110(8):2888–2892.

Dyck V. A., J. Hendrichs, and A. S. Robinson. 2005. *Sterile Insect Technique: Principles and Practice in Area-Wide Integrated Pest Management*. Vienna, Austria: Springer.

Duyck P. F., P. David, and S. Quilici. 2004. A review of relationships between interspecific competition and invasions in fruit flies (Diptera: Tephritidae). *Ecological Entomology* 29:511–520.

Ekesi S., M. K. Billah, P. W. Nderitu, S. A. Lux, and I. Rwomushana. 2009. Evidence for competitive displacement of *Ceratitis cosyra* by the invasive fruit fly *Bactrocera invadens* (Diptera: Tephritidae) on mango and mechanisms contributing to the displacement. *Journal of Economic Entomology* 102:981–991.

Friberg M., O. Leimar, and C. Wiklund. 2013. Heterospecific courtship, minority effects and niche separation between cryptic butterfly species. *Journal of Evolutionary Biology* 26(5):971–979.

Fried M. 1971. Determination of sterile-insect competitiveness. *Journal of Economic Entomology* 64:869–872.

Gröning J., and A. Hochkirch. 2008. Reproductive interference between animal species. *Quarterly Review of Biology* 83:257–282.

Haisch A. 1970. Some observations on decreased vitality of irradiated Mediterranean fruit fly. In *Proceedings, Panel: Sterile-Male Technique for Control of Fruit Flies*. Joint FAO/IAEA Division of Atomic Energy in Food and Agriculture, September 1–5, 1969. SIT/PUB/276. Vienna, Austria: IAEA. pp. 71–75.

Hibino Y., and O. Iwahashi. 1991. Appearance of wild females unreceptive to sterilized males on Okinawa Is. in the eradication program of the melon fly, *Dacus cucurbitae* Coquillett (Diptera: Tephritidae). *Applied Entomology and Zoology* 26:265–270.

Honma A., N. Kumano, and S. Noriyuki. 2019. Killing two bugs with one stone: A perspective for targeting multiple pest species by incorporating reproductive interference into sterile insect technique. *Pest Management Science* 75:571–577.

Itô Y., and H. Kawamoto. 1979. Number of generations necessary to attain eradication of an insect pest with sterile insect release method: A model study. *Researches on Population Ecology* 20:216–226.

Kishi S., and T. Nishida. 2009. Reproductive interference determines persistence and exclusion in species interactions. *Journal of Animal Ecology* 78:1043–1049.

Kitano D., N. Fujii, S. Yamaue et al. 2018. Reproductive interference between two serious pests, oriental fruit flies *Bactrocera carambolae* and *B. dorsalis* (Diptera: Tephritidae), with very wide but partially overlapping host ranges. *Applied Entomology and Zoology* 53:525–533.

Klassen W., and C. F. Curtis. 2005. History of the sterile insect technique. In *Sterile Insect Technique: Principles and Practice in Area-Wide Integrated Pest Management*, eds. V.A. Dyck, J. Hendrichs, and A.S. Robinson. Dordrecht, the Netherlands: Springer.

Konuma J., and S. Chiba. 2007. Ecological character displacement caused by reproductive interference. *Journal of Theoretical Biology* 247:354–364.

Kuno E. 1992. Competitive exclusion through reproductive interference. *Researches on Population Ecology* 34:275–284.

Liu, S. S., P. J. De Barro, J. Xu et al. 2007. Asymmetric mating interactions drive widespread invasion and displacement in a whitefly. *Science* 318 (5857):1769–1772.

McInnis D. O., D. R. Lance, and C. G. Jackson. 1996. Behavioral resistance to the sterile insect technique by Mediterranean fruit fly (Diptera: Tephritidae) in Hawaii. *Annals of the Entomological Society of America* 89:739–744.

Takakura K. I. (2013) Two-way but asymmetric reproductive interference between an invasive *Veronica* species and a native congener. *American Journal of Plant Science* 4:535–542.

Takakura K. I., T. Nishida, and K. Iwao. 2015. Conflicting intersexual mate choices maintain interspecific sexual interactions. *Population Ecology* 57:261–271.

Takakura K. I., T. Nishida, T. Matsumoto, and S. Nishida. 2009. Alien dandelion reduces the seed-set of a native congener through frequency-dependent and one-sided effects. *Biological Invasions* 11:973–981.

Tsubaki Y., and S. Bunroongsook. 1990. Sexual competitive ability of mass-reared males and mate preference in wild females: Their effects on eradication of melon flies. *Applied Entomology and Zoology* 25:457–466.

Vanderplank F. L. 1944. Hybridization between *Glossina* species and a suggested new method of control of certain species of tsetse. *Nature* 154:607–608.

15 Sexual Competitiveness of *Anastrepha ludens* (Diptera: Tephritidae) Males from the Genetic Sexing Strain Tap-7 in the Citrus Region of Montemorelos, Nuevo Leon, Mexico

Patricia López, Juan Heliodoro Luis, Refugio Hernández, and Pablo Montoya*

CONTENTS

Abstract The genetic sexing strain *Anastrepha ludens* (Loew) Tap-7 was developed in the Moscafrut Program for male-only releases. Although successful sexual competitiveness and sterility induction in wild females have been demonstrated for Tap-7 males, evaluations in situ are important to corroborate the compatibility and effectiveness of the males of this strain prior to conducting large-scale releases over wide areas. Sexual competitiveness and capacity to induce sterility by *A. ludens* Tap-7 males were evaluated in field cages in the sweet orange *Citrus aurantium var.* Valencia orchard "Las Parcelas" in Montemorelos, Nuevo Leon, Mexico, with an average temperature of 27.2°C and average relative humidity of 68.40%, in September 2017. Sterile males showed sexual activity synchronously with wild males. The relative

* Corresponding author.

sterility index (RSI) and the Fried competitiveness index of sterile males were 0.61 ± 0.03 and 0.43 ± 0.13, respectively, which are considered as acceptable for this species. These results show that males of the Tap-7 strain were accepted by wild females from Montemorelos in the presence of wild males and induced sterility in the wild population effectively.

15.1 INTRODUCTION

Anastrepha ludens (Loew) is a key pest in commercial citrus-growing regions in Mexico where programs of integrated pest management (IPM) are performed, including, in some cases, the sterile insect technique (SIT) (Gutiérrez et al. 2010). The efficacy of SIT has been improved with the use of genetic sexing strains (GSS), which allows the release of only males (McInnis et al. 1996, Rendón et al. 2004). In SIT programs, an essential parameter requiring evaluation is the sexual competitiveness of sterile males as well as their capacity to induce sterility in wild females (Lance and McInnis 2005, Pérez-Staples et al. 2012). Tap-7 is a genetic sexing strain of *A. ludens* that is currently mass reared in the Moscafrut facility in Chiapas, Mexico, at levels of 25 million pupae per week according to procedures described by Orozco-Dávila et al. (2017). Sterilized pupae are packed and sent to different states of Mexico such as Chiapas, San Luis Potosi, and Tamaulipas for the release of sterile males (Orozco-Dávila et al. 2017, Quintero-Fong et al. 2018). Although initial results indicate that wild populations and this sterile strain are sexually compatible (Orozco et al. 2013), in situ evaluations are required to determine and assure the adequate performance of sterile males before the initiation of sterile male releases (FAO/IAEA/USDA 2014). In addition, these evaluations serve to increase the confidence of the farmers involved in such SIT programs.

The citrus region of Montemorelos, Nuevo Leon, is an important fruit-growing zone where control of *A. ludens* is considered necessary, and it is believed that an SIT program involving the release of *A. ludens* Tap-7 sterile males will greatly contribute to the control of this pest in this region. Based on these considerations, we aimed to determine the sexual competitiveness and sterility induction ability of *A. ludens* Tap-7 sterile males within wild populations of *A. ludens* in Montemorelos, Nuevo Leon.

15.2 MATERIAL AND METHODS

Male sexual competitiveness as well as the capacity to induce sterility was evaluated in field cage tests. The methodology used was based on the manual Food and Agriculture Organization/International Atomic Energy Agency/US Department of Agriculture (FAO/IAEA/USDA, 2014) and Quintero-Fong et al. (2018). The bioassays were carried out from 16:30 to 21:00 h in the citrus orchard "Las Parcelas" (25° 11′ 46.9″ N, −99° 51′ 17.6″ W) during September 2017, with an average temperature of 27.2°C and average relative humidity (RH) of 68.40% (Figure 15.1).

15.2.1 Biological Material

Wild flies were obtained from infested fruit collected in Montemorelos, Nuevo Leon, including fruits of sour orange (*Citrus aurantium* L.), sweet orange Valencia Late variety (*Citrus* × *sinensis* L.), yellow splash (*Sargentia greggii* S. Watts), and grapefruit double red variety (*Citrus* × *paradisi* L.). The sampled fruits were kept in trays with fine vermiculite for 4 days at 22°C ± 1°C to allow larval development. Third instar larvae were extracted via dissection and kept in 13 × 9 cm plastic containers with moist vermiculite as pupation substrate. When pupae were 18 days old, they were placed in groups of 500 pupae per cage (30 × 30 × 30 cm wooden frame with tulle fabric with 0.1 mm mesh). One day after emergence, males and females were separated and kept in groups of 200 adults per cage with sugar and protein (hydrolyzed enzymatic yeast MP Biomedicals, LLC, CA, USA) at a proportion of 1:24 (Quintero-Fong et al. 2018) and pieces of sweet orange Valencia Late variety as a food source. Water was provided in a container with a cotton wick. Both food and water were provided ad libitum. Adults were kept at 25 ± 1°C, 40%–70% RH, and 12:12 h light-to-dark (L:D) photoperiod until they were used for evaluations. When tested, wild flies were 18–22 d old.

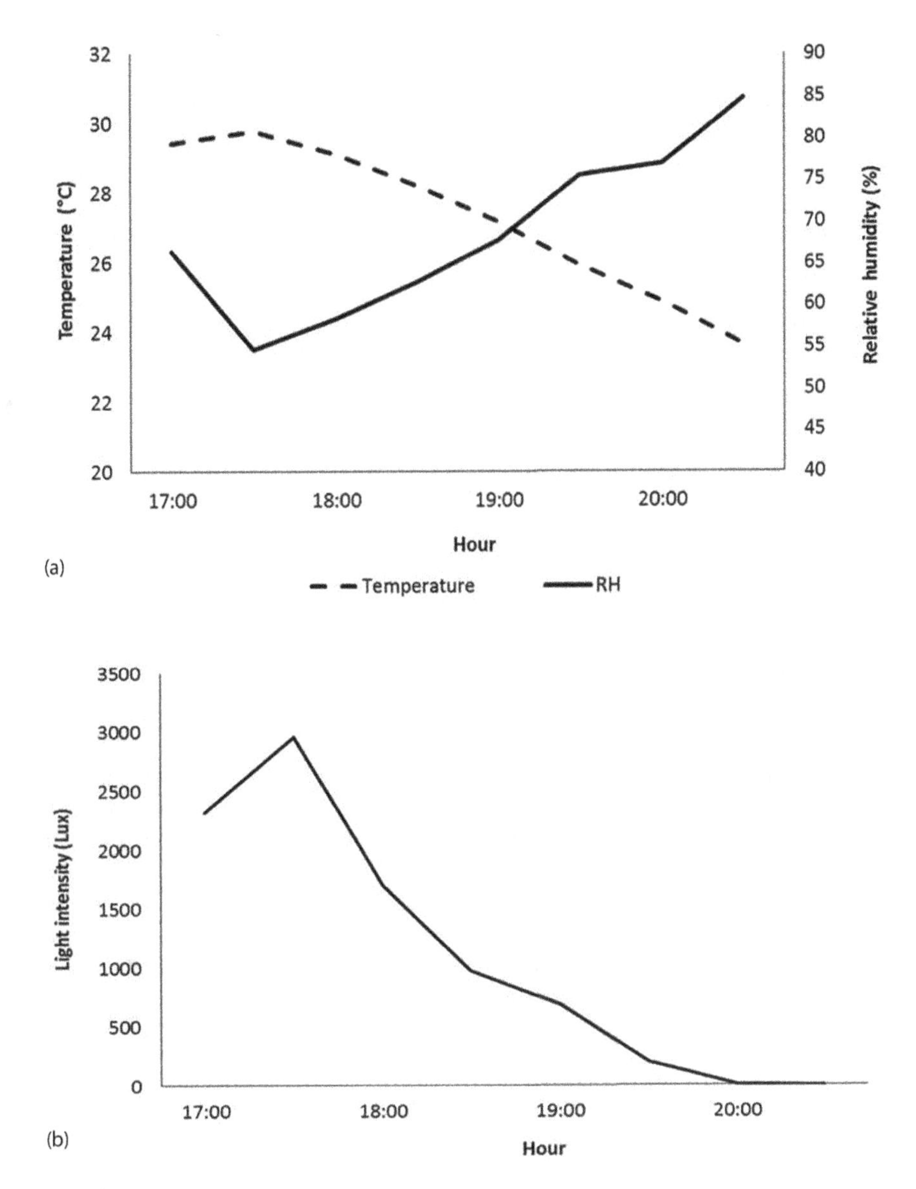

FIGURE 15.1 Environmental conditions registered during field evaluations in Montemorelos, Nuevo Leon: (a) temperature and relative humidity (RH) and (b) light intensity.

Sterile pupae of *A. ludens* Tap-7 were sent by air from the Moscafrut facility located in Metapa de Dominguez, Chiapas in shipments of 1 million pupae per day, with 15–16 h of hypoxia during transportation. Sterile pupae were packed in Mexican towers (Hernández et al. 2010) placing 20,000 pupae per level (~1 fly/cm^2) and maintained at 23°C ± 1°C and 60%–70% RH until the sixth day when adults were chilled and packed for release. Flies were introduced into a cold room when the chamber reached 3°C ± 1°C for a period of 2 h until packing into 20-L cylindrical containers (Cancino et al. 2017). From these packing units, four 60-mL samples of chilled flies (~400 flies) were taken to perform the tests, which were placed separately in the cages previously described with food (sugar-to-protein ratio 1:24), pieces of open sweet orange, and water provided ad libitum. Adults were maintained at 25°C ± 1°C, 40%–70% RH, and L:D period of 12:12 until the day of evaluation. When tested, sterile Tap-7 males were 10–12 d old.

15.2.2 Evaluation

Field cages of 2 m high × 3 m diameter were placed in the sweet orange orchard where also walnut trees (*Juglans regia* L.), pear (*Pyrus communis* L.), and wild vegetation were present. Inside each cage we placed ten 1-m sweet orange trees in pots that were distributed as three in the center and seven in the periphery of each field cage.

To monitor sexual competition, we introduced 30 sterile males, 30 wild males, and 30 wild females per field cage. Males were marked 1 day before the evaluation with different consecutive numbers printed on 1 mm^2 pieces of Bond paper that were glued to the thorax of the flies with Resistol-850® (Meza-Hernández and Díaz-Fleischer 2006). After marking, each group was placed separately in wooden frame cages (10 × 10 × 10 cm) with tulle fabric sides and supplied with food (sugar-to-protein ratio 1:24) and water. To acclimate to outdoor environmental conditions, the small cages with flies were placed for 4 h next to the field cages before the test. Males were released at 16:30 h, and females were released 30min later. The number of calling males was recorded every 30 min, and such males were identified by vigorous flapping, dilated prostiger, and inflated pleural glands (Orozco-Dávila et al. 2007). Matings were scored until 20:30 h, with mating males being identified by number. Groups of 10 field cages were evaluated for 3 consecutive days, including three lots of produced sterile flies, with a total of 30 replicates.

Sterility induction was evaluated using the Fried test (Fried 1971, FAO/IAEA/USDA 2014). We established three treatments: (i) competition, in which 48 sterile males, 16 wild males, and 16 wild fertile females were released per cage; (ii) sterile control, with 40 sterile males and 40 fertile wild females per cage; and (iii) wild control, with 40 wild males and 40 fertile wild females. Adults in each treatment were introduced at 12:00 h and females were recaptured 48 h later. In the cages, flies were supplied with food through fiberglass mesh bags of 10 × 9 cm and water hung on tree branches. Recaptured females (12 ± 0.4 females per cage in the competition treatment and 25 ± 12 sterile control and 30 ± 6 wild control in the control treatments) were placed in laboratory cages of 30 × 30 × 30 cm and maintained at 27°C ± 1°C with a photoperiod of 12:12 L:D. Every 24 h over 5 consecutive days, five spheres (4.5 cm diameter) made of fursellerone painted with green dye and covered with parafilm were offered as artificial hosts (Quintero-Fong et al. 2018). The spheres were smeared with orange natural juice to stimulate oviposition by females. The oviposited eggs were extracted and placed on a black cloth resting on top of a water impregnated sponge in a Petri dish (100 × 15 mm). Incubation units were maintained at 27°C and 40%–70% RH and were observed in a stereoscope after 7 days of incubation. There were nine replicates for the competition treatment and three for the sterile and wild controls each.

15.2.3 Data Analysis

Only replicates with at least 20% of females mating were considered for the analysis of sexual competition (FAO/IAEA/USDA 2014). Number of calling males (total number of males calling per cage per day) and number of matings involving sterile males or wild males were compared with a *t*-student test for paired data ($\alpha = 0.05$). The percentage of calling males and matings before the peak of activity were compared by means of a two-way contingency table chi-square test. We also calculated the relative sterility index (RSI = number of matings with sterile males/total number of matings) (FAO/IAEA/USDA 2014, McInnis et al. 1996).

The Fried competitiveness index was determined according to the following equation:

$$C = W/S*(Hw - Hc)/(Hc - Hs)$$

where W, number of wild males in the competitiveness cage; S, number of sterile males in the competitiveness cage; Hw, percentage of eggs hatched from wild females in the wild control cage; Hc, percentage of eggs hatched from wild females in the competiveness cage; and Hs, percentage of eggs hatched from fertile wild females in the sterile control cages. Values of 1 indicate equal competitiveness between sterile and wild males, replicates with values greater than 1.1 were discarded, and values between 1 and 1.1 were rounded down to 1.0 (FAO/IAEA/USDA 2014).

15.3 RESULTS

15.3.1 Male Calling

Sterile males and wild males began sexual signaling around 17:30 h, showed a peak of activity at 19:30 h, and then declined to complete inactivity at 20:30 h (Figure 15.2). Although there was a trend showing that Tap-7 males comprised most of the males calling early (before the peak time of sexual activity), this difference was not significant ($\chi^2 = 2.5824$; df = 1; $P = 0.1081$). The total number of calling males per cage was significantly higher for Tap-7 strain males compared to wild males (Table 15.1, $t = -5.100$, df = 17, $P < 0.001$).

15.3.2 Matings

Only 18 replicates reaching a mating rate equal or greater than 0.2 were included in the analysis. The first mating by a sterile male occurred at 17:30 h, and the first mating involving a wild male was recorded at 19:00 h. Most of the males mating early (before the peak time of activity) where Tap-7 males, although this was not statistically significant (sexual competitiveness of Tap-7 males, $\chi^2 = 0.96$, df = 1, $P = 0.3272$). The number of matings increased with time with a peak at 19:30 h

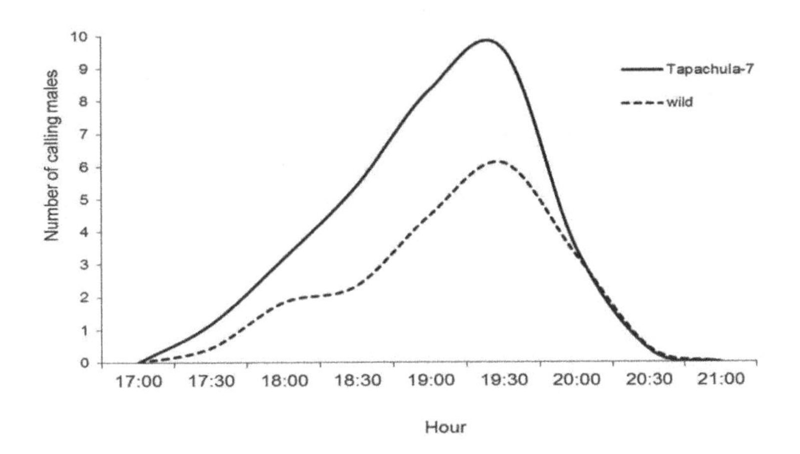

FIGURE 15.2 Mean number of calling males registered during the sexual competitiveness evaluation of sterile Tap-7 and wild males of *Anastrepha ludens* in Montemorelos, Nuevo Leon, Mexico.

TABLE 15.1
Average (± SE) Number of Callings and Matings by Tap-7 and Wild Males Observed per Field Cage during the Sexual Competitiveness Tests. Montemorelos, Nuevo Leon, Mexico (n = 18)

Male Type	Number of Males Calling	Number of Matings
Tap-7	31.56 ± 1.89 b	5.00 ± 0.36 b
Wild	18.94 ± 1.95 a	3.33 ± 0.38 a

Different letters in the same column indicate significant differences, t-student paired data test with $\alpha = 0.05$.

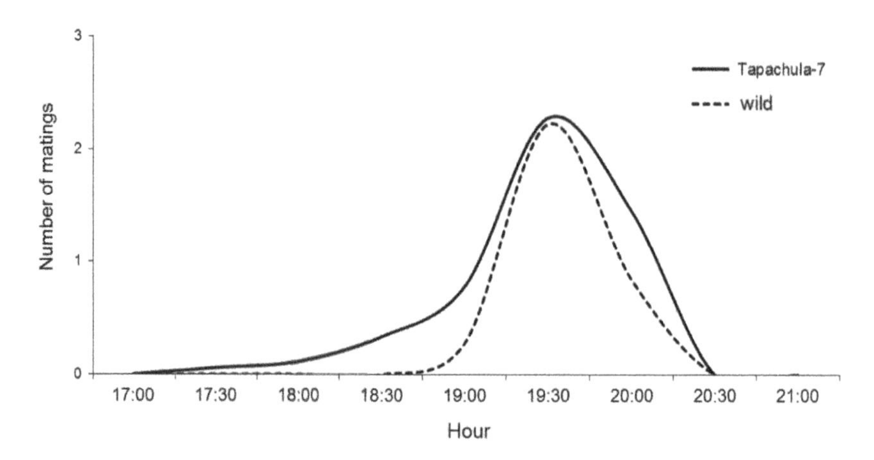

FIGURE 15.3 Mean number of matings registered during the sexual competitiveness evaluation of sterile Tap-7 and wild males of *Anastrepha ludens* in Montemorelos, Nuevo Leon, Mexico.

in both groups (2.28 ± 0.38 for sterile males and 2.22 ± 0.35 for wild males), with the last mating recorded at 20:00 h (Figure 15.3). The total number of matings per cage was significantly higher for males of the Tap-7 strain than for wild males (Table 15.1, $t = -4.366$, df = 17, $P < 0.001$). The RSI was 0.61 ± 0.03, which indicates that sterile males achieved 61% of the copulations.

15.3.3 Induction of Sterility

Egg hatching in the wild control treatment was 75.86% ± 1.12%, 0% in the sterile control treatment, and 39.47% ± 7.41% in the competition treatment. The Fried competitiveness index was 0.43 ± 0.13.

15.4 DISCUSSION

The sexual activity of sterile males observed resembled that reported in previous studies for both bisexual and Tap-7 strains of *A. ludens* (Meza-Hernández and Díaz-Fleischer 2006, Orozco-Dávila et al. 2007, Quintero-Fong et al. 2018). Calling behavior initiated during the first half-hour of observation, with its maximum peak 1 h before the end of daylight. Matings involving sterile males started before those involving wild males. Although this difference was not significant here, it was nonetheless consistent with findings by Quintero-Fong et al. (2018). Such differences may be a result of sterile males invariably subject to constant environmental and confinement conditions in the mass-rearing process, favoring those males that initiate calling and courting earlier (Meza-Hernández and Díaz-Fleischer 2006, Quintero-Fong et al. 2018). A higher proportion of copulations gained by sterile males in this study was also reported in San Luis Potosí (Quintero-Fong et al. 2018); in contrast, Orozco-Dávila et al. (2015) and Quintero-Fong et al. (2016) reported that wild males obtained more copulations than Tap-7 males in two studies performed in Chiapas. It is important to highlight that sterile males and wild males reached the peak of sexual activity at the same time, which is additional evidence of their sexual compatibility. Furthermore, as the number of calling males throughout the study was highly consistent, we consider that it could be used as a reliable estimation of mating success in this strain. The high levels of sexual callings and matings by sterile males indicate that the Tap-7 strain can be successfully used in SIT programs against *A. ludens* pest populations.

This study reaffirms the sexual competitiveness of Tap-7 sterile males and their ability to induce sterility evidenced by an acceptable value of the Fried competitiveness index. It is important to note that environmental conditions were favorable in the citrus orchard because the average temperature and RH prevalent during the bioassays were adequate for the recording of optimal survival and performance of sterile insects. The case of replicates with a mating proportion lower than 0.2 can be considered as the influence of microclimate variations during the bioassays, a situation that deserves further research.

15.5 CONCLUSION

Males of the Tap-7 strain were competitive in the presence of wild males and achieved an acceptable sterility induction in wild *A. ludens* females of Montemorelos, Nuevo Leon, Mexico.

ACKNOWLEDGMENTS

We appreciate the logistic support of Luis Quintero, Dina Orozco, Eduardo Solís, and Salvador Breceda. We also thank Oscar de Leon, Samuel Álvarez, and the technical staff of CESAVENL in Montemorelos, Nuevo Leon, and Carlos Montemayor (owner of the orange orchard) and his staff. We are grateful to Jorge Villalobos, Daniel Gallardo, Alvaro Meza, and Antonio Escobar for their technical support. We also acknowledge the support from Francisco Ramirez y Ramirez, Director of the Fruit Fly National Program.

REFERENCES

Cancino, J., F. López-Arriaga, and P. Montoya. 2017. Packaging conditions for the field release of the fruit fly parasitoid *Diachasmimorpha longicaudata* (Ashmead) (Hymenoptera: Braconidae). *Austral Entomology* 56:261–267.

Food and Agriculture Organization/International Atomic Energy Agency/US Department of Agriculture (FAO/IAEA/USDA). 2014. Manual for Product Quality Control for Sterile Mass-Reared and Released Tephritid Fruit Flies, Version 6.0 May 2014 IAEA, Vienna, Austria. 164pp.

Fried, M. 1971. Determination of sterile insect competitiveness. *Journal Economic Entomolology* 64:869–872.

Gutiérrez, M. J. 2010. El programa moscas de la fruta en México. In *Moscas de la fruta: Fundamentos y procedimientos para su manejo*, eds. J. Toledo, P. Montoya and E. Hernández, 3–10. Mexico, DF: S and G editors.

Hernández, E., A. Escobar, B. Bravo, and P. Montoya. 2010. Chilled packing systems for fruit flies (Diptera: Tephritidae) in the Sterile Insect Technique. *Neotropical Entomology* 39:601–607.

Lance, D. R., and D. O. McInnis. 2005. Biological basis of the sterile insect technique. In *Sterile Insect Technique: Principles and Practice in Area-Wide Integrated Pest Management,* eds. V. A. Dyck, J. Hendrichs, and A. S. Robinson, pp. 69–94. Dordrecht, the Netherlands: Springer.

McInnis, D. O., D. R. Lance, and C. G. Jackson. 1996. Behavioral resistance to the sterile insect technique by Mediterranean fruit fly (Diptera: Tephritidae) in Hawaii. *Annals of the Entomological Society of America* 89:739–744.

Meza-Hernández, J. S., and F. Díaz-Fleischer. 2006. Comparison of sexual compatibility between laboratory and wild Mexican fruit flies under laboratory and field conditions. *Journal of Economic Entomology* 99:1979–1986.

Orozco, D., J. S. Meza, S. Zepeda, E. Solís, and J. L. Quintero. 2013. Tapachula-7, new genetic sexing strain of the Mexican fruit Fly (Diptera: Tephritidae): Sexual compatibility and competitiveness. *Journal of Economic Entomology* 106:735–741.

Orozco-Dávila, D., M. de Lourdes Adriano-Anaya, L. Quintero-Fong, and M. Salvador-Figueroa. 2015. Sterility and sexual competitiveness of Tapachula-7 *Anastrepha ludens* males irradiated at different doses. *PLoS One 10* (8):e0135759.

Orozco-Dávila, D., R. Hernández, J. S. Meza, and J. Domínguez. 2007. Competitiveness and compatibility between mass-reared sterile flies and wild populations of *Anastrepha ludens* (Diptera: Tephritidae) from different regions in Mexico. *Florida Entomologist* 90:19–26.

Orozco-Dávila, D., L. Quintero, E. Hernández, E. Solís, T. Artiaga, R. Hernández, and P. Montoya. 2017. Mass rearing and sterile insect releases for the control of *Anastrepha* spp. (Diptera: Tephritidae) pests in Mexico. *Entomologia Experimentalis et Applicata* 164:176–187.

Pérez-Staples, D., T. E. Shelly, and B. Yuval. 2012. Female mating failure and the failure of "mating" in sterile insect programs. *Entomologia Experimentalis et Applicata* 146:66–78.

Quintero-Fong, L., J. H. Luis, P. Montoya, and D. Orozco-Dávila. 2018. In situ sexual competition between sterile males of the genetic sexing Tapachula-7 strain and wild *Anastrepha ludens* flies. *Crop Protection* 106:1–5.

Quintero-Fong, L., J. Toledo, L. Ruiz, P. Rendón, D. Orozco-Dávila, L. Cruz, and P. Liedo. 2016. Selection by mating competitiveness improves the performance of *Anastrepha ludens* males of the genetic sexing strain Tapachula-7. *Bulletin of Entomological Research* 106:624–632.

Rendón, P., D. McInnis, D. Lance, and J. Stewart. 2004. Medfly (Diptera: Tephritidae) genetic sexing: Large-scale field comparison of males-only and bisexual sterile fly releases in Guatemala. *Journal of Economic Entomology* 97:1547–1553.

16 A New Diet for a New Facility

Development of a Starter-Finalizer Diet System for Rearing Colonies of the Ceratitis capitata *Vienna 8 Strain at a New Facility of Mexico's Moscamed Program*

*Milton Arturo Rasgado-Marroquín, Emmanuel Velázquez-Dávila, José Antonio De la Cruz-De la Cruz, Reynaldo Aguilar-Laparra, Luis Cristóbal Silva Villareal, and Marco Tulio Tejeda**

CONTENTS

* Corresponding author.

Abstract The new mass-rearing facility of Mexico's Moscamed Program, currently under construction, has been projected to produce up to 1000 million sterile male pupae per week of the *Ceratitis capitata* Vienna 8 strain. Given the slower development of females of this strain, the starter-finalizer larval diet system was proposed to sustain the biological amplification of female-male colonies. To start operations in the new facility, a suitable novel larval diet formulation with this system had to be determined. In this study, several starter-finalizer larval diet formulations were designed and evaluated in two stages. In the first stage, eight novel diet formulations were designed using the standard ingredients of mass rearing operations. Larval recovery, larval weight, development speed, and temperature profile were determined in these diets using the male-only stream (thermally treated eggs). The two best diets were further evaluated in the second stage. An additional adjusted diet formulation and the standard mass rearing diet were also included in this stage. In this second stage, larval recovery, larval weight, temperature profile, and microbiological and physicochemical properties of the larval diets were evaluated using the male and female streams. Based on productivity, quality, and development speed, two out of eight larval diets were further evaluated in the second stage. Of the four diets included in the second stage, the formulation "T5-N3" presented the best balance between productivity and quality: a good development time, a stable productivity, an optimal thermal performance, and a relatively better suppression of fungal and bacterial development. The formulation "T5-N3" was determined as the best option for initiating the mass rearing of Vienna 8 colonies at the new facility of Mexico's Moscamed Program. Nevertheless, this formulation still needs to be optimized. Most importantly, mass-rearing biofactories have to be prepared for the continuous improvement of larval diets. Future directions for the optimization and design of larval diets for large-scale operations are discussed.

16.1 INTRODUCTION

Built it in 1979, the current Moscamed biofactory has been the main base for the application of the sterile insect technique (SIT) in Mexico. This facility has been operating for four decades to produce 500 million sterile *Ceratitis capitata* (Wiedemann) pupae per week. Throughout this time, the biofactory has experienced relevant technological advances, such as the transition from bisexual to genetic-sexing strains (GSS; Cáceres 2002, Robinson et al. 1999). Since late 2002, this facility had the purpose of mass rearing only male flies by using a single-diet rearing system. Mixture proportions and ingredients characteristic of this larval diet are recurrently optimized for male-only rearing conditions.

The Moscamed Program in Mexico is currently building a new mass-rearing facility for the regional release of sterile insects. With more than 3 ha of construction distributed as five buildings, this new facility was designed aiming to produce up to 1 billion sterile male pupae per week. To achieve this objective, a biological amplification of several female and male colonies of the *C. capitata* "Vienna 8" GSS has been devised (see Caceres 2002, Caceres et al. 2000, and Fisher and Caceres 2000 for details). Although the Moscamed Program has a long history of mass rearing *C. capitata* (Enkerlin et al. 2015, Schwarz et al. 1985, Villaseñor et al. 2000), the biological amplification of Vienna 8 colonies has been one of the main challenges for the start of operations in the new facility. Females of this strain have never been reared on the ingredients used by the Moscamed facility. Moreover, because Vienna 8 female larvae develop slower than males and have a thermo-sensitive lethal condition (Franz 2005), the male-only rearing conditions established at Moscamed could impose suboptimal to lethal conditions on their development. Thus, an appropriate diet system and setup conditions for rearing females was mandatory and highly needed.

One of the larval rearing systems that have been proposed for the mass rearing of fruit flies is the starter-finalizer diet system. This system consists in using a small diet portion for early larval development stages in combination with a finalizer diet for the later larval development stages (Fay 1988). Compared to a single-diet approach, the starter-finalizer diet system reduces production costs, optimizes mass-rearing spaces, delays microbiological growth, and preserves the physicochemical characteristics of the diet for larval development (Chan and Jang 1995, Economopoulos et al. 1990, Fay 1988, Domínguez et al. 1993, Pinson et al. 1993). Taking into account the specific characteristics of the *C. capitata* Vienna 8 strain, a starter-finalizer diet system for rearing females could be more appropriate than the single-diet approach used for male-only production. In this strain, females develop slower than males because of a pleiotropic effect of the temperature sensitive lethal (*tsl*) gene or to an independent action of a gene close to *tsl* (Caceres 2002, Franz 2005, Salvador Meza personal communication). The slower development of females imposes a longer rearing time compared to that of male rearing. Also, the thermal sensitive lethal condition of females (homozygous for *tsl*-) demands "colder" rearing temperatures than those normally used for male-only production, thus extending further the rearing time. Because of the conditions mentioned, the use of a fresh diet for advanced development stages, as in the starter-finalizer diet system, provides a suitable technical approach to extend the availability of resources needed by females to complete their development.

In this study, several diet formulations were evaluated for the implementation of a starter-finalizer diet system for colonies of the *C. capitata* Vienna 8 strain to start the operations of the new Moscamed facility. As a first exploratory stage of the study, eight diet formulations were evaluated using eggs of the "male-only" stream (eggs on which a specific thermal treatment was applied to kill females) that were available and suitable for this exploratory phase. Two of the most effective diet formulations were then used to initiate a small "Mother colony" using the biological material "Vienna-8 (Toliman), D53-," provided as eggs (without thermal treatment, i.e., viable males and females) by El Pino Facility, Guatemala. In the second stage, the two most effective diets from the first stage and two additional diets were tested using eggs (males and females) from the pre-established Mother colony of Metapa.

16.2 MATERIALS AND METHODS

16.2.1 Insects

The GSS *Ceratitis capitata* "Vienna-8/Toliman $^{D53-}$" was used for the first and second experimental stages (Augustinos et al. 2017). This strain has a reciprocal translocation between the chromosome that determines the male sex (Y) and autosome 5, which carries the wild-type alleles of the genes *tsl* and *white pupa* color (*wp*) (Robinson 2002, Franz 2005). Insects provided by El Pino Facility (generation 73 by January 2015) had been maintained on a larval diet mixture of sugar, yeast, wheat flour, sodium benzoate, hydrochloric acid, water, and sugarcane bagasse (Ramírez-Santos et al. 2016). However, insects in the Metapa facility are reared using different ingredients and mixture proportions (Table 16.1). The second stage of this study was performed with the fourth generation of insects reared on Metapa ingredients.

16.2.2 First Stage

16.2.2.1 Egg Source

The first stage was carried out using the male-only stream produced at El Pino, Guatemala. Eggs were thermally treated to kill females and were transported by land to Metapa, Mexico.

In Metapa, eggs were incubated for 8 h using a 22:1 water-to-egg solution (mL) in constant bubbling (15 lb air pressure) and at a temperature of 26°C. After this time, eggs were suspended in a guar gum solution (5 g/L concentration) with a final volume of 160 mL of guar gum-to-egg solution. For each treatment, 20 mL of guar gum-to-egg solution were added to each starter diet block with the help of a syringe. Egg-seeding density was 4 mL of eggs per 1 kg of starter diet (30,000 eggs/mL × 4 mL ≈ 120,000 eggs, of which ≈ 30,000 were viable males and ≈ 90,000 were nonviable eggs, [see Robinson et al. 1999 for details]).

TABLE 16.1
Percentages of Ingredients in Each Larval Diet Formulation (First Stage)

	Diet Treatments							
Ingredients	**T-1**	**T-2**	**T-3**	**T-4**	**T-5**	**T-6**	**T-7**	**T-8**
Dry ground maize plant fractions	0.00%	31.00%	19.00%	26.00%	15.50%	20.00%	7.75%	3.10%
Wheat bran	31.00%	0.00%	13.35%	5.00%	15.50%	11.00%	23.25%	27.90%
Sugar	12.00%	12.00%	13.72%	12.00%	12.00%	12.00%	12.00%	12.00%
Inactive dry yeast	7.50%	7.50%	7.00%	7.50%	7.50%	7.50%	7.50%	7.50%
Citric acid	1.91%	1.91%	2.11%	1.91%	1.91%	1.91%	1.91%	1.91%
Nipagin (methyl paraben)	0.20%	0.20%	0.44%	0.20%	0.20%	0.20%	0.20%	0.20%
Sodium benzoate	0.30%	0.30%	0.38%	0.30%	0.30%	0.30%	0.30%	0.30%
Water	47.09%	47.09%	44.00%	47.09%	47.09%	47.09%	47.09%	47.09%

16.2.2.2 Formulation and Preparation Methods

Based on the standard diet formulation used in male-only stream production at Metapa, eight different formulations were designed and established as starter-finalizer diet treatments (Table 16.1). These diet formulations were arbitrarily named as "treatment one," "treatment two," and so on until "treatment eight" (T-1 … T-8). In the case of treatment three, the ingredient "fine crushed corn" was sifted using a No. 10 sieve (ASTM), and only large particles were used.

For each treatment, 10 kg of starter diet were prepared with the help of a mixing machine with a capacity of 20 kg. To obtain a homogeneous mixture, the diet was prepared in two steps. First, the ingredients "fine crushed corn," wheat bran, yeast, and preservatives (nipagin and sodium benzoate) were dry-mixed for approximately 2 min. Then, water was added and the mixture was shaken for a total of 15 min. The resulting 10 kg of diet were then distributed in two trays of 77.5 × 40.5 × 9 cm each. Thus, two trays with 5 kg of diet each were obtained per treatment. Subsequently, the starter diet in each tray was divided into five equal diet portions (slices) of approximately 1 kg each and with an approximate size of 15 × 40 × 7 cm.

For each treatment, the finalizer diet formulation was the same as that of its respective starter diet. Forty kilograms of finalizer diet were prepared on two batches using the same procedure described for the starter diet. The finalizer diet was distributed in 10 plastic trays. An empty space was left in the center of each to place the starter diet in that space. One kilogram (slice) of starter diet was then transferred from the starter diet tray to the empty space of the finalizer diet tray. This was repeated for the 10 kg of starter diet and for each treatment.

16.2.2.3 Rearing Procedures

After the eggs were placed on the larval diet, starter trays were covered with a cloth and a plastic bag to maintain the moisture of the diet. The cloth and the plastic bag were removed when the diet had reached a temperature of 29°C. All treatments were maintained in a rearing room at a temperature of 27°C for 96 h. Starter diets were transferred into the finalizer diets once the 96 h had elapsed. Immediately after the starter diet was placed with the finalizer diet, approximately 100 mL of water was added to the starter diet using a garden irrigation tool. The starter-finalizer diet trays were stacked according to treatment and all treatments were transferred to a rearing room at 20°C. Larval collections were initiated around the seventh day after egg seeding (depending on larval maturity) and continued for 3 to 4 days in intervals of

24 h. During larval collections, diets were irrigated approximately three to five times with a mixture of 0.5 g of sodium benzoate per liter of water. Aluminum trays filled with water were used to collect larvae.

16.2.2.4 Measured Parameters

16.2.2.4.1 Larval Weight and Recovery

Larvae were removed daily from the aluminum trays, filtered from the water, dried with wheat bran, and sieved from the wheat bran. Larval production was recorded each day as the total weight of dried larvae. A sample was separated from each larval collection to calculate larval weight. The total number of larvae in approximately 2 g of sample was counted and weighed to the closest 0.1 mg. This was repeated for three samples. The weight of one larva was calculated by interpolation. Larval weight was determined as the mean of the three samples. The number of larvae in each collection was calculated by multiplying larval weight by collection weight. Total larvae produced were calculated as the sum of larval collections.

For every batch of starter-finalizer diet, larval recovery (i.e., egg to larva transformation) was estimated as the percentage of total larvae produced from the number of seeded eggs. Similarly, yield production was obtained as total larvae produced per kilogram of diet used.

16.2.2.4.2 Larval Development Speed

The time between egg seeding and the completion of larval development is a critical parameter for mass-rearing operations. To account for this parameter, development speed allowed by a diet formulation was estimated as the accumulated proportion of larval recovery (0%–100%) at the time of the second larval collection.

16.2.2.4.3 Temperature Profile

For each treatment, diet temperature was registered with a data logger (Onset MX2303) at intervals of 15 min by inserting the probes directly into the diet. Data were recovered after the last larval collection.

16.2.3 Second Stage

16.2.3.1 Egg Source

Between the first and second stages, a small "Mother colony" was established at Metapa. The colony was established using male and female eggs of "Vienna-8/Toliman $^{D53-}$" provided by the Filter colony of El Pino, Guatemala.

The second stage was carried out using nonthermally treated eggs (males and females) of the newly established Mother colony of Metapa, Mexico. At the time of the experiment, insects had been reared for four generations with Metapa ingredients.

Eggs were incubated for 48 h using a 22:1 water-to-egg solution (mL) in constant bubbling (15 lb air pressure) and at a temperature of 26°C. After this time, eggs were suspended in a guar gum solution (5 g/L concentration) with a final volume of 160 mL of guar gum-to-egg solution. For each treatment, 20 mL of guar gum-to-egg solution were added to each starter diet block with the help of a syringe. Egg final density was 1.5 mL of eggs per kilogram of starter diet (30,000 eggs/mL × 1.5 mL ≈ 45,000 eggs; of which ≈ 11,250 were viable males, ≈ 11,250 were viable females, and ≈22,500 were nonviable eggs [see Robinson et al. 1999 for details]).

16.2.3.2 Formulation and Preparation Methods

For this stage, starter and finalizer diet formulations were adjusted and re-evaluated. The two most effective formulations from the first stage, T-5 and T-3, were included in the second stage. Additionally, the proportion of nipagin (methyl paraben) was increased in the T-5 formulation (to reduce microorganism activity). This modified diet was labeled as formulation "T5-N3." The standard male-only mass rearing diet formulation "T21L8N3" was included as an external control. The proportion of each ingredient in all diet formulations evaluated in this stage is shown in Table 16.2.

TABLE 16.2
Percentages of Ingredients in Each Larval Diet Formulation (Second Stage)

	Diet Treatments			
Ingredients	**T5-N3**	**T-3**	**T21L8N3**	**T-5**
Dry ground maize plant fractions	15.50%	19.00%	21.00%	15.50%
Wheat bran	15.50%	13.35%	10.00%	15.50%
Sugar	12.00%	13.72%	12.41%	12.00%
Inactive dry yeast	7.50%	7.00%	8.00%	7.50%
Citric acid	1.91%	2.11%	1.91%	1.91%
Nipagin (methyl paraben)	0.36%	0.44%	0.36%	0.20%
Sodium benzoate	0.34%	0.38%	0.34%	0.30%
Water	47.09%	44.00%	45.98%	47.09%

For each treatment, 10 kg of starter diet was prepared following the process described previously. The starter diet was placed on trays, and egg seeding was performed until the diet reached a temperature of 23°C–24°C.

The finalizer diet was prepared under the same scheme described previously. The starter diet was transferred to the finalizer diet trays 96 h after egg seeding.

16.2.3.3 Rearing Procedures

Once egg seeding was performed, starter diet trays were stacked according to treatment. Each tray stack was covered with a cloth and a plastic bag to maintain the moisture of the diet. Tray stacks were maintained in a rearing room with a temperature of 24 ± 1°C for 5 days. The plastic bag was removed when the diet temperature reached 28°C, and the cloth was removed after the diet temperature reached 29°C. Starter diets were transferred to the finalizer diet trays once the 96 h had elapsed. Immediately after the starter diet was transferred to the finalizer diet trays, approximately 100 mL of water were added to the starter diet with the help of a garden irrigation tool. The starter-finalizer diet trays were stacked according to treatment formulations, and all treatments were transferred to a rearing room at 20°C.

Larval collections were initiated around the seventh day after the eggs were placed on the starter diet (depending on larval maturity) and continued for 8 days in intervals of 24 h. During larval collections, diets were irrigated approximately 7–10 times with a mixture of 0.5 g sodium benzoate per liter of water. Aluminum trays filled with water were used to collect larvae. A total of nine repetitions were performed for each diet treatment formulation.

16.2.3.4 Measured Parameters

16.2.3.4.1 Larval Weight and Recovery

Larvae were removed daily from the aluminum trays, filtered from the water, dried with wheat bran, and sieved from the wheat bran. Daily larval production was recorded as the total weight of dried larvae. A sample was separated from each larval collection to calculate larval weight. The total number of larvae in a sample of approximately 2 g was counted and weighed to the closest 0.1 mg. This was repeated for three samples. The weight of one larva was calculated by interpolation. Larval weight was determined as the mean of the three samples. Number of individuals in each larval collection was calculated by multiplying larval weight by collection weight. Total larvae produced were calculated as the sum of larval collections. For every batch of starter-finalizer diet, larval recovery (i.e., egg to larva transformation) was estimated as the percentage of total larvae produced from the number of seeded eggs. Similarly, yield production was obtained as total larvae produced per kg of diet used.

16.2.3.4.2 Temperature Profile

For each treatment, diet temperature was registered with a data logger (Onset MX2303) at intervals of 15 min by inserting the probes directly into the diet. Data were recovered after the last larval recollection.

16.2.3.4.3 Diet Microbiology

Microbiological evaluations consisted in determining total microbial counts of aerobic mesophilic bacteria per gram of diet and units forming fungi colonies per gram of diet. Diets were sampled several times throughout the rearing process. Following time series data, a starter-finalizer diet batch was determined as the unit of repetition. A starter diet was sampled as (i) fresh diet "0 h" and (ii) 96 h after egg seeding, right after it was transferred to finalizer diet trays. A finalizer diet was sampled as (iii) fresh diet, (iv) 120 h after diet preparation, and (v) 240 h after diet preparation. This scheme was repeated for three batches of each diet treatment.

Samples were prepared using the plate technique. Standard method agar was used as a culture medium for the determination of total counts of aerobic mesophilic bacteria colony-forming unit (CFU) per gram of diet (NOM-092-SSA1-1994). In the case of fungi quantification, Rose Bengal Chloramphenicol agar was used as a selective culture medium for the isolation and enumeration of fungi colonies (NOM-111-SSA1-1994). Incubation times and temperatures for bacteria and fungi were 48 h at 35°C and 96 h at 25°C, respectively.

16.2.3.4.4 Physicochemical Properties of the Diets

Physicochemical evaluations consisted in quantifying the pH and moisture of the diets throughout the rearing process. Similar to diet microbiology, sampling points for the starter diet were considered as fresh diet and 96 h after preparation, whereas the finalizer diet was sampled as freshly prepared diet and 48, 96, 144, and 192 h after preparation. The pH was measured using a Hanna potentiometer (model HI 2216). Diet moisture was determined by the thermogravimetric method, which consisted in placing a 5-g sample in an oven at 110°C for 2 h and calculating moisture percentage as the difference between fresh and dry weight. Both procedures are part of the quality control manual of the Moscamed facilities (Programa Moscamed, 2015).

16.2.4 Statistical Analysis

A batch of starter-finalizer diet was used as an experimental unit. Diet yields, development speed, larval weight, temperature, larval recovery (%), total bacteria count (CFU/g of diet), total fungi count (CFU/g of diet), pH, and diet moisture (%) were used as response variables, whereas diet formulation was used as the predictor, explanatory variable. The relationship of the response variables with the predictor was tested using analysis of variance (ANOVA). For temperature and microbiological and physicochemical variables, full factorial designs, that included time as the explanatory variable and interaction terms, were modeled. Normality and homoscedasticity assumptions were tested with Shapiro-Wilk and Levene's tests, respectively. Because sample repetitions for larval recovery and development speed were relatively small (n = 3–9), significant differences were detected on assumptions. For these cases, power transformations were performed and models were retested. Transformations did not alter the qualitative result of the models. Results are presented on retransformed data. Statistical tests were performed using the software R (R Development Core Team, 2014).

16.3 RESULTS

16.3.1 First Stage

16.3.1.1 Larval Recovery and Weight

Although there was a wide variability in the obtained data, an interdependent relationship among egg-larva recovery, larval weight, and texturizer proportion in the diet (dry ground maize fractions to wheat bran) was observed (Figure 16.1). Egg to larva recovery was negatively correlated with wheat

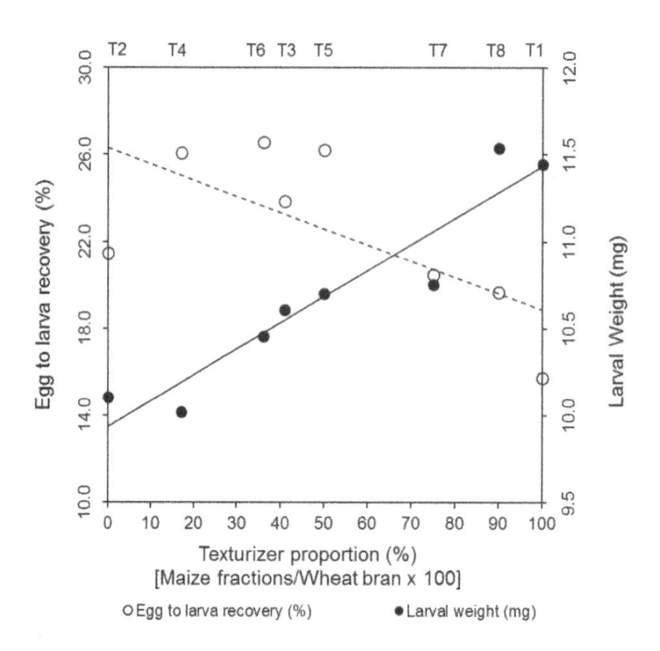

FIGURE 16.1 Larval weight and egg-larva transformations of formulations designed for the starter-finalizer larval diet system. Stage 1 was performed with thermally treated eggs of the Vienna 8 strain, thus thermosensitive females were absent in this stage. Wheat bran (*x*-axis) is indicated as a proportion of texturizer (dry ground maize plant fractions/wheat bran × 100). The dotted line represents the relationship between egg to larva recovery and percentage of wheat bran in the diet. The continuous line represents the relationship between larval weight and percentage of wheat bran in the diet.

bran percentage in the diet (Y1 = 26.3–0.074*wheat, R = 0.46, $P = 0.06$) and with larval weight (Y1 = 74.9–4.9*larva weight, R = 0.49, $P = 0.052$) (Figure 16.1). Larval weight was positively correlated with wheat bran percentage in the diet (Y2 = 9.9 + 0.015*wheatbran, R = 0.89, $P < 0.0001$).

In the case of egg to larva recovery, diets T-1, T-7, and T-8 presented less than 20% mean larval recovery, whereas in diet treatments T-2, T-3, T-4, and T-6 more than 25% larval recovery was obtained. For the quality parameter of larval weight, diet treatments T-1 and T-8 presented a mean weight greater than 11 mg, whereas all the other treatments (T-2 to T-7) were less than this weight.

16.3.1.2 Larval Development Speed Allowed by Diet

For all diet treatments, the accumulated larval recovery at the second larval collection is presented in Figure 16.2. The lowest mean value of accumulated recovery and, thus, the slowest development, was observed in diet treatment T-2. Diet formulations T-1, T-3, and T-4 exhibited an accumulated recovery of approximately 40%. A faster development was observed in diets T-5, T-6, T-7, and T-8, where accumulated recovery was higher than 50%.

16.3.1.3 Temperature Profile

There were significant statistical differences in mean diet temperature between diet treatments (ANOVA, $N = 140$, $F_{7,932} = 10$, $P < 0.01$). Treatments presented particular patterns of thermal variation and thermal limits. Diet treatments T-2, T-4, and T-6 showed the highest variation and exceeded the temperature of 29°C, which may not favor larval recovery of thermal sensitive females. Treatments T-1, T-7, and T-8 presented a mean temperature below 26°C and lower limits below 23°C, which may delay larval development. In contrast, diet treatments T-3 and T-5 presented an average temperature of 26°C and variation within the upper limit of 29°C and the lower limit of 23°C (Figure 16.3).

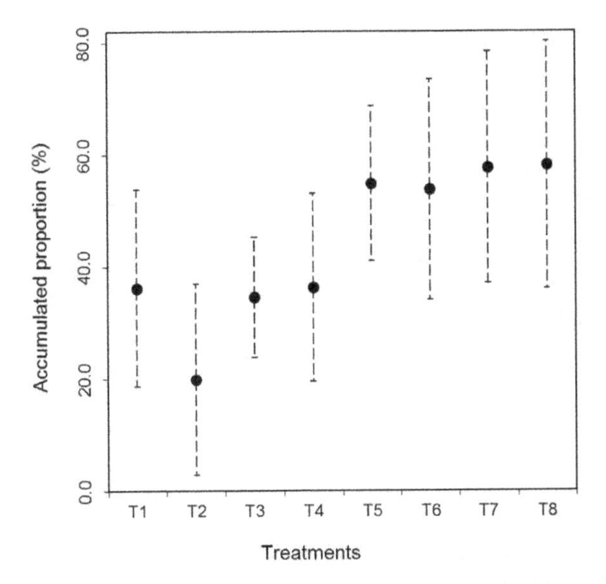

FIGURE 16.2 Larval development speed. The accumulated proportion (0%–100%) of recovered larvae in the second larval collection is presented for each treatment. Each point represents the mean and the standard deviation ($n = 5$).

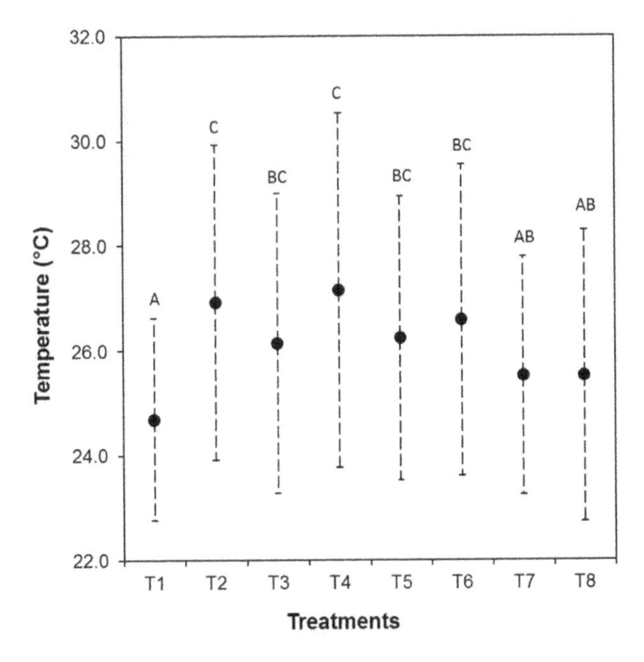

FIGURE 16.3 Temperature profiles in the first stage of the evaluation. The temperature recorded in each larval diet treatment is presented as the pooled data of the starter and finalizer diets. Each point represents the mean and the standard deviation ($N = 941$).

16.3.2 Second Stage

Diet treatments T-3 and T-5 were further evaluated using thermosensitive females of the Vienna 8 strain. Two additional diet treatments were included: the standard mass-rearing diet of the male-only stream ("T21L8N3") and a diet with an increased Nipagin concentration ("T5-N3") (see methods and Table 16.2).

16.3.2.1 Larval Recovery and Diet Yield

When data were pooled across diet treatments, 44.1% mean egg to larva recovery was obtained. No statistical differences were observed in larval recovery between diet treatments ($F_{3,19} = 0.13$, $P = 0.9$). The standard mass-rearing diet of the male-only stream "T21L8N3" generated a larval recovery of 42.9% and a yield production of 3.8 ± 0.7 (mean ± standard deviation) thousand larvae per kilogram of diet; this was the lowest mean value of all treatments. The highest recovery and yield were obtained in the "T5-N3" treatment (45.3% egg to larva recovery and 4.07 ± 0.1 thousand larvae per kilogram of diet), followed by treatments T-3 and T-5 (Figure 16.4).

16.3.2.2 Larval Weight

Larval weight was statistically different between diet treatments ($F_{3,190} = 3.93$, $P < 0.05$) (Figure 16.5). Post hoc analyses revealed that the larval weight of treatment T-5 was significantly higher than the weight obtained for the "T21L8N3" treatment (paired *t*-test, $P < 0.05$). There were no statistical differences in other treatment pair-wise contrasts ($P > 0.05$).

16.3.2.3 Temperature Profile

There were no significant statistical differences in mean diet temperature between diet treatments ($N = 178$, $F_{3,121} = 1.0$, $P = 0.38$). The interaction term treatment-to-time was also not significant ($F_{39,121} = 0.2$, $P > 0.9$). However, diet temperature was significantly different across time ($F_{13,121} = 23.4$, $P < 0.001$). For the first stage of larval development (0–96 h), the starter diet blocks of all treatments registered average temperatures within the range of 25°C–28°C. Additionally, when the starter diet was placed with the finalizer diet and the trays were moved to a rearing room at 20°C (120 h), diet temperature decreased to a range between 18°C and 22°C (Table 16.3). Although there were no statistical differences between treatments when temperature data were pooled, there were slight dissimilarities in temperature ranges of finalizer diets (Figure 16.6). Although temperature values in diets T-5 and "T5N3" were between 18°C and 20°C, formulations "T21L8" and T-3 presented a higher range that varied between 18°C and 22°C.

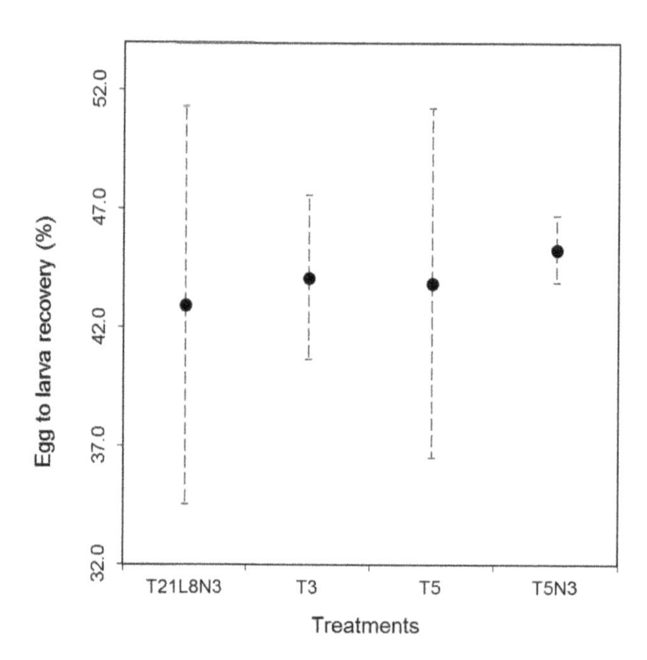

FIGURE 16.4 Egg to larva recovery obtained for each larval diet formulation (treatments) evaluated in the second stage where the female-male stream was used. Each point represents the mean and the standard deviation ($n = 5$).

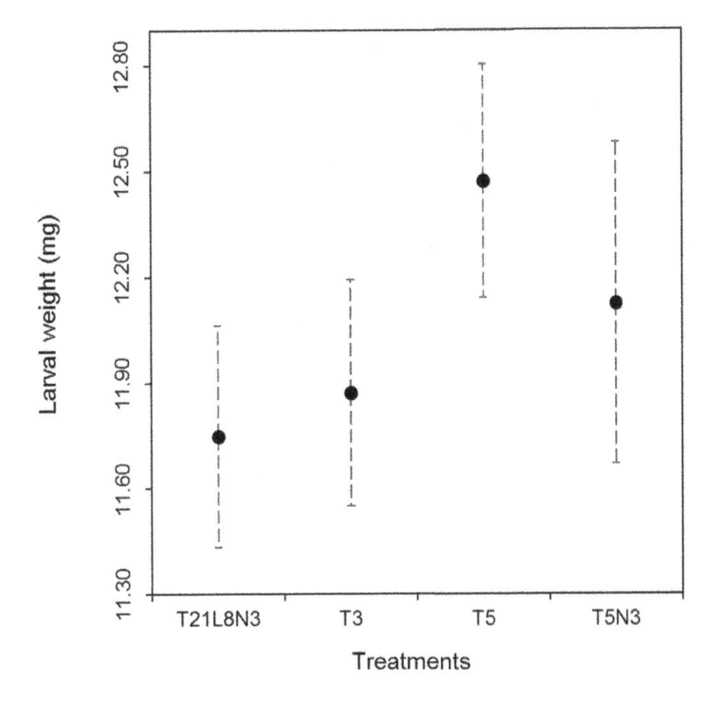

FIGURE 16.5 Larval weight (mg) obtained for each starter-finalizer larval diet formulation (treatments). Each point represents the mean and the standard deviation.

TABLE 16.3
Temperature Profiles of Starter and Finalizer Larval Diets Rearing Female and Male Streams

Rearing Stage	Time (h)	T21 L8 N3 (°C)	T3 (°C)	T5 N3 (°C)	T5 (°C)	Environmental Temperature (°C)
Larval development (1st instar)	24	26.1	25.8	25.8	26.3	24 ± 1
	48	25.7	25.5	26.3	26.3	24 ± 1
	72	25.9	25.6	25.9	26.2	24 ± 1
Transfer of starter diet	96	25.7	25.5	25.9	25.7	24 ± 1
Larval development (2nd instar)	120	23.7	24.0	23.2	23.4	20 ± 1
Larval development (3rd instar)	144	22.6	24.4	23.2	22.5	20 ± 1
1st collection	168	21.3	22.3	20.2	20.9	20 ± 1
2nd collection	192	20.3	22.4	19.5	21.3	20 ± 1
3rd collection	216	19.5	20.7	18.9	20.6	20 ± 1
4th collection	240	20.0	20.3	18.7	20.4	20 ± 1
5th collection	264	18.9	19.6	18.8	20.3	20 ± 1
6th collection	288	18.6	20.1	18.7	20.4	20 ± 1
7th collection	312	18.1	19.3	18.4	20.3	20 ± 1
8th collection	336	19.2	19.8	19.1	21.0	20 ± 1

There were no significant differences.

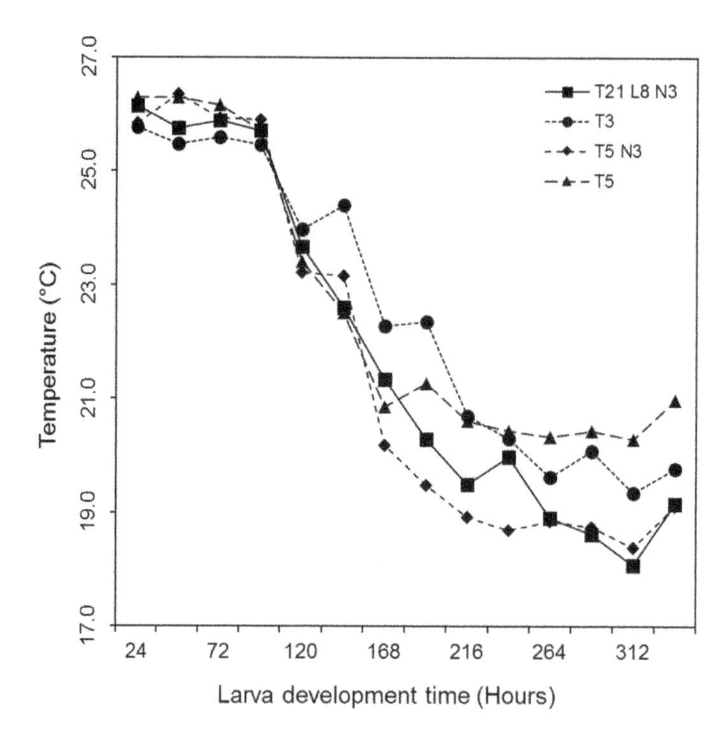

FIGURE 16.6 Detailed temperature profiles in the second stage. The temperature recorded from egg seeding to diet disposal is presented for each larval diet formulation (treatments). Each point represents the mean and the standard deviation.

16.3.2.4 Microbiological Profile

The obtained bacteria and fungi (total unit counts) were not significantly different between diet treatments (ANOVAs, treatment effect, $P > 0.05$). In both starter and finalizer diets, total bacteria and fungi counts increased significantly with time (ANOVAs, time effect, $P < 0.05$). Also, there were marginal differences in the increase of microbiota with time between treatments for the finalizer diets (time*treatment effect $F_{3,52} < 1.3$, $P = 0.3$). This suggests that some diets may control bacteria and fungi development differently across time (Table 16.4).

For bacteria, fresh diets "T21L8N3" and T-3 presented relatively higher values of initial bacteria load than diets T-5 and "T5N3." At the end of the rearing process, relatively higher bacteria counts were observed in finalizer diets "T21L8N3" and T-5 compared to those observed in finalizer diets "T5N3" and T-3.

In the case of fungi unit counts, with the exception of finalizer diet T-3, we observed that CFUs tended to decrease from fresh to 120 h and then increase from 120 to 240 h. This suggests that diets controlled fungi development for a limited time, which did not last for the entire rearing process. Diet "T5N3" was the formulation with the relatively lowest fungi development.

16.3.2.5 Physicochemical Profile

The pH was determined for starter and finalizer diet formulations. Mean values ($n = 5$) ranged from 3.89 to 4.09 for fresh diets and from 3.67 to 3.83 for diets 8 days after their preparation (192 h of rearing process). For both starter and finalizer diets, there were no statistical differences between diet formulations (ANOVAs, treatment effect, starter diet: $F_{3,39} = 1.3$, $P = 0.26$, finalizer diet: $F_{3,79} = 1.2$, $P = 0.3$). Results show that pH significantly decreased with time (starter diet: $F_{3,39} = 4.2$, $P < 0.01$, finalizer

TABLE 16.4
Microbiological Performance of the Evaluated Starter and Finalizer Diet Formulations. Mean Values of Five Repetitions Are Presented

		Starter Diet			Finalizer Diet		
Parameter	**Treatment**	**Fresh**		**96 h**	**Fresh (96 h)**	**216 h**	**336 h**
Bacteria (CFU/g)	T21L8N3	53,200	a	92,600	61,000	195,200	2,490,000
	T3	69,200	a	256,200	57,600	130,400	488,000
	T5N3	41,600	a	176,000	47,400	39,400	952,000
	T5	32,800	a	76,000	41,200	55,800	1,662,000
Fungi (UFC/g)	T21L8N3	682	a	503	192	70	1,150
	T3	624	a	304	487	779	700
	T5N3	772	a	139	186	76	300
	T5	918	a	210	507	167	690

There were no significant differences (diet by time).

TABLE 16.5
Physicochemical Records of the Evaluated Starter and Finalizer Diet Formulations

		Starter Diet				Finalizer Diet									
Parameter	**Treatment**	**Fresh**		**96 h**		**Fresh (96 h)**		**144 h**		**192 h**		**240 h**		**288 h**	
pH	T21L8N3	4.09	a	3.78	a	4.00	a	3.96	a	3.96	a	3.75	a	3.83	a
	T3	4.03	a	3.77	a	3.89	a	3.95	a	4.04	a	3.80	a	3.69	a
	T5N3	3.98	a	3.73	a	4.07	a	3.98	a	4.02	a	3.84	a	3.98	a
	T5	4.02	a	3.73	a	3.97	a	3.99	a	3.87	a	3.65	a	3.67	a
Moisture (%)	T21L8N3	48.10	a	51.09	a	51.15	Aa	50.83	Aa	51.74	Aa	54.45	Aa	57.06	Aa
	T3	47.55	a	50.49	a	47.68	Ba	49.51	Ba	49.40	Ba	52.90	Ba	52.84	Ba
	T5N3	49.62	a	52.90	a	52.47	Aa	50.31	Aa	52.93	Aa	54.52	Aa	56.36	Aa
	T5	49.25	a	52.35	a	51.18	Aa	49.90	Aa	53.34	Aa	54.45	Aa	52.90	Aa

Mean values in each column followed by the same lowercase letter are not significantly different (diet by time, Tukey test, $P > 0.05$).
Mean values in each row followed by the same capital letter are not significantly different (Diet, Tukey test, $P > 0.05$).

diet: $F_{4,79} = 11$, $P < 0.01$). The interaction term diet*time was not statistically significant ($P > 0.3$). Moreover, pairwise contrasts show that there were no differences between diet treatments across rearing time (Table 16.5).

The analysis of moisture percentage in the diets showed mixed results. For starter diets, there were no statistical differences in moisture between treatments, across time, or for the interaction term time*treatment (ANOVA, $P > 0.05$). In contrast, the results for finalizer diets showed significant differences in moisture between diet treatments (ANOVA, $F_{4,79} = 4.8$, $P < 0.01$) and across time of the rearing process ($F_{4,79} = 4.8$, $P < 0.01$), but there were no differences for the interaction term time*treatment ($P > 0.05$). Pos hoc tests for finalizer diets indicated that diet T-3 presented a significantly lower moisture percentage and that moisture percentage in the diets increased with time of the rearing process. Additionally, pairwise contrasts showed that there were no differences between diet treatments across rearing time (Table 16.5).

16.4 DISCUSSION

To initiate the mass rearing of females in the new Moscamed facility, a suitable diet formulation has been established. Using a starter-finalizer diet system, the formulation "T5-N3" presented the best balance between productivity and quality.

For mass-rearing operations, several qualities are desirable in larval diets: high productivity, low cost, fast development, high stability, resilience at large scales, and high quality of produced insects, among others. However, developing a formulation that presents all of these characteristics is challenging, and the best compromise is often used in practice. Of the 10 diets evaluated in this study, the formulation "T5-N3" was selected because it presented a good development time, a stable productivity, and an optimal thermal performance for rearing thermosensitive females. These characteristics, in addition to a relatively better suppression of fungal and bacterial development and a lower initial fungi load, make the formulation "T5-N3" the best option for the biological amplification of Vienna 8 females. This formulation was still considered as the best from all the tested formulations even though it did not result in the highest productivity, the heaviest larvae, or the most ideal performance. However, this diet formulation was regarded as a good starting point for initiating the rearing process of the colonies at the new facility of Mexico's Moscamed Program.

Larval diet formulations can always be optimized. For fruit flies of economic importance, several examples with *Ceratitis*, *Anastrepha*, and *Bactrocera* have shown that diet systems (Chang 2009, Pascacio-Villafán et al. 2018), ingredients of diets (Hernández et al. 2016, Moadeli et al. 2018, Pascacio-Villafán et al. 2015, Rivera et al. 2012), and ingredient proportions (Moadeli et al. 2017, Pascacio-Villafán et al. 2017) can be modified to improve several aspects of mass rearing. In the case of the diet "T5-N3," several aspects can be optimized. For example, even though we used the same formulation for starter and finalizer diets for convenience, it is very likely that an optimal starter formulation will differ from an optimal finalizer diet formulation (e.g., Domínguez et al. 1993). Also, one of the most expensive ingredients, the protein source (inactive dry yeast), could be reduced, especially for the starter diet. Similarly, the proportion of texturizers (dry ground maize plant fractions-to-wheat bran ratio) can be adjusted to reduce the cost and to facilitate larval migration from the starter to the finalizer diet. Moreover, the volume proportion of starter-finalizer diets (e.g., 1:4 kg) and the time to transfer starter diets to finalizer trays can be explored for optimization. Additionally, because some diet ingredients differ between Mexico and Guatemala, it can be expected that the colony recently established in Mexico will undergo a process of adaptation to the new ingredients with a concomitant improvement of the productivity reported here.

The starter-finalizer diet system offers a convenient scheme for the mass rearing of *C. capitata* Vienna 8 colonies. As mentioned before, due to the characteristic slower development of females, larval diets should maintain for an extended time the properties that allow females to complete their development. For mass-rearing operations in Mexico, the starter-finalizer system with meridic diet (containing at least one unknown chemical structure) is operatively straightforward because the diet preparation and rearing methods are not too different from the ones of the diet system currently used. However, the use of other types of promising diet systems for breeding insects, like gel or liquid diets, have not been discarded. In fact, a hybrid diet system is attractive from a technical, operational, and economical point of view. For a starter diet, a gel diet prepared with minimum waste and amount of ingredients and, more importantly, where the growth of bacteria can be controlled more easily could be used. A meridic diet, like those evaluated in this study, could then be used as a finalizer diet due to its relatively lower cost, higher volume, and straightforward logistics in mass rearing.

The design of novel diets involves time, effort, and costs and is generally limited by the volume of the diet logistically viable for evaluation. The accumulated experience of four decades of mass rearing Medflies leads us to believe that even if an ideal diet was developed at a small scale, there is no guarantee that it will work properly at a massive scale. Consequently, mass-rearing biofactories (especially new ones) need to be prepared for the constant improvement of larval diets. When the new Moscamed biofactory was designed, a special area dedicated to this task was included.

The "Experimental diets laboratory" was designed with the conditions normally found in the mass-rearing rooms and at proper scale dimensions. This laboratory will allow the testing of experimental diet batches of around 250 kg without jeopardizing the production line. Therefore, although the selected formulation "T5-N3" will be used to start the operations of the new facility, the optimization of this diet will be carried out simultaneously.

For the new Moscamed facility that will produce up to one billion male pupae per week, larval diets are a fundamental element of the mass-rearing operations. Diet ingredients and formulations will ultimately determine the cost, the quality, and a great part of the logistics of the biofactory. However, it is important to keep in mind that mass rearing is a dynamic constantly changing process, in which the ideal or most suitable larval diet may have a nonstatic formulation. Thus, for the start of operations of the new facility, the diet "T5-N3" can be used to achieve the goals of sustaining the colony and beginning biological amplification. Several aspects of "T5-N3" can be optimized, such as costs, ingredient proportions, productivity, and quality of insects. Nevertheless, and perhaps most importantly, the facility is prepared for the continuous optimization of larval diets in the long term.

ACKNOWLEDGMENTS

We thank the support of Ing. Francisco Ramirez y Ramirez, Director of Fruit Flies National Program at SENASICA, Mexico. We acknowledge the valuable participation of all the technical staff at the Moscamed Biofactory. We are especially grateful to Carlos Cano Vázquez and Meyber Díaz Hernández of the Instituto Tecnológico de Comalapa, Chiapas, for their technical assistance. To Manuel Malo López for the support given to the students for the duration of the study, as well as to Idelma López Alvarez and José Angel Hernandez de la Cruz for performing the physicochemical and microbiological analyses.

REFERENCES

Augustinos, A. A., A. Targovska, E. Cancio-Martinez, E. Schorn, G. Franz, C. Cáceres, A. Zacharopoulou, and K. Bourtzis. 2017. *Ceratitis capitata* genetic sexing strains: Laboratory evaluation of strains from mass-rearing facilities worldwide. *Entomologia Experimentalis et Applicata*, 164(3):305–317. doi:10.1111/eea.12612.

Caceres, C. 2002. Mass rearing of temperature sensitive genetic sexing strains in the Mediterranean fruit fly (*Ceratitis capitata*). *Genetica* 116:107–116.

Cáceres, C., K. Fisher, and P. Rendón. 2000. Mass rearing of the Medfly temperature sensitive lethal genetic sexing strain in Guatemala. In *Area-wide Control of Fruit Flies and Other Insect Pests*, ed. K. H. Tan, 551–558. Penang Malaysia: Penerbit Universiti Sains.

Chan, H. T., and E. B. Jang. 1995. Diet pH effects on mass rearing of Mediterranean fruit fly (Diptera: Tephritidae). *Journal of Economic Entomology* 88(3):569–573.

Chang, C. L. 2009. Fruit fly liquid larval diet technology transfer and update. *Journal of Applied Entomology* 133:164–173.

Domínguez, J. C., J. L. Zavala., P. Liedo., and N. D. Bruzzone. 1993. Implementation of starter diet technique for medfly mass rearing at Metapa, Chiapas, México. In *Fruit Flies: Biology and Management*, ed. M. Aluja, and P. Liedo, 287–280. Springer-Verlag, New York.

Economopoulos, A. P., A. A. Al-Taweel., and N. D. Bruzzone. 1990. Larval diet with starter phase for mass rearing *Ceratitis capitata*: substitution and refinement in the use of yeasts and sugars. *Entomologia Experimentalis et Applicata* 55:239–246.

Enkerlin, W., J. M. Gutiérrez-Ruelas, A. Villaseñor Cortes, et al. 2015. Area freedom in Mexico from Mediterranean fruit fly (Diptera: Tephritidae): A review of over 30 years of a successful containment program using an integrated area-wide SIT approach. *Florida Entomologist* 98(2):665–681.

Fay, H. A. C. 1988. A starter diet for mass rearing larvae of the Mediterranean fruit fly, *Ceratitis capitata* (Wied.). *Journal of Applied Entomology* 105:496–501.

Fisher, K., and C. Caceres. 2000. A filter rearing system for mass reared genetic sexing strains of Mediterranean fruit fly (Diptera: Tephritidae). In *Area-wide Control of Fruit Flies and Other Insect Pests*, ed. K. H. Tan, 543–550. Penang Malaysia: Penerbit Universiti Sains.

Franz, G. 2005. Genetic sexing stains in Mediterranean fruit fly, an example for other species amenable to large-scale rearing for the sterile insect technique. In *Sterile Insect Technique. Principles and practices in Area-Wide Integrated Pest Management*, ed. V.A. Dyck, J. Hendrichs, and A.S. Robinson, 427–451. Dordrecht, the Netherlands: Springer.

Hernández, E., P. Rivera, M. Aceituno-medina, R. Aguilar-Laparra, J. L. Quintero-Fong, and D. Orozco-Dávila. 2016. Yeast efficacy for mass rearing of *Anastrepha ludens*, *A. obliqua* and *Ceratitis capitata* (Diptera: Tephritidae). *Acta Zoológica Mexicana (n.s.)* 32(3):240–252.

Moadeli, T., P. W. Taylor, and F. Ponton. 2017. High productivity gel diets for rearing of Queensland fruit fly, *Bactrocera tryoni*. *Journal of Pest Science* 90(2):507–520.

Moadeli, T., B. Mainali, F. Ponton, and P. W. Taylor. 2018. Evaluation of yeasts in gel larval diet for Queensland fruit fly, *Bactrocera tryoni*. *Journal of Applied Entomology* 142(7):679–688.

NOM-092-SSA1-1994. Norma Oficial Mexicana NOM-092-SSA1-1994, Bienes y Servicios. Método para la cuenta de bacterias aerobias en placa. DOF

NOM-111-SSA1-1994. Norma Oficial Mexicana NOM-111-SSA1-1994, Bienes y Servicios. Método para la cuenta de mohos y levaduras en alimentos. DOF

Pascacio-Villafán, C., A. Birke, T. Williams, and M. Aluja. 2017. Modeling the cost-effectiveness of insect rearing on artificial diets: A test with a tephritid fly used in the sterile insect technique. *PLOS ONE* 12(3):e0173205.

Pascacio-Villafán, C., L. Guillén, T. Williams, and M. Aluja. 2018. Effects of larval density and support substrate in liquid diet on productivity and quality of artificially reared *Anastrepha ludens* (Diptera: Tephritidae). *Journal of Economic Entomology* 111(5):2281–2287.

Pascacio-Villafán, C., T. Williams, J. Sivinski, A. Birke, and M. Aluja. 2015. Costly nutritious diets do not necessarily translate into better performance of artificially reared fruit flies (Diptera: Tephritidae). *Journal of Economic Entomology* 108:53–59.

Pinson E., W. Enkerlin, S. Arrazate, and A. Oropeza.1993. Adaptation of *Anastrepha ludens* (Loew) to an enriched recycled diet, In *Fruit flies Biology and management*, ed. M. Aluja and P. Liedo, 285–287. New York: Springer-Verlag.

Programa Moscamed. 2015. Manual de procedimientos de la planta de producción de moscas estériles del Mediterráneo. SENASICA/DGSV.

R Core Team. 2014. R: A language and environment for statistical computing. *R Foundation for Statistical Computing*, Vienna, Austria. http://www. R-project.org/.

Ramírez-Santos, E. M., P. Rendón, L. Ruiz-Montoya, J. Toledo, and P. Liedo. 2016. Performance of a genetically modified strain of the Mediterranean fruit fly (Diptera: Tephritidae) for area-wide integrated pest management with the sterile insect technique. *Journal of Economic Entomology* 110(1):24–34.

Rivera, J. P., E. Hernández, J. Toledo, M. Salvador, and Y. Gomez-Simuta. 2012. Optimización del proceso de cría de *Anastrepha ludens* Loew (Diptera: Tephritidae) utilizando una dieta larvaria a base de almidón pre-gelatinizado. *Acta Zoológica Mexicana* (*n.s*) 28:102–117.

Robinson, A. S., Franz, G., and K. Fisher. 1999. Genetic sexing strains in the medfly, *Ceratitis capitata*: Development, mass rearing and field application. *Trends in Entomology* 2:81–104.

Schwarz, A. J., A. Zambada, D. Orozco, J. L. Zavala and C. O. Calkins. 1985. Mass production of the Mediterranean fruit fly at Metapa Mexico. *Florida Entomologist* 68:467–477.

Villaseñor, A., J. Carrillo, J. Zavala, et. al. 2000. Current progress in the medfly program Mexico-Guatemala. In *Area-Wide Control of Fruit Flies and Other Insect Pests*, ed. K. H. Tan, 361–368. Penang Malaysia: Penerbit Universiti Sains.

Section VI

Natural Enemies and Biological Control

17 Biological Control of *Anastrepha* Populations in Wild Areas to Strengthen the Commercial Status of Mango Production along the Pacific Coast of Mexico

Jorge Cancino, Arturo Bello-Rivera, Jesús Cárdenas-Lozano, Fredy Gálvez-Cárdenas, Víctor García-Pérez, Eduardo Camacho-Bojórquez, Emiliano Segura-Bailon, Maximino Leyva-Castro, and Francisco Ramírez y Ramírez*

CONTENTS

* Corresponding author.

Abstract The commercial crop of mango has expanded successfully along the Pacific coast of Mexico. The most important pest species of mango in this region is *Anastrepha obliqua* (Macquart), with the exception of Chiapas, where *Anastrepha ludens* (Loew) is considered to be the mango pest of highest priority. Effective integrated management of *Anastrepha* in commercial areas maintains the high-level of mango production. Intense technical activities inside orchards control *Anastrepha* populations. However, a latent problem is the presence of *Anastrepha* populations in the surrounding wild areas. Application of augmentative releases of the Anastrepha spp. parasitoid *Diachasmimorpha longicaudata* (Ashmead) was proposed to reduce pest flies in wild areas. A percentage of parasitism between 30% and 50% reduced the risk of *Anastrepha* populations invading mango orchards. Although the specific objective of the biological control program in each region varies, the general intention is to achieve the control of *Anastrepha* populations in wild fruit hosts in the peripheral areas. The most successful use of augmentative biological control to maintain a low *Anastrepha* prevalence occurred in the producing area of Tecpan de Galeana, Guerrero, where high parasitism levels were found in *A. obliqua*–infested fruits of the genus *Spondias* and creole mango. Parasitism of *Anastrepha* spp. was more important when small fruits were infested, such as species of *Spondias* (>50%), but the effectiveness was decreased when large fruits were infested, such as sour orange (<30%) infested by *A. ludens* in the Coast of Chiapas. Augmentative parasitoid releases contributed to the suppression of fruit fly populations in the surrounding areas of commercial mango orchards. This has become an important technique included in the integrated pest management (IPM) of the National Program of Fruit Flies in Mexico.

17.1 INTRODUCTION

Mango is one of the high-quality fruits exported from Mexico with approximately 1.8 million tons produced annually, from which about 20% are exported (SAGARPA 2017). Mango orchards have been established in tropical areas in approximately 200,000 ha and include varieties such as Ataulfo, Manila, Tommy Atkins, Kent, Headen, and several creole varieties (Monter and Aguilera 2011; SAGARPA 2017). The mango-producing regions with the highest economic significance and production and with modern infrastructure are located along the Pacific Coast (CONASPROMANGO 2012; SAGARPA 2017) (Figure 17.1).

Each region copes with different phytosanitary issues; however, the presence of fruit flies is a common issue. Although there is a tephritid fruit fly species complex in wild areas surrounding production sites, *Anastrepha obliqua* (Macquart) and *Anastrepha ludens* (Loew) are pests. *A. ludens* is limited to the southeast region, whereas *A. obliqua* is a pest throughout the mango production area (Ruiz-Arce et al. 2012, 2015).

Marketing of mango at the export level is achieved through the implementation of efficient integrated management (Aluja 1993, 1994). This has allowed a great extension of the Pacific Coast–producing regions to keep the low prevalence or free area status of *Anastrepha* spp. (FAO 2016). Flies per trap per day (FTD) values below 0.001 and absolute absence of infested fruit in orchards are required during the production period to obtain export permits.

The integrated pest management (IPM) of fruit flies has contributed substantially to reducing pest populations. However, a high risk of introduction of *Anastrepha* spp. populations from wild areas to the commercial orchards remains a problem. The proposal of releasing parasitoids for the control of fly populations in wild areas has great promise and many ecological advantages, particularly because it is an environmentally friendly technique (Montoya et al. 2016).

This chapter analyzes and reviews the current situation of mango production along the Pacific Coast as part of the National Program of Fruit Flies in Mexico. This analysis is reinforced with data of the last 2 years obtained from field operation samplings. The main objective is to show the efficacy of augmentative releases of fruit fly parasitoids in the surrounding areas of mango orchards

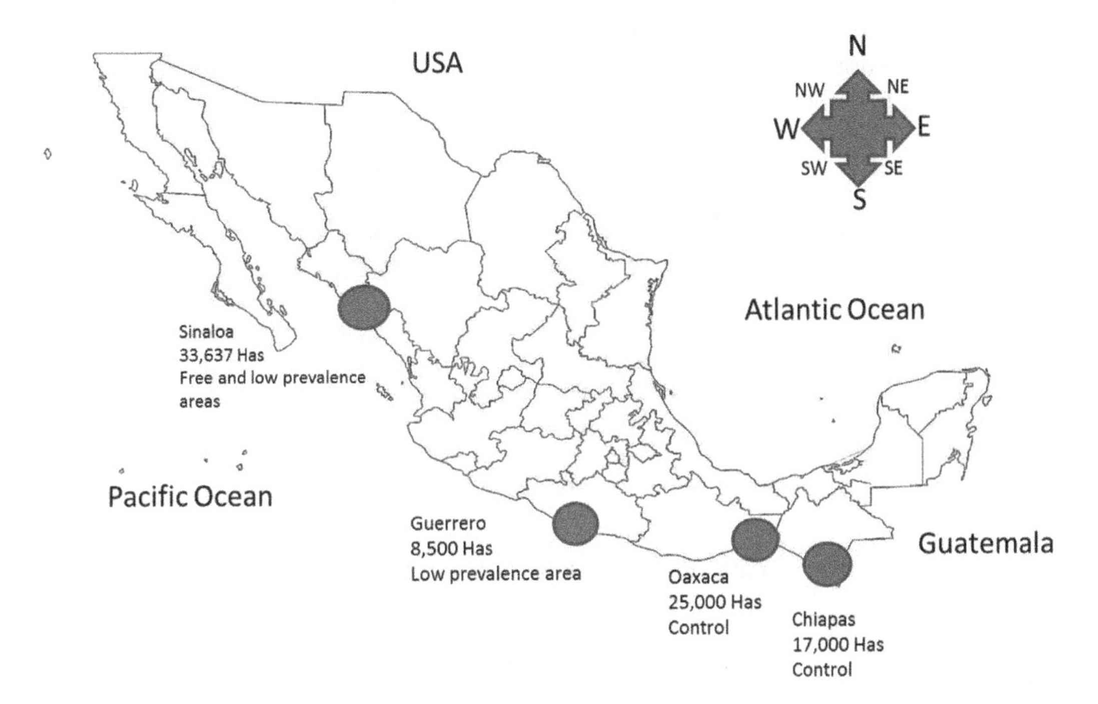

FIGURE 17.1 Economically important mango production regions along the Pacific Coast of Mexico working with integrated pest management techniques against fruit flies, including augmentative biological control.

as part of an IPM strategy to reduce risks of introduction of fly populations from wild areas into mango crops. We organized this chapter in four sections: (i) a brief description of the progressive experiences in the IPM of *Anastrepha* populations and limitations of tactics in wild vegetation, (ii) a discussion of the Mexican biological control program against fruit fly pests, (iii) a specific discussion of biological control as an important tool to reduce fruit fly populations in wild areas of mango production regions of the Pacific Coast, including data on parasitism rates, and (iv) a conclusion with future plans.

17.2 FRUIT FLY SITUATION IN MEXICO

17.2.1 Origin of Fruit Fly Wild Populations and Damage Produced

Anastrepha is a highly diverse neotropical genus, and many species are distributed throughout different regions in Mexico (Hernández-Ortíz 1992; Vanoye-Eligio et al. 2015). Before the introduction of non-native crop fruits, flies originally oviposited and developed as larvae in native fruits (Aluja 1994). According to trapping data, adult populations show seasonal fluctuations associated with the availability of host fruits and environmental conditions (Celedonio-Hurtado et al. 1995; Vanoye-Eligio et al. 2015). Little is known about the native host fruits of these fruit fly species because most of these fruits have little commercial value. The diversity of host fruits is vast and allows the maintenance of fly populations throughout the year. However, fly populations are reduced to low detection levels during periods of host shortages (Aluja 1994). Diapause factors of *Anastrepha* spp. are currently unknown (Aluja et al. 2000).

The most dominant *Anastrepha* pest species of mango in the Pacific Coast is *A. obliqua*, which is associated with fruits of the genus *Spondias* and other native fruits, with guava wild types as important alternative hosts (Aluja et al. 1987; Marsaro et al. 2011). In the southeast region of the Pacific Coast in the state of Chiapas, *A. obliqua* coexists with populations of *A. ludens*, and both species have been identified as important mango pests in this region (Aluja et al. 1987; Celedonio-Hurtado et al. 1995). *A. ludens* is associated with native fruits of the genus *Casimiroa* and *Sargentia greggii* (S. Watson) (both from the Family Rutaceae) (Thomas 2003; Hernandez-Ortíz 2007; Vanoye-Eligio et al. 2015). Both *Anastrepha* spp. increased their host range with the introduction of fruit species after the Spanish conquest (Ramírez et al. 2008). The introduction of mango allowed populations of *A. obliqua* and *A. ludens* to invade new areas and to establish using this fruit as an alternative host for the maintenance of populations. Similarly, both species were able to adapt to other fruits, as is the case of *A. ludens*, which is often a pest of commercial fruits of the genus *Citrus* (Leyva et al. 1991). Marketing of mango in Mexico for exportation is seriously impacted by the presence of *Anastrepha* spp., and the control of these fruit flies is a common requirement in commercial regions. The insects' impact is maximized in areas declared as low prevalence or free areas because the presence of a single specimen in an orchard involves the potential of losing the commercial status of mango exportation throughout the region (NOM023FIT95 1999).

17.2.2 Integrated Management of Fruit Flies in Mango Orchards

Each region of Mexico has adopted different strategies to apply control techniques for fruit flies in mangos, ranging from temporary to permanent efforts. Important technical activities are carried out in mango-growing regions of Mexico that achieve low-prevalence levels of fruit flies (50.25%) and even free-area status (10.44%) in orchards (Senasica 2017). Activities carried out inside orchards depend to a large extent on the application of toxic baits. In the chemical control method, the use of GF-120 (Spinosaid) is increasing (Flores et al. 2011; De los Santos-Ramos et al. 2012), replacing the use of malathion. Cultural activities such as raking, pruning, and fallowing are performed to keep the orchard clean and prevent the presence of pests (Aluja 1993). Mechanical control is used to eliminate infested fruits during the harvest. After the harvest, fruits undergo hydrothermal treatments or irradiation that effectively avoids the risk of eggs or larvae remaining in fruits in case the orchard is infested (Hernández et al. 2012). A highly important activity is the use of the sterile insect technique (SIT), which is applied in orchards and surrounding areas (Orozco-Davila et al. 2017). Strategies are designed based on monitoring with a system of Multilure traps. In addition, regions with low pest population levels are protected with a legal quarantine barrier (NOM023FIT95 1999; Follet and Vick 2002). Yet, the most important question is: What should be done in commercial areas and, mainly, in free or low prevalence regions where orchards are surrounded by native wild plants that are hosts for native fruit fly pest populations?

17.3 BIOLOGICAL CONTROL: PROPOSAL, EXPERIENCE, AND ESTABLISHMENT IN MEXICO

17.3.1 General Concepts

The control of *Anastrepha* populations settled in wild areas is difficult. Many control efforts have deleterious effects for the ecosystem, such as eliminating noncommercial host trees (many of them native ones) or the application of toxic baits that kill a variety of nontarget insects. It is in this context that the use of parasitoids may be the best option. Parasitoids have high levels of specificity and an efficient foraging behavior that allows them to successfully find hosts (Pascal et al. 2017). They have evolved with their hosts developing in native fruits, creating an intrinsic tritrophic interaction that may be exploited for pest control (Henter and Sara 1995; Ovruski et al. 2016). Many native fruits are small and constitute an attractive opportunity for larval parasitoids (the guild most commonly used) to surpass the physical refuge that larger fruits provide to flies, enabling to keep high natural parasitoid population levels (Sivinski 1991; Leyva et al. 1991; Wang et al. 2009). However, these advantageous aspects that occur in nature are reduced by several factors that decrease the intrinsic rate of increasing parasitoid populations; generally, the level of parasitism remains less than 10%, which is considered insufficient to control fruit fly pest populations (López et al. 1999; Ovruski et al. 2000; Montoya et al. 2016).

17.3.2 Augmentative Release as a Strategy

According to fly and parasitoid population parameters and biological control theory, various methods may be used to gain control of pest populations (Liedo and Carey 1994; Vargas et al. 2002). A feasible method to control native fruit fly pests is conducting augmentative releases of established natural enemies. Augmentative biological control involves the artificial mass rearing and release of parasitoid populations to increase parasitism rates and consequently reduce fly pest populations (Knipling 1993). Successful results of fruit fly pest suppression have been achieved in Florida and the southeast of Mexico using augmentative biological control (Sivinski et al. 1996; Montoya et al. 2000).

In relation to an augmentative biological control program for fruit flies in Mexican mango orchards, it is necessary to select a natural enemy species that will be able to control *Anastrepha* spp. populations. Although there are a number of native parasitoids that attack *Anastrepha* spp. in Mexico, none have been studied to develop an artificial mass-rearing technique. However, *Diachasmimorpha longicaudata* (Ashmead) (Hymenoptera: Braconidae), a parasitoid of fruit flies of the genus *Bactrocera*, has been successfully used in augmentative biological control programs. *D. longicaudata* is native to the Indo-Australian region (Dashavant et al. 2018) and was successfully introduced and established in the Americas for the control of fruit fly populations (Ovruski et al. 2000). This parasitoid has a wide host range, including *Anastrepha* spp. (Lawrence et al. 1976; Lawrence 2005; Mierelles et al. 2013). Successful mass-rearing techniques were developed and large numbers of *D. longicaudata* can be reared, packaged, and released (Cancino and Montoya 2007). Studies have been conducted that document the effectiveness of augmentative releases of *D. longicaudata* to suppress several species of fruit fly populations, including *Anastrepha* spp. (Wong et al. 1991; Knipling 1992; Sivinski et al. 1996; Montoya et al. 2000).

17.4 BIOLOGICAL CONTROL OF FRUIT FLIES ALONG THE MANGO PRODUCTION REGIONS OF THE PACIFIC COAST

17.4.1 Shipping, Packing, and Release of Parasitoids

D. longicaudata parasitoids are reared at the Moscafrut Plant, where between 20 and 50 million pupae (60%–70% of parasitoid emergence) are produced weekly (Cancino and Montoya 2006). Pupae are shipped under hypoxia conditions by commercial flights and are packaged in "Arturito" type containers (Cancino and López-Arriaga 2016; Cancino et al. 2017). The prerelease process implicates maintaining the parasitoids at temperatures of 26°C (4 days) during 7 days for complete emergence and at 21°C for 2 days to stimulate the copula, supplying honey as food (Cancino et al. 2017). Each release of parasitoids is conducted on land at a ratio of 1,500 parasitoids per ha per week during the larval infestation period (Montoya et al. 2016).

17.4.2 Methods to Estimate Parasitism Levels in the Field

A formal method was developed for fruit sampling in parasitoid release areas (Figure 17.1) to calculate fruit fly parasitism rates and was used in wild areas of each release region over a 2-year period on a weekly basis.

17.4.2.1 Sampling

Fruit fly host species were randomly collected and fruits collected per species depended on the following fruit size designations: 500 g for small fruits, yellow mombins, plums (*Spondias* spp.), wild guavas (*Psidium* spp.), etc.; 2–3 kg for medium-sized fruits, commercial guavas (*Psidium guajava* L.), creole mangoes (*Mangifera indica* L.), sapodilla (*Manilkara zapota* L.), etc.; and 5 kg for large fruits, Mammee apple (*Pouteria sapota* L.), sour orange (*Citrus aurantium* L.), large mangoes, etc. Fruits were dissected in the laboratory on the same day they were collected and all third stage larvae found in the fruits were placed in cylindrical plastic containers (7 × 5 cm) with vermiculite. Containers with pupae were kept for 15 days at temperatures between 24°C and 26°C until emergence of either a fly or a parasitoid was obtained. Parasitism was calculated using the following formula: % parasitism = (No. of Parasitoids/No. of Parasitoids + No. of Flies) × 100. This parameter is the best indicator of the effect of augmentative releases, and the information is supplemented with other indicators (FTD with trapping, infestation with fruit sampling, etc.).

17.4.2.2 Data Analysis

Parasitism percentages derived from fruit collections were represented graphically on a monthly basis for each region over a 2-year period. Total monthly numbers of larvae collected during 2 years of fruit sampling were analyzed for the Guerrero region at Tecpan de Galeana with a linear relationship to present the respective monthly percentage of parasitism. Percentage of parasitism was also related linearly with the respective monthly FTD over the same 2-year period at the Tecpan de Galeana location. In both cases, a logarithmic data transformation was applied. Four fruits from host plant species from different size classes were sampled randomly each month for a 1-year period in the Chiapas region at Soconusco, and the obtained percentage of parasitism was compared among host species by means of a one-way ANOVA and a Tukey's test.

17.4.3 Results and Discussion

17.4.3.1 Analysis by Region

Mango production information and *D. longicaudata* parasitism rates on fruit flies resulting from the augmentative biological control program are presented for each of the four mango-producing regions along the Pacific Coast region of Mexico.

17.4.3.1.1 Sinaloa

In this state, the commercial mango production covers 33,637 ha. The southern part was deemed as a low prevalence fruit fly area (85%) and the northern part as a fruit fly free area (15%). FTD levels in orchards have been kept between 0 and 0.0031. The vegetation surrounding the orchards was characterized as subdeciduous tropical rainforest and scrubland (Vega-Aviña et al. 2000), with relatively few fruit fly hosts, and the only pest species for mango was *A. obliqua*. Parasitoid releases were basically carried out in wild areas near commercial mango orchards, and the sampled fruit species were mainly wild plums (*Spondias* spp.) and guavas. Parasitism rates in sampled fruits fluctuated slightly on a monthly basis with an average rate of 30% (Figure 17.2). The main mango varieties were Ataulfo, Tommy Atkins, Haden, Malika, and Rio Red. The annual production capacity was of 241,446 tons (2017), from which 77% of the harvest was exported.

17.4.3.1.2 Guerrero

Augmentative parasitoid releases were carried out in the Municipality of Tecpan de Galeana, located in Costa Grande in the state of Guerrero. There are approximately 8,500 ha with mango orchards of the varieties Ataulfo and Manila. The fruit area was adjacent to the coast on the south and surrounded by wild vegetation characterized as medium evergreen tropical forest and low deciduous tropical rainforest (CONAFOR 2015). *Spondias* spp. fruits dominated in the wild areas near the orchards, supporting a large population of *A. obliqua* (Lópezet al. 1999; Montoya et al. 2016) and representing

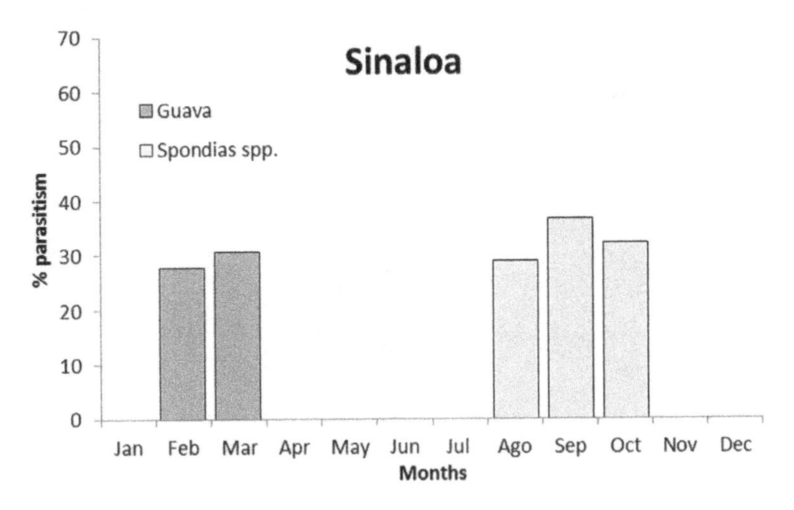

FIGURE 17.2 Monthly percentage of parasitism in guava and fruits of *Spondias* spp. sampled in the mango-producing regions of Sinaloa state (2016–2017).

a high risk of fruit fly dispersion into the mango production areas that were considered to have a low fruit fly prevalence status. Due to favorable weather conditions and the use of bloom stimulants, mango production was constant throughout the year. According to the monthly fruit sampling program of this region, *A. obliqua* populations move into other wild fruit species after the strong fruiting period and infestation in *Spondias* spp. The release of parasitoids was conducted throughout the year in the wild areas, and the resulting monthly parasitism rate of 50% was consistent and considered adequate to keep the population of *A. obliqua* controlled in the wild areas (Figure 17.3). Augmentative parasitoid releases in orchards in Guerrero were the most successful case of fruit fly biological control in Mexico. Thanks to these actions, the level of low fruit fly prevalence was reinforced. An average production of 180,000 tons of mangos are harvested annually, of which 10% are exported.

17.4.3.1.3 Oaxaca Region

The Isthmus of Tehuantepec is the largest mango producing region in Mexico, with 25,000 ha cultivated with the varieties of Ataulfo, Tommy Atkins, and a high-quality Creole mango. Because of

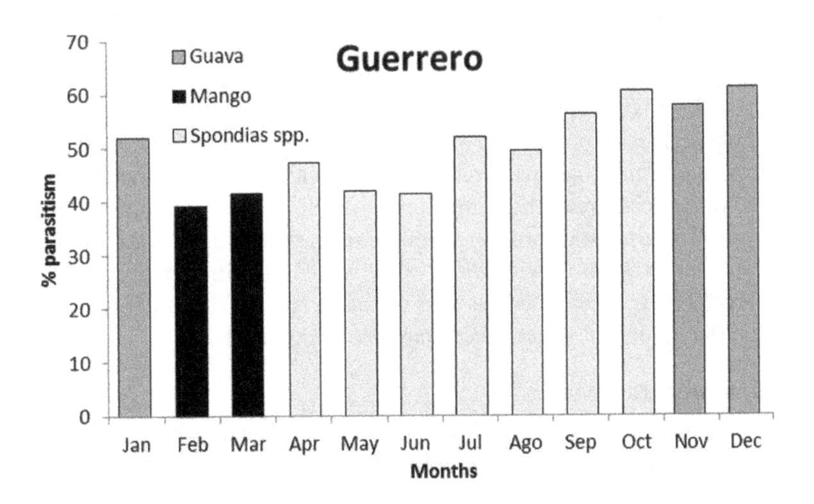

FIGURE 17.3 Monthly percentage of parasitism in guava, creole mango, and fruits of *Spondias* spp. sampled in the mango-producing region of the municipality of Tecpan de Galeana, Guerrero state (2016–2017).

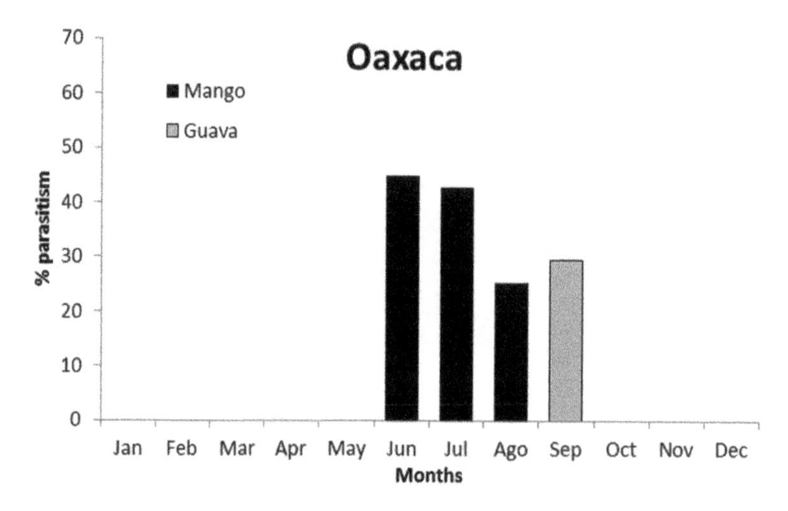

FIGURE 17.4 Monthly percentage of parasitism in creole mango and guava sampled in the mango-producing region of the Tehuantepec Isthmus, Oaxaca state (2016–2017).

the extensive size of the area and a high ecological diversity, it has been difficult to establish a low fruit fly prevalence level in this area, with recorded FTD levels of 0.0403–0.1481. The wild areas are characterized by medium semi-deciduous forest and savanna (Gallardo et al. 2001; González 2011). Even with high ecological diversity, *A. obliqua* was the only species considered as a pest of mango. Parasitoid releases were carried out in orchards adjacent to the wild vegetation. Mangos fruited from January to July, became infested at the end of the fruiting period, and then the populations of *A. obliqua* moved into fruits such as wild guava or fruits of the genus *Spondias* (Aluja and Birke 1993). Obtained parasitism percentages ranged between 30% and 50%, and the primary contribution of augmentative parasitoid releases has been to reduce *Anastrepha* population levels after the fruiting periods, weakening the next fruit fly generation (Figure 17.4). The annual production in this area was of 180,000 tons, of which 57% were exported in 2017.

17.4.3.1.4 Chiapas

The mango producing areas of Chiapas were located in the Soconusco region, which was characterized as an area with the highest *Anastrepha* host plant diversity, surrounded by high evergreen forest with secondary vegetation (Salgado-Mora et al. 2007; Roa-Romero et al. 2009). In this region, *A. obliqua* and *A. ludens* are the two *Anastrepha* pest species present in mango. Ninety percent of the mango production was concentrated on the Ataulfo variety because it is considered more resistant to the two fly pest species (Guillén et al. 2017). Average FTD levels were 0.32 for *A. obliqua* and 0.018 for *A. ludens*. Parasitoids were released in wild areas to minimize the invasion of these fly species in the commercial area. There are many alternating wild hosts for the two fly species in the wild areas (guava, plums, sour orange, creole orange, grapefruit, etc.). Parasitism levels were highly variable, remaining within a wide range of 30%–50% identified in *Spondias* spp. fruits, and at levels of 20%–30% in large fruits, such as sour orange (Figures 17.5 and 17.6). The annual mango production is of 120,000 tons, of which more than 30% is exported.

17.4.3.2 Analysis of Parasitism

It is difficult to estimate the contribution to fruit fly control attributed to augmentative biological control in each mango region presented. However, it does appear that the release of parasitoids had a significant and rapid effect on increased levels of fruit fly parasitism, which were greater than parasitism levels from natural parasitoid populations alone (Ovruski et al. 2000; Montoya et al. 2016). This is supported by previous studies that demonstrated the remarkable level of control that *D. longicaudata*

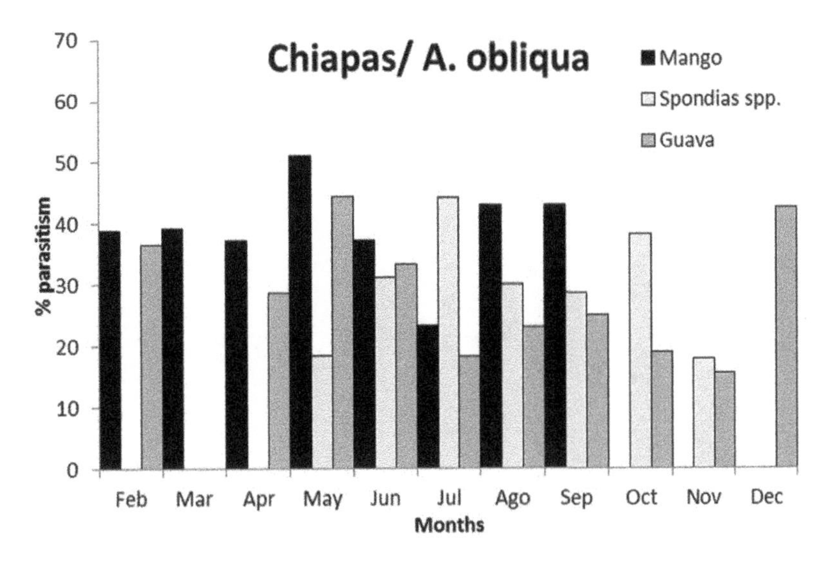

FIGURE 17.5 Monthly percentage of parasitism in creole mango, fruits of *Spondias* spp., and guava sampled in the mango-producing region of the Soconusco, Chiapas state (2016–2017).

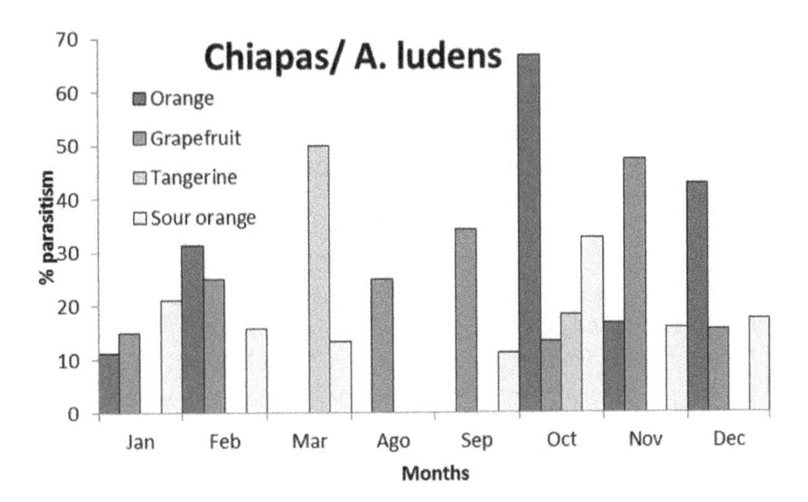

FIGURE 17.6 Monthly percentage of parasitism in orange, grapefruit, tangerine, and sour orange sampled in the mango-producing region of the Soconusco, Chiapas state (2016–2017).

has on *Anastrepha* populations (Sivinski et al. 1996; Montoya et al. 2000). Secondly, the contribution to pest population control in wild areas was evident. There have been points in the surrounding areas with higher percentages of parasitism in wild fruits which are considered as the main factor of the reduction of *Anastrepha* populations in nearby orchards. In general, infestation reduction is consistent for a period of 2 or 3 weeks. The high percentage of parasitism could indicate that parasitoid releases against fruit flies in wild areas prevented the migration of fly populations into commercial areas. Although it was difficult to prove in all cases, the maintenance and control of *A. obliqua* pest populations along the coast of Guerrero was highly effective in reducing the risks of invasion in commercial orchard areas, where the lack of a geographical barrier allowed the spread of flies.

Constant augmentative releases were carried out to obtain the control of pest populations (Knipling 1992; Rossi et al. 2018). Established weekly release intervals were necessary, because in cases where weekly releases were not conducted, the pest population rapidly increased

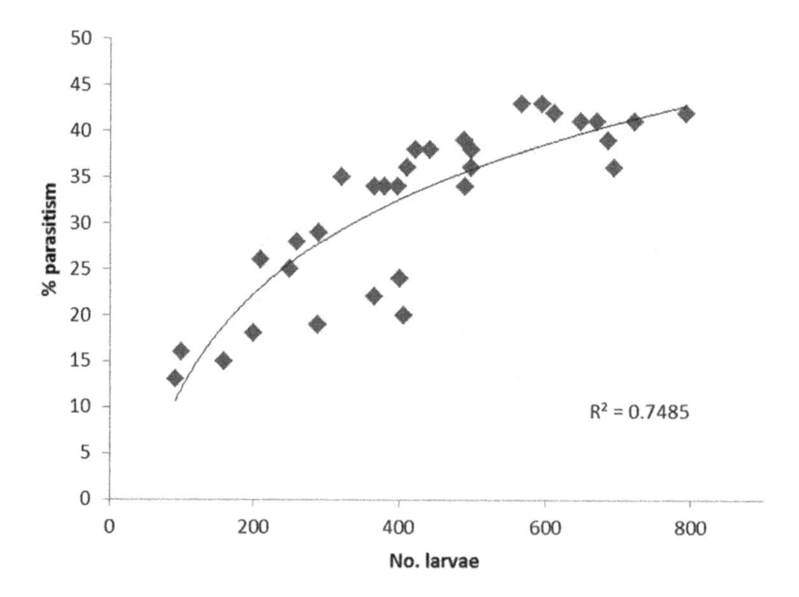

FIGURE 17.7 Relationship between number of larvae and percentage of parasitism obtained monthly over a 2-year period from the mango-producing region in the municipality of Tecpan de Galeana, Guerrero state. Linear regression with logarithmic transformation.

(Cancino et al. 2019). Generally, the targeted release quantity of about 1,000 females per ha was sufficient to maintain a reduced pest population (Montoya et al. 2000).

There was a significant positive relationship between density of pest larvae (measured as number of larvae in sampled fruits) and increased percentage of parasitism (coefficient of correlation (r^2) = 0.74; t-test (t) = 9.45, P <0.0001) (Figure 17.7), and a significant negative relationship between fly density (measured as FTD) and the obtained percentage of parasitism levels (r^2 = 0.59; t = −7.93, P <0.0001) (Figure 17.8). These results obtained in the mango-producing region of the Municipality of Tecpan de Galeana in the state of Guerrero showed a significant effect of parasitoids on the reduction of *Anastrepha* populations. Particularly, the reduction of FTD represents a substantial important result in the indicators of fly pest populations, which contributed directly to the quality improvement

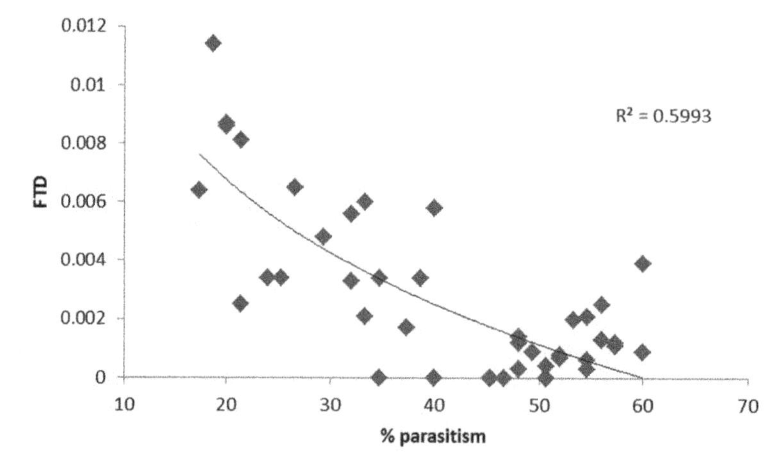

FIGURE 17.8 Relationship between percentage of parasitism and flies/trap/day (FTD) obtained monthly over a 2-year period from the mango-producing region in the municipality of Tecpan de Galeana, Guerrero state. Linear regression with logarithmic transformation.

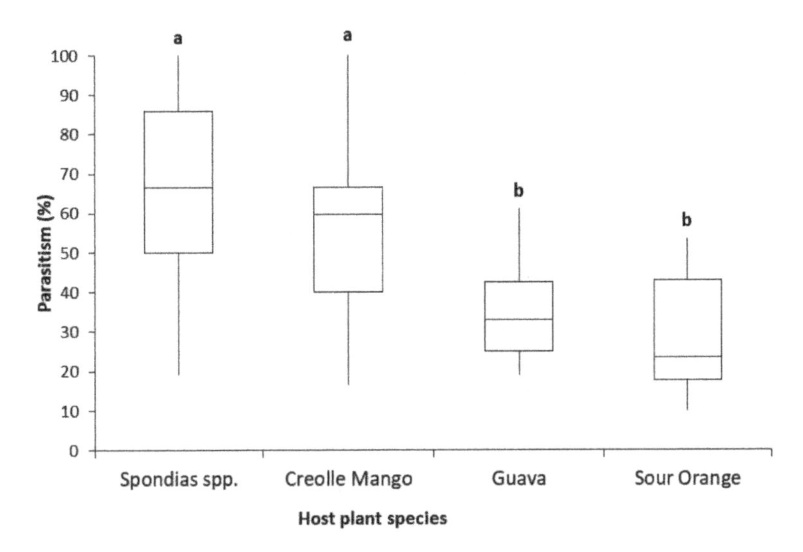

FIGURE 17.9 Average percentage of parasitism obtained in four fruits of host plant species with fruits of different sizes collected monthly over a 1-year period at Soconusco, Chiapas, Mexico. Different letters above boxes indicate statistic differences at $P \leq 0.05$ determined by ANOVA and Tukey's test.

of the fruit produced (Sivinski et al. 1996; Peck and McQuate 2000; FAO 2016). It is important to consider that the use of small fruits as hosts of *Anastrepha* flies was an important opportunity to increase parasitism levels of *D. longicaudata* and reduce fly populations (Leyva et al. 1991; Wang et al. 2009; Montoya et al. 2016). The average percentage of parasitism observed in fruits of the four different size classes from field samplings showed significant differences (ANOVA: $F = 25.44$, d.f. $= 99$, $P < 0.0001$) (Figure 17.9). Fruit flies in *Spondias* spp. fruits, a plant characterized by having small fruits (Hauck et al. 2011), showed an average percentage of parasitism higher than 50%, which is significantly higher than the nearly 30% of parasitism found for fruit flies in larger fruit host plants of sour orange and guava. Strategically, the application of parasitoids should be carried out in areas with an abundance of these fruits. In fact, this seems to be one of the reasons why, in Guerrero, we observed high levels of parasitism in wild areas with concomitant reduced risks of infestation in the adjacent producing area, which was declared as a fruit fly low prevalence area.

Augmentative releases are often criticized for their high costs (Parella et al. 1992). Here we did not conduct any direct cost analyses since this technique is included in the overhead costs of local producer organizations and the national program. This IPM method is also extensively supported by the social sector, unlike insecticide applications that are completely rejected mainly by beekeeping organizations. The inclusion of biological control within the control program has been preferred by producers, giving them the benefits of including their mango products in markets that demand insecticide free commodities. From a general point of view, it can be concluded that the investment cost has been profitable in the commercialization of mango in these regions.

17.5 FUTURE PLANS

There are three different plans to strengthen the application of biological control for *Anastrepha* populations:

1. Develop and implement a packaging method that allows aerial releases. The most viable method involves the use of the "chilled adult" technique. This requires packaging in Mexico-type cages (Leal-Mubarqui et al. 2014) and air release. The application of

this method can be accomplished with manned aircraft or possibly with drones. In both cases, an improved spread of parasitoids and increased access to difficult areas would be beneficial.
2. Reduce the costs of the artificial host diet production that produces host larvae of high quality specially developed for parasitoid rearing. Costs could be reduced by using cheaper ingredients or reducing the amount of expensive ingredients. Currently, the host larval diet component of mass rearing represents about 70% of the production cost of parasitoids.
3. Increase the application of native parasitoid species to control *Anastrepha* spp. Native parasitoid species are often significantly involved in the control of *Anastrepha* populations (Aluja et al. 1987; López et al. 1999; Montoya et al. 2016). For example, *Doryctobracon crawfordi* (Viereck) develops in *A. ludens* feeding in native fruits (López et al. 1999), the more specific parasitoid *Utetes anastrephae* (Viereck) attacks *A. obliqua* developing in *Spondias* spp. fruits, and abundant *Doryctobracon areolatus* (Szépligeti) (Ovruski et al. 2000; Marsaro et al. 2011; Montoya et al. 2016) attack *Anastrepha* spp. populations (Ovruski et al. 2000; Silva et al. 2010; Montoya et al. 2016). Additional research on native parasitoids could give rise to options that may be applied to increase the efficiency of the biological control of pest populations.

ACKNOWLEDGMENTS

The authors are grateful to the respective technical team in each region. Their enthusiastic participation was essential for the information obtained in this work. To the Biological Control Department staff, especially to Cesar Galvez for his help with the statistical analysis. Also, to Stephen Hight for his valuable and helpful comments. Two anonymous reviewers provided important comments to improve the manuscript.

REFERENCES

Aluja, M. 1993. *Manejo integrado de moscas de la fruta*. México, DF: Editorial Trillas.

Aluja, M. 1994. Bionomic and management of *Anastrepha*. *Annual Review of Entomology* 39:155–178.

Aluja, M., and A. Birke. 1993. Habitat use by adults of *Anastrepha obliqua* (Diptera: Tephritidae) in a mixed mango and tropical plum orchard. *Annals of the Entomological Society of America* 86:799–812.

Aluja, M., J. Guillen, G. de la Rosa, M. Cabrera, H. Celedonio, P. Liedo, and J. Hendrichs. 1987. Natural host plant survey of the economically important fruit flies (Diptera: Tephritidae) of Chiapas, Mexico. *Florida Entomologist* 70:329–338.

Aluja, M., J. Piñero, J. Sivinski, F. Diaz-Fleisher, and J. Sivinski. 2000. Behavior of flies in the genus *Anastrepha* (Trypetinae: Toxotrypanini). In *Fruit Flies (Tephritidae) Phylogeny and Evolution of Behavior*, ed. Aluja, M. and A. L. Norrbon, 375–406, Boca Raton, FL: CRC Press.

Cancino, J., and P. Montoya, P. 2006. Advances and perspectives in the mass rearing of fruit fly parasitoids in Mexico. In *Fruit Flies of Economic Importance: From Basic to Applied Knowledge*, ed. Sugayama, R., R. Zucchi, S. Ovruski, and J. Sivinski, 133–142. Salvador Bahia, Brazil: Proceedings of the 7th International Symposium on Fruit Flies of Economic Importance.

Cancino, J., and F. López-Arriaga. 2016. Effect of hypoxia its repercussion in packing pupae of the parasitoid *Diachasmimorpha longicaudata* (Hymenoptera: Braconidae). *Biocontrol Science and Technology* 26:665–677.

Cancino, J., F. López-Arriaga, and P. Montoya. 2017. Packaging conditions for the field releases of the fruit fly parasitoid *Diachasmimorpha longicaudata* (Hymenoptera: Braconidae). *Austral Entomology* 56:261–267.

Cancino, J., C. Gálvez, A. López, U. Escalante and P. Montoya. 2019. Best timing to determine field parasitism by released Diachamimorpha longicaudata (Hymenoptera: Braconidae) against Anastrepha (Diptera: Tephritidae) pest populations. *Neotropical Entomology* 48:143–151.

Celedonio-Hurtado, H., M. Aluja, and P. Liedo. 1995. Adult population fluctuations of *Anastrepha* species (Diptera: Tephritidae) in tropical orchards habitats of Chiapas, Mexico. *Environmental Entomology* 24:861–869.

CONAFOR. 2015. Guerrero. *SEMARNAT*. http://www.conafor.gob.mx/web/ms_gerencias_estatales.

CONASPROMANGO. 2012. Plan rector nacional del sistema producto mango. *Documento validado CONASPROMANGO, A. C.*, Sesión 2012.

Dashavant, A, R. K. Patil, and A. Gupta. 2018. Per cent parasitism of *Diachasmimorpha longicaudata* complex (Hymenoptera: Braconidae) on *Bactrocera* spp. (Diptera: Tephritidae) from North Karnataka. *Journal of Experimental Zoology India* 21:1135–1138.

De los Santos-Ramos, M., A. Bello-Rivera, R. Hernández-Pérez y F. Leal-García. 2012. Efectividad de la estación cebo MS2 y atrayente alimenticio Ceratrap ® como alternativas en la captura de moscas de la fruta en Veracruz, México. *Interciencia* 37:279–283.

FAO, 2016. Establishment of pest free areas for fruit flies (Tephritidae). *ISPM 26, IPPC.* Rome, Italy.

Flores, S., L. E. Gómez, and P. Montoya. 2011. Residual control and lethal concentrations of GF-120 (spinosad) for *Anastrepha* spp. (Diptera: Tephritidae). *Journal of Economic Entomology* 104:1885–1891.

Follet, P. A. y K. W. Vick. 2002. Desarrollo de estrategias de manejo integrado de plagas para eliminar las barreras sanitarias que restringen la exportación de productos agrícolas. *Manejo Integrado de Plagas y Agroecología* 65:43–49.

Gallardo, C., J. Meave y E. Pérez-García. 2001. Vegetación y flora de la región de Nizanda, Istmo de Tehuantepec, Oaxaca, México. *Acta Botánica Mexicana* 56:19–88.

González, A. 2011. Oaxaca 2011: Un diagnóstico breve. Grupo Mesofilo, A. C. Pags: 66.

Guillén, L., R. Adaime, A. Birke, O. Velázquez, G. Angeles, F. Ortega, E. Ruiz, and M. Aluja. 2017. Effect of resin ducts and sap content in infestation and development of immature stages of *Anastrepha obliqua* and *Anastrepha ludens* (Diptera: Tephritidae) in four mango (Sapindales: Anacardiaceae) cultivars. *Journal of Economic Entomology* 110:719–730.

Hauck, J., A. Rosenthal, R. Deliza, R. L. de Oliveira, and S. Pacheco. 2012. Nutritional properties of yellow mombin (*Spondias mombin*) pulp. *Food Research International* 44:2326–2331.

Henter, H. J., and V. Sara. 1995. The potential for coevolution in the host-parasitoid system. I, Genetic variation within an aphid population in susceptibility to a parasitoid wasp. *Evolution* 49:427–438.

Hernández-Ortíz, V. 1992. *El género* Anastrepha *Schiner en México (Diptera: Tephritidae), Taxonomía, distribución y sus plantas huéspedes.* 33, Xalapa, México: Instituto de Ecología, Publicaciones.

Hernandez-Ortíz, V. 2007. Diversidad y biogeografía del género *Anastrepha* en México. In *Moscas de la fruta en Latinoamérica (Diptera: Tephritidae): Diversidad, biología y Manejo*, ed. Hernández-Ortíz, 53–76, México, DF: V. S. G.

Hernández, E., P. Rivera, B. Bravo, J. Toledo, J. Caro-Corrales, and P. Montoya. 2012. Hot-water phytosanitary treatment against *Ceratitis capitata* (Diptera: Tephritidae) in "Ataulfo" mangoes. *Journal of Economic Entomology* 105:1940–1953.

Knipling, E. F. 1992. Principles of insect parasitism analyzed from new perspectives. Practical implications for regulating insect populations by biological means. *USDA-ARS*, Agricultural Handbook No. 693. Washington, DC: United States Department of Agriculture.

Lawrence, P. 2005. Morphogenesis and cytopathic effects of the *Diachasmimorpha longicaudata* entomopoxvirus in host haemocytes. *Journal of Insect Physiology* 51:221–233.

Lawrence, P., R. Baranowski, and P. Greany. 1976. Effect of host size on development of *Biosteres* (=*Opius*) *longicaudatus*, a parasitoid of the Caribbean fruit fly, *Anastrepha suspensa. Florida Entomologist* 59:33–39.

Leal-Mubarqui, R., R. Cano, R. Angulo, J. Zavala, A. Parker, M. Talla, S. Baba, and J. Bouyer. 2014. The smart aerial release machine a universal system for applying the sterile insect technique. *Plos ONE* 97(7):e103077.

Leyva, J., H. W. Browning, and F. E. Gilstrap. 1991. Effect of host fruit species, size, and color of parasitization of *Anastrepha ludens* (Diptera: Tephritidae) by *Diachasmimorpha longicaudata* (Hymenoptera: Braconidae). *Environmental Entomology* 20:1469–1474.

Leyva, J., H. W. Browning, and F. E. Gilstrap. 1991. Development of *Anastrepha ludens* (Diptera: Tephritidae) in several host fruit. *Environmental Entomology* 20:1160–1165.

Liedo, P., and J. Carey. 1994. Mass rearing of *Anastrepha* (Diptera: Tephritidae) fruit flies: A demographic analysis. *Journal of Economic Entomology* 87:176–180.

López, M., M. Aluja, and J. Sivinski. 1999. Hymenopterous larval-pupal and pupal parasitoids of *Anastrepha* flies (Diptera: Tephritidae) in Mexico. *Biological Control* 15: 119–129.

Marsaro, A., R. Adaime, B. Ronchi-Tales, C. Ribeiro, and P. da Silva. 2011. *Anastrepha* species (Diptera: Tephritidae), their hosts and parasitoids in the extreme North of Brazil. *Biota Neotropica* 11:117–123.

Mierelles, R. N., L. Rodriguez, and C. Bernardes. 2013. Comparative biology of *Diachasmimorpha longicaudata* (Hymenoptera: Braconidae) reared on *Anastrepha fraterculus* and *Ceratitis capitata* (Diptera: Tephritidae). *Florida Entomologist* 96:412–418.

Monter, A. V., and A. M. Aguilera. 2011. Avances de la fruticultura en México. *Revista Brasilera de Fruticultura* Vol. Especial: 179–186.

Montoya, P., P. Liedo, B. Benrey, J. Cancino, J. F. Barrera, J. Sivinski, and M. Aluja. 2000. Biological control of *Anastrepha* spp. (Diptera: Tephritidae) in mango orchards through augmentative releases of *Diachasmimorpha longicaudata* (Ashmead) (Hymenoptera: Braconidae). *Biological Control* 18:216–224.

Montoya, P., J. Cancino, M. Zenil, G. Santiago, and J. M. Gutiérrez. 2007. The augmentative biological control component in the Mexican National Campaign against *Anastrepha* spp. fruit flies. In *Area-wide control of insect pest*, ed. M. J. B. Vreysen, A. S. Robinson, and J. Hendrichs, 661–670. Dordrecht, the Netherlands: Springer.

Montoya, P., A. Ayala, P. López, J. Cancino, H. Cabrera, J. Cruz, A. Martínez, I. Figueroa, and P. Liedo. 2016. Natural parasitism in fruit fly (Diptera: Tephritidae) populations in disturbed areas adjacent to commercial mango orchards in Chiapas and Veracruz, Mexico. *Environmental Entomology* 45:1–10.

NOM-023-FITO-95. 1999. Norma oficial Mexicana por la que se establece la Campaña Nacional contra Moscas de la Fruta. *Diario Oficial de la Federación*, Gobierno de México, México D. F.

Orozco-Dávila, D., L. Quintero, E. Hernández, E. Solís, T. Artiaga, R. Hernández, C. Ortega, and P. Montoya. 2017. Mass rearing and sterile insect releases for the control of *Anastrepha* spp. pests in Mexico—A review. *Entomologia Experimentalis et Applicata* 164:176–187.

Ovruski, S., M. Aluja, J. Sivinski, and R. Wharton. 2000. Hymenopteran parasitoids of fruit-infesting Tephritidae (Diptera) in Latin America and the Southern United States: Diversity, distribution, taxonomic status and their use in the fruit fly biological control. *Integrate Pest Management Reviews* 5:81–107.

Ovruski, S., P. Schliserman, and M. Aluja. 2016. Occurrence of diapause in neotropical parasitoids attacking *Anastrepha fraterculus* (Diptera: Tephritidae) in a subtropical rainforest from Argentina. *Austral Entomology* 55:279–283.

Pascal, M. A., A. A. Sanzogan, A. H. Bokonon-Ganta, and M. Karlsoon. 2017. Host species and vegetables fruit suitability and preference by parasitoid wasp *Fopius arisanus. Entomologia Experimentalis et Applicata* 163:70–81.

Parella, M., K. M. Heinz and L. Nunney. 1992. Biological control through augmentative releases of natural enemies: a strategy whose time has come. *American Entomologist* 28:172–179.

Peck, S. L., and G. T. McQuate. 2000. Field test of environmentally friendly malathion replacements to suppress wild Mediterranean fruit fly (Diptera: Tephritidae) populations. *Journal of Economic Entomology* 93:280–284.

Ramírez, B., P. Barrios, J. Castellanos, A. Muñoz, G. Palomeno, and E. Pimienta. 2008. Sistema de producción de *Spondias purpurea* (Anacardacea) en el Centro Occidente de México. *Revista Biología Tropical* 56:675–687.

Roa-Romero, H., M. Salgado-Mora, and J. Alvarez-Herrera. 2009. Análisis de la estructura arbórea del sistema agroforestal cacao (*Theobroma cacao* L.) en el Soconusco, Chiapas, México. *Acta Biológica Colombiana* 14:97–110.

Rossi, M. V., A. Grassi, C. Ioriatti, and G. Anfora. 2018. Augmentative releases of *Trichopria drosophilae* for the suppression of early season *Drosophila suzuki* populations. *BioControl* 64:9–19.

Ruiz-Arce, R., N. B. Barr, C. L. Owen, D. B. Thomas, and B. A. McPheron. 2012. Phylogeography of *Anastrepha obliqua* inferred with mtDNA sequencing. *Journal of Economic Entomology* 105:2147–2160.

Ruiz-Arce, R., C. L. Owen, D. B. Thomas, N. B. Barr, and B. A. McPheron. 2015. Phylogeography structure in *Anastrepha ludens* (Diptera: Tephritidae) populations inferred with mtDNA sequencing. *Journal of Economic Entomology* 108:1324–1336.

SAGARPA. 2017. Mango. *Planificación Agrícola Nacional 2017–2030*. Gobierno de México, México, DF.

Salgado-Mora, M., G. Ibarra-Nuñez, J. Macías-Sámano, and O. López-Baéz. 2007. Diversidad arbórea en cacaotales del Soconusco, Chiapas, México. *Interciencia* 32:363–368.

SENASICA. 2017. Moscas nativas de la fruta. http://www.gob.mx/senasica/documentos/moscas-nativas-de-la-fruta.

Silva, J. G., V. S. Dutra, M. S. Santos, N. M. Silva, D. B. Vidal, R. A. Nink, J. A. Guimaraes, and E. L. Araujo. 2010. Diversity of *Anastrepha* spp. (Diptera: Tephritidae) and associated braconid parasitoids from native and exotic hosts in Southern Bahia, Brazil. *Environmental Entomology* 39:1457–1465.

Sivinski, J. 1991. The influence of host fruit morphology on parasitization rates in the Caribbean fruit fly, *Anastrepha suspensa*. *Entomophaga* 36:447–454.

Sivinski, J., C. Calkins, R. Baranowski, D. Harris, J. Brambila, J. Diaz, E. Burns, T. Holler, and D. Dodson. 1996. Suppression of the Caribbean fruit fly (*Anastrepha suspensa* (Loew) Diptera: Tephritidae) population through releases of the parasitoid *Diachasmimorpha longicaudata* (Ashmead) (Hymenoptera: Braconidae). *Biological Control* 6:177–185.

Thomas, D. B. 2003. Reproductive phenology of the Mexican fruit fly, *Anastrepha ludens* (Loew) (Diptera: Tephritidae) in the Sierra Madre Oriental, Northern Mexico. *Neotropical Entomology* 32:385–397.

Vanoye-Eligiio, V., L. Barrientos-Lozano, G. Gaona-García, and M. Lara-Villalon. 2015. New wild host of *Anastrepha ludens* in Northeastern Mexico. *Southwester Entomologist* 40:435–438.

Vanoye-Eligiio, V., L. Barrientos-Lozano, R. Pérez-Castañeda, and M. Lara-Villalon. 2015. Population dynamics of *Anastrepha ludens* (Loew) (Diptera: Tephritidae) on citrus areas in Southern Tamaulipas, Mexico. *Neotropical Entomology* 44:565–573.

Vargas, R. I., M. Ramadan, T. Hussain, N. Mochizuki, R. C. Bautista, and J. D. Stark. 2002. Comparative demography of six fruit fly (Diptera: Tephritidae) parasitoids (Hymenoptera: Braconidae). *Biological Control* 25:30–40.

Vega-Aviña, R., H. Aguiar-Hernández, J. A. Gutiérrez-García, J. A. Hernández-Vizcarra, I. F. Vega-López, and J. L. Villaseñor. 2000. Endemismo regional presente en la flora del Municipio de Culiacán, Sinaloa, México. *Acta Botánica Mexicana* 53:1–15.

Wang, X., M. Jhonson, K. Daane, and V. Yokoyama. 2009. Large olive fruit size reduces the efficiency of *Psyttalia concolor*, as a parasitoid of the olive fruit fly. *Biological Control* 49:45–51.

Wong, T., M. Ramadan, D. Mcinnis, N. Mochizuki, J. Nishimoto, and J. Herr. 1991. Augmentative releases of *Diachasmimorpha tryoni* (Hymenoptera: Braconidae) to suppress a Mediterranean fruit fly (Diptera: Tephritidae) population in Kula, Maui, Hawaii. *Biological Control* 1:2–7.

18 Use of Entomopathogenic Fungi for the Biological Control of the Greater Melon Fly *Dacus frontalis* in Libya

Esam Elghadi and Gordon Port*

CONTENTS

Abstract The Greater melon fly, *Dacus frontalis* Becker, is one of the most economically damaging pests of cucurbit fruits in Africa. The fly is considered to have a negative impact on food security in the continent. In Libya, a range of major cucurbit crops are attacked, causing extensive yield losses of up to 100%. Direct damage is caused by the larval stage, which decreases quality and quantity of the fruit production, raising concern among growers. Currently, Libyan farmers still rely mainly on extensive application of several insecticides; however, such applications often fail to suppress the fly damage. Information on other management options for the fly is limited. The aim of this chapter is to evaluate entomopathogenic fungi for use against *D. frontalis* and

* Corresponding author.

develop better strategies in using these biological agents for integrated fly management. The pathogenicity of five commercial biopesticides based on several strains of entomopathogenic fungi, *Metarhizium anisopliae*, *Beauveria bassiana*, and *Isaria fumosoroseus*, against larvae, pupae, and adult stages of *D. frontalis* was evaluated by using various inoculation methods under laboratory conditions. The most effective formulation, Met52 Granular biopesticide, based on *M. anisopliae* var *anisopliae* strain F52 (MET52), was selected for further investigations. A dose-response of the target pest to MET52 was examined. Effects of formulation and application time on the efficacy of the fungus were also evaluated. The results revealed that *D. frontalis* adults are more susceptible to the fungal pathogens than pupae. Met52 caused the greatest pathogenicity to the adults ranging from approximately 88% to 100% mortality. Pupal age and increasing rate of MET52 had no effect on pupal mortality. However, MET52 increased mortality of emerging adults by 15% when applied on young pupae. Approximately 10 days were required to get 90% adult mortality when pupae were placed into soil treated with the lowest rate tested. Also, early application of MET52 in granule form caused a significant reduction in adult emergence compared to a drench and untreated control. The effect of MET52 against *D. frontalis* was influenced by application time with the greatest pathogenicity recorded when the treatment occurred 2 weeks before larvae entered the soil, resulting in a 55% reduction in adult emergence rate. This is the first study to demonstrate the susceptibility of *D. frontalis* to entomopathogenic fungi, suggesting that early soil application of MET52 offers a promising biological control for *D. frontalis*.

18.1 INTRODUCTION

Dacus frontalis Becker is one of the economically damaging fruit fly species having a negative impact on food security in Africa (Foottit and Adler 2009). The fly is widely distributed in Africa and some parts of Asia (Steffens 1982; Ba-Angood 1977; Abukhashim et al. 2003a; White 2006; Mwatawala et al. 2010; El-Hawagry et al. 2013; Gameel 2013; De Meyer et al. 2013; Redha 2013; Badii et al. 2015; Hafsi et al. 2015). In Libya, the first observation of the fly was in 1992 in Marzak farms in the south (Ramadan Abdallah 2002). Then, the fly spread across the country but was less abundant in the eastern region (Abukhashim et al. 2003a).

Similar to other tephritid species, direct damage is caused by the larval stage, which decreases quality and quantity of fruit production, making the fruits unmarketable. In Libya, a range of major cucurbit crops are attacked. Results of a survey conducted throughout the country showed that seven cucurbit species, *Cucumis sativus* L., *Cucumis melo* L., *Cucumis melo* var. *flexuosus*, *Cucurbita moschata* Duchesne ex Poir., *Cucurbita pepo* L., *Citrullus lanatus* (Thunb.) Matsum. & Nakai, and *Citrullus colocynthis* (L.) Schrad are reported as fly hosts, and *Solanum melongena* L. (Solanaceae) was reported as a new fly host (Abukhashim et al. 2003a). Infestations of *D. frontalis* caused 100% losses in cucurbit fruit production and raised concern among growers leading to the local authority proposing a national project aimed to study and control the fly in the country. Currently, Libyan farmers still rely mainly on extensive application of several insecticides (Abukhashim et al. 2003b), but they often fail to suppress the fly population, resulting in economic losses. Libyan farmers have little knowledge of the use of developed control agents and strategies to protect their crops (personal observation). For example, although they are aware of the considerable damages caused by the fly, they ignore the application of even the traditional agricultural methods such as field sanitation. They throw infested fruits on the sides of fields and use them as food for agricultural animals instead of collecting and correctly disposing of them. This action might be a reason for the increasing damage and losses of cucurbit fruits in Libya. Thus, safer and more effective approaches are required to suppress damage and losses caused by *D. frontalis*. Biological control is one of the available alternative control strategies to traditional insecticides (Esser and Lemke 1995). Fungal pathogens are valuable biological agents for controlling some agricultural insect pests (Esser and Lemke 1995; Butt 2002; Roy et al. 2010). They are environmentally safe in general (Esser and Lemke 1995; Wraight and Hajek 2009), and some fruit flies are susceptible to fungi (Castillo et al. 2000; Ekesi et al. 2002; Dimbi et al. 2003; Konstantopoulou and Mazomenos 2005; Mochi et al. 2006; Daniel and Wyss 2009; Svedese et al. 2012; Beris et al. 2013; Imoulan and Elmeziane 2014; Gul et al. 2015). Various inoculation approaches have been used to determine the pathogenicity of several entomopathogenic fungi against fruit flies. In this context, soil application of insect fungal pathogens has been suggested as a strategy to reduce emergence rates and induce postemergence mortality of adults (Ekesi et al. 2005; Garrido-Jurado et al. 2011a). Insect pests have varying susceptibilities to different strains of entomopathogenic fungi (Butt et al. 1995). Therefore, investigation of pathogenicity is an essential step to select appropriate fungal strains. To date, using such fungal pathogens against *D. frontalis* has not been studied. The aim of this study was to investigate the susceptibility of different life stages of *D. frontalis* to commercial biopesticides based on different species of entomopathogenic fungi when applied by different methods under controlled conditions. Also, a dose-response of the target pest to selected pathogens was examined. Effects of formulation and application time on the efficacy of the fungus were also evaluated. This may help to find an effective pathogen and strategy for the biological control of the fly.

18.2 METHODS

18.2.1 Insect Culture

About 400 *D. frontalis* pupae were obtained from the Biotechnology Research Centre in Libya. Mass rearing was maintained at 25°C, 50%–55% relative humidity (RH), and 14:10 hour light-dark (L/D) photoperiod. Adults were kept in transparent perspex cages (25 cm × 25 cm × 25 cm) covered with gauze on one side for ventilation. The cages were supplied with water and adult artificial diet consisting of 1:3 ratio of yeast hydrolysate enzymatic (MP Biomedicals, France) and sucrose. Eggs were collected by introducing whole fresh squashes into the cage, which were replaced regularly. Larvae were fed on squash in plastic containers (20 cm × 30 cm × 15 cm) filled with sterilized soil where full-grown larvae could pupate.

18.2.2 Bioinsecticides

Five commercial bioinsecticides were tested in this study. The products depended on different strains and isolates of entomopathogenic fungi (Table 18.1). Fungal pathogens were kept at 4°C in a refrigerator until used.

TABLE 18.1
Sources and Isolates of Entomopathogenic Fungi Tested against *D. frontalis*

Commercial Name	Strain*	Supplier	Recommended Rate	Concentration
Met52® Granular (MET52)	*Metarhizium anisopliae* var *anisopliae* strain F52	Fargro® Ltd, West Sussex UK	0.5 kg m^{-3} of growing media or 122 kg ha^{-1} for open ground use	9.0×10^8 (CFU) g^{-1}
Bio-Magic	*Metarhizium anisoplae* (Metchnikoff) Sorokin	T. Stanes & Company limited, India	4 kg ha^{-1} in 500 L of water	1×10^8 (CFU) mL^{-1}
Bio-Power	*Beauveria bassiana* (Balsamo) Vuillemin	T. Stanes & Company limited, India	4 kg ha^{-1} in 500 L of water	1×10^8 (CFU) mL^{-1}
Bio-Catch	*Isaria fumosoroseus* (Wize) Brown and Smith	T. Stanes & Company limited, India	4 kg ha^{-1} in 500 L of water	1×10^8 (CFU) mL^{-1}
Naturalis-L®	*Beauveria bassiana* strain ATCC 74040	Belchim Crop Protection	3 L in 1000 L of water	2.3×10^7 (CFU) mL^{-1}

* Only MET52 and Naturalis-L had information on the strain. CFU, colony-forming units; ha, hectare.

18.2.3 Pupal and Adult Experiment

Thirty plastic cups (4 cm height × 4 cm diameter) were filled with 30 g of sterilized sandy clay loam soil (65% sand, 12 silt, and 23% clay) (autoclaved at 1.5 bar, 123°C for 25 minutes) obtained from Cockle Park, Morpeth, United Kingdom. Six treatments, the five bioinsecticides and one control, were prepared as follows: Soil was inoculated with 1.5 g of MET52. For other inoculated treatments, 2 mL suspensions of Bio-Power, Bio-Magic, Bio-Catch or Naturalis-L® were applied to the soil in the cups. Two mL of sterilized distilled water were added to the control and all the fungal treatments. For all treatments, the fungus was mixed with the soil and 20 pupae (2 days old) of *D. frontalis* were buried at a depth of 2 cm. Soil moisture content in all treatments was maintained at 35% water-holding capacity (WHC) daily until adult emergence. Treatment cups were covered with cups of the same size, inverted and perforated at the top for air flow. Cups of each replicate were sealed together at the sides with Parafilm M® (VWR, UK) and kept in an incubator at 25°C, 60%–70% RH and 14:10 L/D. Five replicates were done for each treatment. Nine days later (2 to 3 days before emergence), cups were placed in transparent plastic cages 10 cm × 10 cm × 10 cm to assess emerging adult mortality. Cages covered with gauze on one side were supplied with artificial diet and water as previously described. After 14 days, the number of emerging flies in the treatments was assessed. With the total of emerged adults, for each treatment, four cages with 15 flies were arranged. Adult cages were kept in the same conditions described previously for the pupal cups. Dead flies were collected daily from the cages and assessed over a period of 2 weeks. To confirm a fungal infection, pupae that failed to produce adults and dead adult flies were individually sterilized with 70% ethanol followed by three rinses with sterile distilled water. Samples were placed in Petri dishes with moist sterile filter papers. The Petri dishes were incubated at 25°C in the dark. Insect samples were subjected to microscopic observation every 24 h for a week to 10 days. Only pupae and adults covered with fungal mycelium were considered as hosts to fungi.

18.2.4 Larval Experiment

Six treatments, one control and five bioinsecticides (Table 18.1), were prepared and inoculated following the same process described previously. Twenty third-instar larvae of *D. frontalis* were released in the treatments. Cups were kept by following the same process and conditions described previously. Five replicates were done for each treatment. After 1 week, pupae were sieved from the soil and examined under a microscope to determine if any growth of mycelium was apparent. To evaluate emerging adult mortality, all pupae recovered from the treatments were placed in Petri dishes and kept at 25°C, 60%–70% RH, and 14:10 L/D until adult emergence. After 10 to 11 days, emergence rates were assessed. Then, 19 adults (0 to 1 day old) were placed in adult cages to assess mortality. Cages were kept as previously described for adults in experiment one. Four replicates were done for each treatment. Dead flies were collected daily from the cages over a period of 12 days. Fungal infection in adults was investigated by following the same process described in experiment one.

18.2.5 Pupal Age Experiment

As the results of the experiments (larval, pupal, and adult) indicated that MET52 caused the highest mortality against *D. frontalis* (Figure 18.1 and Table 18.2), the product was selected for further investigations.

The effect of pupal age (2 and 8 days old from pupation) on susceptibility to MET52 was examined. A number of cups filled with soil inoculated with 1.5 g of MET52 were prepared as previously described. Four mL of sterilized distilled water was added to the untreated control and the MET52 treatments. The cups were kept by following the same process and conditions as described

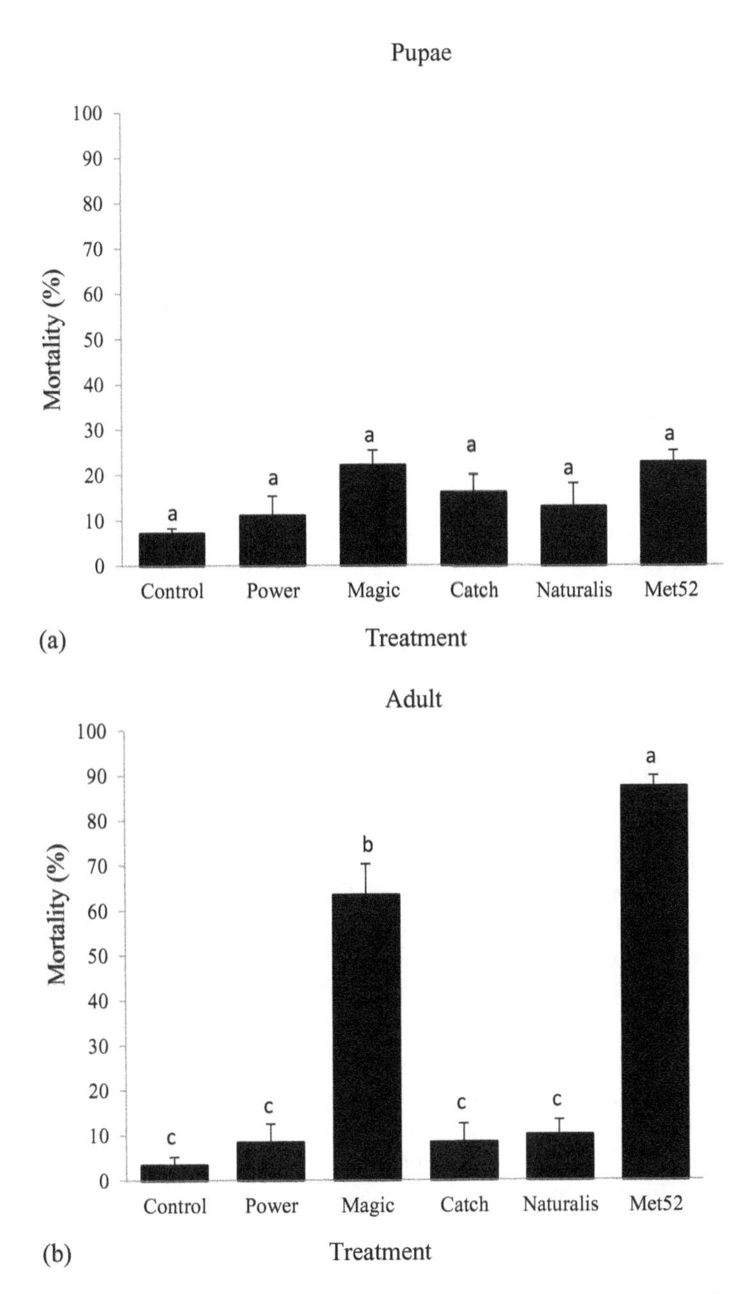

FIGURE 18.1 (a) Mean (% + SE) percentage of pupal mortality, $n = 5$. (b) Mean (% ± SE) subsequent adult mortality (2 weeks after emergence), $n = 4$ of *Dacus frontalis* treated with different biopesticides at recommended doses in 30 g of soil. Bars with different letters have significantly different means based on Tukey's HSD test ($P < 0.05$) after ANOVA.

in experiment one. Five replicates with 20 pupae were done for each treatment. After emergence, the number of dead pupae was assessed. Then, 20 adults (0 to 1 day old) were transferred to adult cages and kept as described previously to assess mortality. Four replicates were done for each treatment. Dead flies were collected daily from the cages over a period of 12 days. Fungal infection in dead pupae and adults was investigated by following the same process described in experiment one.

TABLE 18.2
Median Mortality of Pupae and Emerging Adults (12 Days after Emergence) of *Dacus frontalis* Following Larval Treatment with Different Entomopathogenic Fungi at Tested Doses, $n = 5$

Treatment	% Pupal mortality	Median[a]	% Adult mortality	Median[b]
Control	1	0	1	1
MET52	4	5	29	29[c]
Bio-Magic	4	5	0	0
Bio-Power	3	0	0	0
Naturalis L	4	0	0	0
Bio-Catch	1	0	3	3

[a] Kruskal–Wallis: Difference not significant.
[b] Kruskal–Wallis: Difference is significant.
[c] Mann–Whitney test: MET52 treatment showed significant difference compared to the untreated control ($P < 0.05$).

18.2.6 Rate Effect

In a similar procedure as previously described in experiment one, MET52 was applied to 30 g of soil at different rates (1.5, 1.0, 0.75, 0.5, 0.25, 0.125 g). Four mL of sterilized distilled water were added to the untreated control and the six fungal treatments. Fifteen 2-day-old pupae of *D. frontalis* were used in each replicate, and there were five replicates for each treatment. Cups were then kept by following the same process and conditions as described in the experiment one. Ten to 11 days after application, adult emergence was assessed. Then, 12 adults (0 to 1 day old) were transferred to and kept in adult cages to assess mortality following the same process and conditions previously described. Four replicates were done for each treatment. Dead flies were collected daily from the cages and assessed for 5 to 9 days. Fungal infection in adults was investigated by following the same process described in experiment one.

18.2.7 Formulation

The efficacy of two formulations (granule and drench) of *Metarhizium anisopliae* var *anisopliae* strain F52 was evaluated in reducing emergence rate and emerging adults of *D. frontalis*. For the granule treatment, cups filled with 30 g soil were prepared and inoculated with 1.5 g of MET52, as previously described. In the drench application, a suspension of 1.5 g of the fungus was prepared in 2 mL of water and the soil in the cups was drenched in it. Both treatments were applied 1 week prior to placing 20 third-instar larvae in the cups. Four mL of sterilized distilled water were added to the untreated control and the two fungal treatments. The cups were then kept following the same process and conditions described previously. Five replicates were done for the two fungal treatments and the control. After emergence, 15 adults (0 to 1 day old) were transferred to and kept in adult cages as described. Four replicates were done for each treatment. Dead flies were collected daily from the cages and assessed over a period of 2 weeks. Fungal infection in dead pupae and adults was investigated by following the same process described in experiment one.

18.2.8 Application Time

The result of the formulation experiment indicated that MET52 caused a high reduction in adult emergence of *D. frontalis* when applied as granule 1 week prior to releasing larvae in the soil (Figure 18.4). Thus, the effect of MET52 applied at different times prior to releasing larvae in the soil on emergence rate and adult mortality was investigated. Cups were filled with 30 g soil inoculated with 1.5 g of MET52 as previously described. The fungus was applied 2, 4, 6, 8, and 10 weeks before 20 third-instar larvae were introduced in each treatment. Four mL of sterilized distilled water were added to the untreated control and the other treatments. Cups were kept following the same process and conditions described previously. Five cups per treatment were used. After emergence, 15 adults (0 to 1 day old) were placed in cages and kept as described in experiment one. Four replicates were done for each treatment. Dead flies were collected daily from the cages and assessed over a period of 2 weeks. Fungal infection in adults was investigated by following the same process previously described in experiment one.

18.2.9 Statistical Analysis

Percentages of dead pupae, adult emergence rate, and emerging adults were arcsine transformed and analyzed by appropriate analysis of variance (ANOVA) tests. Mean differences among the treatments were then compared with a Tukey's test ($P < 0.05$). If data were not normally distributed, a nonparametric analysis was performed with a Kruskal–Wallis test. A Mann–Whitney test was then used to compare the differences between the treatments. A probit analysis was performed to calculate LT_{50} and LT_{90}. All statistical analyses were performed in Minitab 16 Statistical Software.

18.3 RESULTS

18.3.1 Pupal and Adult Experiment

Mycoses of the fungi applied in the soil were observed growing on dead pupae and emerging adults. None of the biopesticides used in the present study caused a significant increase in mortality of *D. frontalis* pupae ($F = 2.43$; $df = 5, 24$; $P > 0.05$). Figure 18.1a shows that MET52 and Bio-Magic biopesticides caused increasing mortality of pupae compared to the other treatments and untreated control, resulting in approximately 22% of pupal mortality. Two weeks after emergence, mortality of adults in the fungal treatments ranged from approximately 8% to 88% compared to the untreated control with approximately 3% (Figure 18.1b). There was a significant difference in adult mortality between treatments ($F = 74.67$; $df = 5, 18$; $P < 0.002$). MET52 showed the greatest pathogenicity against the fly, inducing approximately 88% of mortality (Figure 18.1b). Mortality of flies treated with MET52 was significantly greater than that of those treated with Bio-Magic, and both treatments had significantly greater mortality than the other treatments (Figure 18.1b).

18.3.2 Larval Experiment

Larvae of *D. frontalis* pupated normally in all the treatments. The results showed that larvae were not susceptible to any of the tested products, with no significant differences in pupal mortality between the treatments and untreated control (Kruskal Wallis test statistic (H) $= 4.57$; $df = 5$; $P > 0.05$). The percentage pupal mortality ranged from 1% to 4% (Table 18.2). Visible mycelium was detected growing around pupae recovered only from the soil treated with MET52. Twelve days after emergence, significant mortality was observed in the MET52 cages compared to the other treatments ($H = 17.41$; $df = 5$; $P < 0.005$). Adult mortality was 29% compared to the untreated control, with approximately 1% adult mortality. No infected adults emerged from the other fungi-treated soil and adult mortality varied from 0% to nearly 3% (Table 18.2).

18.3.3 Pupal Age Experiment

The age of the pupae did not affect their susceptibility to MET52. No significant difference in mortality was observed between 2- and 8-day-old treated pupae and the untreated control ($F = 0.64$; $df = 3, 16$; $P > 0.05$). Pupal mortality ranged from 2% to 15% in the treatments (Figure 18.2). Mortality of adults emerged from 2- and 8-day-old inoculated pupae was significantly higher than

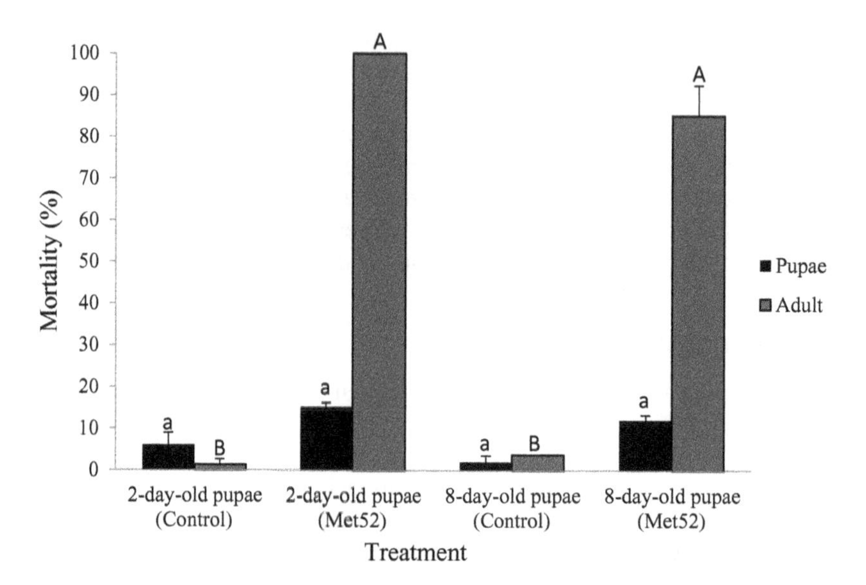

FIGURE 18.2 Mean (+ SE) percentage of pupae mortality (n = 5) and subsequent adult mortality after 12 days (n = 4) of *Dacus frontalis* pupae treated at different ages (2- and 8-day-old) with MET52 at recommended doses. Bars for different life stages with different letters represent significantly different means based on Tukey's HSD test (P <0.05) after ANOVA.

the mortality in the untreated control ($F = 545.68$: $df = 3, 12$; $P < 0.001$). The greatest pathogenicity was found in the 2-day-old pupae treatment, with 100% adult mortality compared to the 8-day-old treatment with 85% mortality, but there was no significant difference between these treatments (Figure 18.2). In the untreated control, approximately 1% and 4% adult mortality were found in the 2- and 8-day-old treatments, respectively.

18.3.4 Rate Effect

Areas of green fungal vegetative growth were visible on the soil surface two weeks after application of MET52 at different rates. There was no significant effect of applying MET52 at the tested rates on *D. frontalis* pupae ($F = 1.76$; $df = 6, 28$; $P > 0.05$). Mortality of pupae ranged from approximately 7% to 20% (Figure 18.3a). No pupal mortality was recorded in the untreated control. Adult mortality increased with the application rate of the fungal pathogen (Figure 18.3). At 5 days from emergence, the fungus induced significant mortality in adults from the inoculated soil compared to the untreated control ($F = 5.43$; $df = 6, 21$; $P < 0.01$), with no significant differences between the fungal treatments (Figure 18.3b). The greatest pathogenicity occurred when the highest rate (1.5 g) of the fungus was applied, giving slightly more than 70% adult mortality compared to the untreated control, in which the adult mortality was 2%. At 9 days from emergence, adult mortality in the fungal treatments increased significantly compared to the untreated control ($F = 30.4$; $df = 6, 21$; $P < 0.001$), with no significant differences found between fungal treatments (Figure 18.3b). Adult mortality ranged from approximately 79% to 100% in the fungal treatments and 6% in the untreated control. The lethal time to 50% adult mortality (LT_{50}) in the fungal treatments ranged from 5.5 days to 7.7 days. The shortest LT_{50} was found when pupae had been exposed to 1.5 g of MET52, whereas approximately 10 days were required to get 90% of adult mortality when pupae were placed in the soil treated with the lowest rate (0.125 g) (Table 18.3).

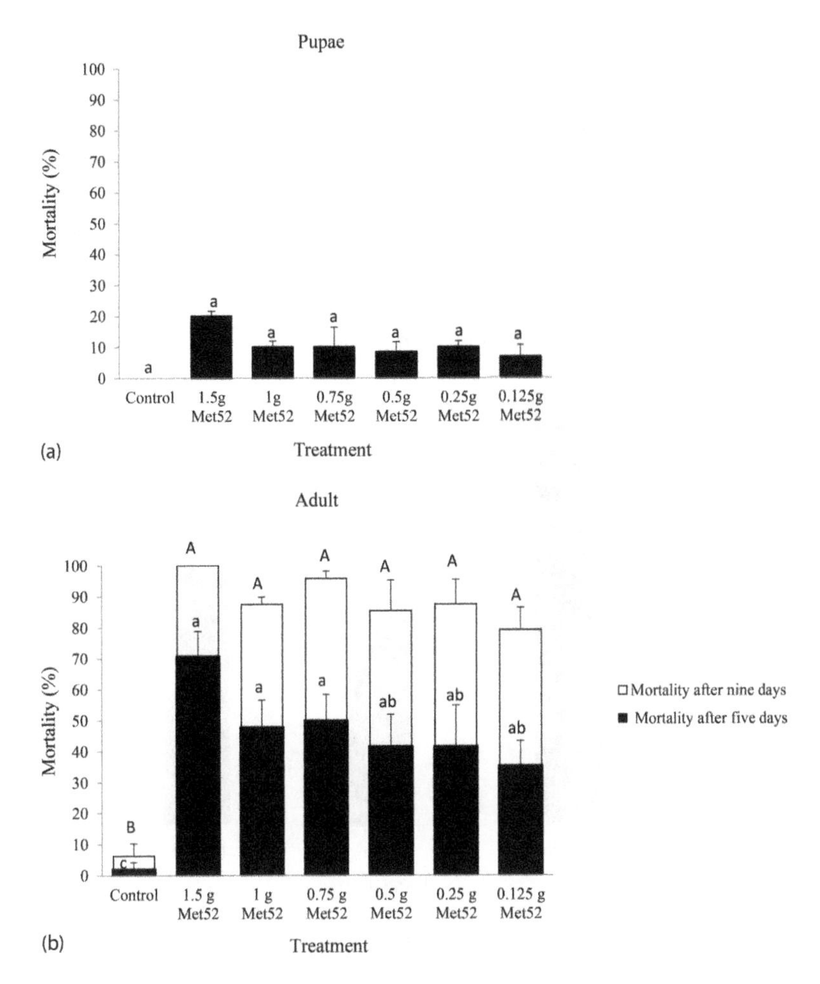

FIGURE 18.3 Mean (+ SE) percentage of pupal mortality, $n = 5$ (a) and subsequent adult mortality (5 and 9 days), $n = 4$ (b) after *Dacus frontalis* pupae were treated with different rates of MET52 (1.5, 1, 0.75, 0.5, 0.25, 0.152 g) in 30 g of soil. Bars within treatments with different letters represent significantly different means based on Tukey's HSD test ($P < 0.05$) after ANOVA.

TABLE 18.3
Mean Lethal Time (LT_{50} and LT_{90}) of Different Rates of MET52 Applied against *Dacus frontalis* Adults

Treatment	LT_{50}(Days)	LT_{90} (Days)
1.5 g	5.5	7.5
1.0 g	5.8	8.1
0.75 g	6.1	8.2
0.5 g	7.1	9.8
0.25 g	6.7	9.3
0.125 g	7.1	9.9

The values were calculated by Probit analysis.

18.3.5 Formulation

Results indicated that applying *M. anisopliae* var *anisopliae* strain F52 in a granular form caused a significant reduction in adult emergence compared to the drench and untreated treatments (F = 41.63; df = 2, 12; P <0.001). As shown in Figure 18.4a, adult emergence reached 52% in the granule treatment compared to the drench treatment, with 91%, and the untreated control, with 97%. In the case of emerging adults, both application methods significantly reduced the number of adults compared to the untreated

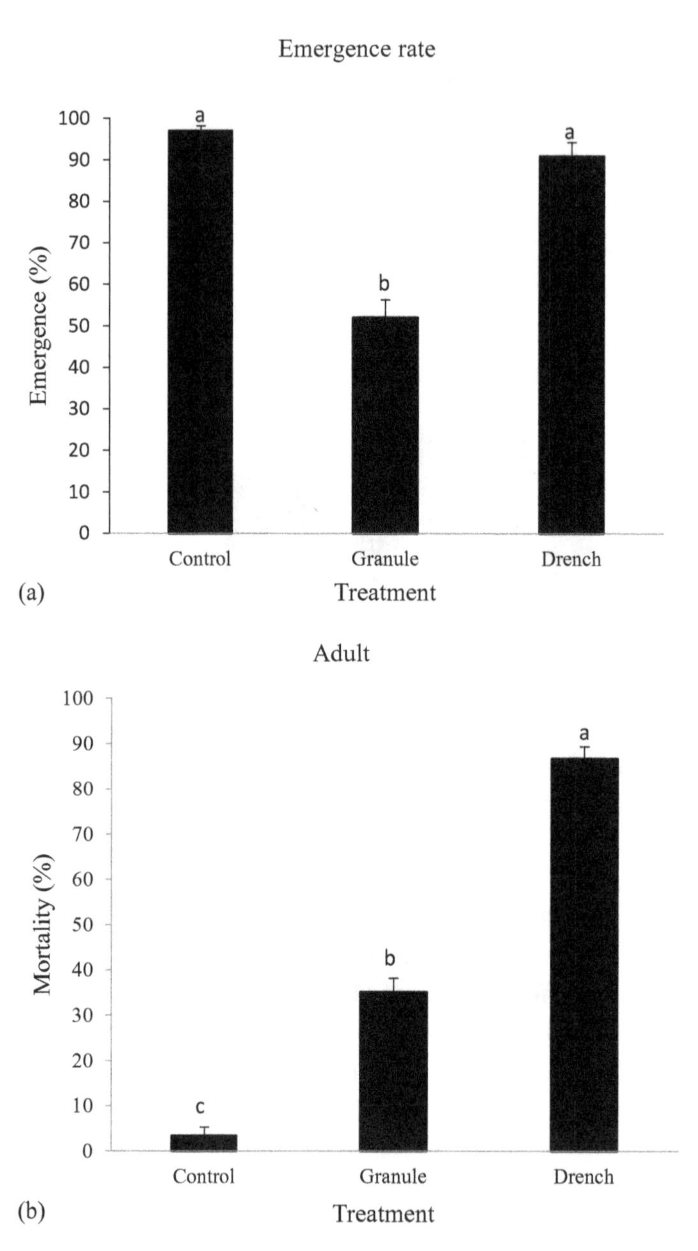

FIGURE 18.4 Mean (+ SE) percentage of emergence rate, n = 5 (a) and subsequent adult mortality (two weeks after emergence) n = 4 (b) after *Dacus frontalis* larvae were released in soil treated with different formulations (granule and drench) of *M. anisopliae* var anisopliae strain F52 one week earlier. Bars with different letters represent significantly different means based on Tukey's HSD test (P < 0.05) after ANOVA.

control (F = 529; df = 2, 9; P <0.001). Adult mortality was low (35%) when treated with granules compared to the drench treatment (87%) 2 weeks after emergence, with a significant difference between both treatments (Figure 18.4b). Only approximately 3% of adult mortality was found in the untreated control.

18.3.6 Application Time

The pathogenicity of MET52 against *D. frontalis* was influenced by application time. Results showed that adult emergence was significantly lower in inoculated soils (except in the 4-week treatment) than in untreated control treatments (F = 111.44; df = 5, 24; P <0.001). The greatest pathogenicity was recorded in the 2-week treatment with a 55% reduction in adult emergence rate. Adult emergence reduction in the other fungal treatments ranged from 16% to 40%, with significant differences between treatments (Figure 18.5a). A high adult emergence (97%) was found in the untreated control. After emergence, the number of dead flies was significantly higher in the inoculated treatments

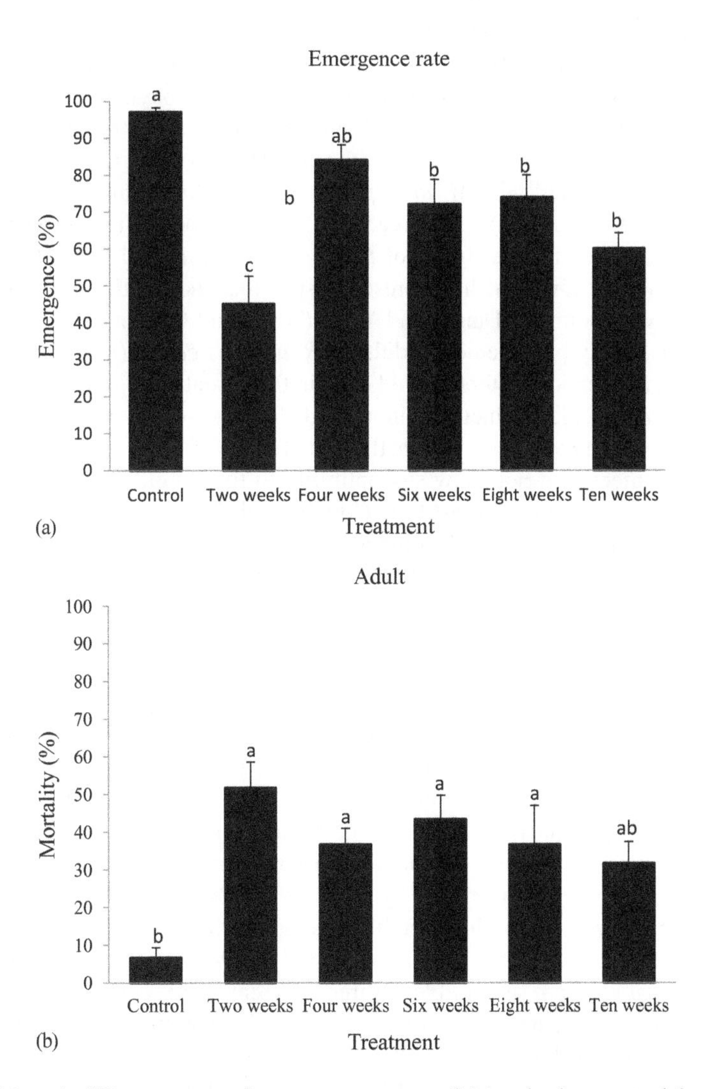

FIGURE 18.5 Mean (+ SE) percentage of emergence rate, $n = 5$ (a) and subsequent adult mortality (2 weeks from adult emergence), $n = 4$ (b) after *Dacus frontalis* larvae were released in 30 g of soil treated with 1.5 g of MET52 at different times. Bars with different letters represent significantly different means based on Tukey's HSD test (P <0.05) after ANOVA.

than in the untreated control, with the exception of the 10-week treatment ($F = 98.02$; $df = 5, 18$; $P < 0.01$) (Figure 18.5b). Pathogenicity ranged from approximately 32% to 52% of adult mortality over the period of 2 weeks from emergence. The greatest pathogenicity against the fly was induced when the fungus was applied 2 weeks after the larvae were released, with no significant differences between treatments (Figure 18.5b). Adult mortality in the untreated control was 6.6%.

18.4 DISCUSSION

The present investigation compared five commercial entomopathogenic fungi products to explore their efficacy against *D. frontalis* under laboratory conditions. This study demonstrated the susceptibility of *D. frontalis* to entomopathogenic fungi. The susceptibility of other species of fruit flies to entomopathogenic fungi inoculated through different methods has been confirmed before (De la Rosa et al. 2002; Ekesi et al. 2002; Dimbi et al. 2003; Sookar et al. 2008; Daniel and Wyss 2009; Cossentine et al. 2010; Goble et al. 2011).

In the adult experiment, the product MET52, based on *M. anisopliae* var *anisopliae* strain F52, was the most pathogenic against *D. frontalis* adults and reached 100% mortality when applied at 1.5 g (9.0×10^8 CFU g^{-1}). The LT_{50} ranged from 4 to 5 days throughout the experiments. Other authors (Ekesi et al. 2002, 2005; Mochi et al. 2006) have obtained 100% mortality of adults of *Ceratitis capitata* (Wiedemann), *Ceratitis fasciventris* (Bezzi), and *Ceratitis cosyra* (Walker) that emerged from sand treated with *M. anisopliae*. Also, Sookar et al. (2008) reported mortality of *Bactrocera zonata* (Saunders) adults reaching 98% after being treated with *M. anisopliae*. The same study indicated that some strains of *Beauveria bassiana* and *Isaria fumosoroseus* have low pathogenicity, and this agrees with the present results that showed that both fungi caused only 8% of adult mortality. In contrast, Daniel and Wyss (2009) and Cossentine et al. (2010) reported that *B. bassiana* was highly pathogenic to adults of *Rhagoletis cerasi* (L.). Different target insect species and different application methods could explain these contrasting findings. Gul et al. (2015) indicated that different inoculation methods induced different susceptibility levels in larval, pupal, and adult stages of *B. zonata* to three insect pathogenic fungi.

Larvae in all treatments did not show susceptibility to the fungi. This was probably because exposure to the pathogens was for a short time (Mochi et al. 2006). Another possible explanation is that tephritid larvae have soft cuticle, lacking any hairs, which could limit the number of conidia that can be attached. Mochi et al. (2006) indicated that the E9 isolate of *M. anisopliae* had no effect on the larval stage of *C. capitata*. Also, De la Rosa et al. (2002) found that *B. bassianna* caused low mortality against larvae of the Mexican fruit fly, *Anastrepha ludens* (Loew). Moreover, larvae of *R. cerasi* were found to not to be susceptible to entomopathogenic fungi (Daniel and Wyss 2009).

None of the tested strains caused significant mortality in *D. frontalis* pupae, although fungal mycelia was observed growing inside and outside the pupal cuticle. Additionally, pupal age and increasing rate of MET52 had no effect on pupal mortality. This could be due to pupae having a thick cuticle that might prevent the penetration of conidia spores. However, the fungus induced higher mortality in emerging adults when applied in young pupae. Similar results were obtained by Beris et al. (2013), who found low mortality, from 19% to 24%, in *C. capitata* pupae when they were exposed to three fungi species. However, higher mortality rates were induced after emergence. Previous studies by De la Rosa et al. (2002) and Daniel and Wyss (2009) indicated that pupae of *A. ludens* and *R. cerasi* were not susceptible to three different fungi species applied to the soil, and there were also no effects on emerging adults, suggesting that the pupal stage is not susceptible to fungal infection. In contrast, Ekesi et al. (2002) indicated that different isolates of *M. anisopliae* and *B. bassiana* caused great reduction in adult emergence of three tephritid fruit fly species in an experiment in Petri dishes. The authors found that adult emergence decreased when old pupae were used, inducing mortality in emerging adults. Differences in inoculation methods and the expected high humidity level in the Petri dish experiment might be the reason for the high mortality.

The application time experiment showed that early application of MET52 reduced adult emergence to 45%. This was probably due to increasing conidia density of MET52 in the soil over the time of the experiment. A possible explanation is that the granule form promotes the growth of the pathogen, thus increasing the concentration of conidia. Another possible explanation for the reduction in emergence is that released larvae could have ingested some conidia spores before entering into the pupal stage. The same results were obtained by Ekesi et al. (2002), who observed that prophylactic application with *M. anisopliae* was more effective than a curative treatment in reducing adult emergence of three species of fruit flies.

The results of the present study also revealed that *M. anisopliae* var *anisopliae* strain F52 can greatly reduce adult emergence when it is applied early as granules rather than mixed in a suspension and drenched. However, after adult emergence, mortality was greater in the drenched treatment, with 52%. This could be due to a greater adhesion of conidia to the emerging adult cuticle in the drenched treatment than in the granular treatment. In future investigations, it might be useful to use the fungus in both forms of the tested formulations at the same time for a better control; however, the cost of this should be analyzed. Our microscopic observations showed that large parts of the emerging adults were covered with greenish dry conidia in the granule application but not in the drenched treatment. Similarly, Ekesi et al. (2005) found that a granule form of another strain of *M. anisopliae* was more effective in reducing adult emergence of *C. capitata*, *C. fasciventris*, and *C. cosyra* than a suspension and drench treatment.

A granule formulation of some species, isolates and strains of *Metarhizium* fungi, has been successfully used as a pathogen against other agricultural insect pests (Moorhouse et al. 1993; Bruck and Donahue 2007; Ansari and Butt 2013; Arthurs et al. 2013; Mauchline et al. 2013; Williams et al. 2013).

In the current study, a sandy clay loam soil (65% sand, 12% silt, and 23% clay) was used, but very few studies have investigated the effect of soil type on the efficacy of entomopathogenic fungi. Garrido-Jurado et al. (2011b) found that soil proprieties had no effect on the pathogenicity of *M. anisopliae* EAMa 01/58-Su and *B. bassiana* EABb 01/110-Su against soil stages of *C. capitata*. Further investigations focused on the type of soil are required.

18.5 CONCLUSIONS

In conclusion, *D. frontalis* adults are highly susceptible to some fungal pathogens. Our results suggest that applying entomopathogenic fungi as granules to the soil could be a promising biological control, reducing adult emergence and causing high mortality in emerging adults. This mortality strategy could provide some benefits for *D. frontalis* control because dead pupae and adults could serve as future infection sources against new fly offspring in the soil (Ekesi et al. 2002). Also, soil provides a good opportunity for recycling and protecting pathogens, which may help increase conidia density and spread in the environment. To improve our understanding of how to maximize the effect of *M. anisopliae*, a next step is to focus on the effect of abiotic factors such temperature, humidity, and soil-moisture content on the efficacy of MET52. In addition, studying the combined use of the product with other biological control agents could increase larval mortality. The current market price of MET52 suggests that it will not be economically feasible compared to chemical pesticides because applying it at the recommended rate will cost an excess of $5000/ha.

REFERENCES

Abukhashim, N. K., Abdussalam, A. M., Ashleeb, M. A., Alshareef, S. K., Ben Husine, T. O., Elghadi, E. O., Albakkoush, F. E., Abdulmalek, H. M. and Ahmadeh, A. M. 2003a. Survey for distribution and host range of the Greater melon fly *Dacus frontalis* (Becker) in Libya. *Eighth Arab Congress of Plant Protection*. El-Beida, Libya, 22-E–23-E.

Abukhashim, N. K., A. M. Abdussalam, T. O. Ben Husine, S. K. Alshareef, and E. O. Elghadi. 2003b. *The Greater Melon fly Dacus frontalis (Becker) and Its Management*. Tripoli, Libya: Biotechnology Research Center Newsletter 1–40.

Ansari, M., and T. Butt. 2013. Influence of the application methods and doses on the susceptibility of black vine weevil larvae *Otiorhynchus sulcatus* to *Metarhizium anisopliae* in field-grown strawberries. *BioControl* 58: 257–267.

Arthurs, S. P., L. F. Aristizábal, and P. B. Avery. 2013. Evaluation of entomopathogenic fungi against chilli thrips, *Scirtothrips dorsalis. Journal of Insect Science* 13: 1–16.

Ba-Angood, S. A. S. 1977. Control of the melon fruit fly, *Dacus frontalis* Becker (Diptera: Trypetidae), on cucurbits. *Journal of Horticultural Science and Biotechnology* 52: 545–547.

Badii, K. B., M. Billah, K. Afreh-Nuamah, D. Obeng-Ofori, and G. Nyarko. 2015. Review of the pest status, economic impact and management of fruit-infesting flies (Diptera: Tephritidae) in Africa. *African Journal of Agricultural Research* 10: 1488–1498.

Beris, E. I., D. P. Papachristos, A. Fytrou, S. A. Antonatos, and D. C. Kontodimas. 2013. Pathogenicity of three entomopathogenic fungi on pupae and adults of the Mediterranean fruit fly, *Ceratitis capitata* (Diptera: Tephritidae). *Journal of Pest Science* 86: 275–284.

Bruck, D. J., and K. M. Donahue. 2007. Persistence of *Metarhizium anisopliae* incorporated into soilless potting media for control of the black vine weevil, *Otiorhynchus sulcatus* in container-grown ornamentals. *Journal of Invertebrate Pathology* 95: 146–150.

Butt, T. 2002. Use of entomogenous fungi for the control of insect pests. In *Agricultural Applications*, ed. F. Kempken, 111–134. Berlin, Germany: Springer.

Butt, T., L. Ibrahim, S. Clark, and A. Beckett. 1995. The germination behaviour of *Metarhizium anisopliae* on the surface of aphid and flea beetle cuticles. *Mycological Research* 99: 945–950.

Castillo, M. A., P. Moya, E. Hernández, and E. Primo-Yúfera. 2000. Susceptibility of *Ceratitis capitata* Wiedemann (Diptera: Tephritidae) to entomopathogenic fungi and their extracts. *Biological Control* 19: 274–282.

Cossentine, J., H. Thistlewood, M. Goettel, and S. Jaronski. 2010. Susceptibility of preimaginal western cherry fruit fly, *Rhagoletis indifferens* (Diptera: Tephritidae) to *Beauveria bassiana* (Balsamo) Vuillemin Clavicipitaceae (Hypocreales). *Journal of Invertebrate Pathology* 104: 105–109.

Daniel, C. and E. Wyss. 2009. Susceptibility of different life stages of the European cherry fruit fly, *Rhagoletis cerasi*, to entomopathogenic fungi. *Journal of Applied Entomology* 133: 473–483.

De la Rosa, W., F. L. Lopez, and P. Liedo. 2002. *Beauveria bassiana* as a pathogen of the Mexican fruit fly (Diptera: Tephritidae) under laboratory conditions. *Journal of Economic Entomology* 95: 36–43.

De Meyer, M., I. M. White, and K. F. M. Goodger. 2013. Notes on the frugivorous fruit fly (Diptera: Tephritidae) fauna of western Africa, with description of a new Dacus species. *European Journal of Taxonomy* 50: 1–17.

Dimbi, S., N. K. Maniania, S. A. Lux, S. Ekesi, and J. K. Mueke. 2003. Pathogenicity of *Metarhizium anisopliae* (Metsch.) *Sorokin* and *Beauveria bassiana* (Balsamo) *Vuillemin*, to three adult fruit fly species: *Ceratitis capitata* (Weidemann), *C. rosa* var. *fasciventris* Karsch and *C. cosyra (*Walker) (Diptera:Tephritidae). *Mycopathologia* 156: 375–382.

Ekesi, S., N. K. Maniania, and S. A. Lux. 2002. Mortality in three African tephritid fruit fly puparia and adults caused by the entomopathogenic fungi, *Metarhizium anisopliae* and *Beauveria bassiana. Biocontrol Science and Technology* 12: 7–17.

Ekesi, S., N. K. Maniania, S. A. Mohamed, and S. A. Lux. 2005. Effect of soil application of different formulations of Metarhizium anisopliae on African tephritid fruit flies and their associated endoparasitoids. *Biological Control* 35: 83–91

El-Hawagry, M., M. Khalil, M. Sharaf, H. Fadl, and A. Aldawood. 2013. A preliminary study on the insect fauna of Al-Baha Province, Saudi Arabia, with descriptions of two new species. *ZooKeys* 274: 1–88.

Esser, K., and P. A. Lemke. 1995. *The Mycota*. Berlin, Germany: Springer.

Foottit, R. G., and P. H. Adler. 2009. *Insect Biodiversity: Science and Society*. Chichester, UK: Blackwell publishing Ltd.

Gameel, S. M. M. 2013. Species composition of piercing-sucking arthropod Pests and associated natural enemies inhabiting cucurbit fields at the New Valley in Egypt. *Egyptian Journal of Biological Pest Control* 6: 73–79.

Garrido-Jurado, I., F. Ruano, M. Campos, and E. Quesada-Moraga. 2011a. Effects of soil treatments with entomopathogenic fungi on soil dwelling non-target arthropods at a commercial olive orchard. *Biological Control* 59: 239–244.

Garrido-Jurado, I., J. Torrent, V. Barrón, A. Corpas, and E. Quesada-Moraga. 2011b. Soil properties affect the availability, movement, and virulence of entomopathogenic fungi conidia against puparia of *Ceratitis capitata* (Diptera: Tephritidae). *Biological Control* 58: 277–285.

Goble, T. A., J. F. Dames, M. P. Hill, and S. D. Moore. 2011. Investigation of native isolates of entomopathogenic fungi for the biological control of three citrus pests. *Biocontrol Science and Technology* 21: 1193–1211.

Gul, H. T., S. Freed, M. Akmal, and M. N. Malik. 2015. Vulnerability of different life stages of *Bactrocera zonata* (Tephritidae: Diptera) against entomogenous fungi. *Pakistan Journal Zoology* 47: 307–317.

Hafsi, A., K. Abbes, A. Harbi, S. Ben Othmen, E. Limem, M. Elimem, M. Ksantini, and B. Chermiti. 2015. *The Pumpkin Fly Dacus Frontalis (Diptera: Tephritidae): A New Pest of Curcubits in Tunisia.* Bulletin OEPP/EPPO Bulletin 45: 1–5.

Imoulan, A., and A. Elmeziane. 2014. Pathogenicity of *Beauveria bassiana* isolated from Moroccan Argan forests soil against larvae of *Ceratitis capitata* (Diptera: Tephritidae) in laboratory conditions. *World Journal of Microbiology and Biotechnology* 30: 959–965.

Konstantopoulou, M. A., and B. E. Mazomenos. 2005. Evaluation of B*eauveria bassiana* and *B. brongniartii* strains and four wild-type fungal species against adults of *Bactrocera oleae* and *Ceratitis capitata. BioControl* 50: 293–305.

Mauchline, N.A ., K. A. Stannard, and S. M. Zydenbos. 2013. Evaluation of selected entomopathogenic fungi and bio-insecticides against *Bactericera cockerelli* (Hemiptera). *New Zealand Plant Protection* 66: 324–332.

Mochi, D. A., A. C. Monteiro, S. A. De Bortoli, H. O. S. Doria, and J. C. Barbosa. 2006. Pathogenicity of *Metarhizium anisopliae* for *Ceratitis capitata* (Wied.) (Diptera: Tephritidae) in soil with different pesticides. *Neotropical Entomology* 5: 382–389.

Moorhouse, E. R., A. T. Gillespie, and A. K. Charnley. 1993. Application of *Metarhizium anisopliae* (Metsch.) *Sor.* conidia to control *Otiorhynchus sulcatu*s (F.) (Coleoptera: Curculionidae) larvae on glasshouse pot plants. *Annals of Applied Biology* 122: 623–636.

Mwatawala, M., A. P. Maerere, R. Makundi, and M. De Meyer. 2010. Incidence and host range of the melon fruit fly *Bactrocera cucurbitae* (Coquillett) (Diptera: Tephritidae) in Central Tanzania. *International Journal of Pest Management* 56: 265–273.

Ramadan Abdallah, A. 2002. The first record of *Dacus frontalis* (Becker) and *Dacus longistylus* (Wiedemann) in Libya. *Arab and Near East Plant Protection Newsletter* 34: 29–30.

Redha, S. A. L. J. 2013. Relative incidence of the Greater melon fly, *Dacus frontalis* Becker and cucurbit fly, *Dacus ciliates* Loew on cucumber *Cucumis sativus* (L.). *Mesopotamia Journal of Agriculture*, 41(*A special international conference of the Department of Plant Protection*) 44, 88–94.

Roy, H. E., E. L. Brodie, D. Chandler, M. S. Goettel, J. K. Pell, E. Wajnberg, and F. E. Vega. 2010. Deep space and hidden depths: understanding the evolution and ecology of fungal entomopathogens. *Biocontrol* 55: 1–6.

Sookar, P., S. Bhagwant, and E. A. Ouna. 2008. Isolation of entomopathogenic fungi from the soil and their pathogenicity to two fruit fly species (Diptera: Tephritidae). *Journal of Applied Entomology* 132: 778–788.

Steffens, R. J. 1982. Ecology and approach to integrated control of *Dacus frontalis* on the Cape Verde Islands. *The International Symposium on Fruit Flies of Economic Importance.* Athens, Greece. A. A. Balkema, 632–638.

Svedese, V. M., A. da Silva, R. D. Lopes, J. F. dos Santos, and E. Lima. 2012. Action of entomopathogenic fungi on the larvae and adults of the fig fly *Zaprionus indianus* (Diptera: Drosophilidae). *Ciencia Rural* 42: 1916–1922.

White, I. M. 2006. *Taxonomy of the Dacina (Diptera: Tephritidae) of Africa and the Middle East.* African Entomology: Memoir 2: Special Issue 2: 1–156. Hatfield, South Africa: Entomological Society of Southern Africa.

Williams, C. D., A. B. Dillon, C. D. Harvey, R. Hennessy, L. M. Namara, and C. T. Griffin. 2013. Control of a major pest of forestry, Hylobius abietis, with entomopathogenic nematodes and fungi using eradicant and prophylactic strategies. *Forest Ecology and Management* 305: 212–222.

Wraight, S. P., and A. E. Hajek. 2009. Manipulation of arthropod pathogens for IPM. In: *Integrated Pest Management: Concepts, Tactics, Strategies and Case Studies*, ed. E. B. Radcliffe, W. D. Hutchison, and R. E. Cancelado, 131–150. New York: Cambridge University Press.

19 Natural Parasitism and Parasitoid Releases to Control *Anastrepha obliqua* (Diptera: Tephritidae) Infesting *Spondias* spp. (Anacardaceae) in Chiapas, Mexico

Patricia López[*], *Jorge Cancino, and Pablo Montoya*

CONTENTS

Abstract *Anastrepha obliqua* (Macquart) is a fruit fly pest in the Neotropical region associated with fruits of the genus *Spondias* spp. (Anacardiaceae) as natural hosts, where a conspicuous guild of native parasitoids has been identified. This species is also reported as the main pest in the mango (*Mangifera indica* L.) (Anacardiaceae) growing zones in Mexico, the Caribbean, and several countries of Central and South America, which demands the use of different control strategies to produce healthy fruits. Literature describing the association between native parasitoids and *A. obliqua* in *Spondias* fruits in Mexico and Central and South America is reviewed. We also provide results of field releases of the hymenopteran parasitoid *Diachasmimorpha longicaudata* (Ashmead), introduced in Mexico in the 1970s, and the native *Utetes anastrephae* (Viereck) in Spondias fruits infested with *A. obliqua* as a preliminary evaluation to determine the potential of native parasitoids in the control of *A. obliqua*. The association of *A. obliqua* with *Spondias* spp. and the guild of native parasitoids as a

[*] Corresponding author.

natural resource for the management of this pest are described. Our results indicate that the native parasitoid *U. anastrephae* has the potential to control pest populations of *A. obliqua*. The management of *A. obliqua* with parasitoids, whether native or introduced, seems to be suitable and ecologically convenient. High levels of parasitism can be achieved depending on the size, shape, and structure of *Spondias* fruits, thus diminishing the numbers of *A. obliqua* flies that later will invade mango commercial areas at a larger scale.

19.1 INTRODUCTION

Anastrepha obliqua, the West Indian fruit fly, is an economically important Tephritid pest species in the Western Hemisphere (López-Guillén et al. 2009; Leite et al. 2017; Montoya et al. 2016). This species is distributed mainly in the tropical regions of the Americas, from the north of Mexico to the southeast of Brazil, including the Caribbean Islands and, occasionally, the south of the United States (Steck 2001; Silva et al. 2010; Garcia and Ricalde 2013; Sousa et al. 2017), being dominant in tropical regions with high temperature and humidity (CABI 2017).

According to different surveys, *A. obliqua* is mainly associated with *Spondias* spp. (Anacardaceae) (Aluja et al. 2001), but the list of host fruits include the families Annonaceae (custard apple, anona), Myrtaceae (guava), Oxalidaceae (carambola), Passifloraceae (granadilla), and Sapotaceae (mamey, chicozapote) (Mangan et al. 2011; Taira et al. 2013; Montoya et al. 2017). The current extension of the mango (*Mangifera indica* L.) growing zone in several countries of the Americas makes this exotic species a preferential host for *A. obliqua*, which highlights the economic importance of this pest (Sivinski et al. 2001; Jenkins and Goenaga 2008).

Different authors proposed that the relation between *A. obliqua* and *Spondias* fruits is an intrinsic association in different American locations (Hernández-Ortiz and Pérez-Alonso 1993; Silva et al. 2010; Jesus-Barros et al. 2012). The diversity of *Spondias* fruits is broad, with some species adapted to subtropical and even temperate regions of America (Arce-Romero et al. 2017). The commercialization of these fruits is variable; in some areas they are of marketable interest but in others they lack economic importance (Ramírez et al. 2008). However, extensive commercial areas are scarce and production is limited to a small scale or backyard orchards (Hauck et al. 2011; Alia-Tejacal 2012). Under this scenario, the importance of *Spondias* spp. as preferential hosts of *A. obliqua* is twofold: It shows the direct effect of *A. obliqua* limiting the commercialization of cultivars of interest, and it highlights their role as reservoir fruits maintaining pest populations that later will invade mango commercial orchards (Aluja and Birke 1993). Both cases represent an ideal model to study different fruit-host-parasitoid relationships, which are characterized by highly dynamic and complex structures (López et al. 1999; Silva et al. 2010).

Here, we describe the distribution of *A. obliqua* in relation to its main hosts, the control tactics used against it, and the natural parasitism exerted by native parasitoids associated with this species infesting *Spondias* spp. in Mexico. We also provide results of simultaneous releases of *Utetes anastrephae* (Viereck) and *Diachasmimorpha longicaudata* (Ashmead) to control this important pest.

19.2 MANAGEMENT OF *ANASTREPHA OBLIQUA* PEST POPULATIONS

The management of *A. obliqua* in mango commercial areas has been based on the application of toxic baits focused on pest populations (Aluja 1994; Díaz-Fleischer et al. 2017). Currently, the Spinosad mixture, derived from the metabolism of the soil bacterium *Saccharopolyspora spinosa* Mertz & Yao (Flores et al. 2011), is increasingly applied with food attractants obtained from vegetal protein. Pest management in commercial orchards is complemented with activities such as tracking, pruning, and fertilization, as well as the elimination of infested fruits (Peña et al. 1998). Other activities such as postharvest treatments by means of hydrothermal treatments or fruit radiation are complementary procedures for fruits destined for the export market (Neven 2010).

In 2002, the Mexican campaign against fruit flies initiated the release of sterile *A. obliqua* insects in different states of the country. This species is currently produced and sterilized at the Moscafrut Facility-SAGARPA-SENASICA, located in Metapa, Chiapas, Mexico, at a scale of 60 million sterile pupae per week (Orozco-Dávila et al. 2017). The insects are then sent for subsequent releases into mango-producing regions in northwestern Mexico, mainly in the states of Sinaloa and Nayarit. Release densities are ca. 2,000–2,500 sterile adults per ha (SAGARPA/SENASICA 2012). Currently, the north of the state of Sinaloa is an internationally recognized fruit fly-free zone of mango (Flores et al. 2017), and this state has become the main exporter of mango in the country (PNMF/DGSV 2017).

Another strategy to control *A. obliqua* populations is augmentative biological control (ABC). This strategy is used in Mexico since the 1990s in regions where agro-ecological conditions are adequate (Montoya et al. 2007). Augmentative releases of the exotic parasitoid *D. longicaudata* to control *Anastrepha* pest populations were evaluated by Montoya et al. (2000), reaching pest suppression of around 70%. After this, ABC has been successfully applied in different states of Mexico (Montoya et al. 2007), focusing parasitoid releases on marginal areas adjacent to commercial zones to reduce the movement of wild flies to commercial orchards. This strategy was used to achieve a fruit fly-low prevalence zone located in the state of Guerrero, Mexico, through the release of *D. longicaudata* against *A. obliqua* populations in marginal areas with high density of hog plums (*Spondias mombin* L.) (Segura et al. 2016). Hog plums are small fruits with a central seed that limit the escape of fruit fly larvae once they have been detected by the female parasitoid (Montoya et al. 2016), thus providing important opportunities for successful wasp oviposition.

The microbiological control of *A. obliqua* by means of applications of the fungus *Beauveria bassiana* (Bals) (Toledo et al. 2007) is another novel option that recently emerged through the use of dissemination of conidia devices baited with fruit fly attractants (Campos-Carbajal 2017). It has been used at a density of 10 devices per ha in the field.

19.3 DISTRIBUTION OF *A. OBLIQUA* IN *SPONDIAS* SPP. IN MEXICO

Anastrepha obliqua is widely distributed in Mexico; in most cases, it is associated with *Spondias* spp. (López et al. 1999; Murillo et al. 2015; Montoya et al. 2017) as natural hosts, whose presence is also widespread along both coastal lines of Mexico. In the Pacific coastal region, the genus *Spondias* is distributed from northern Sinaloa (next to the Gulf of California) to the border with Guatemala. In the slope of the Gulf of Mexico, its presence is concentrated in the states of Veracruz, Campeche, and Yucatan (Arce-Romero et al. 2017). As mentioned previously, the distribution of *A. obliqua* is associated with mango-growing zones, and it shares this niche with *Anastrepha ludens* (Loew) in the state of Chiapas (Montoya et al. 2000). The importance of *A. obliqua* as a pest of *Spondias* spp. is overlapped with its presence in mango orchards, either by direct fruit damage caused by larval infestation or just by adult trap captures in areas of low prevalence (Gutiérrez-Ruelas 2009; Santiago-Martínez 2008; Montoya et al. 2016).

There is a broad guild of native parasitoids closely associated with *A. obliqua* in *Spondias* spp.: *Utetes anastrephae* (Viereck), *Doryctobracon areolatus* (Szepligeti) (Braconidae), *Aganaspis pelleranoi* (Brèthes) (Eucolidae), and *Odontosema anastrephae* Borgmeier (Figitidae) are the most conspicuous species in this guild (Ovruski et al. 2000; Sivinski et al. 2000). However, there are no detailed reports on population variations of the native species that integrate this guild, which becomes a subject requiring further studies that would be beneficial for a better control of this pest. The preference of *A. obliqua* for infesting *Spondias* fruits suggests that these trees could be used as "trap crops" to protect commercial areas because they could be an important source of native parasitoids, thus contributing to biological control and conservation. Nevertheless, there is no empirical evidence supporting this strategy (Aluja et al. 2014).

19.4 NATIVE PARASITOIDS ASSOCIATED WITH *A. OBLIQUA* IN *SPONDIAS* SPP.

19.4.1 *Utetes anastrephae* (Viereck) (Hymenoptera: Braconidae)

This species is a synovigenic, koinobiont, larval-pupal endoparasitoid (Stul and Sivinski 2012). Adult size can vary depending on the host species. The length of its ovipositor and its host searching behavior allow to consider this parasitoid as specific to the tritrophic relation with *A. obliqua* and *Spondias* spp. (Sivinski et al. 2001). This species also has the ability to enter into diapause inside its hosts, a capacity that is influenced mainly by environmental conditions (Aluja et al. 1998).

Utetes anastrephae is native to the neotropical region (Ovruski et al. 2000), and it attacks larvae of *Anastrepha* spp., *Rhagoletis pomonella* (Walsh), and the exotic *Ceratitis capitata* (Wied). In this region, it coexists with the rest of the native parasitoid species and also with exotic species such as *D. longicaudata* and *Aceratoneuromyia indica* (Silvestri) (Rull et al. 2009; Silva et al. 2010) (Table 19.1). Its first larval stage possesses large jaws, providing a competitive advantage under

TABLE 19.1
Fruit Fly Species and Native and Exotic* Parasitoid Species Associated with *Spondias* Spp. Taken from References Published for the Neotropical Region

Fruit	Fly	Parasitoid	Country	References
S. cytherea	Ao	Asa, Da, Ua	Brazil	Silva et al. (2010)
S. dulcis	AN	Da	Brazil	Leal et al. (2009)
S. jambo	Af	Da, Dc, Dl*, Ua	Mexico	Sivinski et al. (2000)
S. malaccense	Af	Asa, Da, Ua	Brazil	Silva et al. (2010)
S. mombin	Ao	Da, Dl, Ua	Mexico	Sivinski et al. (2000)
	Ao	Da, Ua	Mexico	Murillo et al. (2015)
	Ao	Ua	Puerto Rico	Jenkins and Goenaga (2008)
	AN	Da, Ua	Domin. Rep.	Serra et al. (2011)
	Ao	Da, Ua, Dl	Mexico	López et al. (1999)
	Ao	Asa, Da, Ua	Brazil	Silva et al. (2010)
S. purpurea	Ao	Da, Dl*, Ua, Ai, Ch, Pv*	Mexico	Sivinski et al. (2000)
	Ao	Da	Brazil	Bittencourt et al. (2011)
	Af, Aso	Da	Brazil	Leal et al. (2009)
	AN, Af, Ao	Da	Domin. Rep. Rep.	Serra et al. (2011)
	Aso	Asa, Da, Ua	Brazil	Silva et al. (2010)
	Ao	Da, Ua	Mexico	López et al. (1999)
S. radkolferi	Ao	Da, Dl, Ua	Mexico	Sivinski et al. (2000)
Spondias spp.	Af, Ao	Da	Brazil	Silva et al. (2010)
	Ao, Af, Al, Ad	Da, Ua, Ap Oa, Dl	Mexico	Montoya et al. (2017)

Ad, *Anastrepha distincta* Greene; Af, *Anastrepha fraterculus* Wiedemann; Al, *Anastrepha ludens* Loew; AN, *Anastrephas* spp.; Ap, *Aganaspis pelleranoi*; Ao, *Anastrepha obliqua*; Asa, *Asobara anastrephae*; Aso, *Anastrepha sororcula* Zucci; Ch, *Coptera haywardi* Oglobin; Da, *Doryctobracon areolatus*; Dl = *Diachasmimorpha longicaudata*; Oa, *Odontosema anastrephae*; Pv, *Pachycrepoideus vindemiae*; Ua, *Utetes anastrephae*.

intrinsic competition against other first instar parasitoid larvae of species such as *D. areolatus* and *D. longicaudata* (Murillo et al. 2016). In Mexico, it is considered as a potential biocontrol agent, given that it can be mass reared using irradiated third instar larvae of *A. ludens* as hosts (Cancino et al. 2009).

19.4.2 *Doryctobracon areolatus* (Szépligeti) (Hymenoptera: Braconidae)

Doryctobracon areolatus is a synovigenic and koinobiont endoparasitoid of tephritid larvae-pupae (Stul and Sivinski 2012; Garcia and Ricalde 2013; Garcia et al. 2017). It is the native species with the highest natural parasitism in *Anastrepha* flies (Table 19.1) in the neotropics (Ovruski et al. 2000, Garcia and Ricalde 2013; García et al. 2017). This species also parasitizes early host life stages such as eggs and first instar larvae (Murillo et al. 2015). Its larval development is completed in 22–25 days; however, it takes longer when it attacks younger host stages. Diapause is common, but it can vary in relation to the host fruit of origin (Aluja et al. 1998; Carvalho 2005).

Doryctobracon areolatus is distributed from Florida, United States, to the north of Argentina, and it coexists with *U. anastrephae*, *Doryctobracon brasiliensis* (Szépligeti), *Opius bellus* (Gahan), *A. pelleranoi*, *O. anastrephae*, and the exotic parasitoids *D. longicaudata* and *A. indica*. The species that it attacks most frequently are *A. obliqua* and *A. fraterculus*, but it also parasitizes *C. capitata* and *R. pomonella* (Table 19.1). Although its reproduction has been documented in irradiated larvae of *A. suspensa* (Palenchar et al. 2009) and *A. ludens* (Aluja et al. 2009), there is no economically feasible method for its mass production.

19.4.3 *Aganaspis pelleranoi* (Brèthes) (Hymenoptera: Eucolidae)

This species is a koinobiont endoparasitoid whose females differentiate from males by having shorter antennae (Guimarães et al. 2000). Its average life span is of approximately 20 days at 25°C, and it is able to reproduce from the first day of emergence, producing an average of two offspring per day (Ovruski 1994a). This parasitoid also has the capacity of entering into diapause (Carvalho 2005). Females oviposit directly on the hosts. They enter the fruit by perforating the epi- and mesocarp with their mandibles or by taking advantage of holes originated by other factors (Ovruski 1994b).

Along the neotropical region (Table 19.1), this species attacks lonchaeidae and tephritid flies (Wharton 1998) such as *A. obliqua*, *A. ludens*, *Anastrepha fraterculus* (Wiedemann), *Anastrepha striata* (Schiner), and *C. capitata*. *Aganaspis pelleranoi* can be easily reared in the laboratory using third stage larvae of *A. ludens*, *C. capitata*, and *Aganaspis fraterculus* (Ovruski 1994a; Cancino et al. 2009).

19.4.4 *Odontosema anastrephae* Borgmeier (Hymenoptera: Figitidae)

Odontosema anastrephae is a solitary koinobiont, ecto-endoparasitoid with arrhenotokous populations; however, thelytokous populations have been recorded (Ramirez-Romero et al. 2011; Copeland et al. 2010). Females of both types of populations respond to guavas with larvae of *A. ludens* and exhibit similar patterns of host searching and choosing behaviors (Ramirez-Romero 2011). Under laboratory conditions, the development of immatures takes 30 days in larvae of *A. ludens* (Cancino et al. 2009). Immatures can enter diapause as third instar larvae in host pupae (Aluja et al. 1998). This species has been reproduced in larvae of *A. ludens* by exposing the hosts in open units to favor the contact between foraging females and available hosts. The development of *O. anastrephae* in irradiated hosts is not suitable yet (Cancino et al. 2009).

19.5 CASE STUDY: RELEASES OF *U. ANASTREPHAE* AND *D. LONGICAUDATA* TO CONTROL *A. OBLIQUA* PEST POPULATIONS IN *SPONDIAS MOMBIN* L.

We evaluated the impact of *U. anastrephae* and *D. longicaudata* releases on *A. obliqua* populations infesting *S. mombin* in Tuzantan, Chiapas, Mexico (15° 05′N and 92° 26′W, altitude of 81 m.a.s.l., average temperature of 27.6°C, and precipitation of 3,304 mm) (CLIMATE-DATA.ORG). The main objective was to determine which parasitoid species is more suitable to control *A. obliqua* populations infesting *Spondias* fruits. The release zone was located near a mango-growing zone *cv* Ataulfo in Huehuetan, Chiapas. The evaluation was conducted from June to September 2013, during the fructification season of this fruit.

Utetes anastrephae was reared in the Biological Control laboratory of the Moscafrut Program SADER-IICA, located in Metapa de Dominguez, Chiapas, Mexico, according to Cancino et al. (2009). Adults of *D. longicaudata* were provided by the Moscafrut facility where they are produced at a rate of 20 million parasitized pupae per week (Cancino et al. 2010). For both species, 9-day-old *A. ludens* larvae, previously irradiated at 4.5 Krads, were used as hosts (Cancino et al. 2009).

The treatments were: (i) release of *U. anastrephae* at a density of approximately 1,500 adults per ha, (ii) release of *D. longicaudata* at the same density as *U. anastrephae*, (iii) concurrent releases of both parasitoid species, and (iv) a control treatment without parasitoid releases. The experimental unit was a tree of *S. mombin* chosen at random and separated by at least 500 m from the rest of the trees included in the experiment. Each treatment consisted of two trees. Parasitoids were packed for release using plastic containers of 19 L, where 2,500 pupae close to adult emergence were placed with food consisting of honey mixed with soft paper (Montoya et al. 2012). We obtained close to 1,500 parasitoids (c.a. two females to 1 male) that were kept at 25 ± 1°C and 70%–80% relative humidity (RH) until females reached 7 days of age to be released in the field. Containers with parasitoids were transported to the field in a van at 20 ± 2°C. Parasitoids were released in the drop zone of trees according to each treatment between 7:00 and 9:00 am. Releases were performed weekly during the months of *Spondias* spp. fructification.

Four days after the releases, we sampled approximately 300 g of fruit per tree. Fruits were taken to the laboratory to obtain second and third instar larvae, which were kept in cylindrical plastic containers (3.3 cm high × 8 cm d) with humid vermiculite at 25 ± 1°C and 70%–80% RH until adult emergence. A total of 28 replicates were carried out. A replicate consisted of a parasitoid release plus the respective fruit sampling (see Cancino et al. 2018). We registered the number of adult flies and parasitoids that emerged to obtain emergence and parasitism percentages.

Estimations of parameters were as follows: *Total emergence percentage* = [(number of parasitoids + number of flies)/number of larvae] × 100; *Parasitism percentage* = [number of parasitoids/(number of parasitoids + number of flies)] × 100; *Fly emergence percentage* = [number of flies/(number of parasitoids + number of flies)] × 100.

19.5.1 Data Analysis

Data were analyzed using one-way analysis of variance (ANOVA). Number of larvae per fruit and number of larvae per kilogram of fruit were previously square-root transformed, and emergence percentage and parasitism emergence were arc-sine transformed to satisfy the assumptions of normality. Means were compared using Tukey's test with a significance level of 0.05. We also calculated indexes of relative abundance and richness (diversity) of species determined by the Shannon index (Ho) using the formulas described by Smith and Smith (2001).

19.5.2 Results and Discussion

The number of sampled fruits and the number of larvae extracted from dissected fruits are shown in Table 19.2. *Anastrepha obliqua* was the only fruit fly species that emerged from the sampled fruits, with a notable natural parasitism in the control treatment (Table 19.3). *Utetes anastrephae*

TABLE 19.2
Number of Sampled Fruits, Weight of Samples, and Number of *Anastrepha obliqua* Larvae Obtained during Releases of *Utetes anastrephae* and *Diachasmimorpha longicaudata* to Control *A. obliqua* Populations in *Spondias mombin.* Tuzantan Chiapas, June–September 2013

Treatments	Number of Fruits	Weight of Sample (g)	Number of Larvae
U. anastrephae	2717	15299	1517
D. longicaudata	1835	10916	2043
D. longicaudata + *U. anastrephae*	1694	9350	1035
Control	1792	9780	1386

TABLE 19.3
Number of Larvae per Fruit, Number of Larvae per Kilogram of Fruit, Parasitism Percentage (Mean ± SE), and Diversity Index during Releases of *Utetes anastrephae* and *Diachasmimorpha longicaudata* to control populations of *Anastrepha obliqua* in *Spondias mombin.* Tuzantan Chiapas, June–September 2013

Released Species	Number of Larvae/Fruit	Number of Larvae/Kg Fruit	Parasitism Percentage	Diversity (H′)
U. anastrephae	0.71 ± 0.06 AB	153.86 ± 16.79 AB	50.87 ± 5.5 AB	1.1034
D. longicaudata	1.18 ± 0.18 A	194.8 ± 25.18 A	63.86 ± 5.6 A	0.9496
D. longicaudata + *U. anastrephae*	0.59 ± 0.09 B	111.73 ± 16.90 B	50.76 ± 5.9 AB	1.2557
Control	0.73 ± 0.13 AB	133.27 ± 26.26 AB	38.53 ± 5.9 B	1.0056

Different letters between rows indicate significant differences among treatments. ANOVA and Tukey tests with 95% significance. ANOVA, analysis of variance.

was the native species with the highest relative abundance index (Figure 19.1), coexisting with *D. areolatus*, *A. pelleranoi*, and the exotic *D. longicaudata*. Species richness was not affected by the releases of each species separately, but there was an effect when they were released together (treatment 3). *longicaudata* was the species with the greatest impact on the abundance of *A. obliqua* (Figure 19.1). The relative abundance of *D. areolatus* was lower when *U. anastrephae* was released; whereas *A. pelleranoi*, although present in all treatments, decreased with the release of *U. anastrephae* and was not recovered when *D. longicaudata* and *U. anastrephae* were released together.

Total emergence percentage was not different between treatments ($F_{3,110} = 0.5817$, $P = 0.6283$). The number of larvae per fruit and the number of larvae per kilogram of fruit were higher in the release zones of both parasitoids (number of larvae/fruit, $F_{3,110} = 4{,}062$, $P < 0.0001$; number of larvae per kilogram, $F_{3,110} = 3.5455$, $P = 0.0169$) (Table 19.3). The emergence of *A. obliqua* was reduced numerically in the three treatments with parasitoid releases, but it was significant only when *D. longicaudata* was released by itself ($F_{3,110} = 2.944$, $P = 0.0321$) (Table 19.3). Similarly, parasitism was greater in treatments with parasitoid releases, but it was significant only when *D. longicaudata*

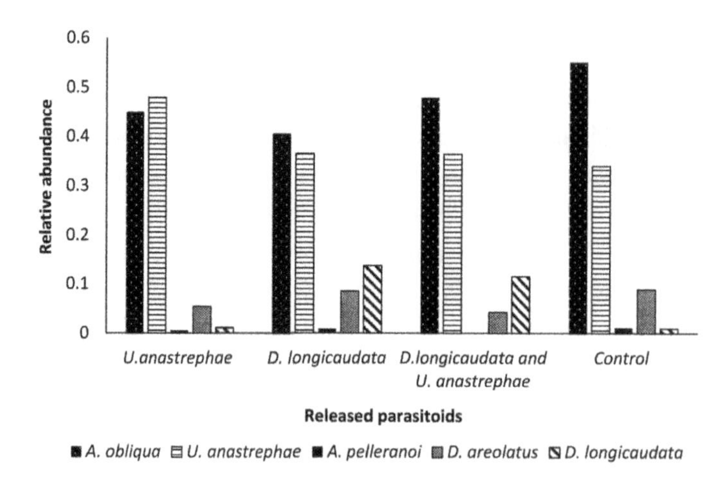

FIGURE 19.1 Relative abundance of parasitoid species during the releases of *Utetes anastrephae* and *Diachasmimorpha longicaudata* in *Anastrepha obliqua* populations in *Spondias mombin,* June–September 2013, Tuzantan Chiapas, Mexico.

was released by itself ($F_{3,110} = 2.9444$, $P = 0.0362$) (Table 19.3). In the control treatment, the most abundant parasitoid in the study region was *U. anastrephae*, attacking *A. obliqua* in *S. mombin*; although in other reports, the presence of *D. areolatus* was reported as predominant (Murillo et al. 2015; Montoya et al. 2017). The release of *U. anastrephae* increased the parasitism but not in a significant way, and it negatively influenced the abundance of *D. areolatus*. The concurrent release of both species of parasitoids seems debatable since it did not increase the percentage of parasitism compared to *D. longicaudata* acting by itself. Only the release of *D. longicaudata* by itself caused a significant reduction in the population of *A. obliqua*.

It is important to note that there was a high parasitism percentage observed in the control treatment, as well as a high diversity of species of parasitoids registered, which agrees with previous reports (e.g., Aluja et al. 2014; Montoya et al. 2016) that highlight the importance that *S. mombin* trees may have as reservoirs of native parasitoids of fruit flies.

19.6 CONCLUSIONS

The association of *A. obliqua* with *Spondias* fruits along natural settings strongly influences the presence of this pest in commercial areas of mango and other marketable fruits. Fortunately, the management of this pest with parasitoids, whether native or introduced, seems to be a suitable and ecologically convenient strategy. High levels of parasitism can be achieved in *Spondias* spp. hosts to diminish the numbers of *A. obliqua* flies that later will invade mango commercial areas. Studies on the population dynamics of *A. obliqua* and its associated native parasitoids in *Spondias* spp. fruits can improve the design of biological control strategies.

ACKNOWLEDGMENTS

We are grateful to Jorge Villalobos and Fredi Gálvez (CESAVE, Chiapas) who contributed with suggestions and activities during field sampling. We also thank Cesar Galvez, Velisario Rivera, and Patricia Rosario for technical support, and Salvador Flores (Programa Moscafrut, SADER-SENASICA) for statistical advice.

REFERENCES

Alia-Tejacal, I., Y. I. Astudillo-Maldonado, C. A. Nuñez-Colín, L. A. Valdez-Aguilar, S. Bautista-Baños, E. García-Vázquez, R. Ariza-Flores, and F. Rivera-Cabrera. 2012. Caracterización de frutos de ciruela mexicana (*Spondias purpurea* L.) del Sur de México. *Revista Fitotecnia Mexicana* 35: 21–26.

Aluja, M. 1994. Bionomics and management of *Anastrepha*. *Annual Review Entomology* 39: 155–178.

Aluja, M., and A. Birke. 1993. Habitat use by adults of *Anastrepha obliqua* (Diptera: Tephritidae) in a mixed mango and tropical plum orchard. *Annals of the Entomological Society of America* 86: 799–812.

Aluja, M., F. Díaz-Fleisher, D. R. Papaj, G. Lagunes, and J. Sivinski. 2001. Effect of age, diet, female density, and the host resource on egg load in *Anastrepha ludens* and *Anastrepha obliqua* (Diptera: Tephritidae). *Journal of Insect Physiology* 47: 975–988.

Aluja, M., M. López, and J. Sivinski. 1998. Ecological evidence for diapause in four native and one exotic species of larval-pupal fruit fly (Diptera: Tephritidae) parasitoids tropical environments. *Annals of the Entomological Society of America* 91(6): 821–833.

Aluja, M., J. Sivinski, S. Ovruski, L. Guillen, M. Lopez, J. Cancino, A. Torrez-Anaya, G. Gallegos-Chan, and Ruiz, L. 2009. Colonization and domestication of seven species of native New World hymenopterous larval-prepupal and pupal fruit fly (Diptera: Tephritidae) parasitoids. *Biocontrol Science and Technology* 19: 49–79.

Aluja, M., J. Sivinski, R. Van Driesche, A. Anzures-Dadda, and L. Guillén. 2014. Pest management through tropical tree conservation. *Biodiversity and Conservation* 23: 831–853.

Arce-Romero, A. R., A. I. Monterroso-Rivas, J. D. Gómez-Díaz, and A. Cruz-León. 2017. Mexican plums (*Spondias* spp.): their current distribution and potential distribution under climate change scenarios for México. *Revista Chapingo Serie Horticultura* 13: 5–19.

Bittencourt, M. A. L., A. C. M. Da Silva, V. E. S. Silva, Z. V. Bomfim, J. A. Guimares, M. F. de Souza Filho, and E. L. Arajo. 2011. Moscas-das-frutas (Diptera: Tephritidae) e seus parasitoides (Hymenoptera: Braconidae) associados às plantas hospedeiras no sul da Bahia. *Neotropical Entomology*, 40(3): 405–406.

CABI. 2017. Invasive species compendium. www.cabi.org/isc/datasheet/5659.

Campos-Carbajal, S. E. 2017. Infección de adultos de *Anastrepha ludens* y *A. obliqua* con diseminadores de conidios de *Beauveria bassiana* en campo. Tesis de Maestría, El Colegio de la Frontera Sur, Tapachula, Chiapas, México, 42p.

Cancino, J., C. Gálvez, A. López, U. Escalante, and P. Montoya. 2018. Best timing to determine field parasitism by released *Diachasmimorpha longicaudata* (Hymenoptera: Braconidae) against *Anastrepha* (Diptera: Tephritidae) pest populations. *Neotropical Entomology* 48: 143–151.

Cancino, J., L. Ruiz, P. López, and F. M. Moreno. 2010. Cría masiva de parasitoides. In: *Moscas de la fruta: Fundamentos y procedimientos para su manejo*, ed. P. Montoya, J. Toledo, and E. Hernández, 291–306. Mexico: S y G Editores.

Cancino, J., L. Ruiz, J. Sivinski, F. O. Gálvez, and M. Aluja. 2009. Rearing of five hymenopterous larval-prepupal (Braconidae, Figitidae) and three pupal (Diapriidae, Chalcidoidea, Eurytomidae) native parasitoids of the genus *Anastrepha* (Diptera: Tephritidae) on irradiated *A. ludens* larvae and pupae. *Biocontrol Science and Technology* 19: 193–209.

Carvalho, R. D. S. 2005. Diapause in fruit fly parasitoids in the Recôncavo Baiano, Brazil. *Neotropical Entomology* 34: 613–618.

CLIMATE-DATA-ORG. 2018. https://es.climate-data.org/america-del-norte/ mexico/ chiapas/ tuzantan-359392 /#climate-graph.

Copeland C. S., M. A. Hoy, A. Jeyaprakash, M. Aluja, R. Ramirez-Romero, and J. M. Sivinski. 2010. Genetic characteristics of bisexual and female-only populations of *Odontosema anastrephae* (Hymenoptera: Figitidae). *Florida Entomologist* 93: 437–443.

Díaz-Fleischer, F., D. Pérez-Staples, H. Cabrera-Mireles, P. Montoya, and P. Liedo. 2017. Novel insecticides and bait stations for control of *Anastrepha* fruit flies in mango orchards. *Journal of Pest Science* 90(3): 865–872.

Flores, S., L. Gomez, and P. Montoya. 2011. Residual control and lethal concentrations of GF-120 (spinosad) for *Anastrepha* spp. fruit flies (Diptera: Tephritidae). *Journal of Economic Entomology* 104: 1885–1891.

Flores, S., E. Gómez, S. Campos, F. Gálvez, J. Toledo, P. Liedo, R. Cardoso, and P. Montoya. 2017. Evaluation of mass trapping and bait stations to control *Anastrepha* fruit flies (Diptera: Tephritidae) in mango Francis orchards of Chiapas, Mexico. *Florida Entomologist* 100: 358–365.

Garcia, F., and M. Ricalde. 2013. Augmentative biological control using parasitoids for fruit fly management in Brazil. *Insects* 4(1): 55–70.

Garcia, F., A. Brida, L. Martins, L. Abeijon, and J. Lutinski. 2017. Biological control of fruit flies of the genus *Anastrepha* (Diptera: Tephritidae): current status and perspectives. In: *Biological Control: Methods, Applications and Challenges*, org. L. Davenport, 29–71. Hauppauge, NY: Nova Science.

Guimarães, J. A., N. B. Diaz, and R. A. Zucchi. 2000. Parasitóides (Figitidae: Eucoilinae). In *Moscas-das-frutas de importância econômica no Brasil: Conhecimento básico e aplicado*, ed. A. Malavasi and R.A. Zucchi, 127–134. Ribeirão Preto, Brazil: Holos Editora.

Gutiérrez-Ruelas, J. M. 2009. Programa Nacional contra Moscas de la Fruta (Situación actual y perspectivas). In: *Memorias del XVIII Curso Internacional sobre Moscas de la Fruta, 27 Agosto-14 Septiembre 2009, Centro Internacional de Capacitación en Moscas de la Fruta, Programa Moscamed-Moscafrut*, 1–9. Metapa de Domínguez, Chiapas, México: DGSV-SENASICA-SAGARPA.

Hauck, J., A. Rosenthal, R. Deliza, R. L. de Oliveira, and S. Pacheco. 2011. Nutritional properties of yellow mombin (*Spondias mombin* L.) pulp. *Food Research International* 44: 2326–2331.

Hernández-Ortiz, V., and R. Pérez-Alonso. 1993. The natural host plants of *Anastrepha* (Diptera: Tephritidae) in a tropical rain forest of Mexico. *Florida Entomologist* 76: 447–460.

Jenkins, D. A., and R. Goenaga. 2008. Host breadth and parasitoids of fruit flies (*Anastrepha* spp.) (Diptera: Tephritidae) in Puerto Rico. *Environmental Entomology* 37: 110–120.

Jesus-Barros, C. R., R. Adaime, M. N. Oliveira, W. R. Silva, S. V. Costa-Neto, and M. F. Souza-Filho. 2012. *Anastrepha* (Diptera: Tephritidae) species, their hosts and parasitoids (Hymenoptera: Braconidae) in five municipalities of the State of Amapá, Brazil. *Florida Entomologist* 95: 694–705.

Leal, M. R., S. A. D. S. Souza, E. D. L. Aguiar-Menezes, M. Lima Filho, and E. B. Menezes. 2009. Diversidade de moscas-das-frutas, suas plantas hospedeiras e seus parasitóides nas regiões Norte e Noroeste do Estado do Rio de Janeiro, Brasil. *Ciência Rural*, 39(3).

Leite, S. A., M. A. Castellani, A. E. L. Ribeiro, D. R. D. Costa, M. A. L. Bittencourt, and A. A. Moreira. 2017. Fruit flies and their parasitoids in the fruit growing region of Livramento de Nossa Senhora, Bahia, with records of unprecedented interactions. *Revista Brasileira de Fruticultura* 39(4): 592–602.

López, M., M. Aluja, and J. Sivinski. 1999. Hymenopterous larval-pupal and pupal parasitoids of *Anastrepha* flies (Diptera: Tephritidae) in México. *Biological Control* 15: 119–129.

López-Guillén, G., A. Virgen, and J. Rojas. 2009. Color preference of *Anastrepha obliqua* (McQuart) (Diptera: Tephritidae). *Revista Brasileira de Entomología* 53: 157–159.

Mangan, R. L., D. B. Thomas, A. T. Moreno, and D. Robacker. 2011. Grapefruit as a host for the West Indian fruit fly (Diptera: Tephritidae). *Journal of Economic Entomology* 104: 54–62.

Murillo, F. D., H. Cabrera-Mireles, J. F. Barrera, P. Liedo, and P. Montoya. 2015. *Doryctobracon areolatus* (Hymenoptera, Braconidae) a parasitoid of early developmental stages of *Anastrepha obliqua* (Diptera: Tephritidae). *Journal Hymenoptera Research* 46: 91–115.

Murillo F. D., P. Liedo, M. G. Nieto-López, H. Cabrera-Mireles, J. F. Barrera, and P. Montoya. 2016. First instar larvae morphology of Opiinae (Hymenoptera: Braconidae) parasitoids of *Anastrepha* (Diptera: Tephritidae) fruit flies. Implications for interspecific competition *Arthropod Structure & Development* 45: 294–300.

Montoya, P., P. Liedo, B. Benrey, J. Cancino, J. F. Barrera, J. Sivinski, and M. Aluja. 2000. Biological control of *Anastrepha* spp. (Diptera: Tephritidae) in mango orchards through augmentative releases of *Diachasmimorpha longicaudata* (Ashmead) (Hymenoptera: Braconidae). *Biological Control* 18: 216–224.

Montoya, P., J. Cancino, M. Zenil, G. Santiago, and J. M. Gutiérrez. 2007. The augmentative biological control component in the Mexican Campaign against *Anastrepha* spp. fruit flies. In: *Area-Wide Control of Insect Pests: From Research to Field Implementation*, ed. M. J. B. Vreysen, A. S. Robinson, and J. Hendrichs, 661–670. Dordrecht, the Netherlands: Springer.

Montoya, P., J. Cancino, and L. Ruiz. 2012. Packing of fruit fly parasitoids for augmentative releases *Insects* 3: 889–899.

Montoya, P., A. Ayala, P. López, J. Cancino, H. Cabrera, J. Cruz, A. M. Martínez, I. Figueroa, and P. Liedo. 2016. Natural parasitism in fruit fly populations in disturbed areas adjacent to commercial mango orchards in Chiapas and Veracruz, Mexico. *Environmental Entomology* 42: 328–337.

Montoya, P., P. López, J. Cruz, F. López, C. Cadena, J. Cancino, and P. Liedo. 2017. Effect of *Diachasmimorpha longicaudata* releases on the native parasitoid guild attacking *Anastrepha* spp. larvae in disturbed zones of Chiapas, Mexico. *BioControl* 62: 581–593.

Neven, L. G. 2010. Postharvest management of insects in horticultural products by conventional and organic means, primarily for quarantine purpose. *Steward Postharvest Review* 4: 1–11.

Orozco-Dávila, D., Quintero, L., Hernández, E., Solís, E., Artiaga, T., Hernández, R., Ortega, C. and Montoya, P. 2017. Mass rearing and sterile insect releases for the control of Anastrepha spp. pests in Mexico–a review. *Entomologia Experimentalis et Applicata* 164(3): 176–187.

Ovruski S. M. 1994a. Immature stages of *Aganaspis pelleranoi* (Hymenoptera: Cynipoidea: Eucolidae), a parasitoid of *Ceratitis capitata* (Wied) and *Anastrepha* spp. (Diptera: Tephritidae). *Journal of Hymenoptera Research* 3: 233–239.

Ovruski S. M. 1994b. Comportamiento en la detección del huésped de *Aganaspis pelleranoi* (Hymenoptera:Eucolidae), parasitoide de larvas de *Ceratitis capitata* (Diptera: Tephritidae). *Revista de la Sociedad Entomológica Argentina* 53: 121–127.

Ovruski, S., M. Aluja, J. Sivinski, and R. Wharton. 2000. Hymenopteran parasitoids on fruit-infesting Tephritidae (Diptera) in Latin America and the southern United States: diversity, distribution, taxonomic status and their use in fruit fly biological control. *Integrated Pest Management Reviews* 5: 81–107.

Palenchar, J., T. Holler, A. Moses-Rowley, R. McGovern, and J. Sivinski. 2009. Evaluation of irradiated Caribbean fruit fly (Diptera: Tephritidae) larvae for laboratory rearing of *Doryctobracon areolatus* (Hymenoptera: Braconidae). *Florida Entomologist* 92: 535–537.

Peña, J. E., A. I. Mohyuddin, and M. Wysoki. 1998. A review of the pest management in mango agroecosystem. *Phytoparasitica* 26:129.

PNMF-DGSV (Programa Nacional Moscas de la Fruta-Dirección General de Sanidad Vegetal). 2017. Informe del Programa de Exportación de Mango a los Estados Unidos con Tratamiento Hidrotérmico, *Dirección del Programa Nacional de Moscas de la Fruta*, SENASICA-SAGARPA, México DF, 46p.

Ramírez, B. C., P. Barrios, J. Z. Castellanos, A. Muñoz, G. Palomino, and E. Pimienta. 2008. Sistemas de producción de *Spondias purpurea* (Anacardiaceae) en centro-occidente de México. *Revista de Biología Tropical* 58: 675–687.

Ramirez-Romero, R., J. Sivinski, C. S. Copeland, and M. Aluja. 2011. Are individuals from thelytokous and arrhenotokous populations equally adept as biocontrol agents? Orientation and host searching behavior of a fruit fly parasitoid. *BioControl* 57: 427–440.

Rull, J., R. Wharton, J. L. Feder, L. Guillén, J. Sivinski, A. Forbes, and M. Aluja. 2009. Latitudinal variation in parasitoid guild composition and parasitism rates of North American hawthorn infesting Rhagoletis. *Environmental Entomology* 38: 588–599.

SAGARPA/SENASICA. (2012). Informe mensual de la Campaña Nacional contra Moscas de la Fruta del mes de diciembre de 2012. file:///D:/Downloads/12_INFORME_CNCMF_DICIEMBRE_2012.pdf.

Santiago-Martínez, G. 2008. Manejo de áreas libres de plagas. In: *Manejo Integrado de Plagas*, ed. J. Toledo and F. Infante, 227–236. Mexico: Ed. Trillas.

Segura, E., A. Hernández-Tinoco, M. Leyva, J. Chavero, and J. Cancino. 2016. Control biológico de *Anastrepha obliqua* (McQuart) en *Spondias* spp. silvestre como soporte del status de baja prevalencia en el área de producción de mango en Tecpan, Guerrero en la Costa Pacifico de México. Book of Abstracs, *9th Meeting of Tephritid Workers of the Western Hemisphere*. Buenos Aires, Argentina, p. 77.

Serra, C. A., M. Ferreira, S. García, L. Santana, M. Castillo, C. Nolasco, P. Morales, T. Holler, A. Roda, M. Aluja, and J. Sivinski. 2011. Establishment of the West Indian Fruit Fly (Diptera: Tephritidae) Parasitoid Doryctobracon areolatus (Hymenoptera: Braconidae) in the Dominican Republic. *Florida Entomologist*, 94(4): 809–817.

Silva, J. G., V. S. Dutra, M. S. Santos, N. M. O. Silva, D. B. Vidal, R. A. Nink, J. A. Guimarães, and E. L. Araujo. 2010. Diversity of *Anastrepha* spp. (Diptera: Tephritidae) and associated braconid parasitoids from native and exotic hosts in Southeastern Bahia, Brazil. *Environmental Entomology* 39: 1457–1465.

Sivinski, J., J. Pinero, and M. Aluja. 2000. The distributions of parasitoids (Hymenoptera) of *Anastrepha* fruit flies (Diptera: Tephritidae) along an altitudinal gradient in Veracruz, Mexico. *Biological Control* 18: 258–269.

Sivinski, J., K. Vulinec, and M. Aluja. 2001. Ovipositor length in a guild of parasitoids (Hymenoptera: Braconidae) attacking *Anastrepha* spp. fruit flies (Diptera: Tephritidae) in southern Mexico. *Annals of the Entomological Society of America* 94: 886–895.

Smith R.L., and T. M. Smith. 2001. *Ecología*. Madrid: Pearson Educación S.A.

Sousa, L. D. S., P. R. R. Silva, M. P. P. Nascimento, S. Franca, and A. A. R. Araujo. 2017. Fruit flies (Diptera: Tephritidae) and their parasitoids associated with different hog plum genotypes in Teresina, Piauí. *Revista Brasileira de Fruticultura* 39(4).

Steck, G. J. 2001. Concerning the occurrence of *Anastrepha obliqua* (Diptera: Tephritidae) in Florida. *Florida Entomologist* 84: 320–321.

Stul C. J., and J. Sivinski. 2012. Featured Creatures. *Utetes anastrephae* (Viereck) (Insecta: Hymenoptera: Braconidae). University of Florida, USA. http://entnemdept.ufl.edu.

Taira, T. L., A. R. Abot, J. Nicácio, M. A. Uchôa, S. R. Rodrigues, and J. A. Guimarães. 2013. Fruit flies (Diptera, Tephritidae) and their parasitoids on cultivated and wild hosts in the Cerrado-Pantanal ecotone in Mato Grosso do Sul, Brazil. *Revista Brasileira de Entomologia* 57(3): 300–308.

Toledo, J., S. E. Campos, S. Flores, P. Liedo, J. F. Barrera, A. Villaseñor, and P. Montoya. 2007. Horizontal transmission of *Beauveria bassiana* in *Anastrepha ludens* (Diptera: Tephritidae) under laboratory and field cage conditions. *Journal Economic Entomology* 100: 291–297.

Wharton R. 1988. Manual para los géneros de la familia Braconidae (Hymenoptera) del nuevo mundo. Subfamilia Opiinae. Ed. Wharton A, Robert; Marsh M, Paul; Sharkey J, Michael, (Ed. en Español por I. Mercado). *International Society Hymenopterist*. Washington.

Section VII

Area-Wide Integrated Pest Management and Action Programs

20 Holistic Pest Management

Juan F. Barrera[*]

CONTENTS

Abstract This chapter introduces holistic pest management (HPM), a decision-making system for pest management based on a holistic approach. The concept of HPM is presented and some of its methods are described. Unlike integrated pest management (IPM), the strategy of HPM is to reduce both the population densities of the pest and the vulnerability of the farmer and his farm, while it works to increase the producer's response capacity. As an example, the holistic management of coffee leaf rust (*Hemileia vastatrix*) is cited.

20.1 INTRODUCTION

How do you kill a werewolf? ... Of course! Mythology teaches that the only reliable way to kill a werewolf is with a silver bullet. They say that a wooden stake to the heart or direct sunlight works on vampires but not on werewolves. Weapons that are lethal to other creatures have no effect. The deaths of countless brave souls who confronted this mythic creature without the aid of silver bullets are the anecdotal evidence that supports this truth. Although preventing werewolf attacks could be as simple as not going outside during a full moon, or a full lunar cycle quarantine of all animal bite victims, costly silver bullets sell very well in a crisis. In a crisis, we want a guaranteed solution.

However, ignoring all the causes of a crisis and putting blind faith in traditional solutions developed in another era is unwise. The "silver bullet" is a metaphor that I will use to address the weaknesses of popular methods of pest management. Currently, pest management is strongly influenced by a reductionist approach (i.e., Lewis et al. 1997). That is, we reduce everything to a silver bullet. The consequence of this reductionist approach is that many pests continue to cause significant losses to producers as they did decades ago (Oerke 2006, Dhaliwal et al. 2015).

About 15 years ago, we wondered why control of the coffee berry borer, *Hypothenemus hampei*, an important pest of coffee worldwide, was so difficult. We also asked ourselves why the methods developed by the research institutions for the management of this pest were rarely used by producers. In other words, why were they ignoring our advice? In the case of Mexico, the majority of coffee producers are small landholders, and they do not have the resources to implement the methods (Segura et al. 2004). When we analyzed the situation, we could see that this problem was shared by many coffee-producing countries. Strictly speaking, small landholders have high vulnerability and poor response capacity to the threat of pests.

[*] Corresponding author.

From this diagnosis (Barrera et al. 2004), we understood the urgent need to change the approach to control pests. In particular, it seemed obvious to us that conventional pest control, called integrated pest management (IPM) was not the most appropriate approach. This assertion was even truer in the case of developing countries, where IPM is not used very often (Morse and Buhler 1997, Morse 2009, Parsa et al. 2014). When reviewing the literature on the subject, we found that our concern was not new or unique. Even some researchers, like Lester E. Ehler and Dale G. Bottrell (2000), were very critical of IPM. The agreement of these authors with our ideas reinforced our efforts to find an alternative approach to pest management. Various alternative approaches like ecological (NRC 1996), agroecological (Altieri et al. 1983), biointensive (Frisbie and Smith 1991), and total system management (Lewis et al. 1997) had already been suggested.

For this reason, we have been developing an alternative to IPM that we have called holistic pest management (HPM). As the name implies, the most particular characteristic of this new paradigm is its holistic approach. To report what it means and how it is implemented, this chapter has a dual objective: It will present the concept and methods of HPM, and it will stimulate reflection on how we have been dealing with pest management.

20.2 AN INTRODUCTION TO HPM

The holistic approach is based upon two central ideas. First, pest management actions must put the producer at the center of the system. Second, pest management must consider not only the pests but also the other important components of the system in question. This approach, based on the producers and the system in which they are immersed, is called holistic pest management, or HPM. What does this term mean? The term "holistic," first proposed by Jan Christiaan Smuts in 1926 (Smuts 1936), comes from the Greek word *holos* which means "everything, whole, or entire." Holistic is the idea that systems and their properties should be viewed as wholes and not just as a collection of parts. HPM is defined as a participatory regional system of decision making in pest management. It is aimed at the well-being of the human population. It implements low impact and quality processes and products for self-consumption and competition in the market. These processes and products are generated from integral production systems. They are managed by the producer as a strategy to focus on the causes of pest outbreaks. The producer implements these tactics to minimize the economic, environmental, and social costs derived from pest outbreaks and their mismanagement (Barrera 2006, 2007).

What do I mean by mismanagement? I will start by exposing some characteristics of IPM that limit its effectiveness and justify the search for an alternative approach (see Table 20.1 for a comparison between IPM and HPM approaches). First, IPM is limited in its effectiveness because the producer is not the priority. My experience has taught me that the fundamental thing must be to satisfy the needs of the producer so that he lives well. Because IPM does not care about him, he does not care about IPM, and so he is not careful in its implementation. Second, the producer is not the decision maker who defines the needs, generates the information and, with that knowledge, implements the programs. Third, the old paradigm is essentially reactive because it emerged as a response to reduce the excessive use and misuse of pesticides. This characteristic has led to give little attention to prevention, a basic aspect in pest management. Fourth, decision making in IPM is based on action thresholds. Decision making is based almost entirely on the population densities of pests and the damage they cause. This approach overlooks other important components of the system. Therefore, the IPM approach is reductionist. Fifth, IPM relies on silver bullets, which will provide guaranteed solutions without regard to circumstance. However, as silver bullets focus on the symptoms and not on the causes, they rarely solve the problems. Therefore, silver bullets trigger never-ending processes such as the substitution of pesticides. Sixth, IPM is oriented to situations of intensive agriculture and production for the market. Its methods do not integrate well with agroecological actions. Farmers focused on intensive production do not focus on production for self-consumption. And seventh, IPM operates mainly at a small scale like a farm, making it difficult to implement it at a community or regional scale.

TABLE 20.1
Comparison between IPM and HPM Approaches in Agricultural Production Systems

Approach	IPM	HPM
Priority	Management focused on the pest. IPM emphasizes the management of pests through harmonious integration of various control methods with the least economic, environmental, and social impact.[1] Even the "biointensive IPM," whose approach is more holistic, complex, and sensitive to change, is based on understanding the ecology of the pest.[2] However, by favoring attention to pests, attention to the farmer usually becomes secondary; in general, this situation is taken to the extreme in area-wide management programs,[3] particularly in cases of pest eradication[4] where people living in the region where these programs operate are rarely taken into account.[5]	Management focused on improving the well-being of the farmer. The ultimate purpose should not be the pest and not even the crop, but the viability of rural life as a whole.[6] The root of HPM comes from "holistic management," which proposes to manage natural resources, promote biodiversity, improve production, and generate financial strength to improve the quality of life while conserving the environment.[7] For this, it is necessary to resolve economic, social, and environmental aspects that limit the ways of life[8] of farmers. As a consequence of improving the well-being of the farmer (i.e., increased income), the conditions for carrying out pest management are improved.[9]
Participation	The farmer almost always acts as a receiver. The farmer is a user of information on pest management. In general, processes follow a top-down (i.e., government, technicians to farmers) pathway.[10] Typically, technicians employed by the government bring technological packages to farmers.[11]	The farmer receives and contributes. The farmer participates in the information-generation process for pest management.[12] Emphasis is placed on action-research processes[13] and training through Farmer Field Schools.[14] Knowledge and perceptions of farmers play an important role in pest management.[15]
Response	Reactive response. It is based on management of symptoms caused by pests (i.e., damage, population outbreaks). The focus on symptoms leads to this question: What should I do to control the pest?[16] Answering this question promotes the use and excessive use of pesticides (i.e., chemical, biological, genetically modified organisms).[17] A reactive response leads to a reductionist approach.	Preventive response. It is based on managing the causes that trigger pests. Understanding the causes that give rise to a problem of pests leads to the question: Why is the pest a pest?[18] Answering this question requires knowing the structure and understanding the function and dynamics of the system, very much in the manner proposed by ecologically based pest management.[19] A preventive response leads to a holistic approach.
Decision making	Decision making is based on action thresholds (Economic Injury Level and Economic Threshold). The Economic Injury Level (EIL), defined as the lowest pest population density that will cause economic damage, is calculated with information on cost of control (C), value of the product in the market (V), losses attributable to the pest (I, D), and the cost of pest control (K), through this equation: $EIL = C/VIDK$. The Economic Threshold (ET) is the density at which control measures should be determined to prevent an increasing pest population from reaching the EIL. In other words, ET is the operational criterion that determines whether a management action against a pest is required. These thresholds, most useful in the application of therapeutic practices, are widely recognized as the most important concepts of IPM.[20]	Decision making is based on risk. The risk, which can be defined as the likelihood of an adverse event and the magnitude of the consequences,[21] is obtained through the Holistic Risk Index (HRI). The components of HRI are threat (A), vulnerability (V), and response capacity (C), and is calculated with $HRI = (A + V)/C$.[22] The threat may be the occurrence of one (or more) pest that acts on certain conditions of vulnerability. Vulnerability is integrated with the characteristics of producers and farms, which determine their degree of exposure to a threat. Response capacity refers to the attributes and mechanisms of producers (and other stakeholders) to reduce the risks and survive, resist, and recover from the damage. In this approach, the action thresholds are one of several indicators used to calculate the components of HRI.

(*Continued*)

TABLE 20.1 (*Continued*)
Comparison between IPM and HPM Approaches in Agricultural Production Systems

Approach	IPM	HPM
Strategy	Search for silver bullet products and tactics. This begins with the idea that each problem has a unique and effective solution. These solutions are temporary. When the solution fails, the replacement is sought. This strategy leads to input substitution, that is, to permanently search for the silver bullet.[23]	Increase resilience. This approach promotes the health of the system. To a large extent, the resilience of the system is increased by reducing the risk of the threat, while reducing vulnerability and increasing the response capacity of the farmer and other stakeholders involved in the system.[24]
Production system	Modern industrial agriculture. IPM tends to be more efficient in monoculture production systems, generally for export or conventional markets,[25] a situation that occurs mostly in developed countries (IPM emerged in the context of US agriculture).[26] These systems depend, to a large extent, on using large quantities of external inputs (pesticides and fertilizers) to produce large volumes of products per unit area. Therefore, IPM is considered as an essential tool to reduce dependence on pesticides.[27]	Agroecological agriculture. HPM tends to be more efficient in diversified systems, either for the export of products in special markets (i.e., organic, fair trade) or for self-consumption. These systems with greater diversification of species—which are less prone to pest invasion and pest outbreaks[28]—make use of few external inputs because they strengthen natural processes (i.e., water and nutrient cycles, beneficial fauna and flora) and prioritize the conservation of natural resources and quality over quantity.[29] The ultimate goal is not to eliminate the pests but to keep them below the level of economic damage.[30]
Scale	From small to large scale. However, the farm is almost always the unit in which decisions are made.[31] In some cases, it can be implemented in large areas with the participation of many individual farms (area-wide approach,[32] i.e., Mediterranean fruit fly program[33]).	From small to large scale. The size of the scale does not matter in the implementation. However, to be implemented in large areas, it is desirable to do it in coordination with farmer organizations.[34]

HPM, holistic pest management; IPM, integrated pest management.
[1]Bottrell (1979), Kogan (1998), Norris et al. (2003), Prokopy and Kogan (2003), Gray et al. (2009); [2]Benbrook (1996); [3]Lindquist (2000), Hendrichs et al. (2007); [4]Walters et al. (2009); [5]Klassen (2000); [6]Levins (2007); [7]Savory and Butterfield (1999); [8]Herrera et al. (2019); [9]Barrera (2006); [10]Thrupp (1996); [11]Morse and Buhler (1997); [12]Thrupp (1996); [13]van de Fliert and Braun (2002); [14]Braun et al. (2000); [15]Bentley and Andrews (1996), Segura et al. (2004), Liebig et al. (2016), Munyuli et al. (2017); [16]Lewis et al. (1997); [17]Hokkanen (2015); [18]Lewis et al. (1997); [19]NRC (1996), Menalled et al. (2004); [20]Stern et al. (1959), Higley and Pedigo (1996), Higley and Peterson (2009); [21]Griffin (2012); [22]Barrera et al. (2007), Barrera et al. (2011); [23]Altieri and Nicholls (2005); [24]Barrera et al. (2007), Montalba et al. (2013), Henao-Salazar (2013), Machado-Vargas et al. (2018); [25]Morse (2009); [26]Parsa et al. (2014); [27]Benbrook (1996), Aselage and Johnson (2009); [28]Levins (2007); [29]Nicholls et al. (2016); [30]Aselage and Johnson (2009); [31]Levins (2007); [32]Lindquist (2000), Hendrichs et al. (2007); [33]Enkerlin et al. (2017); [34]Barrera et al. (2017).

In contrast to the weaknesses of IPM, the holistic approach is a better solution (Table 20.1). At best, it can result in a participatory and regional system for decision making for pest management that results in a better solution for integrated and agroecological systems of production. It also increases the prosperity of producers, those who mainly produce for the market and those who produce for self-consumption (Barrera 2006).

Who is behind all these ideas? The theory behind holistic management of pests has its basis in three schools of thought. The first is the "Holistic Management" of systems, led by Dr. Allan Savory (Savory and Butterfield 1999). The second is "Agroecology," led by Dr. Miguel A. Altieri (1995). Agroecology means production that results in low environmental impacts. The third is

"Complex Thought," developed by Dr. Edgar Morin (2000). Morin's writings suggest focusing analyses on the interactions between elements in a system instead of analyzing the elements disconnected from their system. HPM, which integrates these three schools of thought (holistic management, agroecology, and complex thought), gives us the tools to manage complex systems like we see in agriculture.

Getting people to work together long term is important. The holistic approach assumes that it will always be a challenge to maintain a permanent dialogue between professionals in different disciplines (Barrera 2009). The biologist, agronomist, and entomologist have different paradigms than the sociologist and economist. However, permanent dialogue is essential, and it leads to inter and transdisciplinary processes that address the true complexity of the problems. As Odum suggested many years ago when he defined the term "ecosystem" (Odum 1986), the holistic approach with its more complete data inputs generates more reliable results. For that reason, the holistic approach relies on the work of inter- and transdisciplinary groups. We suggest creating a working group composed of producers, researchers, and facilitators to perform a participatory diagnosis. Some members of this group will be devoted to maintaining the channels of communication, which is essential to the success of the working group. We also suggest the creation of collaborative networks composed of these working groups. These networks should follow the methodology of farmer field schools, which emphasize "learning by doing" (Jarquín-Gálvez 2003).

If we want to apply the holistic approach, it is necessary to have tools to study and manage complex systems. Defining the system to be studied is the first step in HPM. The limits of the system, its main components, and the goal must be defined in dialogue with the producers and other stakeholders. We define the system and its components in meetings with the working groups. For example, the "coffee system" includes elements that go from production and transformation of coffee beans to their commercialization (Barrera et al. 2004). One way to know the most important elements of the system is through knowing the problems that affect it. For this, the working groups identify the most outstanding problems that limit the system and prioritize them by importance. Determining the interactions that exist between the components of the system is equally important. A tool that helps us in this process is structural analysis. This analysis allows us to read reality as a system, a structure, and a complex phenomenon (Mojica 1991, 2004). The results of the structural analysis are represented in a graph with four zones or areas that relate interdependence and mobility among the components of the system (Figure 20.1). The "interdependence" indicates the strength of association between components, that is, the degree to which a problem needs other components to change to be solved. "Mobility" refers to the change that a graphed component experiences as a result of solving another or other problems. Interdependence and mobility are determined by establishing the lack of influence (value 0) or the direct influence (value 1) of one component over another. The four zones of the graph are determined by allowing the axes of the graph to intersect in the averages of interdependence and mobility values. The power zone (low interdependence and high mobility) groups the most important problems of the system. These elements influence the majority and depend little on them. The conflict zone (high interdependence and high mobility) groups the problems that are characterized by being very influential on the other components but which are also highly dependent on them. The exit zone (high interdependence and low mobility) groups the problems that are the product of the previous zones. The autonomous zone (low interdependence and low mobility) groups problems that neither influence the other components nor are affected by them. Because our priority is the prosperity of the producers, good solutions fall in the highly mobile and noninterdependent part of the graph or "power zone." In contrast, the element labeled "pests" falls in the immobile and interdependent part of the graph or "exit zone." With so many variables, this analysis helps to prioritize and plan effective solutions. In other words, the solution of a pest problem depends on solving other problems. In the case study of pests and diseases of coffee in Chiapas (Barrera et al. 2004, 2013), the organization of producers, the industrialization of coffee, and the access to markets with better prices (Figure 20.1) result in more income, which allows other problems to be solved, like managing the pests.

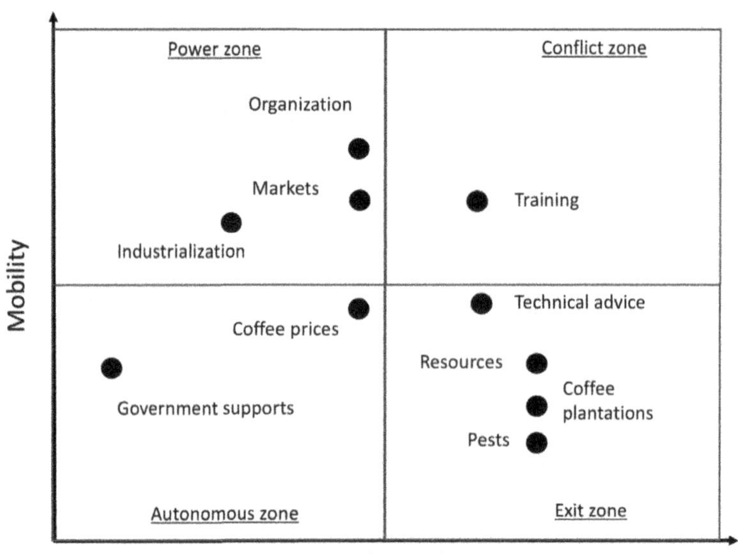

FIGURE 20.1 Relationships of interdependence and mobility among the 10 most important components of the coffee system in Chiapas, Mexico, elaborated by means of a structural analysis. (From Barrera, J. F. et al., *Plan Estatal de Manejo Agroecológico del Café en Chiapas: Guía hacia una cafeticultura sustentable*, Comisión para el Desarrollo y Fomento del Café de Chiapas y El Colegio de la Frontera Sur, Tapachula, México, 2004.)

In contrast with the holistic approach, decisions made with the old paradigm rely on action thresholds that do not provide sufficient information for making good decisions. Pest-management decisions must consider the complexity of agroecosystems. Analysis of complexity is the strength of the HPM approach. Decisions made with this approach are based on the concept of managing risk (Barrera et al. 2007). The risk can be defined in several ways, but for the purpose of this chapter we will define risk as "the possibility that something bad will occur" (i.e., a pest and its consequences).

A lot of information about risk management has been developed in the area of natural disasters. From there (ITDG 2002), we have taken the formula $R = (T + V)/C$, which allows us to estimate the risk (R, or holistic risk index, HRI) from three components: the threat (T), the vulnerability (V), and the response capacity (C). The "threat" is the infestation of a pest. The "vulnerability" is those characteristics of producers, and their farms that determine the degree of exposure to a pest. And the "response capacity" is composed of both those attributes and mechanisms of both producers and other stakeholders that reduce risks and increase the ability to survive. Both resistance and recovery from the damage that the pests cause define the response capacity. The formula allows us to consider one or several pests as a threat. Vulnerability can be integrated with information from the producer and the farm. Information from the producer may include age, sex, health, or schooling, and information from the farm may include altitude, soil fertility, or crop production cost. Examples of variables related to the response capacity may be knowledge about pest management, the level of producer organization, and access to technical advice, among others. The risk formula mentioned previously allows us to estimate the risk towards pests taking into account other components of the system that are usually ignored by the action thresholds used in the IPM. In other words, the assessment of risk is the essential tool for decision making in the holistic approach. This is a stark contrast to the tunnel vision that action thresholds inspire.

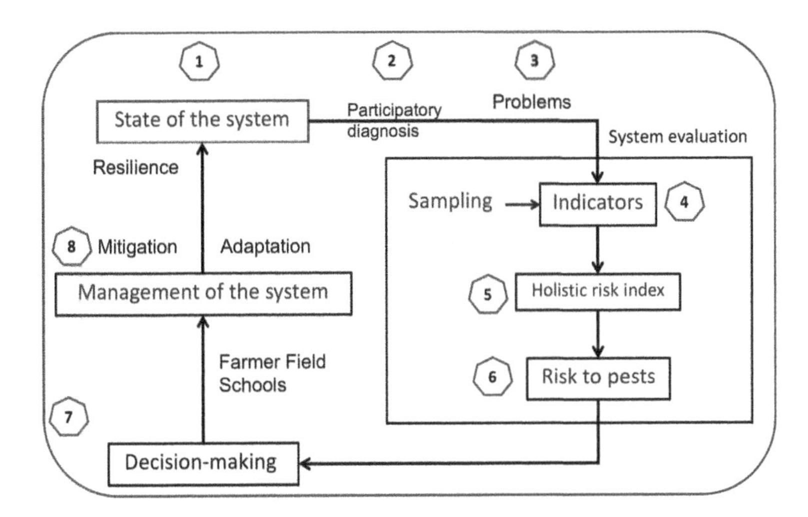

FIGURE 20.2 Process for decision-making in a system under the holistic approach. (1) The decision-making process begins by evaluating the state that the system is in. (2) For this, a diagnosis of the system is carried out with the participation of producers and other stakeholders. (3) The purpose of the diagnosis is to know the main problems of the system. (4–6) Once the problems have been defined, the variables and indicators must be identified and sampled to generate the information to estimate the risk toward the threats (e.g., pests). The risk is estimated as: HRI = (T + V)/C, where HRI = Holistic Risk Index, T = Threat, V = Vulnerability, and C = Response capacity. (7–8) The risk is key to decision making of system management. Before starting to manage the system, it is important to train producers and technical personnel in Farmer Field Schools. The training should provide the concepts, methods, and tools to carry out adaptation and mitigation actions aimed at reducing the risk of threats to increase the system's resilience. The system must be evaluated periodically to determine if the goals committed to were met. (From Barrera, J. F. et al., *Manejo holístico de plagas en zonas cafetaleras: Concepto y método*, Primera edición, El Colegio de la Frontera Sur, Tapachula, México, Folleto Técnico Núm. 15, 32 p., 2017.)

Figure 20.2 shows the process for decision making under the HPM approach (Barrera et al. 2017). The path starts with a participatory diagnosis to determine the state of the system. We identify the main problems through this diagnosis and we propose variables and indicators to estimate the risk. The risk, which is estimated with the risk formula [HRI = (T + V)/C], provides the necessary information to make decisions in the management of the system. Some actions will be to adapt to circumstances (adaptation), and some actions will be to change the circumstances (mitigation). We recommend to periodically evaluate the system following this process. This evaluation will determine both the control of pests and the improvement of the system. In other words, by reducing vulnerability and risk, and increasing the producer's response capacity, HPM aims to improve the resilience of the system.

Risk and resilience are related (Barrera et al. 2017). Response capacity is an estimate of resilience. In the risk equation, we will replace the response capacity with the letter "E," an estimate of resilience. Subsequently, the variable "E" is moved to the other side of the equation: E = (T + V)/R. If we plot the risk with the resilience estimate, we observe a curvilinear relationship like a power equation (Figure 20.3), which is contrary to our expectation of a linear relationship between these variables. According to this power relation, risk values that are greater than 3.0 indicate systems with low resilience. Those systems have little capacity to recover after suffering the impact of a threat. I should point out that the possibility of estimating resilience through the response capacity opens a window of opportunity in the study and management of pests with the holistic approach. This opportunity only exists with HPM.

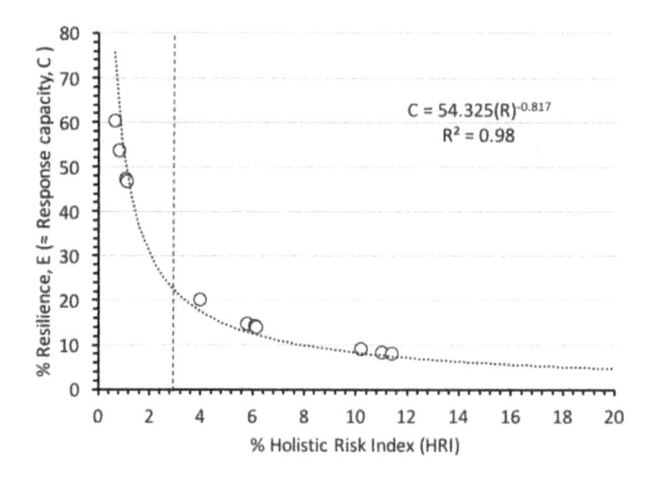

FIGURE 20.3 Power relationship between risk of coffee rust (*Hemileia vastatrix*) and resilience of coffee farms in Soconusco, Chiapas, Mexico. Risk values were calculated with R = (T + V)/C, and resilience values were estimated with E = (T + V)/R, where E ≈ C (see text for explanation). Risk values that are greater than 3.0 indicate that coffee farms have low resilience. (From Barrera, J. F. et al., *Manejo holístico de plagas en zonas cafetaleras: Concepto y método*, Primera edición, El Colegio de la Frontera Sur, Tapachula, México, Folleto Técnico Núm. 15, 32 p., 2017.)

20.3 SOME METHODS TO IMPLEMENT THE HOLISTIC APPROACH

We suggest carrying out a "rapid agroecological sampling" to obtain information about the vulnerability and response capacity of the system. From this sampling, we will estimate the risk. This sampling consists of two parts. One part is to gather socioeconomic information from producers and other stakeholders through interviews and surveys. The other part is to sample the agricultural system to determine the level of infestation by pests and other characteristics such as soil fertility, plant diversity, production, geographic location, and weather. As an example, Figure 20.4 illustrates the variables that make up each of the components to estimate the risk of coffee rust, *Hemileia*

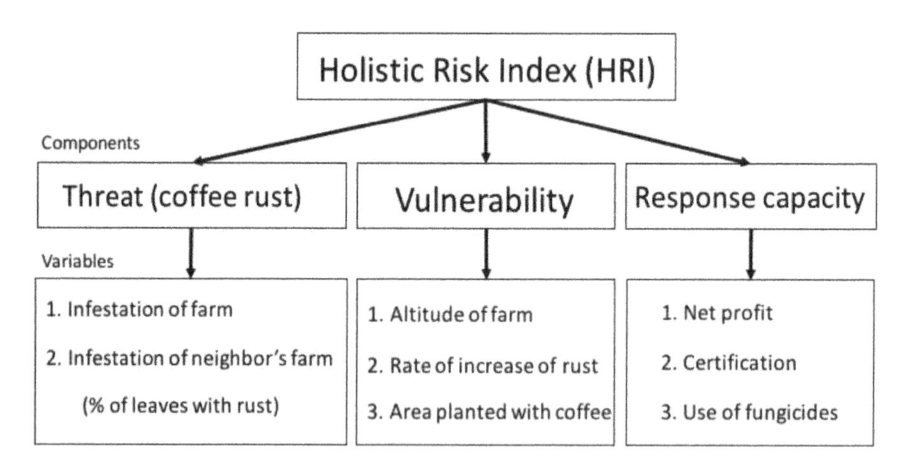

FIGURE 20.4 Variables to determine the components of risk of coffee rust (*Hemileia vastatrix*). (From Barrera, J. F. et al., *Manejo holístico de plagas en zonas cafetaleras: Concepto y método*, Primera edición, El Colegio de la Frontera Sur, Tapachula, México, Folleto Técnico Núm. 15, 32 p., 2017.)

vastatrix (Barrera et al. 2017). The threat is represented by the percentage of coffee leaves with rust. Vulnerability is a function of three variables: altitude of the farm, increase rate of rust, and size of the area planted with coffee. Response capacity is a function of three variables: net profit from sale of coffee, participation of the producer in an organic certification program, and application of fungicides.

Once the variables have been defined, it is necessary to carry out the following procedure to calculate the risk (HRI) value for each producer: (1) collect the data from the field; (2) determine the weight of the variables (e.g., factorial analysis of the data); (3) establish the relationship between the variables and the threat; (4) standardize the data to a scale of 1–100; (5) add the values of the variables to calculate each component of the risk; (6) calculate the percentage that corresponds to each risk component (the sum of the three components must be equal to 100); and (7) apply the risk formula.

With respect to number three of the previous procedure, which refers to establishing the relationship between the variables and the threat, Figure 20.5 shows these relationships in the case of coffee rust (threat). These relationships are established by using information from the literature (e.g., Avelino and Rivas 2013) or through empirical knowledge. Hypothetical relationships can also be established for those who want to model or make predictions without real data. The variables are measured in ordinal scales; for example, in the case of rust infestation, the following scale is used: without rust, very low infestation (<3%), low infestation (3.1%–5%), medium infestation (5.1%–20%), high infestation (20.1%–60%), and very high infestation (>60%). As it can be seen,

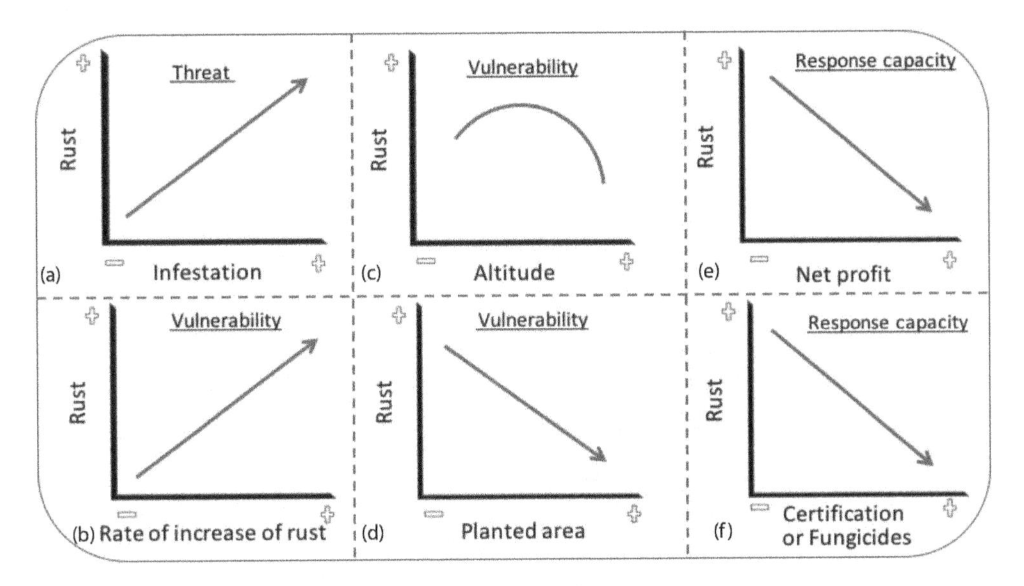

FIGURE 20.5 Relations between components of risk and their variables and coffee leaf rust (*Hemileia vastatrix*) in coffee plantations in Chiapas, Mexico. Coffee rust is the threat. The variables of vulnerability are increased rate of rust, altitude of coffee plantation, and size of coffee plantation (planted area). The variables of response capacity are net profit per sale of coffee, whether coffee production is certified as organic, and whether the farmer uses fungicides to control the rust. (a) Relationship between rust infestation and rust damage; (b) relationship between increased rate of rust and its damage; (c) relationship between coffee plantation altitude and rust damage; (d) relationship between size of planted area and coffee and between size of planted area and rust damage; (e) relationship between net profit and rust damage; and (f) relationships between both certified and uncertified organic coffee and rust damage, and between fungicide use and rust damage. (From Barrera, J. F. et al., *Manejo holístico de plagas en zonas cafetaleras: Concepto y método*, Primera edición, El Colegio de la Frontera Sur, Tapachula, México, Folleto Técnico Núm. 15, 32 p., 2017.)

some relationships are positive linear, that is, there is more rust as the values of those variables increase. For example, with respect to the threat (rust), as the infestation of rust increases, its damage increases (Figure 20.5a); or, with respect to vulnerability, as the rate of increase in rust increases (shorter spore production cycles), an increase in damage also occurs (Figure 20.5b). In contrast, in other cases, the relationship is negative (Figure 20.5d–f). Only in the case of rust and altitude (Figure 20.5c) do we establish a curvilinear relationship because the conditions of low and high altitudes are less favorable for this coffee disease. In this case, the ordinal scale used is: <600 m above sea level (masl), regular infestation of rust; from 601 to 1000 masl, very high infestation (maximum vulnerability); from 1001 to 1200 masl, low infestation; and >1200 masl, very low infestation.

Several procedures can be performed to analyze the risk. We propose at least three types of analyses: holistic risk triangle, radial graphs, and geographical analysis. The holistic risk triangle is a ternary or triangle plot (equilateral triangle) that is generated from the sum of the proportions of three variables. This means that each point of the graph results from the combination of the three variables, whose sum is a constant (e.g., 100%). The ternary plots are useful to identify different regions or phases in the plot (phase diagrams). These plots have applications in several fields such as glass compositions, refractories, aluminum alloys, stainless steels, solder metallurgy, or game theory (Selvaduray 2004, Ponsen et al. 2009). In the case of coffee producers from Soconusco, Chiapas, Mexico, the three variables on the graph are the risk components of each studied producer: the threat, the vulnerability, and the response capacity, as it is shown in Figure 20.6. This figure is divided into areas or phases that indicate the level of risk, ranging from very low risk (<1.0; Figure 20.6) in light

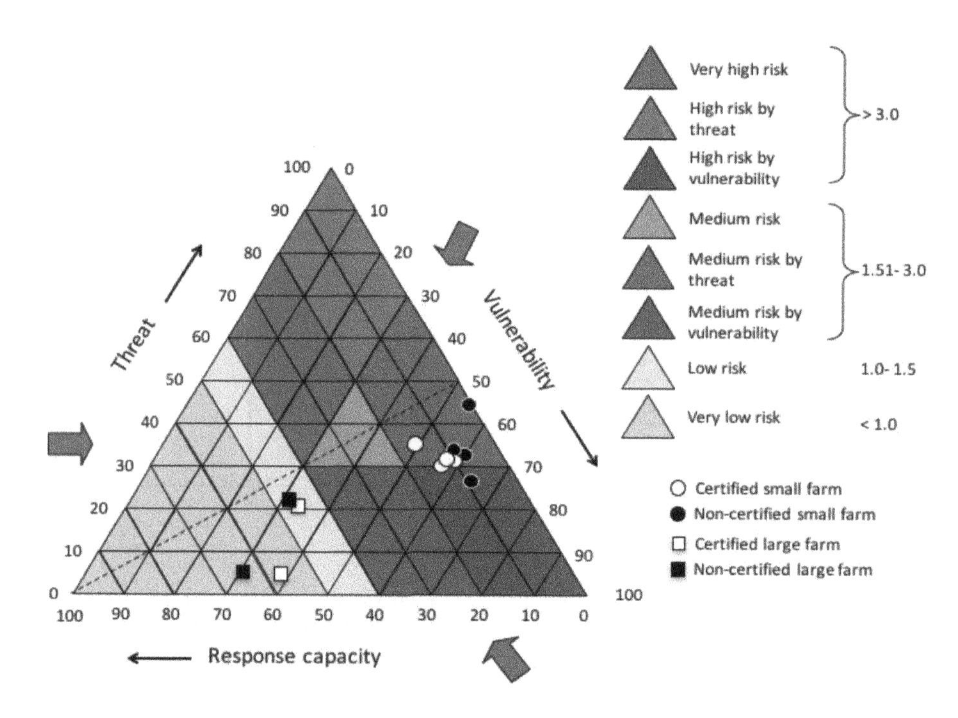

FIGURE 20.6 Risk analysis (holistic risk triangle) of coffee leaf rust, *Hemileia vastatrix*, for certified and noncertified organic coffee producers in Soconusco, Chiapas, Mexico (2013). (From Barrera, J. F. et al., *Manejo holístico de plagas en zonas cafetaleras: Concepto y método*, Primera edición, El Colegio de la Frontera Sur, Tapachula, México, Folleto Técnico Núm. 15, 32 p., 2017.)

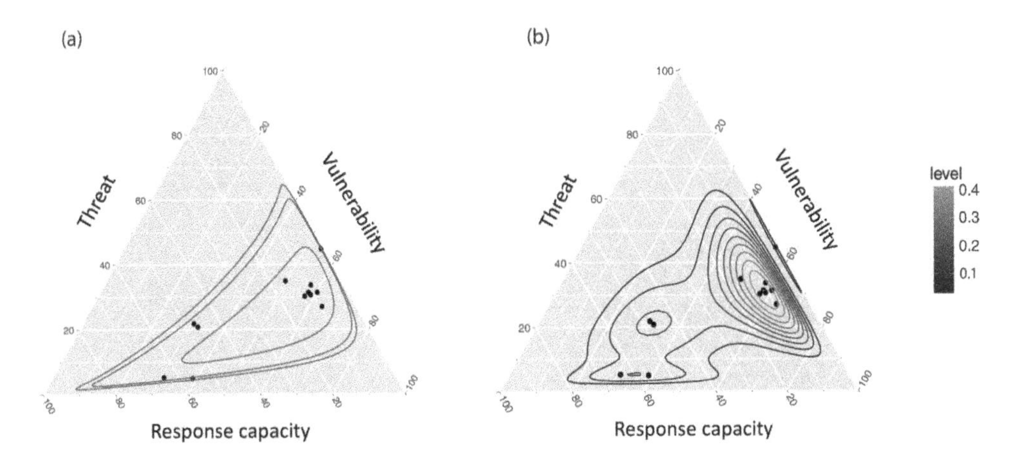

FIGURE 20.7 Risk analysis (holistic risk triangle) of coffee leaf rust, *Hemileia vastatrix*, for certified and noncertified organic coffee producers from Soconusco, Chiapas, Mexico (2013). (a) Confidence intervals at 50%, 90%, and 95%; (b) estimate of probability density. (From Barrera, J. F. et al., *Manejo holístico de plagas en zonas cafetaleras: Concepto y método*, Primera edición, El Colegio de la Frontera Sur, Tapachula, México, Folleto Técnico Núm. 15, 32 p., 2017.)

colors to very high risk (>3.0; Figure 20.6) in dark colors, such as the red area that indicates the highest risks. Figure 20.6 shows that large producers have a lower risk of rust in comparison with small producers. Large producers are less vulnerable to rust (V = 30.8%–38.8%) than small producers (V = 49.6%–63.6%), and they have a greater response capacity to reduce the damage that this coffee disease causes in their plantations (C = 46.9%–64.0%) than small producers (0.7%–15.6%). This results in lower infestation rates of rust in large farms (4.9%–21.7%) as opposed to small farms (27.0%–44.8%). We also see that certification does not influence the rates of rust infestation in large farms. However, Figure 20.6 does show that certified small producers have a better response capacity to deal with rust (C = 11.0%–15.6%) than noncertified small producers (0.7%–9.4%).

Also, as shown in Figure 20.7, triangle plots can include confidence intervals of the values (Cornell 2011, Lawson and Willden 2016, Hamilton and Ferry 2018). These are statistical measurements of certainty. These confidence intervals are used to infer the risk of rust that neighboring producers not considered in the study may have (Figure 20.7a). According to the information obtained from sampled producers who participated in the study (n = 12), confidence intervals show that 95% of all producers in the studied region would be concentrated in the medium- and high-risk areas of the plot. It is also observed that part of a producers' population, mainly represented by large farms, would be dispersing toward the low-risk area of the plot (Figure 20.7a).

Likewise, Figure 20.7 includes a nonparametric estimate of the probability density function (Figure 20.7b), that is, a plot that shows the region in which the observations are distributed. As it can be seen, the analysis indicates that there are two clearly recognizable producer groups. One of the groups (small producers) is concentrated in the part of the graph with the highest risk of rust and the other (large producers) is dispersed in the graph from the medium-risk area to the low-risk area (Figure 20.7b).

In Figure 20.8 we show how radial graphs are also used in risk analysis. These graphs allow us to analyze the variables that make up each component of the risk in such a way that the causes associated with the risk can be identified with greater precision. In the case study of coffee producers in Chiapas, which included data from 12 producers grouped in six pairs, the fourth pair of producers is comprised

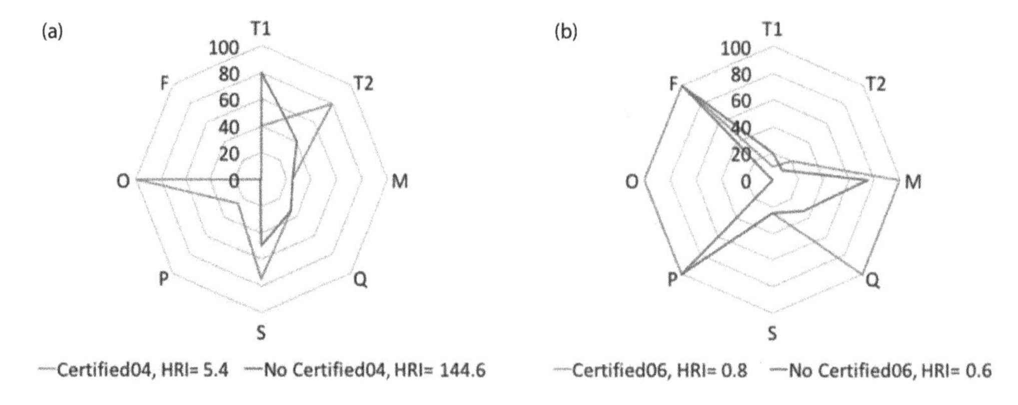

FIGURE 20.8 Comparison of variables that integrate the components of the risk of coffee leaf rust, *Hemileia vastatrix*, in certified and noncertified organic coffee producers from Soconusco, Chiapas, Mexico (2013). Each radial graph represents a pair of producers, one certified and the other not certified, whose farms are relatively close. (a) Pair 4, small farms (up to 3.5 ha); (b) pair 6, large farms (up to 300 ha). T1 = rust infestation in the studied farm (threat); T2 = rust infestation in neighboring farm (threat); M = altitude of the farm (vulnerability); Q = increase rate of rust (vulnerability); S = area size of the farm planted with coffee (vulnerability); P = net profit of coffee sale (response capacity); O = organic coffee certification (response capacity); F = use of fungicides (response capacity). (From Barrera, J. F. et al., *Manejo holístico de plagas en zonas cafetaleras: Concepto y método*, Primera edición, El Colegio de la Frontera Sur, Tapachula, México, Folleto Técnico Núm. 15, 32 p., 2017.)

of two small producers, one is certified organic coffee, with HRI = 5.4, and the other is not certified, with HRI = 144.6. The radial graph shows that the producer with the highest risk of rust is characterized by having more rust infestation (threat). This producer was cultivating coffee in a smaller area (vulnerability) and did not make a profit by selling coffee (response capacity). This producer did not participate in a certification program (response capacity) and did not use fungicides to control rust (response capacity) (Figure 20.8a). With regard to the sixth pair of producers (Figure 20.8b), it is comprised of two large producers, one is certified organic coffee, with HRI = 0.8, and the other is not certified, with HRI = 0.6. Although both producers have similar rust risk values, the certified producer is more vulnerable to the disease in terms of farm altitude and intrinsic rate of rust. This result indicates that cultivation of organic coffee (e.g., more shade trees, use of organic fertilizers, better price of coffee) gives the producer greater response capacity to the threat that the rust represents.

If there is a database with a large number of producers available, we can do a geographical analysis of the risk. For example, the maps in Figure 20.9 show the geographical distribution of approximately 2,500 coffee producers in central and southern Chiapas with respect to the risk of rust. Apart from the risk (Figure 20.9a), these maps also show its components: threat (Figure 20.9b), vulnerability (Figure 20.9c), and response capacity (Figure 20.9d). As in the case of the graphic risk analysis mentioned previously, the separation of risk into its components allows us to know with more detail the factors associated with the risk. In this case, there is the advantage that the data are presented at a geographical scale. Also, as in the previously mentioned graphic analysis, the risk of neighboring producers, who do not participate in the study, can be inferred from the sampled producers. As expected from this type of analysis, it is possible to make inferences about the rust situation at a regional level to issue, for example, early warning messages.

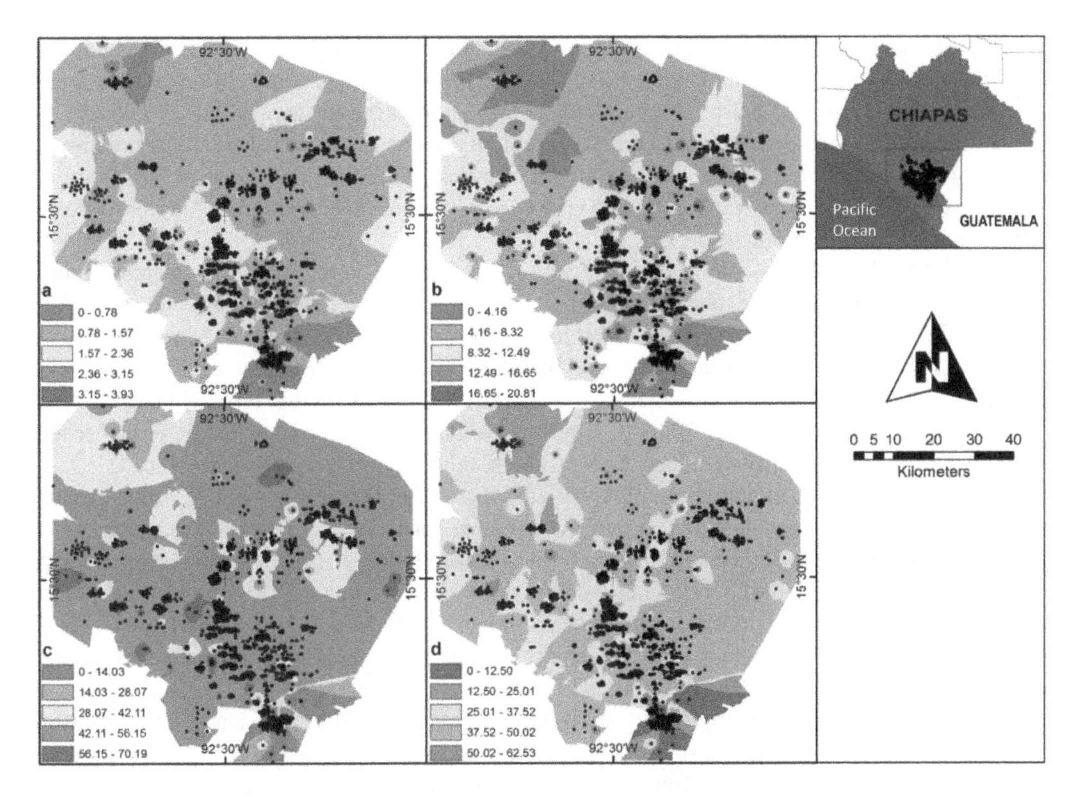

FIGURE 20.9 Geographical analysis of risk of coffee leaf rust (*Hemileia vastatrix*) and its components. The data correspond to approximately 2,500 coffee producers from Chiapas, Mexico. Inverse distance weighting (IDW) was used as an interpolation method. (a) Holistic Risk Index (HRI) calculated with HRI = (T + V)/C; (b) threat (T) is the rust infestation in the farm; (c) vulnerability (V) is the altitude of the farm; and (d) response capacity (C) is the net profit from the sale of coffee. Green colors in a, b, and c indicate low values, and in d indicate high values. (From Barrera, J. F. et al., *Manejo holístico de plagas en zonas cafetaleras: Concepto y método*, Primera edición, El Colegio de la Frontera Sur, Tapachula, México, Folleto Técnico Núm. 15, 32 p., 2017.)

HPM gives priority to prevention over cure. Prevention means that producers have to produce without causing conditions that encourage pests. For this strategy to work, preventive methods must be compatible with the processes and cycles of nature. These methods must conserve and increase the action of natural enemies. They must also increase the resistance or tolerance of crops toward pests and other environmental stresses. To a large extent, this involves cultivating plants in diversified systems with low use of pesticides to promote mechanisms of self-regulation or "autonomous control" of pests. This strategy is suggested by the group led by Vandermeer et al. (2010). As you may suppose, conservation biological control has a primordial place in the holistic approach.

If it is necessary to carry out curative actions, HPM suggests using mainly biological methods like the release of agents of biological control, sterile insects, or odors that modify the behavior of pests and their natural enemies. Under these strategies, chemical pesticides must always be the last resort of defense against pests. These strategies are the opposite of aiming at a moving target in a crisis and relying on silver bullets.

What lessons have been learned when applying holistic management to coffee leaf rust? In our case studies, the concept and methods of HPM have allowed us to propose a different solution to the rust problem. We have focused on promoting a decision-making strategy based on reducing the risk of this disease. This is a contrast to the old paradigm that focuses primarily on reducing disease infestation. The infestation can be controlled with chemical fungicides and resistant coffee varieties. However, these products, fungicides and special varieties, are not available to any producer or they are not what producers want.

Our risk-based strategy is a holistic approach to the problem. The reduction of rust infestation is just one of several actions to be taken to manage it. In effect, reducing risk also means taking actions to reduce the vulnerability of the producer and his farm, as well as to increase his response capacity to the rust problem. To reduce vulnerability and increase response capacity to rust, it is necessary for the producer to have sufficient material and social resources. For this reason, HPM aligns all actions to improve the producer's income. Without a substantive improvement in income, the producer will not have the response capacity to reduce the infestation, much less to reduce vulnerability.

Through participatory diagnostics, we have found that to improve the income of coffee producers it is necessary to establish formal organizations or at least working groups. The organization of producers is a fundamental requirement to have access to better prices for coffee beans, such as certified organic and fair-trade coffee markets (Bara and Pérez-Akaki 2015). Also, organizations are essential to industrialize coffee to give added value. In Mexico, and particularly in Chiapas, successful producer organizations grow coffee organically and export it to the United States and Europe (Folch and Planas 2019).

HPM is based on the cultivation of plants through agroecological production systems (Barrera 2006). The cultivation of coffee through these systems and with access to special or certified markets (organic coffee, fair trade, coffee under shade, bird-friendly, etc.) not only obtains a better price (Arana-Coronado et al. 2019) but also contributes to improving soil fertility through organic fertilizers. Furthermore, it increases production through renewal and pruning of coffee trees. Also, these production systems increase biodiversity by promoting coffee under shade trees, with some producers not using any synthetic agrochemicals and others using much less than before (Escamilla et al. 2005). All these aspects favor more vigorous plants, more diversified production systems, and plants less prone to pest outbreaks. These are desirable characteristics to manage rust more effectively (Avelino and Rivas 2013, Barrera 2018).

All this complexity, which is difficult to deal with in the old paradigm, can be addressed through risk analysis. Thus, the HRI has been a practical and useful tool for decision making aimed at reducing vulnerability and increasing response capacity. The periodic monitoring of the HRI and its analysis by means of the graphic methods presented here allow us to evaluate the effectiveness of the management actions implemented by the producers in reducing the risk of rust. According to our analysis, risks equal to or greater than 3.0 represent systems with very low resilience. Therefore, as risk decreases, the system will be moving toward a more resilient state.

HPM is a new paradigm whose adoption will take time. We need successful case studies, which include other systems besides coffee and other countries besides Mexico. In this regard, it is worth mentioning that in addition to the Mexican experience with coffee pests that is referred to here, we are investigating the application of HPM with coffee producers in Honduras. We have also initiated research to apply HPM in pests of flower crops in central Mexico and in fruit flies (*Anastrepha ludens*) infesting mango in the southeast of Mexico. HRI, the cornerstone of HPM, has generated interest in other contexts and with other threats (Table 20.2), which is an indication of the utility of this index for the management of complex systems.

TABLE 20.2
Some Cases That Have Used the Holistic Risk Index (HRI)[a]—the Decision-Making Tool in HPM—in Different Contexts and Threats

Context	Threat(s)	Country	Reference(s)
Resilience, water resources, rural and indigenous communities	• Drought intensity • Drought frequency • Loss of yields or productivity due to drought	Chile	Montalba et al. (2013, 2015)
Resilience, socio-ecological systems, extreme weather events	• Climate events • Intensity • Duration • Frequency • Damage levels	Colombia	Henao-Salazar (2013)
Social and ecological adaptation strategies, climate change, farmers	• Floods • Heavy rains • Landslides • Droughts	Ecuador	Carpio Sacoto and Carpio Sacoto (2014)
Cattle producers, protected areas, estuaries	• Decrease in grazing lands • Negative socioeconomic and ecological impact on the basin • Insecurity for livestock farming	Uruguay	Gazzano Santos (2014), Gazzano et al. (2015, 2016), Gazzano and Achkar (2015, 2016)
Agroecological sustainability, agricultural production systems	• Climate events • Intensity • Frequency • Level of damage	Mexico	Álvarez Morales (2015)
Agroecological management of the ground pearl in blackberry	Infestation by *Eurhizococcus colombianus* Jakubsky (Hemiptera: Margarodidae)	Colombia	Meneses-Ospina (2015)
Holistic pest management in coffee plantations, coffee rust	Infestation by *Hemileia vastatrix* Berk. & Br. (Basidiomycota: Uredinales)	Mexico	Barrera et al. (2017)
Socioecological resilience of small coffee production	• Water availability due to rainfall • Coffee price fluctuations	Colombia	Machado-Vargas (2017), Machado-Vargas et al. (2018)
Repellence and attraction in the coffee berry borer	Infestation by *Hypothenemus hampei* (Ferrari) (Coleoptera: Curculionidae)	Colombia	Castro-Triana (2018)

[a] *Source:* Barrera, J. F. et al., Riesgo-vulnerabilidad hacia la broca del café bajo un enfoque de manejo holístico, in *La Broca del Café en América Tropical: Hallazgos y Enfoques*, ed. J.F. Barrera, A. García, V. Domínguez and C. Luna, pp. 131–141, Sociedad Mexicana de Entomología y El Colegio de la Frontera Sur, Mexico, 2007; Barrera, J.F. et al., Método holístico para la toma de decisiones en el manejo de plagas, in *Simposio Estado del arte del Manejo Ecológico de Plagas en América Latina, Tercer Congreso Latinoamericano de Agroecología*, August 17–19, Oaxtepec, Mexico, 2011.

20.4 CONCLUSION

From what you have seen, it is obvious that the holistic approach results in better decisions for pest management. This approach improves the prosperity of producers by providing them with greater response capacity. This puts producers in a better position to reduce their vulnerability to pests and increase the resilience of their production systems. Changing the old paradigm to the HPM paradigm implies thinking, and even more importantly, acting holistically. It is time for people working with the old paradigm to declare "mission accomplished." The 60-year-old paradigm that solved some important problems has revealed its limits and weaknesses. Our challenge today is to be more precise and effective in our management of pests. Paraphrasing Dr. Keith Andrews, a former proponent of IPM, this requires a new paradigm that is informed by methods of the past but that transcends it. HPM is that paradigm!

ACKNOWLEDGMENTS

I would like to thank the organizing committee for the invitation to open the 10th International Symposium on Fruit Flies of Economic Importance with the conference that gave rise to this paper. Particularly, I would like to thank Pablo Liedo, a scientist passionate about fruit flies of economic importance, with whom I have had the opportunity to share my ideas and advances in HPM. The data from certified and noncertified organic coffee producers were obtained in Soconusco, Chiapas, in 2013 with Inter-American Institute for Global Change Research (IAI) financing. We also received financing from the MT project of GIEZCA of ECOSUR. I would like to thank Javier Valle Mora and José Higinio Lopez Urbina from ECOSUR for their advice on statistical and geographical data analysis, respectively. I am grateful for the careful editing of Esteban McAndrew and for his help in the writing of this manuscript in English.

REFERENCES

Altieri, M. A. 1995. *Agroecology: The Science of Sustainable Agriculture*, 2nd ed. Boulder, CO: Westview Press.

Altieri, M. A., P. B. Martin, and W. J. Lewis. 1983. A quest for ecologically based pest management systems. *Environmental Management* 7: 91–100.

Altieri, M. A., and C.I. Nicholls. 2005. *Agroecology and the Search for a Truly Sustainable Agriculture*, 1st ed. Mexico: United Nations Environment Programme, Environmental Training Network for Latin America and the Caribbean.

Álvarez Morales, Y. 2015. Evaluación de indicadores de sustentabilidad agroecológica en sistemas de producción agrícola de Baja California Sur, México. *Doctoral thesis*. Centro de Investigaciones Biológicas del Noroeste, S.C. La Paz, Baja California Sur, Mexico.

Arana-Coronado, J. J., C. O. Trejo-Pech, M. Velandia, and J. Peralta-Jimenez. 2019. Factors influencing organic and fair trade coffee growers level of engagement with cooperatives: The case of coffee farmers in Mexico. *Journal of International Food & Agribusiness Marketing* 31(1): 22–51.

Aselage, J., and D. T. Johnson. 2009. From IPM to organic and sustainable agriculture. In *Integrated Pest Management: Concepts, Tactics, Strategies and Case Studies*, ed. E. B. Radcliffe, W. D. Hutchinson and R. E. Cancelado, pp. 489–505. Cambridge, UK: Cambridge University Press.

Avelino, J., and G. Rivas. 2013. *La roya anaranjada del cafeto*. https://hal.archives-ouvertes.fr/hal-01071036/ (accessed March 23, 2019).

Bara, C., and P. Pérez-Akaki. 2015. Status quo, challenges and opportunities of alternative coffee that is produced in Mexico and consumed in Germany. *Agricultura Sociedad y Desarrollo* 12: 59–86.

Barrera, J. F. 2006. Manejo holístico de plagas: Hacia un nuevo paradigma de la protección fitosanitaria. In *El cafetal del futuro: Realidades y Visiones*, ed. J. Pohlan, L. Soto and J. Barrera, pp. 63–82. Aachen, Germany: Shaker Verlag.

Barrera, J. F. 2007. Manejo Holístico de Plagas: Más allá del MIP. In *XXX Congreso Nacional de Control Biológico*, November 11–15, Mérida, Mexico.

Barrera, J. F. 2009. La necesidad del enfoque holístico en el manejo de plagas. In *Enfoques y temáticas en entomología*, ed. J. C. Arrivillaga, M. El Souki, and B. Herrera, pp. 1–28. XXI Congreso Venezolano de Entomología, Julio 2009. Caracas, Venezuela: Sociedad Venezolana de Entomología.

Barrera, J. F. 2018. Insect pests of coffee and their management in nature-friendly production systems. In *Handbook of Pest Management in Organic Farming*, ed. V. Vacante and S. Kreiter, pp. 477–501. Wallingford, UK: CAB International.

Barrera, J. F., W. Gamboa, J. Gómez, and J. Valle. 2011. Método holístico para la toma de decisiones en el manejo de plagas. In *Simposio Estado del arte del Manejo Ecológico de Plagas en América Latina, Tercer Congreso Latinoamericano de Agroecología*, August 17–19, Oaxtepec, Mexico.

Barrera, J. F., J. Gómez-Ruiz, M. R. Parra-Vázquez, G. Mercado-Vidal, and T. Williams. 2013. Análisis de problemas ocasionados por plagas del café bajo un enfoque holístico. *Entomología Mexicana* 12(2): 1128–1133.

Barrera, J. F., J. Herrera, and J. Gómez. 2007. Riesgo-vulnerabilidad hacia la broca del café bajo un enfoque de manejo holístico, In *La Broca del Café en América Tropical: Hallazgos y Enfoques*, ed. J. F. Barrera, A. García, V. Domínguez, and C. Luna, pp. 131–141. Mexico: Sociedad Mexicana de Entomología y El Colegio de la Frontera Sur.

Barrera, J. F., M. Parra Vázquez, O. B. Herrera Hernández, R. Jarquín Gálvez, and J. Pohlan. 2004. *Plan Estatal de Manejo Agroecológico del Café en Chiapas: Guía hacia una cafeticultura sustentable.* Tapachula, México: Comisión para el Desarrollo y Fomento del Café de Chiapas y El Colegio de la Frontera Sur.

Barrera, J. F., J. Valle, J. Gómez, J. Herrera, E. López, and J. de la Rosa. 2017. *Manejo holístico de plagas en zonas cafetaleras: Concepto y método.* Primera edición. Tapachula, México: El Colegio de la Frontera Sur. Folleto Técnico Núm. 15, 32 p.

Benbrook, C. M. 1996. *Pest Management at the Crossroads.* New York: Consumer Union.

Bentley, J., and K. Andrews. 1996. *Through the Roadblocks: IPM and Central American smallholders. Sustainable Agriculture and Rural Livelihoods Programme.* London, UK: International Institute for Environment and Development. Gatekeeper Series No. 56.

Bottrell, D.G . 1979. *Integrated Pest Management: Council of Environmental Quality.* Washington, DC: U.S. Government Printing Office.

Braun, A. R., G. Thiele, and M. Fernandez. 2000. *Farmer Field Schools and Local Agricultural Research Committees: Complementary Platforms for Integrated Decision-Making in Sustainable Agriculture.* London, UK: Agricultural Research and Extension Network, ODI. AgREN Network Paper No. 105.

Carpio Sacoto, J. E., and L.M. Carpio Sacoto. 2014. Determinación de estrategias sociales y ecológicas de adaptación al cambio climático implementadas por los agricultores de las cuatro zonas agroecológicas de la Parroquia San Joaquín. *MSc thesis.* Universidad Politécnica Salesiana, Cuenca, Ecuador.

Castro-Triana, A. M. 2018. Estudio sobre la repelencia y la atracción en la broca del café como herramienta para el manejo agroecológico en los cafetales colombianos. *Doctoral thesis.* Universidad de Antioquia, Medellín, Colombia.

Cornell, J. 2011. *A Primer on Experiments with Mixtures.* Hoboken, NJ: John Wiley & Sons.

Dhaliwal, G. S., V. Jindal, and B. Mohindru. 2015. Crop losses due to insect pests: Global and Indian scenario. *Indian Journal of Entomology* 77(2): 165–168.

Ehler, L. E., and D. G. Bottrell. 2000. The illusion of integrated pest management. *Issues in Science and Technology*, http://www.issues.org/16.3/index.html (accessed March 23, 2019).

Enkerlin, W. R., J. M. Gutiérrez Ruelas, R. Pantaleon, C. Soto Litera et al. 2017. The Moscamed Regional Programme: Review of a success story of area-wide sterile insect technique application. *Entomologia Experimentalis et Applicata* 164: 188–203.

Escamilla, P. E., R. O. Ruiz, P. G. Díaz, S. C. Landeros, R. D. E. Platas, C. A. Zamarripa, and H. V. A. González. 2005. El agroecosistema café orgánico en México. *Manejo Integrado de Plagas y Agroecología* 76: 5–16.

Folch, A., and J. Planas. 2019. Cooperation, fair trade, and the development of organic coffee growing in Chiapas (1980–2015). *Sustainability* 11(2): 357.

Frisbie, R. E., and J. W. Smith Jr. 1991. Biologically intensive integrated pest management: The future, In *Progress and Perspectives for the 21st Century*, ed. J. J. Menn and A. L. Steinhauer, pp. 151–164. Entomological Society of America Centennial Symposium, Lanham, MD, Entomological Society of America.

Gazzano, I., and M. Achkar. 2015. Amenaza, Vulnerabilidad y Riesgo: estrategias de respuesta de ganaderos familiares en el área protegida Esteros de Farrapos – Uruguay. In *Memorias del V Congreso Latinoamericano de Agroecología*, La Plata, Argentina.

Gazzano, I., and M. Achkar. 2016. Conflictos de las transformaciones territoriales: Ganaderos frente a la intensificación agraria en Esteros de Farrapos Uruguay. *Revista Iberoamericana de Economía Ecológica* 26: 109–121.

Gazzano, I., M. Altieri, M. Achkar Borrás, and J. Burgueño. 2015. Holistic risk index: A case study of cattle producers in the protected area of Farrapos Estuaries-Uruguay. *Agroecology and Sustainable Food Systems* 39(2): 209–223.

Gazzano, I., M. Altieri, M. Achkar Borrás, and J. Burgueño. 2016. Riesgo y resiliencia de productores ganaderos familiares en el área protegida Parque Nacional Esteros de Farrapos, Uruguay. *Agrociencia Uruguay* 20(1): 51–60.

Gazzano Santos, M. I. 2014. Viabilidad de la ganadería familiar en áreas protegidas de humedales, en un contexto sinérgico de intensificación agraria e inundaciones: Parque Nacional Esteros de Farrapos-Uruguay. *Doctoral thesis*. Universidad de Córdoba, España.

Gray, M. E., S. T. Ratcliffe, and M. E. Rice. 2009. The IPM paradigm: Concepts, strategies and tactics. In *Integrated Pest Management: Concepts, Tactics, Strategies and Case Studies*, ed. E. B. Radcliffe, W. D. Hutchinson, and R. E. Cancelado, pp. 1–13. Cambridge, UK: Cambridge University Press.

Griffin, R. 2012. Basic concepts in risk analysis. In: *Plant Pest Risk Analysis: Concepts and Application*, ed. C. Devorshak, pp. 7–18. Wallingford, UK: CAB International.

Hamilton, N. E., and M. Ferry. 2018. ggtern: Ternary diagrams using ggplot2. *Journal of Statistical Software, Code Snippets* 87(3): 1–17.

Henao-Salazar, A. 2013. Propuesta metodológica de medición de la resiliencia agroecológica en sistemas socio-ecológicos: Un estudio de caso en los Andes colombianos. *Agroecología* 8(1): 85–91.

Hendrichs, J., P. Kenmore, A. S. Robinson and M. J. B. Vreysen. 2007. Area-wide integrated pest management (AW-IPM): Principles, practice and prospects. In *Area-Wide Control of Insect Pests*, ed. M.J.B. Vreysen, A.S. Robinson, and J. Hendrichs, pp. 3–33. Dordrecht, the Netherlands: Springer.

Herrera, O. B., M. Parra, I. Livscovsky, P. Ramos, and D. Gallardo. 2019. Lifeways and territorial innovation: Values and practices for promoting collective appropriation of territory. *Community Development Journal*, 54(3): 427–445, doi:10.1093/cdj/bsx052.

Higley, L. G., and L. P. Pedigo. 1996. The EIL concept. In *Economic Thresholds for Integrated Pest Management*, ed. L. G. Higley, J. G. and L. P. Pedigo, pp. 9–21. Lincoln, NE: University of Nebraska Press.

Higley, L. G., and R. K. D. Petersen. 2009. Economic decision rules for IPM. In *Integrated Pest Management: Concepts, Tactics, Strategies and Case Studies*, ed. E. B. Radcliffe, W. D. Hutchinson, and R. E. Cancelado, pp. 25–32. Cambridge, UK: Cambridge University Press.

Hokkanen, H. M. T. 2015. Integrated pest management at the crossroads: Science, politics, or business (as usual)? *Arthropod-Plant Interactions* 9: 543–545.

Intermediate Technology Development Group [ITDG]. 2002. *El fenómeno de El Niño y la gestión de riesgo de desastres*. Taller de capacitación. Proyecto Moquegua. Ministerio Alemán- Agro Acción Alemana. Soluciones Prácticas ITDG-CEOP-ILO. Moquegua, Peru.

Jarquín-Gálvez, R. 2003. Las ECEAs: Base para la implementación de proyectos de desarrollo autogestionarios en zonas cafetaleras. *LEISA* 19: 33–36.

Klassen, W. 2000. Area-wide approaches to insect pest management: History and lessons. In *Proceedings: Area-Wide Control of Fruit Flies and Other Insect Pests. International Conference on Area-Wide Control of Insect Pests, and the 5th International Symposium on Fruit Flies of Economic Importance*, ed. K.H. Tan, pp. 21–38. May 28–June 5, 1998, Penang, Malaysia. Pulau Pinang, Malaysia: Penerbit Universiti Sains Malaysia.

Kogan, M. 1998. Integrated pest management: Historical perspectives and contemporary developments. *Annual Review of Entomology* 43: 243–270.

Lawson, J., and C. Willden. 2016. Mixture experiments in R using mixexp. *Journal of Statistical Software, Code Snippets* 72(2): 1–20.

Levins, R. 2007. From simple IPM to the management of agroecosystems. In *Perspectives in Ecological Theory and Integrated Pest Management*, ed. M. Kogan and P. Jepson, pp. 45–64. Cambridge, UK: Cambridge University Press.

Lewis, W. J., J. C. van Lenteren, S. C. Phatak, and J. H. Tumlinson III. 1997. A total system approach to sustainable pest management. *Proceeding of the National Academy of Science USA* 94: 12243–12248.

Liebig, T., L. Jassogne, E. Rahn, P. Läderach et al. 2016. Towards a collaborative research: A case study on linking science to farmers' perceptions and knowledge on Arabica coffee pests and diseases and its management. *PLoS ONE* 11(8): e0159392.

Lindquist, D. A. 2000. Pest management strategies: Area-wide and conventional. In *Proceedings: Area-Wide Control of Fruit Flies and Other Insect Pests. International Conference on Area-Wide Control of Insect Pests, and the 5th International Symposium on Fruit Flies of Economic Importance*, ed. K. H. Tan, pp. 13–19. May 28–June 5, 1998, Penang, Malaysia. Pulau Pinang, Malaysia: Penerbit Universiti Sains Malaysia.

Machado-Vargas, M. M. 2017. Marco conceptual para evaluar los niveles de resiliencia socioecológica, aplicación en estudios de caso en pequeños productores de café. In *Nuevos caminos para reforzar la resiliencia agroecológica al cambio climático*, ed. C.I. Nicholls and M.A. Altieri, pp. 46–53. Berkeley, CA: Sociedad Científica Latinoamericana de Agroecología (SOCLA) and Red Iberoamericana de Agroecología para el Desarrollo de Sistemas Agrícolas Resilientes al Cambio Climático (REDAGRES).

Machado-Vargas, M. M., C. I. Nicholls-Estrada, and L. A. Ríos-Osorio. 2018. Social-ecological resilience of small-scale coffee production in the Porce river basin, Antioquia (Colombia). *Idesia* 36(3): 141–151.

Menalled, F. D., D. A. Landis, and L. E. Dyer. 2004. Research and extension supporting ecologically based IPM systems. *Journal of Crop Improvement* 11(1–2): 153–174.

Meneses-Ospina, E. 2015. Bases para el manejo agroecológico de Eurhizococcus colombianus Jakubski (Hemiptera: Margarodidae) en cultivos de mora del Oriente Antioqueño. *Doctoral thesis.* Universidad Nacional de Colombia, Facultad de Ciencias Agrarias, Medellín, Colombia.

Mojica, F. J. 1991. *La prospectiva técnica para visualizar el futuro.* Bogotá, Colombia: Editorial Legis.

Mojica, F. J. 2004. *El modelo prospectivo llevado a la práctica.* Cartagena, Colombia: Convenio Andrés Bello.

Montalba, R., M. García, M. Altieri, F. Fonseca, and L. Vieli. 2013. Utilización del Índice Holístico de Riesgo (IHR) como medida de resiliencia socioecológica a condiciones de escasez de recursos hídricos. Aplicación en comunidades campesinas e indígenas de la Araucanía, Chile. *Agroecología* 8(1): 63–70.

Montalba, R., F. Fonseca, M. García, L. Vieli, and M. Altieri. 2015. Determinación de los niveles de riesgo socioecológico ante sequías en sistemas agrícolas campesinos de La Araucanía chilena. Influencia de la diversidad cultural y la agrobiodiversidad. *Papers Revista de Sociología* 100(4): 607–624.

Morin, E. 2000. *La mente bien ordenada: Pensar la reforma, reformar el pensamiento.* Barcelona, Spain: Ed. Siex Barral S.A.

Morse, S. 2009. IPM: Ideals and realities in developing countries. In *Integrated Pest Management: Concepts, Tactics, Strategies and Case Studies*, ed. Radcliffe, E. B., W. D. Hutchinson, and R. E. Cancelado, pp. 458–470. Cambridge, UK: Cambridge University Press.

Morse, S., and W. Buhler. 1997. *Integrated Pest Management: Ideals and Realities in Developing Countries.* Boulder, CO: Lynne Rienner Publishers.

Munyuli, T. K., Cihire, D. Rubabura, K. Mitima et al. 2017. Farmers' perceptions, believes, knowledge and management practices of potato pests in South-Kivu Province, eastern of Democratic Republic of Congo. *Open Agriculture* 2: 362–385.

National Research Council [NRC]. 1996. *Ecologically Based Pest Management: New Solutions for a New Century.* Washington, DC: National Academy Press.

Nicholls, C. I., M. A. Altieri. and L. Vazquez. 2016. Agroecology: Principles for the conversion and redesign of farming systems. *Journal of Ecosystem & Ecography* S5: 010.

Norris, R. F., E. P. Caswell-Chen, and M. Kogan. 2003. *Concepts in Integrated Pest Management.* Upper Saddle River, NJ: Prentice Hall.

Odum, E. P. 1986. *Fundamentos de ecología.* Mexico: Nueva Editorial Interamericana, S.A. de C.V.

Oerke, E. 2006. Crop losses to pests. *The Journal of Agricultural Science* 144(1): 31–43.

Parsa, S., S. Morse, A. Bonifacio, T. C. B. Chancellor et al. 2014. Obstacles to integrated pest management adoption in developing countries. *Proceeding of the National Academy of Science USA* 111(10): 3889–3894.

Ponsen, M., K. Tuyls, M. Kaisers, and J. Ramon. 2009. An evolutionary game-theoretic analysis of poker strategies. *Entertainment Computing* 1: 39–45.

Prokopy, R. J., and M. Kogan. 2003. Integrated pest management. In *Encyclopedia of Insects*, ed. V.H. Resh and R.T. Cardé, pp. 4–9. New York: Academic Press.

Savory, A., and J. Butterfield. 1999. *Holistic Management, a New Framework for Decision Making*, 2nd ed. Washington, DC: Island Press.

Segura, H. R., J. F. Barrera, H. Morales, and A. Nazar. 2004. Farmers' perceptions, knowledge, and management of coffee pests and diseases and their natural enemies in Chiapas, Mexico. *Journal of Economic Entomology* 97(5): 1491–1499.

Selvaduray, G. 2004. *Ternary Phase Diagrams: An Introduction.* San Jose, CA: San Jose State University.

Smuts, J. C. 1936. *Holism and Evolution*, 3rd ed. London, UK: Macmillan and Co., 358 p.

Stern, V. M., R. F. Smith, R. van der Bosch, and K. S. Hagen. 1959. The integrated control concept. *Hilgardia* 29: 81–101.

Thrupp, L. A. 1996. Overview. In *New Partnerships for Sustainable Agriculture*, ed. L. A. Thrupp, pp. 1–38. Washington, DC: World Resources Institute.

van de Fliert, E., and A.R. Braun. 2002. Conceptualizing integrative, farmer participatory research for sustainable agriculture: From opportunities to impact. *Agriculture and Human Values* 19: 25–38.

Vandermeer, J., I. Perfecto, and S. M. Philpott. 2010. Ecological complexity and pest control in organic coffee production: Uncovering an autonomous ecosystem service. *BioScience* 60: 527–537.

Walters, M. L., R. Sequeira, R. Staten, O. El-Lissy, and N. Moses-Gonzales. 2009. Eradication: Strategies and tactics. In *Integrated Pest Management: Concepts, Tactics, Strategies and Case Studies*, ed. Radcliffe, E. B., W. D. Hutchinson, and R. E. Cancelado, pp. 298–308. Cambridge, UK: Cambridge University Press.

21 Area-Wide Management of *Anastrepha grandis* in Brazil

Márcio Alves Silva, Gerane Celly Dias Bezerra Silva, Joseph Jonathan Dantas de Oliveira, and Anderson Bolzan*

CONTENTS

Abstract The occurrence of and the damage caused by the South American cucurbit fruit fly (SACFF), *Anastrepha grandis* (Macquart) (Diptera: Tephritidae), are major limiting factors in the production and commercialization of Cucurbitaceae worldwide. Brazil has endemic populations of SACFF in the south-central region. The north and northeast regions do not have endemic populations of the SACFF, allowing the establishment of a SACFF Pest Free Area (PFA) in Ceara and Rio Grande do Norte State. There are also some areas where SACFF is present at low prevalence, allowing the implementation of a SACFF Systems Approach. The areas are maintained using the principles of area-wide integrated pest management (AW-IPM). AW-IPM of SACFF has been practiced in Brazil for many years to protect the phytosanitary status of the areas. Recent research has looked more specifically at the bioecology of SACFF, especially its thermal development requirements. The success of AQ-IPM programs is highly dependent on effective fruit fly trapping, appropriate and quick response to incursions, and an active participation by all growers and the rest of the community and stakeholders in the area under the AW-IPM program. This chapter describes AW-IPM schemes and measures currently being used in Brazil to maintain and expanded the SACFF PFA and SACFF Systems Approach. These phytosanitary schemes are key to maintain Brazil's status as one of the main world producers and exporters of melons and cucurbits.

21.1 INTRODUCTION

Brazil presents areas of cucurbit (family Cucurbitaceae) production in expansion, with emphasis on the cultivation of melon (*Cucumis melo* L.), watermelon (*Citrullus* spp.), cucurbita and squash (*Cucurbita* spp.), and cucumber (*Cucumis sativus* L.). Cucurbits have high commercial value and are appreciated in all continents by people of different cultures; they are traded on a global

* Corresponding author.

scale and represent 20% of the total production of oleraceous products. In the period from 1990 to 2016 (after the establishment of the fruit fly pest-free area [PFA]), Brazil generated a revenue of more than US$2 billion only in melon and watermelon exports (FAOSTAT 2018). In 2016 alone, exports of melon and watermelon were valued at US$180 million (FAOSTAT 2018). The exported Cucurbitaceae are mainly produced in the South American cucurbit fruit fly (SACFF), *Anastrepha grandis* (Macquart) (Diptera: Tephritidae), PFA, located in Ceara and Rio Grande do Norte State, in the Brazilian semiarid region. However, Brazil also has areas with presence of SACFF where a pest risk-mitigation system (or systems approach) is applied for the export of cucurbits.

The occurrence and the damage caused by fruit flies are major limiting factors in the production and commercialization of Cucurbitaceae worldwide (Bolzan et al. 2015, 2017; Meyer et al. 2015; Dominiak and Worsley 2018). The geographical range of fruit flies has increased in recent years due to globalization, including an increase in international trade and globalization (Jiang et al. 2018). More than 10 fruit fly species are known to be able to infest cucurbits around the world, particularly species of the genus *Anastrepha* Schiner (Bolzan et al. 2015, 2017), *Bactrocera* Macquart (Dominiak and Worsley 2018), and *Zeugodacus* Hendel (Meyer et al. 2015). The most important fruit flies affecting Cucurbitaceae in the world are the SACFF (Bolzan et al. 2015, 2017) and the melon fruit fly *Zeugodacus cucurbitae* (Coquillett) (Doorenweerd et al. 2018). The management of SACFF represents a critical factor for success, especially in export crops. The presence of SACFF and its respective surveillance and control methods are subject to several International Standards for Phytosanitary Measures (ISPMs) of the International Plant Protection Convention (IPPC). Among the most relevant and specific are: ISPMs No. 26 "Establishment of pest free areas for fruit flies (Tephritidae)" and ISPM No. 35 "Systems approach for pest risk management of fruit flies (Tephritidae)," which provide the phytosanitary framework under which cucurbits are produced and exported. These international standards include aspects related to pest risk management such as the establishment and maintenance of PFAs, buffer zones, detection and monitoring surveys, and quarantine measures. In this context, this chapter reviews the Brazilian experience, describing bioecological, behavioral, and management aspects of *A. grandis* in an area-wide approach.

21.2 BIOECOLOGY OF *A. GRANDIS*

21.2.1 Biogeography

Anastrepha grandis is a fruit fly pest of quarantine significance native to the American neotropics. Its geographic distribution includes countries of South and Central America, such as Brazil (Silva and Malavasi 1996), Paraguay (Arias et al. 2014), the Andean Mountain Chain from Bolivia to Venezuela (Cabanilla and Escobar 1993; Birke et al. 2013), and Panama (North American Plant Protection Organization [NAPPO] 2009) (Figure 21.1a). In Brazil, the pest has a south-central distribution (except north and northeast), being recorded in the states of central (Mato Grosso do Sul, Mato Grosso, and Goias), southeast (São Paulo, Rio de Janeiro, Espirito Santo and south of Minas Gerais), and southern (Rio Grande do Sul, Santa Catarina and Parana) regions (Figure 21.1b). A single manuscript cites the presence of *A. grandis* in the semiarid region of Bahia in the last century (Bondar 1950). However, there is no evidence of resident populations in the Caatinga biome, a Brazilian semi-arid region.

21.2.2 Damage and Hosts

SACFF is one of the main pests of native and non-indigenous cucurbit species (Norrbom 2000) and attacks may occur at different stages of fruit development (Bolzan et al. 2016; Machado Junior et al. 2017). Females puncture the skin of fruits using their ovipositor and lay eggs in fruits at different stages of development (Figure 21.2a) (Dhillon et al. 2005; Birke et al. 2013; Bolzan et al.

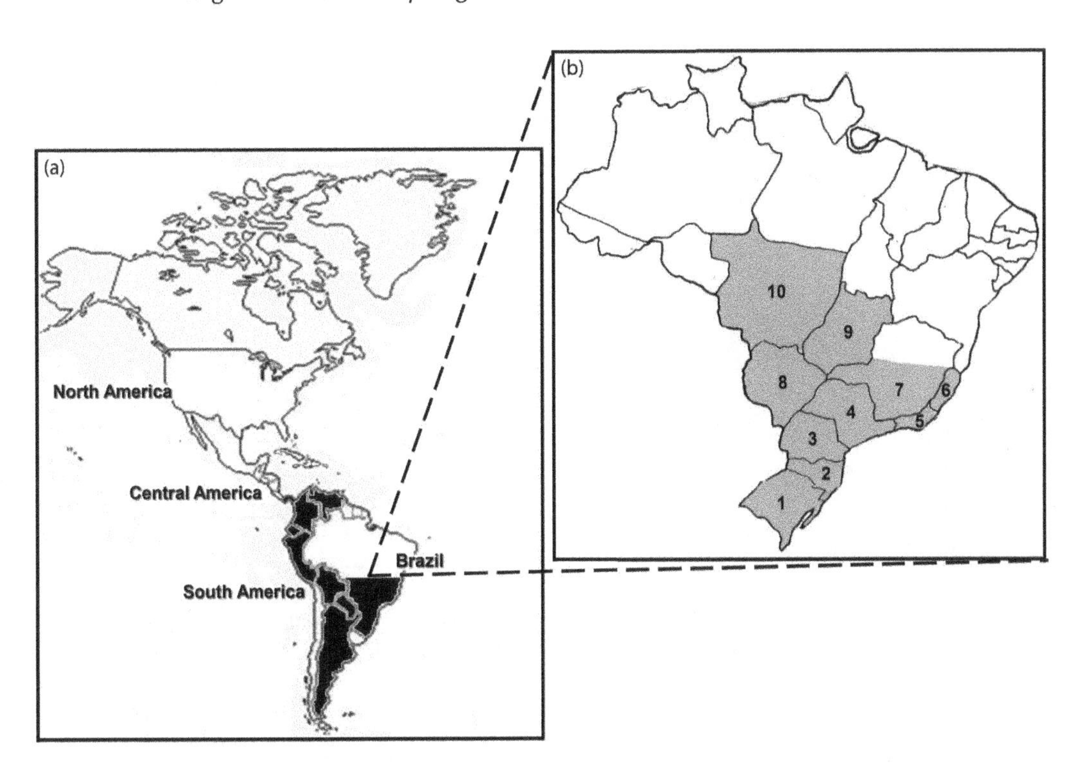

FIGURE 21.1 Biogeography of *Anastrepha grandis*. (a) Occurrence of *A. grandis* in the Americas. White shading indicates countries free of *A. grandis*. Black shading indicates the occurrence of *A. grandis* in Argentina, Brazil, Bolivia, Colombia, Ecuador, Paraguay, Peru, Venezuela, and Panama. (From Cabanilla, G. C. and Escobar, J., Free Zone Program of *Anastrepha grandis* in Ecuador, in *Fruit Flies: Biology and Management*, ed. M. Aluja and P. Liedo, pp. 443–447, Springer-Verlag, New York, 1993; Silva, J. G. and Malavasi, A., Life Cycle of *Anastrepha grandis*, in *Fruit Fly Pests: A World Assessment of Their Biology and Management*, ed. B. A. McPheron and G. J. Steck, pp. 347–351, St Lucie Press, Delray Beach, FL, 1996; North American Plant Protection Organization [NAPPO], Phytosanitary Alert System, Outbreak of *Anastrepha grandis* (South American cucurbit fruit fly) in Panama, http://www.pestalert.org/viewNewsAlert.cfm?naid=76, 2009; Birke, A. et al., Fruit Flies, *Anastrepha ludens* (Loew), *A. obliqua* (Macquart) and *A. grandis* (Macquart) (Diptera: Tephritidae): Three Pestiferous Tropical Fruit Flies that Could Potentially Expand Their Range to Temperate Areas, in *Potential Invasive Pests of Agricultural Crops*, ed. J. Peña, pp. 192–2013, CABI International, Boca Raton, FL, 2013; Arias, O. R. et al., *J. Insect Sci.*, 14, 1–9, 2014.) (b) Gray shading defines the natural occurrence of *A. grandis* in Brazilian States, 1—Rio Grande do Sul, 2— Santa Catarina, 3—Parana, 4—São Paulo, 5—Rio de Janeiro, 6—Espirito Santo, 7—Minas Gerais, 8—Mato Grosso do Sul, 9—Goias, 10—Mato Grosso. (From Uchôa, M. A. and Nicácio, J., *Ann. Entomol. Soc. Am.*, 103, 723–733, 2010; Rabelo, L. R. S. et al., *Arq. Inst. Biol.*, 80, 223–227, 2013.)

2015, 2017). Larvae hatch, feed on the pulp, create galleries, and allow the entrance of microorganisms leading to fruit rot, making them unsuitable for consumption, trade, and industrialization (Figure 21.2b) (Malavasi and Barros 1988; Dhillon et al. 2005; Birke et al. 2013; Bolzan et al. 2015, 2017). However, despite the direct damage to the fruit, the major importance of this pest is associated with economic embargoes imposed by importing countries free of the pest. For example, to import Brazilian Cucurbitaceae, some countries require a certificate of origin indicating that the fruit comes from a PFA or from an area where a systems approach against SACFF is in place.

SACFF has many reported hosts, such as cucurbits of the genera *Cucurbita* (Costa Lima 1926; Malavasi et al. 1980), *Cucumis* (Costa Lima 1926; Silva et al. 1968; Silva and Malavasi 1993a; Oliveira et al. 2012), *Citrullus* (Costa Lima 1926; Korytkowski and Ojeda-Peña 1968), *Lagenaria* (Oliveira et al. 2012; Baldo et al. 2017), and *Sechium* (Silva et al. 1968) (Table 21.1). Species

FIGURE 21.2 Damage caused by *Anastrepha grandis* in Cucurbitaceae. (a) Female inserting the ovipositor into the fruit. (b) Squash cut in half showing larvae feeding and galleries. The circle shows a larva and the arrows indicate the presence of galleries. (From Bolzan, A. et al., *Anastrepha grandis*: Bioecologia e Manejo. Embrapa Clima Temperado [Documentos/Embrapa Clima Temperado, 404], 2016.)

TABLE 21.1
Host Plant Species of *Anastrepha grandis*

Host		
Common Name	**Species**	**References**
Pumpkin	*Cucurbita* sp.	Costa Lima (1926); Malavasi et al. (1980)
Zucchini	*Cucurbita* sp.	Costa Lima (1926); Malavasi et al. (1980)
	Cucurbita pepo L.	Korytkowski and Ojeda-Peña (1968); Baldo et al. (2017)
Squash	*Cucurbita* sp.	Costa Lima (1926); Malavasi et al. (1980)
Melon	*Cucumis* sp.	Costa Lima (1926); Silva et al. (1968); Silva and Malavasi (1993a); Oliveira et al. (2012)
	Cucumis melo L.	Korytkowski and Ojeda-Peña (1968)
Cucumber	*Cucumis* sp.	Costa Lima (1926); Silva et al. (1968); Silva and Malavasi (1993a); Oliveira et al. (2012)
	Cucumis sativus L.	Korytkowski and Ojeda-Peña (1968)
Gherkin	*Cucumis* sp.	Costa Lima (1926); Silva et al. (1968); Silva and Malavasi (1993a); Oliveira et al. (2012)
Watermelon	*Citrullus* sp.	Costa Lima (1926)
	Citrullus vulgaris Schrad	Korytkowski and Ojeda-Peña (1968)
Gourd	*Lagenaria* sp.	Oliveira et al. (2012)
Bottle gourd	*Lagenaria siceraria* (Mol.) Standl.	Baldo et al. (2017)
Chayote	*Sechium edule* (Jacq.) Swartz	Silva et al. (1968)

of the genus *Cucurbita* are primary hosts that provide better fitness for SACFF. Bolzan et al. (2015) evaluated the effect of different hosts on the biology of SACFF; the hosts of *Cucurbita* and *Cucumis* genera allowed complete development of the SACFF. On the other hand, watermelon (*Citrullus lanatus* [Thunb.] Matsum and Nakai) and chayote (*Sechium edule* [Jacq.] Swartz) did not allow the development of SACFF (Bolzan et al. 2015). It was also reported that hosts of the genus *Cucurbita* provided higher fecundity and number of insects per fruit than other genera (i.e., *Cucumber, Citrullus* and *Cucumis*) (Bolzan et al. 2015). A hypothesis has been suggested stating that the neotropical origin of the *Cucurbita* genus and SACFF allowed a coevolution over a long period. This association did not happen between SACFF and other genera such as *Cucumber, Citrullus*, and *Cucumis* (Bolzan et al. 2015). In addition, the chayote *S. edule* is native to Central America and does not appear to favor the development of SACFF (Bisognin 2002; Kokubo 2012; Bolzan et al. 2015).

There are reports of hosts from other plant families. The guava (*Psidium guajava* L.) was reported by Fischer (1934) as a host, but Norrbom and Kim (1988) classified it as an incidental record because only one specimen of *A. grandis* was bred from a fruit of guava tree in the middle of a pumpkin field. There are also reports of passion fruit (*Passiflora alata* Dryand.) and *Citrus* sp. (Costa Lima 1934; Oakley 1950), but these reports are considered doubtful. The record on citrus was classified as questionable by Norrbom and Kim (1988).

21.2.3 Biological, Morphological, and Behavioral Parameters

SACFF is one of the species of the *Anastrepha* genus that lays a large number of eggs per clutch, reaching up to 110 eggs per clutch (Silva and Malavasi 1993b). Females can lay more than 500 eggs during their lifetime (Bolzan et al. 2015). The eggs are white and crescent-shaped, their length ranges from 2.06 to 2.25 mm, and their width is about 0.20 mm (Steck and Wharton 1988). The duration of the egg stage ranges from 21 d at 15°C to 6.9 d at 30°C, the duration was 7.3 d at 25°C (Table 21.2, obtained from Bolzan et al. 2017). In the third instar, larval length ranges from 6.6 to 17.0 mm and the width of the sixth abdominal segment ranges from 1.6 to 2.7 mm. Larvae exhibit cream color and an elongated shape, with a truncate caudal segment and tapered thoracic segments (Steck and Wharton 1988). When pumpkins are host fruits, larval development time at 25°C is of 17.7 d (Silva and Malavasi 1996). At the end of the larval stage, larvae leave the fruit and go into the ground to pupate. Pupae can range from 8 to 9.1 mm in length and a maximum width of 3.2–3.7 mm. The color of integumental areas is golden-brown with a small blackened area around the oral opening (Steck and Wharton 1988). The duration of the pupal stage ranges from 52.3 d at 15°C to 16.5 d at 30°C, the duration was around 20 d at 25°C (Table 21.2, obtained from Bolzan et al. 2017). Female longevity ranged from 119.9 to 9.1 d when temperatures ranged from 15°C to 35°C, longevity was 75.5 d at 25°C (Bolzan et al. 2017). Silva (1991) demonstrated that males present greater longevity than

TABLE 21.2
Biological Parameters of *Anastrepha grandis* at Different Temperatures

	Egg		Egg to Pupa	Pupa		Egg to Adult
Temperatures (°C)	Duration[a] (Days)	Survived[b] (%)	Duration[a] (Days)	Duration[a] (Days)	Survived[b] (%)	Duration[b] (Days)
15	21.0 ± 0.4a	12.2 ± 6.0c	41.0 ± 0.1a	52.3 ± 0.3a	89.4 ± 1.6a	93.3 ± 2.0a
20	10.4 ± 0.1b	53.9 ± 5.7b	23.6 ± 0.1b	29.1 ± 0.1b	95.6 ± 1.1a	52.7 ± 1.0b
25	7.3 ± 0.1c	91.7 ± 3.7a	19.0 ± 0.1d	20.3 ± 0.1c	96.1 ± 1.6a	39.3 ± 0.9c
30	6.9 ± 0.1c	56.1 ± 14.2b	22.6 ± 0.2c	16.5 ± 0.2d	43.0 ± 4.0b	39.1 ± 1.1c
35	—	0.0 ± 0.0d	—	—	0.0 ± 0.0c	—

Source: Bolzan, A. et al., *Crop Prot.*, 100, 38–44, 2017.

[a] Values represent the survival curves that do not differ according to the log rank test when followed by the same letter in the column.

[b] Values followed by the same letter in the column do not differ according to Tukey's test ($P < 0.05$).

females. The adult morphology of *A. grandis* is characterized by a rather large size and the coloration is mostly orange to red-brown with setae usually moderately red-brown (Norrbom 1991). Mesonotum length ranges from 2.88 to 4.22 mm, with distinct medial and lateral stripes of yellow and dark-brown colors (Norrbom 1991). The length of the wings varies from 7.95 to 10.30 mm, with the proximal arm of the V-band ending at M or extending anteriorly (often faintly) to R_{4+5} to fuse with the S-band (Norrbom 1991). The color of the abdomen is yellowish to orange (Norrbom 1991). The seventh abdominal segment (oviscape) can reach 6.28 mm in length and the aculeo can reach 6.18 mm in length, it is often as long as or longer than the seventh abdominal segment (Norrbom 1991).

SACFF completes its developmental cycle (egg-adult) at temperatures between 15 and 30°C, and immatures do not survive at temperatures from 35°C (Table 21.2). The highest fecundity and fertility of *A. grandis* were found at 25°C (Bolzan et al. 2017). At mild temperatures (15°C and 20°C), this pest presented greater fecundity and a larger number of individuals developed per fruit than at higher temperatures (30°C) (Bolzan et al. 2017). The pre-oviposition period also varies according to temperature (the shortest period was 27.3 d at 30°C), demonstrating that the pre-imaginal period and ovarian maturation are influenced by temperature (Bolzan et al. 2017). The lower temperature threshold and thermal constant for *A. grandis* to complete the egg adult cycle are 5.2°C and 858.7 degree days, respectively (Bolzan et al. 2017).

After emergence, females look for protein-rich foods and then are able to reproduce (Cresoni and Zucoloto 2012). Reproduction begins with the copula; the calling behavior of males begins around 5:00 p.m. and ceases around 8:00 p.m., when it is already dark (Silva and Malavasi 1993b). Courtship behavior starts when males present puffed pleural abdominal glands and an everted anal pouch, then they walk, rotating 180° or 360° and touching the substrate with the anal pouch (Silva and Malavasi 1993b). After this, they form leks of 4–10 flies; in these clusters, males initiate sequences of calling behaviors (sex pheromone release and vibration of wings producing sounds), attracting sexually receptive females (Prokopy 1980; Silva and Malavasi 1993b). Immediately after, females arrive and usually approach the males on a face to face orientation before copulation (Silva and Malavasi 1993b). The average duration of the copula is approximately 4 ½ h (Silva and Malavasi 1993b). Oviposition activity of the SACFF occurs from 8:00 a.m. to 4:00 p.m., with greater activity from 11:00 a.m. to 2:00 p.m. (Silva and Malavasi 1993b). Oviposition activity can be divided into four phases: foraging, oviposition, cleaning, and dragging (Silva and Malavasi 1993b; Silva et al. 2012). Mean oviposition time is approximately 45 min (Silva and Malavasi 1993b). Ovipositor dragging after puncturing is an indicative of oviposition (Silva and Malavasi 1993b; Silva et al. 2012).

The knowledge of biological, morphological, and behavioral aspects is a fundamental tool for pest management. Already published bioecology studies provide basic information on the development of *A. grandis*. However, studies on the influence of other factors that affect the biology of this pest are still needed for a better understanding of the population dynamics, spatial distribution, and presence/absence of *A. grandis* in a given environment.

21.3 FRUIT FLY PFAs IN NORTHEAST BRAZIL

The requirements for the establishment of PFAs are available in ISPM No. 26. The ISPMs are recognized as the basis for the application of phytosanitary measures by World Trade Organization (WTO) members in compliance with Article 6 of the Agreement on the Application of Sanitary and Phytosanitary Measures (SPS Agreement) and are adopted by member countries through the Commission on Phytosanitary Measures (CPM) of the IPPC (Matyak 2011). They are drafted by the IPPC Secretariat and used as a guideline by Member Countries of the Food and Agriculture Organization (FAO) of the United Nations and the WTO, as well as by other stakeholders. ISPMs contain harmonized phytosanitary measures to facilitate trade and avoid unwarranted use of trade barriers (Matyak 2011). ISPMs by themselves are not legal instruments; however, they gain this status when countries use them as the basis for setting requirements in their national plant protection legislations. According to ISPM No. 5, PFA is defined as "an area in which a specific pest is absent as demonstrated by scientific evidence and in which, where appropriate, this condition is being officially maintained" (FAO 2018b). Among the general requirements described in ISPM No. 26 (FAO 2016), which deals with the establishment of PFAs for fruit flies (Tephritidae), are public awareness, documentation and record-keeping, and supervision, corrective action plans, loss of status and reinstatement, and others.

In Brazil, the only officially recognized SACFF free area is located in the northeast region. The area is composed of 20 municipalities, seven belonging to the Ceara State (CE) and 13 to the Rio Grande do Norte State (RN). The SACFF free area is about 14,000 km^2 in size and has the Atlantic Ocean as a natural barrier to the north, and to the other directions, there is a buffer zone of approximately 14,700 km^2 (Figure 21.3). Efforts to establish a SACFF free area began in 1985, due to the demand from Açu River Valley farmers in Rio Grande do Norte (RN). Farmers aspired to export cucurbits to the United States; however, based on ISPM No. 11 "Pest risk analysis for quarantine pests," additional studies and adjustments to the quarantine were required (FAO 2017a). The SACFF free area was recognized only in 1990 through a bilateral agreement (Araujo et al. 2000). Before the SACFF PFA, only the PFA in Sonora Mexico had been officially recognized as such (Gutiérrez-Ruelas et al. 2013). In the past, it was common to consider the entire country as being free of a particular pest, making the establishment of a PFA unfeasible from an economic and technical point of view. The fruit fly PFA in Mexico and SACFF PFA in Brazil changed this trend. The Ministry of Agriculture, Livestock and Food Supply (MALFS) of Brazil issued normative acts in 2003 recognizing and delimiting the SACFF PFA, which included other municipalities in the state of Rio Grande do Norte and Ceara, which is the territorial extent that remains at present (Braga Sobrinho et al. 2002; Brasil 2003a, 2003b).

According to the regulatory framework of the SACFF PFA in Brazil, the measures to establish a free area are: (1) The location of the proposed area in the federation unit (free area and buffer zone); (2) identification of the production transport routes of the exit point (i.e., port and airport) to the external market; (3) location of fruit fly monitoring points; and (4) location of cucurbit

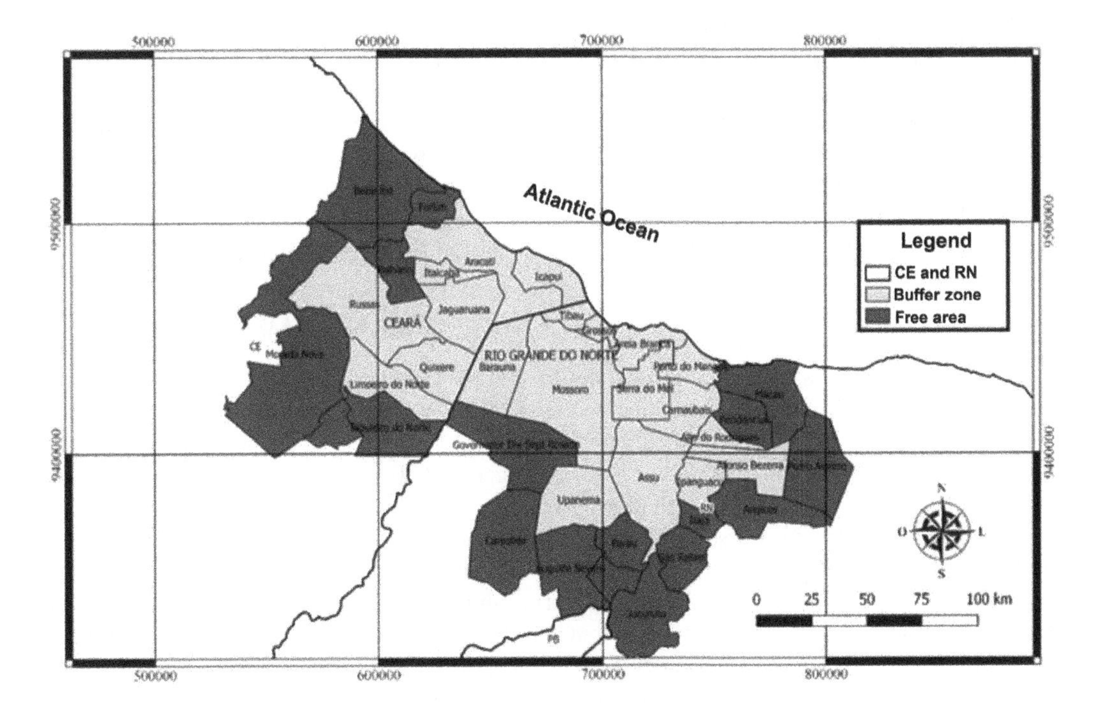

FIGURE 21.3 Free area for *Anastrepha grandis* in Brazil.

crops and geographical factors of the area, such as description, extension, and location of the production sites with indication of the isolation conditions of the area and the existence of possible natural barriers for the pest (Brasil 2006a).

After establishing the free area, it is necessary to adopt a series of measures to maintain this condition. Among these, trapping is important because it is essential in the establishment, verification, and subsequent maintenance of a phytosanitary condition (Jang et al. 2014). Trapping in the SACFF PFA area is carried out continuously using McPhail traps baited with hydrolyzed protein diluted to 5% in water (food attractant). McPhail traps are strategically distributed in the area at critical locations to increase the probability of SACFF detection, such as fruit supply centers and highways through which horticultural products are transported into the free area and buffer zone. Farmers in the SACFF PFA who export cucurbits also need to carry out trappings in their properties. In the farms, traps are installed 35 days after sowing and distributed in a proportion of one for every 5 ha. Trapping and identification of fruit flies is carried out by companies accredited by the MALFS (Brasil 2006a). In the event of a possible detection of the SACFF, the free-area condition would be suspended. Consequently, the export would also be suspended for countries that require the PFA condition. Concomitantly, a contingency plan for eradication would be initiated with the aim of containing and eradicating the SACFF by means of a systematic set of actions (Brasil 2006a). The plan takes into account the dispersal capacity of the SACFF to delimit an area for trapping and eradication actions around the detection area. The SACFF has never been detected in the SACFF PFA in Brazil.

Another fundamental measure for the maintenance of the SACFF PFA is phytosanitary surveillance in fixed and mobile quarantine checkpoints in the main access routes. This action restricts the entry of cucurbits coming from other Brazilian states, allowing the transit of consignments only from another SACFF PFA or from SACFF under a systems approach recognized by MALFS. To ensure the traceability of the production and to be able to transit with loads of cucurbits across the SACFF PFA, each farmer should keep a record of the production units with the State Plant Protection Agency (SPPA) and the transport of the consignment must be accompanied by a certificate of origin issued by a qualified accredited technician.

The SPPA frequently conducts phytosanitary education activities with the sectors involved. It includes training and qualifying the professionals that work in the area, instructing farmers, and raising the awareness of the region's supermarket chains. The SACFF PFA is of great importance for the region because it reduces the cost of agricultural production through the absence of the pest and ensures the preferential access to markets of other countries with quarantine restrictions (Weldon et al. 2014). The National Program to Combat Fruit Flies has been currently established in Brazil, which ensures greater financial and personnel resources, among other aspects, for programs involving the prevention, control, and eradication of *A. grandis*. The MALFS has analyzed the possibility of expanding the SACFF PFA toward the northeast of Brazil (Brasil 2015; MAPA 2017).

21.4 SYSTEMS APPROACH FOR *A. GRANDIS* RISK MANAGEMENT IN BRAZIL

Systems approach is a pest risk-management option that integrates different measures, at least two of which act independently with a cumulative effect (FAO 2018b). An advantage of the systems approach is the ability to address variability and uncertainty by modifying the number and strength of measures to meet phytosanitary import requirements (FAO 2018a). Measures used in a systems approach may be applied pre- or postharvest wherever National Plant Protection Organizations (NPPOs) have the ability to oversee and ensure compliance with phytosanitary procedures (FAO 2017b). Thus, a systems approach may include measures applied in the place of production, during the postharvest period, at the packing house, or during shipment and distribution of the commodity (FAO 2017a). Cultural practices, crop treatment, postharvest disinfestation, inspection, and other procedures may be integrated in a systems approach (FAO 2017b). Risk-management measures designed to prevent contamination or reinfestation are generally included in a systems approach (e.g., maintaining the integrity of lots, requiring pest-proof packaging, and screening packing areas)

(FAO 2017a). Likewise, procedures such as pest surveillance, trapping, and sampling can also be components of a systems approach (FAO 2017b). Measures that do not kill pests or reduce their prevalence but reduce their potential for entry or establishment (safeguards) can be included in a systems approach as well (FAO 2017b). Examples include designated harvest or shipping periods, restrictions on the maturity, color, hardness, or other condition of the commodity, the use of resistant hosts, and limited distribution or restricted use at the destination (FAO 2017b).

The systems approach can be used as an option for pest risk mitigation by farmers of Cucurbitaceae (*C. melo*, *Citrullus* spp., *Cucurbita* spp., and *C. sativus*) who wish to export to countries that demand that fruits do not present the quarantine risk of *A. grandis*. In Brazil, according to Normative Instruction 16/2006, farmers who wish to adopt the systems approach should express their interest to the SPPA. The SPPA shall prepare and submit a project requesting recognition of the SACFF Systems Approach to the MALFS, which will formalize the process containing the following items: (1) Description of the proposed area, geographic extension, and georeferenced location of the Cucurbitaceae cultivation; (2) regulations and control procedures; (3) detection trapping start date and listing and location of traps; and (4) situation of the Cucurbitaceae cultivation in the federation unit: commercial production area in ha, common and scientific names of species and cultivars cultivated, estimated yield per species in tonnes, information on estimated export volume per cultivar and identification of the production transport routes of the exit point (i.e., port and airport) to the external market, cultivation systems, harvest and postharvest procedures, and other pests associated with Cucurbitaceae.

Farmers who join the SACFF Systems Approach must carry out the recommended phytosanitary actions and guarantee the identity, traceability, and phytosanitary compliance of the products from the areas registered in the SACFF Systems Approach, which will be verified by a survey carried out by MALFS (Brasil 2006b). The farmer joining the SACFF Systems Approach should monitor this pest in its production unit, following the guidelines of the technician responsible for the phytosanitary certificate of origin, under the supervision of MALFS (Brasil 2006b).

The SACFF Systems Approach has already been adopted in seven federation units (States) and 38 municipalities in Brazil (Table 21.3). The SACFF Systems Approach in Brazil is

TABLE 21.3
Municipalities with Systems Approach for *Anastrepha grandis* in Brazil

State	Municipalities	Legislation
Rio Grande do Sul	Bagé, Dom Pedrito and Herval	Normative Instruction 35/2008
Paraná	Santa Izabel do Ivaí	Resolution 1/2012
São Paulo	Mesópolis, Paranapuã, Urânia and Presidente Bernardes	Normative Instruction 42/2006
	Tarabai	Normative Instruction 37/2007
	Regente Feijó	Normative Instruction 32/2008
	Rinópolis	Resolution 2/2009
	Indiana	Normative Instruction 9/2014
Goiás	Carmo do Rio Verde, Itapuranga, Jaraguá and Uruana	Normative Instruction 41/2006
	Rio Verde, Maurilândia and Santa Helena	Normative Instruction 22/2008
	Cristalina and Ipameri	Resolution 1/2009
	Goianésia and São Miguel do Araguaia	Normative Instruction 4/2013
	Edealina	Normative Instruction 32/2017
Minas Gerais	Paracatu, João Pinheiro, Unaí, Uberlândia, Jaíba, Matias Cardoso and Manga	Normative Instruction 29/2007
	Luz	Resolution 3/2009
Bahia	Ribeira do Amparo	Normative Instruction 23/2012
	Curaçá	Normative Instruction 22/2017
Rio Grande do Norte	Macau and Jandaíra	Resolution 1/2011
	Apodi and Governador Dix-Sept Rosado	Normative Instruction 21/2016

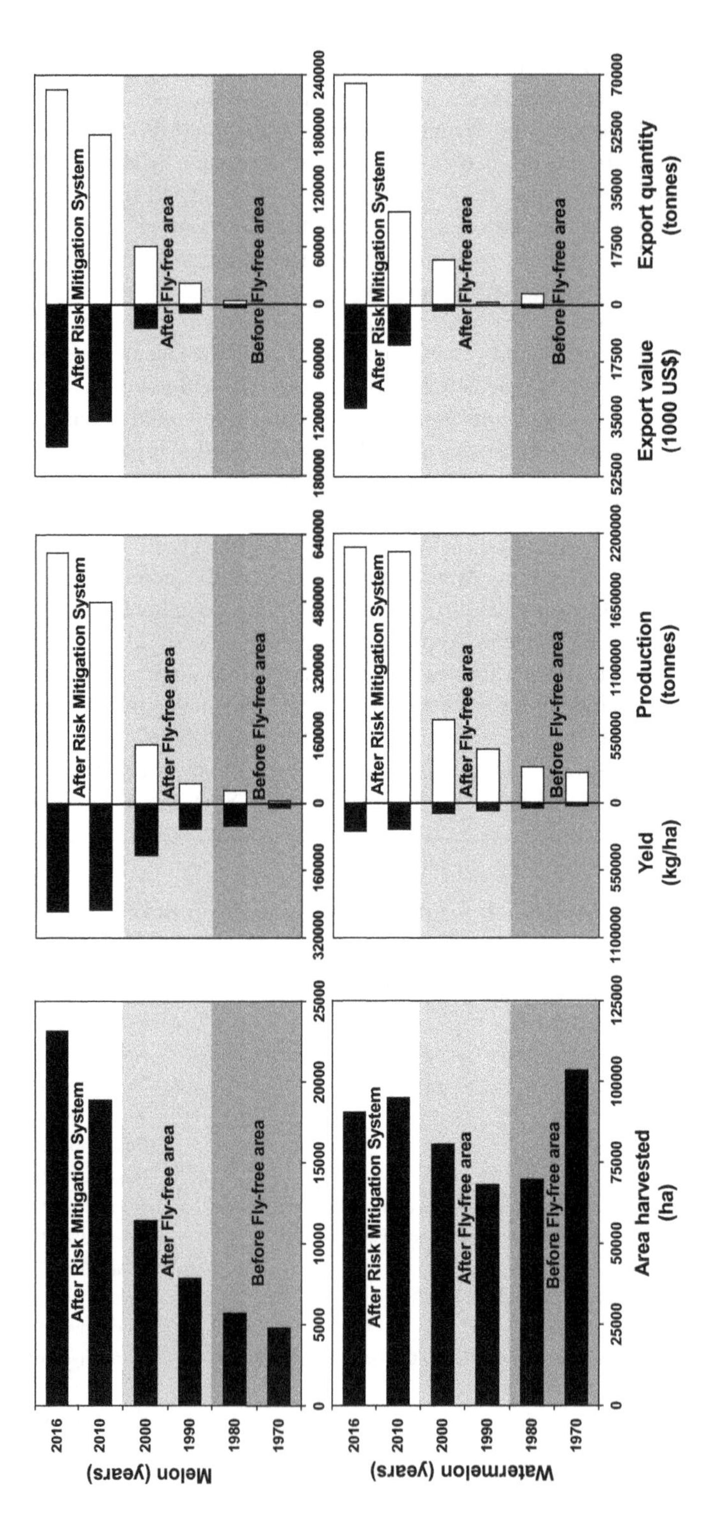

FIGURE 21.4 Production of (a) melon and (b) watermelon in Brazil. (Data compiled from Food and Agriculture Organization of the United Nations [FAOSTAT], Food and Agriculture Data, 2018, http://www.fao.org/faostat/en/#data.)

mainly based on a continuous low pest prevalence and field certification and management (e.g., inspection with traps and pesticides).

Surveys for the detection, delimitation, or trapping of *A. grandis* should be carried out using McPhail traps with hydrolyzed protein diluted to 5% in water (Brasil 2006b). The traps must be installed within 35 days of cultivation (starting from the date of sowing, even for transplanting) and must remain as long as there are cultural remains after harvest (Brasil 2006b). In areas under detection or delimitation, as well as in production units under trapping, traps should be distributed from the periphery to the center of the crop area to cover the whole area (Brasil 2006b).

One trap must be installed every 5 ha or fraction (Brasil 2006b). In the case of the flies per trap per day (FTD) index, it should be kept at 0.1 or less. If the index is greater than 0.1 but less than 0.4 during the weekly trapping period, production from the respective production unit will be prevented from being certified until the pest control plan is implemented and the FTD index is again less than or equal to 0.1 (Brasil 2006b). In the case that the FTD index for *A. grandis* is greater than 0.4 during the weekly trapping period, production from the respective production unit will be prevented from being certified for export in the current harvest (Brasil 2006b).

The Brazilian production of Cucurbitaceae increased sharply after the establishment of the SACFF PFA and SACFF Systems Approach (Figure 21.4). Productivity has increased exponentially due to the increase in the level of technology and the intensification of activities, based mainly on the significant increase in irrigated production systems. Brazil has become a major world producer and exporter of Cucurbitaceae, mainly melon (Figure 21.4).

21.5 FINAL THOUGHTS

The SACFF is an important pest in many countries because of its potential to cause damage to cucurbit fruits and restrict access to international markets. This review expanded the current knowledge on the SACFF area-wide management, initially presenting aspects of taxonomy, morphology, ecology, biology, and behavior. Brazil became a large producer and exporter of Cucurbitaceae since the establishment of the SACFF PFA and use of the Systems Approach for SACFF risk management. The opening of new markets was only possible in the 1990s, when the absence of the pest was officially recognized in the Rio Grande do Norte and Ceara States in Brazil. Evidence of the absence of *A. grandis* was possible through the deployment of a high-sensitivity detection system. This allowed to verify with certainty the absence of the pest in melon-producing areas and its accidental entry into the area. Early detection of pest entries allows the immediate application of eradication measures following a contingency plan. The establishment, maintenance, and expansion of SACFF PFA are an indispensable condition for keeping Brazil's status as one of the main world producers and exporters of melons and other cucurbits.

REFERENCES

Araujo, E. L., F. A. M. Lima, and R. A. Zucchi. 2000. Rio Grande do Norte. In *Moscas-das-frutas de importância econômica no Brasil: Conhecimento básico e aplicado*, ed. A. Malavasi, and R. A. Zucchi, pp. 223–226. Ribeirão preto, Brazil: Holos Editora.

Arias, O. R., N. L. Fariña, G. N. Lopes, K. Uramoto, and R. A. Zucchi. 2014. Fruit Flies of the Genus *Anastrepha* (Diptera: Tephritidae) From Some Localities of Paraguay: New Records, Checklist, and Illustrated Key. *Journal Insect Science* 14:1–9.

Baldo, F. B., L. H. C. Berton, L. J. Gistoli, and A. Raga. 2017. New Records of *Anastrepha grandis* (Diptera, Tephritidae) and *Neosilba zadolicha* (Diptera, Lonchaeidae) in Cucurbitaceae Species in Brazil. *Journal of Agriculture and Ecology Research International* 13:1–7.

Birke, A., L. Guillén, D. Midgarden et al. 2013. Fruit Flies, *Anastrepha ludens* (Loew), *A. obliqua* (Macquart) and *A. grandis* (Macquart) (Diptera: Tephritidae): Three Pestiferous Tropical Fruit Flies That Could Potentially Expand Their Range to Temperate Areas. In *Potential Invasive Pests of Agricultural Crops*, ed. J. Peña, pp. 192–2013. Boca Raton, FL: CABI International.

Bisognin, D. A. 2002. Origin and Evolution of Cultivated Cucurbits. *Ciência Rural* 32:715–723.

Bolzan, A., G. I. Diez-Rodríguez, F. R. M. Garcia, and D. E. Nava. 2016. *Anastrepha grandis*: Bioecologia e Manejo. Embrapa Clima Temperado (Documentos/Embrapa Clima Temperado, 404).

Bolzan, A., D. E. Nava, F. R. M. Garcia, R. A. Valgas, and G. Smaniotto. 2015. Biology of *Anastrepha grandis* (Diptera: Tephritidae) in Different Cucurbits. *Journal of Economic Entomology* 108:1034–1039.

Bolzan, A., D. E. Nava, G. Smaniotto, R. A. Valgas, and F. R. M. Garcia. 2017. Development of *Anastrepha grandis* (Diptera: Tephritidae) Under Constant Temperatures and Field Validation of a Laboratory Model for Temperature Requirements. *Crop Protection* 100:38–44.

Bondar, G. 1950. Moscas de Frutas na Bahia. *Boletim de Campo* 6:13–15.

Braga Sobrinho, R., R. N. Lima, M. A. Peixoto, and A. L. M. Mesquita. 2002. South American Cucurbit Fruit Fly-Free Area in Brazil. In *Proceedings of the 6th International Symposium on Fruit Flies of Economic Importance*, ed. B. N. Barnes, pp. 173–177. Stellenbosch, South Africa, May 6–10.

Brasil. Ministério da Agricultura, Pecuária e Abastecimento. Secretaria de Defesa Agropecuária. 2003a. Instrução Normativa nº 7, de 27 de janeiro de 2003. Brasília: Diário Oficial da União, January 20. http://sistemasweb.agricultura.gov.br/sislegis/action/detalhaAto.do?method=consultarLegislacaoFederal (accessed July 20, 2018).

Brasil. Ministério da Agricultura, Pecuária e Abastecimento. Secretaria de Defesa Agropecuária. 2003b. Portaria nº 150, de 1 de dezembro de 2003. Brasília: Diário Oficial da União, december 2. http://sistemasweb.agricultura.gov.br/sislegis/action/detalhaAto.do?method=consultarLegislacaoFederal (accessed July 20, 2018).

Brasil. Ministério da Agricultura, Pecuária e Abastecimento. Secretaria de Defesa Agropecuária. 2006a. Instrução Normativa nº 13, de 31 de março de 2006. Brasília: Diário Oficial da União, April 13, 2006. file:///C:/Users/gerente/Downloads/INSTRU%C3%87%C3%83O%20NORMATIVA%20132006%20%20Procedimentos%20para%20a%20%C3%A1rea%20livre%20de%20Anastrepha%20grandis.pdf (accessed July 19, 2018).

Brasil. Ministério da Agricultura, Pecuária e Abastecimento. Secretaria de Defesa Agropecuária. 2006b. Instrução Normativa nº 16, de 05 de março de 2006. Brasília: Diário Oficial da União, April 13, 2006. http://www.editoramagister.com/doc_833886_INSTRUCAO_NORMATIVA_N_16_DE_5_DE_MARCO_DE_2006.aspx (accessed July 19, 2018).

Brasil. Ministério da Agricultura, Pecuária e Abastecimento. Gabinete da Ministra. 2015. Instrução Normativa nº 24, de 8 de setembro de 2015. Brasília: Diário Oficial da União, September 9. http://sistemasweb.agricultura.gov.br/sislegis/action/detalhaAto.do?method=consultarLegislacaoFederal (accessed July 19, 2018).

Cabanilla, G. C., and J. Escobar. 1993. Free Zone Program of *Anastrepha grandis* in Ecuador. In *Fruit Flies: Biology and Management*, ed. M. Aluja and P. Liedo, pp. 443–447. New York: Springer-Verlag.

Costa Lima, A. 1926. Sobre as moscas das frutas que vivem no Brasil. *Chácaras e Quintais* 34:20–24.

Costa Lima, A. 1934. Moscas de frutas do genero *Anastrepha* Schiner. *Memórias do Instituto Oswaldo Cruz* 28:487–575.

Cresoni, C. P., and F. S. Zucoloto. 2012. Fruitflies. In *Insect Bioecology and Nutrition for Integrated Pest Management*. ed. A. R. Panizzi and J. R. P. Parra, pp. 451–472. New York: CRC Press.

Dhillon, M. K., R. Singh, J. S. Naresh, and H. C. Sharma. 2005. The Melon Fruit Fly, *Bactrocera cucurbitae*: A Review of Its Biology and Management. *Journal of Insect Science* 5:40–56.

Dominiak, B. C., and P. Worsley. 2018. Review of Cucumber Fruit Fly, *Bactrocera cucumis* (French) (Diptera: Tephritidae: Dacinae) in Australia: Part 1, Host Range, Surveillance and Distribution. *Crop Protection* 106:79–85.

Doorenweerd, C., L. Leblanc, A. L. Norrbom, M. S. Jose, and D. Rubinoff. 2018. A Global Checklist of the 932 Fruit Fly Species in the Tribe Dacini (Diptera, Tephritidae). *ZooKeys* 730:17–54.

Fischer, C. R. 1934. Variação das cerdas frontais e outras notas sobre duas espécies de *Anastrepha* (Diptera: Trypetidae). *Revista de Entomologia* 4:18–22.

Food and Agriculture Organization of the United Nations [FAO]. 2016. Establishment of Pest Free Areas for Fruit Flies (Tephritidae). International Standards for Phytosanitary Measures (ISPM 26), International Plant Protection Convention (IPPC). https://www.ippc.int/en/core-activities/standards-setting/ispms/ (accessed July 20, 2018).

Food and Agriculture Organization of the United Nations [FAO]. 2017a. Pest Risk Analysis for Quarantine Pests (ISPM 11). International Plant Protection Convention (IPPC). https://www.ippc.int/static/media/files/publication/en/2017/05/ISPM_11_2013_En_2017-05-25_PostCPM12_InkAm.pdf (accessed September 2018).

Food and Agriculture Organization of the United Nations [FAO]. 2017b. The Use of Integrated Measures in a Systems Approach for Pest Risk Management. International Standards for Phytosanitary Measures (ISPM 14), International Plant Protection Convention (IPPC). http://www.fao.org/3/a-y4221e.pdf (accessed August 10, 2018).

Food and Agriculture Organization of the United Nations [FAO]. 2018a. Systems Approach for Pest Risk Management of Fruit Flies (Tephritidae) (ISPM 35), International Plant Protection Convention (IPPC). https://www.ippc.int/static/media/files/publication/en/2018/10/ISPM_35_2012_En_FF_Post-CPM-13_InkAm_2018-10-01.pdf (accessed November 4, 2018).

Food and Agriculture Organization of the United Nations [FAO]. 2018b. Glossary of Phytosanitary Terms. International Standards for Phytosanitary Measures (ISPM 05), International Plant Protection Convention (IPPC). https://www.ippc.int/static/media/files/publication/en/2018/06/ISPM_05_2018_En_Glossary_2018-05-20_PostCPM13_R9GJ0UK.pdf (accessed July 20, 2018).

Food and Agriculture Organization of the United Nations [FAOSTAT]. 2018. Food and Agriculture Data. http://www.fao.org/faostat/en/#data (accessed July 31, 2018).

Gutiérrez-Ruelas, J. M., G. S. Martínez, A. Villaseñor Cortes, W. R. Enkerlin, and F. Hernández López. 2013. *Los Programas de Moscas de la Fruta en México. Su Historia Reciente*. Mexico: Talleres de S y G Editores.

Jang, B. E., W. Enkerlin, C. E. Miller et al. 2014. Trapping Related to Phytosanitary Status and Trade. In *Trapping and the Detection, Control, and Regulation of Tephritid Fruit Flies*, ed. T. Shelly, N. Epsky, E. B. Jang, J. Reyes-Flores, and R. Vargas, pp. 589–608. Dordrecht, the Netherlands: Springer.

Jiang, F., L. Liang, Z. Li, Y. Yu, J. Wang, Y. Wu, and S. Zhu. 2018. A Conserved Motif Within *cox 2* Allows Broad Detection of Economically Important Fruit Flies (Diptera: Tephritidae). *Scientific Reports* 8:1–7.

Kokubo, M. C. C. 2012. Aspectos bioecológicos sobre a mosca-das-cucurbitáceas-sul-americana *Anastrepha grandis* (Diptera: Tephritidae). 74p. Master's dissertation, Instituto Biológico, São Paulo.

Korytkowski, C., and D. Ojeda-Peña. 1968. Espécies del genero *Anastrepha* Schiner 1868 en el noroeste Peru. *Revista Peruana de Entomologia* 11:32–70.

Machado Junior, R., R. S. Gomes, C. F. Almeida, and F. M. Alves. 2017. Incidência de moscas-das-frutas em lavouras de cucurbitáceas. *Revista Campo & Negócios HF* 145:50–52.

Malavasi, A., and M. D. Barros. 1988. Comportamento sexual e de oviposição em moscas-das-frutas (Tephritidae). In *Moscas-das-frutas no Brasil*, ed. H. M. L. de Souza, pp. 22–53. Campinas, Brazil: Fundação Cargil.

Malavasi, A., J. S. Morgante, and R. A. Zucchi. 1980. Biologia de "moscas-das-frutas" (Diptera: Tephritidae): lista de hospedeiros e ocorrência. *Revista Brasileira de Biologia* 40:9–16.

Matyak, E. 2011. Sistema de mitigação de risco de *Anastrepha grandis* (Macquart, 1846) (Diptera: Tephritidae) em cucurbitáceas, no município de Santa Isabel do Ivaí – PR. 170 p. Specialization's monograph, Setor de Ciências Agrárias, Universidade Federal do Paraná.

Meyer, M., H. Delatte, M. Mwatawala, S. Quilici, J. F. Vayssières, and M. Virgilio. 2015. A Review of the Current Knowledge on *Zeugodacus cucurbitae* (Coquillett) (Diptera: Tephritidae) in Africa, with a List of Species Included in *Zeugodacus.ZooKeys* 540:539–557.

Ministério da Agricultura, Pecuária e Abastecimento [MAPA]. 2017. Planejamento Estratégico 2017–2018. http://www.agricultura.gov.br/assuntos/camaras-setoriais-tematicas/documentos/camaras-tematicas/suasa/2017/05a-ro/1-planej-estrat-sda-2017-2018.pdf (accessed July 20, 2018).

Norrbom, A. L. 1991. The Species of *Anastrepha* (Diptera: Tephritidae) with a *Grandis*-type Wing Pattern. *Proceeding of the Entomological Society of Washington* 93:101–124.

Norrbom, A. L. 2000. *Host Plant Database for* Anastrepha *and* Toxotrypana *(Diptera: Tephritidae: Toxotrypani), Diptera Data Dissemination Disk 2*. Washington, DC: USDA-APHIS.

Norrbom, A. L., and K. C. Kim. 1988. *A List of the Reported Host Plants of the Species of Anastrepha (Diptera: Tephritidae)*. Washington, DC: USDA-APHIS.

North American Plant Protection Organization [NAPPO]. 2009. Phytosanitary Alert System. Outbreak of *Anastrepha grandis* (South American cucurbit fruit fly) in Panama. http://www.pestalert.org/viewNewsAlert.cfm?naid=76 (accessed June 21, 2018).

Oakley, R. G. 1950. Fruit Flies (Tephritidae). In *Manual of Foreign Plant Pests for Fruit Flies, vol 3*. Bureau of Entomology and Plant Quarantine, pp. 168–248. Washington, DC: USDA-Division of Foreign Plant Quarantine.

Oliveira, A. S., M. F. Souza-Filho, A. Raga, A. M, Almeida, J. A. Azevedo-Filho, M. J. D. M. Garcia. 2012. Levantamento de hospedeiros da mosca-das-cucurbitáceas-sul-americana *Anastrepha grandis* (Macquart, 1846) (Diptera: Tephritidae) no Estado de São Paulo. In: *10º Congresso de Iniciação Científica em Ciências Agrárias, Biológicas e Ambientais*. O Biológico (Impresso). São Paulo, Brazil: Instituto Biológico, v. 74. p. 46.

Prokopy, R. J. 1980. Mating Behavior of Frugivorous Tephritidae in Nature. In *Proceedings of the International Congress of Entomol Symposium of Fruit Fly Problems*, pp. 37–46. XVI International Congress of Entomology. Kyoto: A Ntl. Inst. Agric. Sci. Yatabe.

Rabelo, L. R. S., V. R. S. Veloso, A. D. F. Rios, C. S. Queiroz, and F. H. S. Meshima. 2013. Moscas-das-frutas (Diptera, Tephritidae) em municípios com sistema de mitigação de risco para *Anastrepha Grandis* Macquart. *Arquivos do Instituto Biológico* 80:223–227.

Silva, A. G. D'A., C. R. Gonçalves, D. M. Galvão et al. 1968. *Quarto catálogo dos insetos que vivem nas plantas do Brasil, seus parasitos e predadores*. Tomo 1. Parte 2. Rio de Janeiro, Brazil: Ministério da Agricultura.

Silva, J. G. 1991. Biologia e Comportamento de *Anastrepha grandis* (Macquart, 1846) (Diptera: Tephritidae). 135 p. Master's dissertation, Instituto de Biociências, Universidade de São Paulo.

Silva, J. G., and A. Malavasi. 1993a. The Status of Honeydew Melon as a Host of *Anastrepha grandis* (Diptera: Tephritidae). *Florida Entomologist* 76:516–519.

Silva, J. G., and A. Malavasi. 1993b. Mating and Oviposition Behavior of *Anastrepha grandis* Under Laboratory Conditions. In *Fruit Flies: Biology and Management*, ed. M. Aluja, and P. Liedo, pp. 181–184. New York: Springer-Verlag.

Silva, J. G., and A. Malavasi. 1996. Life Cycle of *Anastrepha grandis*. In *Fruit Fly Pests: A World Assessment of Their Biology and Management*, ed. B. A. McPheron and G. J. Steck, pp. 347–351. Delray Beach, FL: St Lucie Press.

Silva, M. A., G. C. D. Bezerra-Silva, and T. Mastrangelo. 2012. The Host Marking Pheromone Application on the Management of Fruit Flies—A Review. *Brazilian Archives of Biology and Techonology* 55:835–842.

Steck, G. J., and R. A. Wharton. 1988. Description of Immature Stages of *Anastrepha interrupta, A. limae, A. grandis* (Diptera: Tephritidae). *Annals of the Entomological Society of America* 81:994–1003.

Uchôa, M. A., Nicácio, J. 2010. New Records of Neotropical Fruit Flies (Tephritidae), Lance Flies (Lonchaeidae) (Diptera: Tephritoidea), and Their Host Plants in the South Pantanal and Adjacent Areas, Brazil. *Annals of the Entomological Society of America* 103:723–733.

Weldon, C. W., M. K. Schutze, and M. Karsten. 2014. Trapping to Monitor Tephritid Movement: Results, Best Practice, and Assessment of Alternatives. In *Trapping and the Detection, Control, and Regulation of Tephritid Fruit Flies*, ed. T. Shelly, N. Epsky, E. B. Jang, J. Reyes-Flores, and R. Vargas, pp. 175–217. Dordrecht, the Netherlands: Springer.

22 Eradication of an Outbreak of *Bactrocera carambolae* (Carambola Fruit Fly) in the Marajo Archipelago, State of Para, Brazil

Maria Julia S. Godoy, Wilda S. Pinto, Clara A. Brandão, Clóvis V. Vasconcelos, and José M. Pires*

CONTENTS

Abstract Results of the eradication program of the fruit fly pest *Bactrocera carambolae* (carambola fruit fly [CFF]) in the municipalities of Portel and Curralinho in the Marajó Archipelago, Pará, Brazil, are presented in this chapter. Fly population were monitored by using Methyl eugenol poisonous baits in a ratio of 6:1 in Jackson-type traps and, McPhail-type traps baited with three torula tablets per trap. Control methods such as insecticide bait applications, male annihilation technique (MAT), mechanical control, phytosanitary regulations, and, phytosanitary education actions were used. During the two years of activities, a total of 164 specimens, 59 males and 105 females, were captured in Curralinho from March to December of 2014. The last capture recorded in Curralinho was on May 2014. In Portel, a total of 493 specimens, 263 males and 230 females, were captured from April to December 2014. In the period from January to December 2015, 263 specimens, 138 males and 125 females, were captured. The last capture recorded in this town was on September 2015. Among the adopted practices, fruit collection and burial (mechanical control) were particularly effective and contributed significantly to the eradication of the pest.

* Corresponding author.

22.1 INTRODUCTION

The National Eradication Program for *Bactrocera carambolae* (carambola fruit fly [CFF]) (Drew & Hancock) (Diptera: Tephritidae) (*PBc*), which is coordinated by the Ministry of Agriculture, Livestock and Food Supply (MAPA) through the Department of Plant Health and the Brazilian NPPO, is responsible for carrying out control actions to eradicate the pest species *B. carambolae* from the Brazilian territory. The eradication of this pest will prevent economic and environmental damages, as well as phytosanitary restrictions imposed by countries that import Brazilian fresh fruit. In 1989, this pest was initially detected in French Guiana and was found to be dispersed throughout its territory. In 1996, *B. carambolae* had reached Brazil, being first detected at the bordering municipality of Oiapoque, in the state of Amapa (Godoy et al., 2011a).

Since 1996, the CFF has been included in the list of current quarantine pests and is under official control in the state of Amapa (van Sauers-Muller, 1991, 2005; White and Elson-Harris, 1992). Despite continuous control actions throughout these years in Amapa, the lack of control actions in French Guiana has been detrimental to the eradication actions carried out in Brazil because of continuous pest pressure. Phytosanitary measures have been carried out to prevent the dispersion of the CFF to the neighboring state of Para, located to the south of Amapa. However, new outbreaks of the pest have been detected in Para, in municipalities of the Marajo Archipelago along the boat route that links the cities of Macapa, capital of Amapa, and Belem, capital of Para. The first outbreaks were detected in the towns of Curralinho on March 2014 and Portel on April 2014.

The state of Para, in the Brazilian Amazon, shares a border with Suriname and the state of Amapa to the north. There is a dense range of tropical rain forest between Para and Amapa, a natural buffer zone against the dispersal of the CFF. However, the wide local waterway network has intense boat flow among cities, towns, and villages of these two states. This hinders the legal control of CFF host fruits that are transported by passengers on routes passing by Marajo.

Marajo, with 49,602 km^2, is the largest fluvial-maritime archipelago on the planet. It extends from the mouth of the Amazon River to the Atlantic Ocean. It is located between the Equator and 01°33′00″S, and between 47° and 53°W. The climate is hot and humid, with an average temperature of 30°C. The vegetation comprises large savanna on the east side, seasonally flooded varzea forest on the west side, and flooded palm swamps on the north side. The Archipelago has sixteen municipalities located between 00°40′00″N and 01°50′00″S, and between 48°10′00″ and 51°13′00″ W (Source: www.sectam.pa.gov.br). The municipality of Curralinho is 127 km away from the city of Belem. It is located at 01°48′54″S and 49°47′45″W and has an altitude of 15 meter above sea level (masl). It has an estimated population of 32,881 in a total area of 3,617.25 km^2 (FAPESPA, 2016a). However, the downtown area, considered as the target area in the present work, covers only 1.35 km^2; the remaining area is covered by natural vegetation (FAPESPA, 2016a) with some settlements distributed along river banks (Figure 22.1).

The first detection of *B. carambolae* in the Marajo Archipelago occurred on March 7, 2014, in the town of Curralinho. Nine male specimens were captured, distributed in four Jackson-type traps. The second detection occurred in the town of Portel, located at 01°55′45″S and 50°49′15″W. It has an altitude of 19 masl, a population of 59,322, and a total area of 25,384.96 km^2 (FAPESPA, 2016b). However, the working area for CFF control was established as 2 km^2. In Portel, the occurrence of *B. carambolae* was recorded on April 23, 2014, with the capture of two male specimens in one Jackson-type trap (Figure 22.2).

FIGURE 22.1 Location and aerial view of the city of Curralinho, Para, Brazil.

FIGURE 22.2 Location and aerial view of the city of Portel, Para, Brazil.

22.2 MATERIALS AND METHODS

CFF detection surveys in the state of Para are currently carried out by the Para State Agency of Agriculture and Livestock Health and Inspection (ADEPARA). Because the Marajo Archipelago was identified as the main risk route for the introduction of the CFF between the states of Amapa and Para in 2007, 8 Jackson- and 2 McPhail-type traps were installed in Curralinho and Portel. The initial objectives were: to determine the size of the working areas, to increase the number of traps at the Curralinho and Portel working areas, to verify the characteristics of the CFF population, to evaluate control measure effectiveness, and to increase monitoring efforts to delimit the infested area. Methyl eugenol poisonous baits were used in a ratio of 6:1 in Jackson-type traps, and three torula tablets per trap were used in McPhail-type traps. Taking into account the outbreaks in Curralinho and Portel, adapted strategies based on the new information and a CFF Emergency Plan for Corrective Actions were elaborated for each municipality according to their specificities, and contemplating monitoring surveys as control actions. Trap density was determined based on the Trapping Guidelines for Area-wide Fruit Fly Programmes (IAEA, 2003).

Therefore, areas near the outbreaks, in the urban area of the municipalities and in neighboring communities, with installed traps were considered as a risk areas. In both Portel and Curralinho, most of the rural area is comprised of small coastal villages that are flooded daily by tides and, therefore, are accessed by boats. The working area for Curralinho was of 135 ha (1.35 km^2) where 54 Jackson-type and 32 McPhail-type traps were installed in the urban area, and 21 Jackson-type traps were installed in the rural area. The working area in Portel was of 200 ha (2.00 km^2) with 80 Jackson-type and 40 McPhail-type traps in the urban area and 17 Jackson-type traps in the rural area. Host fruits were dissected to check for the occurrence of *B. carambolae* larvae in the infested working area. The aim was to determine the location of the pest before it was captured in the trap and to break its cycle by the immediate collection of infested fruits.

22.2.1 Phytosanitary Procedures Used in Management Strategies for CFF Eradication

22.2.1.1 Insecticide Bait Application (Bait Station)

This technique, based on the principle of attract and kill, is widely used in the integrated pest management (IPM) of fruit flies because of its minimum impact on the environment. The lure attracts both sexes of the CFF; however, it attracts a higher number of females because they have a greater requirement for protein ingestion to achieve sexual maturity and ensure a good fertilization (Godoy et al., 2011b). The solution composed by the lure, insecticide, and water was sprayed on the underside of the leaves of the host plant, covering around 1 m^2 of canopy. The solution was sprayed on 3,478 and 4,371 plants in the working areas of Curralinho and Portel, respectively.

22.2.1.2 Male Annihilation Technique (MAT)

This control method reduces the male population of *B. carambolae* using a solution of methyl eugenol (male lure) combined with insecticide. Bait blocks were soaked in this toxic solution and attached to host plants, with a total of 23,500 blocks in Curralinho and 71,302 blocks in Portel.

22.2.1.3 Mechanical Control by Elimination of Host Fruits—Fruit Collection and Burial

This method consists of collecting most of the fruits, especially those well ripened, from host plants of the CFF, particularly *Averrhoa carambola* (star fruit). This procedure aims to minimize the proliferation of the CFF by the interruption of its biological cycle. The presence of fallen fruits of host plants on the soil leads to large populations of the pest because larvae move from fallen fruits into the soil to initiate the pupal stage (Godoy et al., 2011b). Host fruits collected from trees and the ground were placed in strong plastic bags and were exposed to the sun

TABLE 22.1
Phytosanitary Measures Applied in the Eradication Plan for *Bactrocera carambolae* in the Municipalities of Curralinho and Portel from March 2014 to December 2015

	Curralinho		Portel		
Treatment	**2014**	**2015**	**2014**	**2015**	**Total**
Toxic bait spray (No. of sprayed plants)	67.760	69.365	284.115	240.723	661.963
Male Annihilation Technique—(No. of block units)	23.300	2.700	39.120	6.200	71.320
Collected host fruits (kg)	7.165	6.545	5.060	22.265	41.035

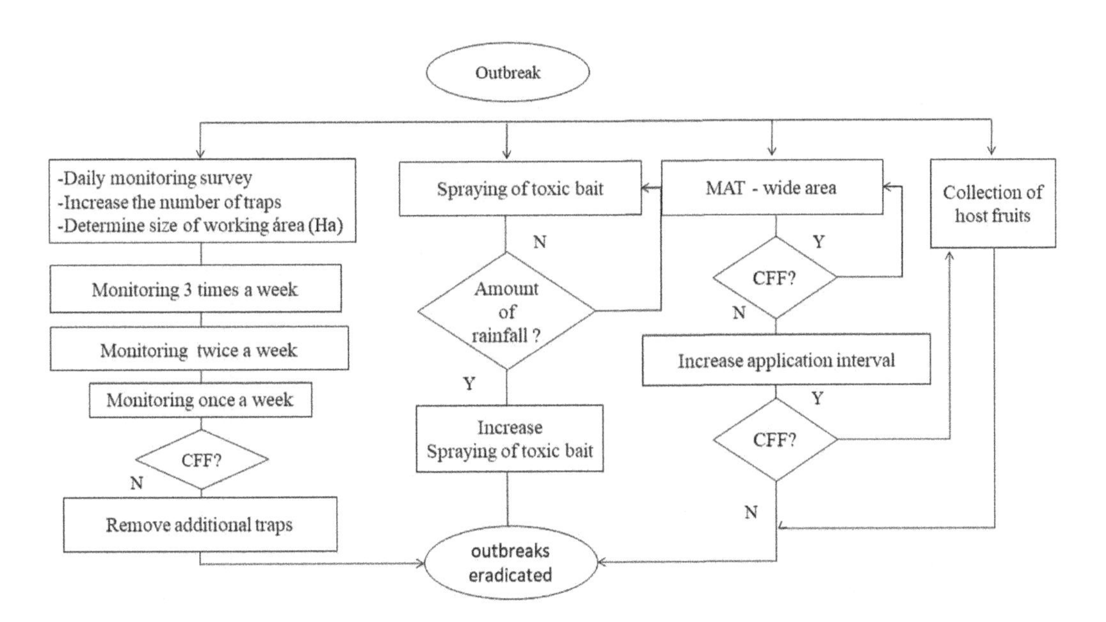

FIGURE 22.3 Control flow diagram of the *Bactrocera carambolae* eradication plan (corrective actions) applied in Curralinho and Portel. CFF, carambola fruit fly; MAT, male annihilation technique.

for 7 days before they were buried. The total number of discarded fruits in Curralinho from March to December 2014 was 13,710 kg and in Portel from April 2014 to December 2015 was 27,325 kg. Phytosanitary measures were applied in a "wide area" and quantitative measures are described in Table 22.1.

The entire sequence of actions that were performed is described in the flow diagram shown in Figure 22.3 (control flow diagram of the Emergency Plan for Corrective Actions for the eradication of *B. carambolae*).

22.2.1.4 Phytosanitary Regulations

The Marajo Archipelago has an extensive range of forest acting as a buffer zone between CFF infested and noninfested municipalities. Nevertheless, there is still a high local risk for the dispersal of this pest because host fruits infested by the CCF can be transported by passengers or cargo travelling by boat lines in the region. The Portaria SFA-PA nº 55-2014 legislation recognized Curralinho and Portel as "areas under quarantine." Even though there has not been

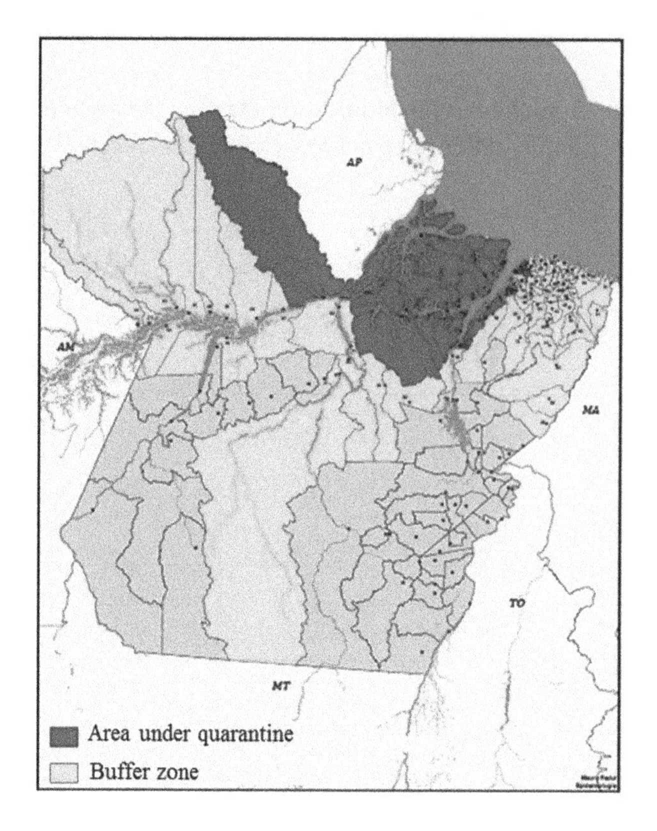

FIGURE 22.4 Area under quarantine and buffer zone established in the state of Para in April 2014.

any recorded occurrence of the CFF in the other 14 Marajo municipalities and 11 neighboring municipalities in Northeast Para, the transit and commercialization of CFF host fruits were forbidden because of weak regional phytosanitary security conditions. As part of the strategies in the present work, a buffer zone, covering 54 municipalities located on the routes of the Lower Amazon river, the Xingu river, and from Belem to the border with the state of Maranhão, was established (Figure 22.4).

22.2.1.5 Phytosanitary Education Actions

Phytosanitary education actions were fundamental to support the execution of all other actions carried out in Curralinho and Portel, as well as to maintain the status of areas "without occurrence of CFF." Education actions were based on general information about CFF biology, the problems generated by commercialization and transportation of CFF host fruits, and the socioeconomic impacts generated by the pest on the fruit business at state and national levels.

22.3 RESULTS

A total of 164 specimens—59 males and 105 females—were captured in Curralinho from March to December 2014 (Figure 22.5). The last capture recorded in Curralinho was on May 14, 2014.

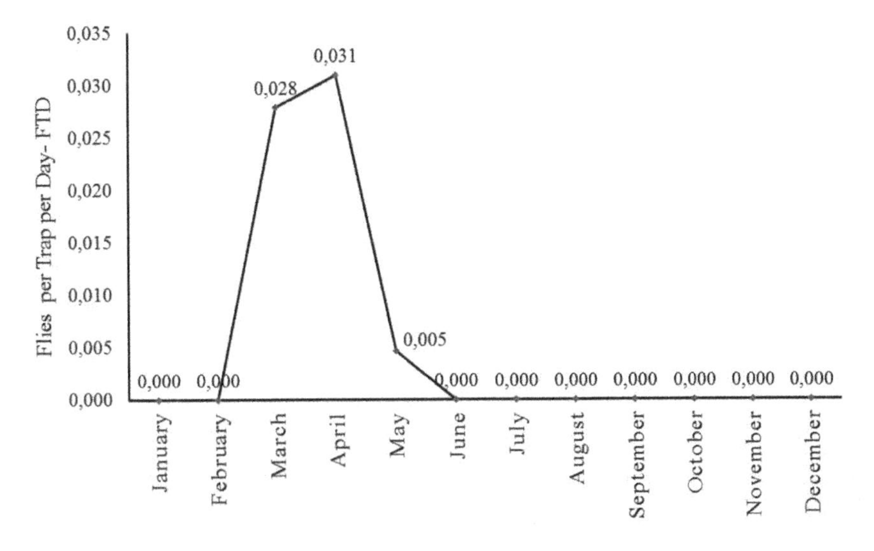

FIGURE 22.5 Population fluctuation of *Bactrocera carambolae* in Curralinho, Para, Brazil, from January to December 2014.

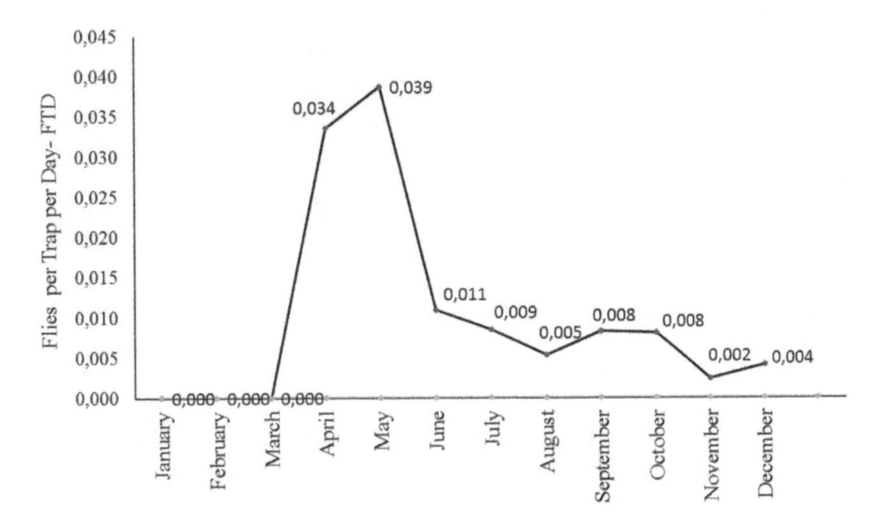

FIGURE 22.6 Population fluctuation of *Bactrocera carambolae* in Portel, Para, Brazil, from April to December 2014.

A total of 493 specimens—263 males and 230 females—were captured in Portel from April to December 2014. In the period from January to December 2015, 263 specimens—138 males and 125 females—were captured (Figures 22.6 and 22.7). The last capture recorded in Portel was on September 16, 2015.

The number of days without captures of *B. carambolae* in the municipalities of Curralinho and Portel from the date of the last capture until the end of June 2018 was 1,508 and 1,018, respectively.

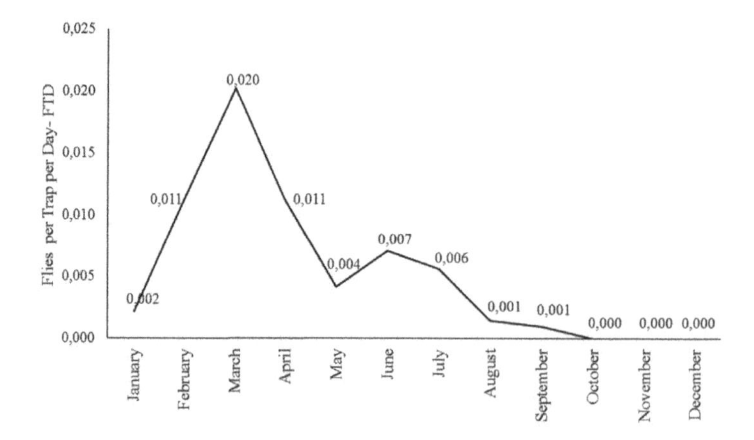

FIGURE 22.7 Population fluctuation of *Bactrocera carambolae* in Portel, Para, Brazil, from January to December 2015.

22.4 CONCLUSION

The phytosanitary procedures used in the management strategies for the eradication of the CFF allowed to counteract the main biological characteristics of the pest that naturally hinder its control (e.g., a high capacity for reproduction, longevity, and dispersion). These procedures also had an important effect on the relationship of the pest with biotic factors (e.g., availability and maturation stages of host fruits). Among the adopted practices, fruit collection and burial were particularly effective and contributed significantly to the eradication of the pest.

According to the International Standards for Phytosanitary Measures for the Establishment of Pest Free Areas for Fruit Flies (Tephritidae), an outbreak is declared to be eradicated after three life cycles of the species (FAO, 2006). The period corresponding to three complete life cycles of the species was considered based on the survival time of 126 days, and the pests in Curralinho and Portel were eradicated 378 days after the last specimens were captured. However, later studies showed that *B. carambolae* has a long oviposition period because females can oviposit until up to 117 days of life, and they also observed individuals that lived up to 150 days (Jesus-Barros et al., 2017).

The implementation of an Emergency Plan for Corrective Actions and a Phytosanitary Education Action Plan within 48 h after the detection of the focus of the outbreak, the rapid publication of the host fruit traffic restriction legislation by MAPA, and the immediate action by control task forces successfully led to obtain the status of "eradicated area" in a short period of time (Figure 22.8).

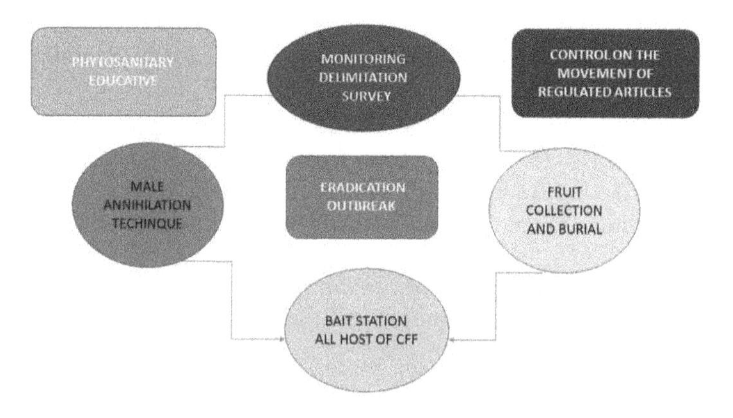

FIGURE 22.8 Phytosanitary procedures used in management strategies for the eradication of the carambola fruit fly.

REFERENCES

FAO. 2006. *Normas Internacionales para Medidas Fitosanitárias*. Rome, pp. 331–341. (Publicaciones, n. 26).

Fundação Amazônia de Amparo a Estudos e Pesquisas [FAPESPA]. 2016a. Estatísticas Municipais Paraenses: Portel./Diretoria de Estatística e de Tecnologia e Gestão da Informação. – Belém, 2016. 60f.: il. Semestral, n. 1, jul./dez. 1. Perfil Municipal – Portel. 2. Aspectos Socioeconômicos – Portel. 3. Dados Históricos – Portel. I. FAPESPA. II. Diretoria de Estatística e de Tecnologia e Gestão da Informação. III. Título. CDD: 23 ed.

Fundação Amazônia de Amparo a Estudos e Pesquisas [FAPESPA]. 2016b. Estatísticas Municipais Paraenses: Curralinho/Diretoria de Estatística e de Tecnologia e Gestão da Informação. – Belém, 2016. 56f.: il. Semestral, n. 1, jul./dez. 1. Perfil Municipal – Curralinho. 2. Aspectos Socioeconômicos – Curralinho. 3. Dados Históricos – Curralinho. I. FAPESPA. II. Diretoria de Estatística e de Tecnologia e Gestão da Informação. III. Título. CDD: 23 ed.

Godoy, M. J. S., W. S. P. Pacheco, R. R. Portal, J. M. Pires Filho, and L. M. M. Moraes. 2011a. Programa Nacional de Erradicação da Mosca-da-Carambola. In: *Moscas-das-frutas na Amazônia brasileira: diversidade, hospedeiros e inimigos naturais*, ed. R. A. Silva, W. P. Lemos, and R. A. Zucchi. Macapá, Brazil: Embrapa Amapá.

Godoy, M. J. S., W. S. P. Pacheco, J. M. Pires Filho et al. 2011b. Erradicação da mosca-da-carambola (Bactrocera carambolae) no Vale do Jari, Amapá-Pará (2007–2008). In: *Moscas-das-frutas na Amazônia Brasileira: diversidade, hospedeiros e inimigos naturais*, ed. R. A. Silva, W. P. Lemos, and R. A. Zucchi, pp. 159–172. Macapá, Brazil: Embrapa Amapá.

International Atomic Energy Agency. 2003. Trapping Guidelines for Area-wide Fruit Fly Programmes, IAEA, Vienna, Austria.

Jesus-Barros, C. R., L. O. Mota Junior, A. L. Lima, A. S. Costa, J. Pasinato, and R. Adaime. 2017. Fecundidade e longevidade de Bactrocera carambolae Drew & Hancock (Diptera: Tephritidae). *Revista Biotemas* 30(4): 7–13.

van Sauers-Muller, A. 1991. An overview of the carambola fruit fly, *Bactrocera* species (Diptera: Tephritidae), found recently in Suriname. *Florida Entomologist* 74(3): 432–440.

van Sauers-Muller, A. 2005. Host plants of the carambola fruit fly, *Bactrocera carambolae* Drew & Hancock (Diptera: Tephritidae), in Suriname, South America. *Neotropical Entomology* 34(2): 203–214.

White, I. M., and M. Elson-Harris. 1992. *Fruit Flies of Economic Significance: Their Identification and Bionomics*. Melksham, UK: CAB International, Redwork Press Ltd.

23 Use of the Sterile Insect Technique in an Area-Wide Approach to Establish a Fruit Fly-Low Prevalence Area in Thailand

Suksom Chinvinijkul[*], *Wanitch Limohpasmanee, Thanat Chanket, Alongkot Uthaitanakit, Puttipong Phopanit, Weerawan Sukamnouyporn, Chanon Maneerat, Weera Kimjong, Phatchara Kumjing, and Naowarat Boonmee*

CONTENTS

Abstract The Mangosteen of Trok Nong subdistrict, Khlung district, Chanthaburi province in Thailand, represents a marketable production area of tropical fruits that has been faced with an exportation trade barrier because of the presence of and infestation by *Bactrocera dorsalis,* among other fruit fly species. Awareness of the degree of damage caused by fruit flies resulted in the creation of the fruit fly control group. After applications of the male annihilation technique (MAT), the aim of this group was to implement an area-wide integrated pest management (AW-IPM) program using the sterile insect technique (SIT) to establish a low-prevalence area of fruit flys. The control group was developed in Trok Nong in 2005. A MAT fruit fly management initiative was applied in 2006, followed by a SIT-AW-IPM research project during 2007–2012. A geographical information system (GIS) was used to map and delineate the action area, guide the release of sterile flies, and design a trapping network system. The white-striped strain of *B. dorsalis* was developed by the Thailand Institute of Nuclear Technology (TINT) in 2007, and sterile flies were released every 28 weeks over 2,590 hectares of tropical fruit plantations during 2008–2013. Quality control of sterile flies and trap inspections were

[*] Corresponding author.

carried out weekly, and fruit sampling was conducted twice a month. In addition, orchard sanitation was carried out, alternative and wild hosts were regularly removed, and mass trapping and interception traps were applied as recommended. In 2013, four organizations supported the establishment of the low prevalence area under a SIT-AW-IPM program, which is a requirement for fruit export. A participatory action plan for fruit fly control was designed and supported financially, and technical backstopping was tasked to the relevant stakeholders. The wild strain of sterile *B. dorsalis* was then continuously released for 20 weeks each year. From 2015 to 2017, the action site was put under an International Atomic Energy Agency (IAEA) technical cooperation program to enhance agricultural productivity by supporting the production of commodities free of fruit flies that meet international standards. The genetic sexing strain (GSS) of *B. dorsalis* is currently under developing process by three collaborators: the Department of Agricultural Extension (DOAE), the TINT, and the International Atomic Energy Agency. The development of this strain will improve the efficiency of the SIT-AW-IPM program. The SIT, integrated with other environmental-friendly control techniques in an AW-IPM approach, has the potential to suppress *B. dorsalis* populations, while increasing the S/N ratio. Further efforts following the International Standards for Phytosanitary Measures could lead the area to reach the goal of obtaining a low prevalence status of fruit flies under National Plant Protection Organization certification.

23.1 INTRODUCTION

The Trok Nong subdistrict (Khlung district, Chanthaburi province) of Thailand is located at 12° 27′ 17″ N and 102° 13′ 17″ E. It has a total area of 44 km^2, of which 21.6 km^2 are conserved forest and 22.4 km^2 are croplands. The area is subdivided into six villages (Figure 23.1). Agriculture is the main practice in this subdistrict, with 80% of the population being crop growers. Mangosteen (*Garcinia mangostana* Linn. [Clusiaceae]), durian (*Durio zibethinus* Murray [Bombacaceae]), rambutan (*Nephelium lappaceum* Linn. [Sapindaceae]), longong, (*Lansium domesticum* Corr. [Meliaceae]), and salak (*Salacca edulis* [Arecaceae]) are the main crops grown in the area. Other soft fruits are

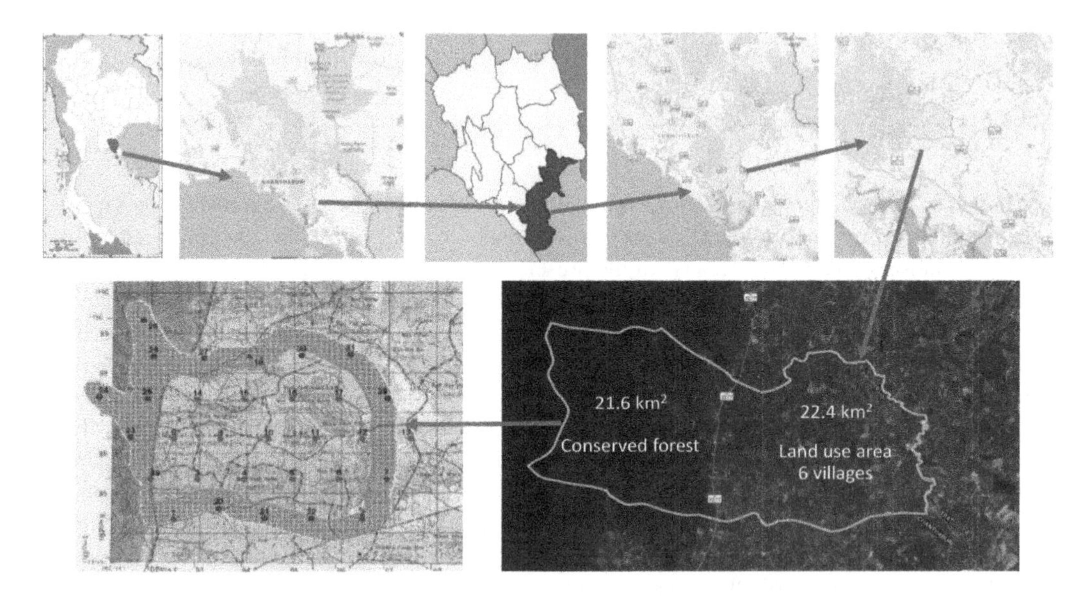

FIGURE 23.1 Location of the Trok Nong subdistrict (in dark red) within Thailand. The red line indicates zooming from the whole country of Thailand, to provinces, then to districts, and finally to the implemented area, where the distribution of the 31 monitoring traps is shown.

also common in the region, particularly guava, which is a key host for *Bactrocera dorsalis* (Hendel). *Bactrocera dorsalis*, like many other fruit flies, is a pest of quarantine nature and is responsible for hindering fruit trade in countries where it has been reported.

Because of the presence of tephritid fruit flies and the extreme marketable quality of the mangosteen produced in the Trok Nong subdistrict, a fruit fly control group was established to implement environmentally friendly control measures, such as the male annihilation technique (MAT). Following periods of implementation of the MAT, the bait application technique (BAT) and orchard sanitation (OS), integrated with the sterile insect technique (SIT) were applied as part of an area-wide integrated pest management (AW-IPM) program. We hoped that the AW-IPM would enhance the management of fruit flies and thus achieve an area of low pest prevalence for fruit flies in the Trok Nong subdistrict. The present work shows how the SIT program was implemented within the AW-IPM in the Trok Nong subdistrict.

23.2 MATERIALS AND METHODS

23.2.1 Background and Implementation Area

A *Trok Nong fruit fly control group* was formed in 2005, initially by a few communities and grower leaders that were trained in fruit fly control measure applications, mainly to implement a basic MAT to control fruit fly populations with the support of the Trok Nong subdistrict administrative organization (SAO).

In 2006, grower leaders requested support from the Thailand Institute of Nuclear Technology (TINT) for a more efficient application of the SIT. Thus far, the governor's office and the Trok Nong SAO had been financially supporting the application of environmentally friendly control methods to reduce *B. dorsalis* populations to affordable levels using the SIT. In 2007, a SIT-AWIPM research program that consisted of the MAT, the BAT, OS, and the release of sterile *B. dorsalis* (SIT) was established by the TINT in collaboration with the Khlung district Agricultural Extension office, Burapha University, the Trok Nong SAO, and grower leaders. In 2013, this program was financially supported by the National Bureau of Agriculture Commodity and Food Standards (BACFS) through the Department of Agricultural Extension (DOAE), with the growers' intention of establishing the first area of low pest prevalence for fruit flies in Thailand. A participatory action plan was designed by the DOAE in cooperation with stakeholders in the Trok Nong region. Special technical issues were transferred, and a SIT AW-IPM fruit fly control campaign was launched. Subsequently, this area was further promoted by the DOAE, and during the period of 2015–2017, the program received further technical support from the International Atomic Energy Agency (IAEA) technical cooperation program (IAEA-TC project (THA5052)). This program is aimed at enhancing agricultural productivity by supporting the production of fruit fly-free commodities that meet international standards. All implementations were supported by the governor's office in constant cooperation with the Trok Nong SAO and the DOAE.

The SIT AW-IPM project covered 25.9 km^2, which included the entire Trok Nong subdistrict cropping area of mangosteen, durian, rambutan, longong, and salak, and some parts of the conserved forest that included 660 fruit-grower households out of a total of 825 households. The SIT AW-IPM program comprised a core area of 15.7 km^2 surrounded by a buffer zone of 10.2 km^2. With the aid of a global positioning system (GPS) and a geographic information system (GIS), fixed sterile fly release points and a trapping network system were marked out (Figure 23.1).

Under the IAEA-TC project, it was suggested that the buffer area around the treated area should be expanded to cover the flight distance of *B. dorsalis*. The new action area would extend to cover 640 ha in all directions (Figure 23.2). Growers in the new area had to agree to participate by preserving the local farming culture and using the SIT to ensure the effectiveness of the expansion.

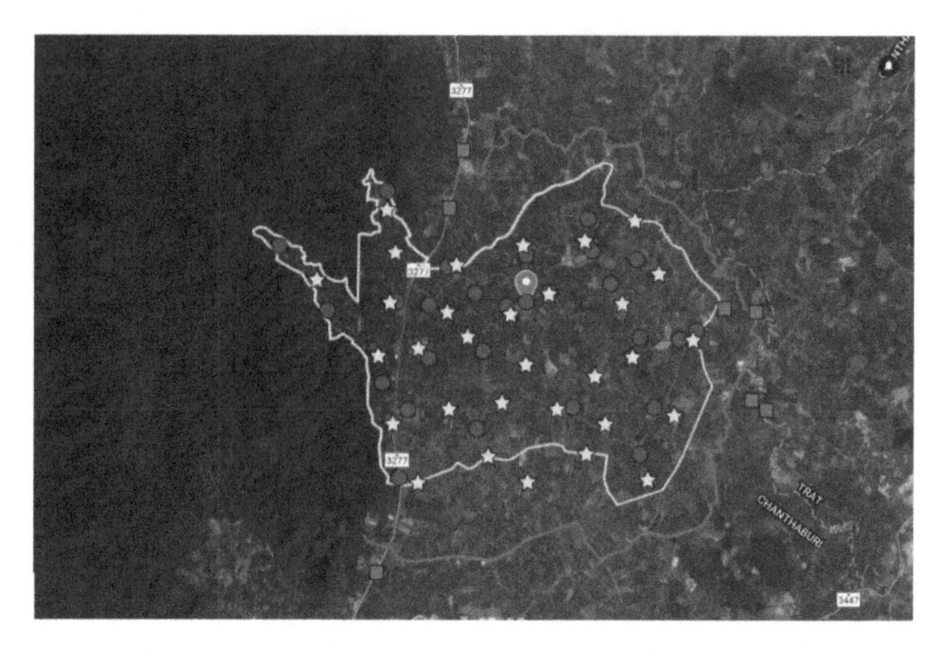

FIGURE 23.2 The new buffer zone of the implemented area, Trok Nong subdistrict. Geographic information system (GIS) was used to re-establish the edge of the buffer zone to a distance of 1 km from the core area as indicated by the blue line.

23.2.2 Integrated Pest Management Program

The SIT AW-IPM program established the diversity and abundance of tephritid fruit flies and host plants. Management options consisting of MAT+BAT+OS+SIT were implemented when alternative and wild-hosts were removed. From the start of the program in 2007, growers were involved along with the SAO and the DOAE's local pest management officers. OS was applied twice per month with a recycle-reuse system, in which damaged or remnant fruits were composted and used as bio-fertilizers. Soil pH was measured to monitor soil status over the period of the implementation. Alternative and wild hosts were removed from the whole area three times per year and were replaced with nonhost plants.

MAT and BAT traps fabricated from local materials were applied prior to SIT releases. MAT traps measuring 5 × 5 cm, made out of fiber blocks or modified recycled water bottles, were dip-soaked in a mixture of methyl eugenol, molasses, and Malathion® and were used for mass trapping at 50-m intervals within the core area during two 3-month cycles. Liquid traps, modified by using recycled water bottles, consisted of 150 cc of total volume. These traps were baited with a mixture of methyl eugenol, protein, and Malathion®, and were placed at 25-m intervals in the buffer area to intercept males and females three times a year.

In 2013, as part of the DOAE's strategies, a community pest management center (CPMC) was formed in the Trok Nong subdistrict. The CPMC consists of a growers' committee and mostly involves the same crop members who manage and make decisions on pest management by themselves. The DOAE and other related organizations support technical knowledge exchange, pest identification, IPM application, parasitoid production, and pest surveillance and monitoring. Fruit fly control activities, including sterile fly releases, were carried out by members of the CPMC in cooperation with the Trok Nong SAO and the DOAE's local officers.

23.2.3 Sterile Male Releases and Surveillance

To safeguard volunteer growers against allergies and environmental pollution caused by pupal fluorescent powder markers in SIT programs (FAO/IAEA/USDA, 2014), the white-striped back strain of *B. dorsalis*, developed by the TINT in 2007 (Boonsirichai et al., 2011), was used. Sterile males were mass produced and released at a rate of 5 million per week in the core area from March to September in each of the 5 years of the project (2008–2012). The white-striped back *B. dorsalis* strain was subjected to quality control measures (FAO/IAEA/USDA, 2014) in a weekly manner. During the same period, the SIT was integrated with other control techniques. In 2013, the same activities were supported by the BACFS and the DOAE.

In 2014, the responsibility of mass production of sterile *B. dorsalis* flies was entrusted to the DOAE. The wild strain of *B. dorsalis* was used due to proprietary issues with the white-striped back strain. Sterile flies were released at the same rate of 5 million per week only from April to August due to budget constraints. Releases were performed at ground level by participating growers (CPMC) and SAO volunteers. Since 2017, releases were adjusted to operate from January to June due to the low population period of wild fruit flies.

A surveillance/monitoring system consisting of a trapping network and fruit sampling was established. Modified Steiner traps distributed as 31 in the core area and 10 in the neighboring area were inspected weekly. Fruit sampling was carried out twice a month for each fruit variety.

Budget and SIT technologies were provided by the TINT during 2007–2012, by the BACFS and the DOAE in 2013, and by the DOAE since 2014 and continuously cooperating with the governor's office and the Trok Nong SAO. Under the IAEA-TC project, the genetic sexing strain (GSS) of *B. dorsalis* has been under a development process, in cooperation with the DOAE, the TINT, and the IAEA, using white pupae selected from the wild strain of the DOAE's mass-rearing facility and the white-striped back strain from the TINT.

23.3 RESULTS AND DISCUSSION

Bractrocera dorsalis is a destructive fly species native to tropical Asia. It has spread around the globe and is one of the most invasive tephritid pest species. In other countries it was synonymized as *Bactrocera invadens* (Schutze et al., 2015), a species with strong quarantine measures that prevent the free movement of fruits between infested countries and even within countries. Such is the case presented here, in Thailand and the Trok Nong subdistrict, with its great production of mangosteen, durian, rambutan, and longong. For this reason, the Thai authorities established a participatory *B. dorsalis* control program engaging national and regional institutions and growers.

23.3.1 *Bractrocera dorsalis* IPM: Host Fruits and Sanitation Practices

Fourteen out of 18 tested fruit species in the Trok Nong subdistrict were potential hosts for *B. dorsalis*; however, guava (*Psidium guajava* (L.) Kunze 1898) was preferred by this species. The preference for guava by *B. dorsalis* was also reported by Goergen et al. (2011). Based on this information, guava, mamiew pomerac, wild banana, java apple, mango, jujube, and star fruit trees, which are alternative hosts of *B. dorsalis*, were eliminated by growers and the CPMC as recommended by the international standards for phytosanitary measures (FAO, 2012).

OS involved the conversion of fallen ripe and damaged fruits into bio-fertilizers that were used to fertilize the soil; this improved the soil quality of 320 ha. These bio-fertilizers raised the pH by approximately one point, which indicated that the soil could maintain its own organic matter mineralization process, and that beneficial bacteria would increase their activity, thus enhancing crop yields.

23.3.2 *Bractrocera dorsalis* AW-IPM Program with a SIT Component

Approximately 200 million *B. dorsalis* sterile flies were released in the target area of the Trok Nong subdistrict (over 25.9 km^2) during 7 months of 2007–2013, and approximately 100–120 million sterile flies were released during 5–6 months from 2014 to 2018. Average S/N sterile-to-wild or sterile-to-native (S/N) ratios and fly per trap per day (FTD) during 2013, 2014, 2015, 2016, 2017, and 2018 compared to those of 2012 are shown in Figures 23.3 and 23.4.

The application of an AW-IPM program using the SIT as a main component resulted in a reduction of longong fruit damage caused by *B. dorsalis*, from 30% in 2005 to 5% in 2013, 0% in 2016, 2% in 2017, and 1% in 2018, along with a reduction of chemical fertilizer costs of about US$406 per ha.

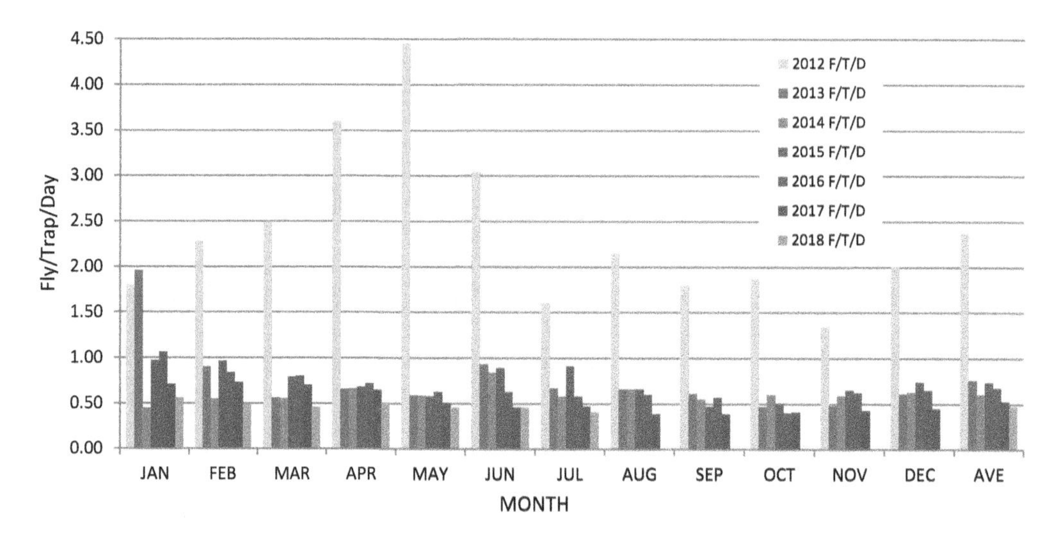

FIGURE 23.3 Average sterile-to-wild or sterile-to-native (S/N) ratios during 2013, 2014, 2015, 2016, 2017, and 2018 compared to those of 2012.

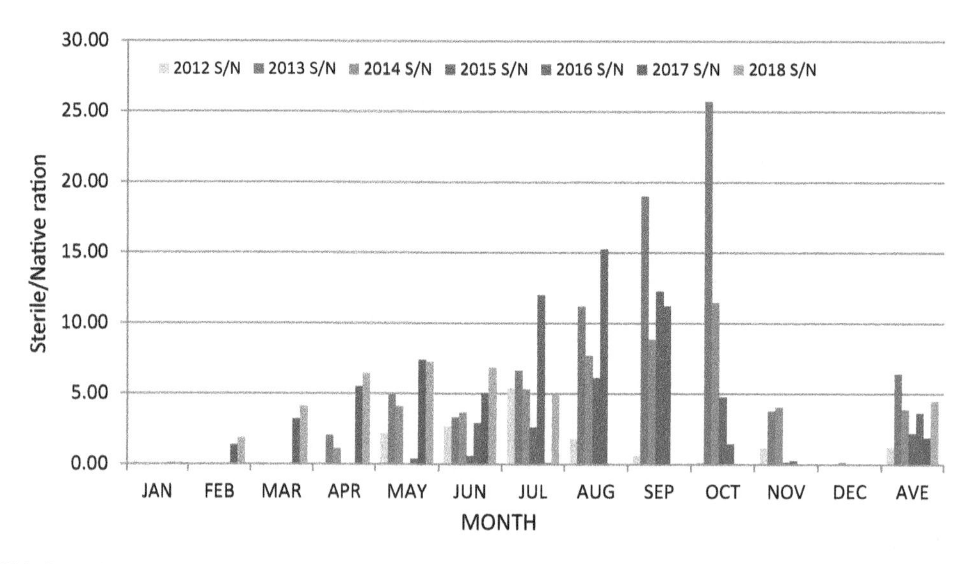

FIGURE 23.4 Average fly per trap per day (FTD) during 2013, 2014, 2015, 2016, 2017, and 2018 compared to those of 2012.

Furthermore, the *B. dorsalis* SIT AW-IPM program resulted in increased market values. The market value of longong was increased to US$83 per ton, and of mangosteen to US$100–167 per ton, or about US$57,500 and US$850,000 per year, respectively. This is a large increase compared to the neighboring control orchards that were not subjected to AW-IPM with SIT under farmer practice.

Similarly to the Mexican AW-IPM program (Salcedo Baca et al., 2010), the Tronk Nong subdistrict SIT project is having a high economic impact as a result of low fruit infestation because of a significantly reduced prevalence of *B. dorsalis* in the region. Fruit prices have increased and hence net income has increased. The orchards under SIT AW-IPM have experienced remarkably lower chemical applications and overall production costs. The treated area has been considered eco-friendly as no pesticides have been applied after the SIT approach. High-quality fruits, especially mangosteen, produced in Trok Nong can be exported in amounts of approximately 4,000 tons each year and have access to markets they could not get into before.

The *B. dorsalis* sterile-to-wild ratio, even if the maximum average was ≈26 for the entirety of the years (the trend indicated in Figure 23.3), showed an increased progression within the sterile fly release period. When the number of released sterile males is constant, the sterile-to-wild ratio starts to increase, which is a direct measure of the reduction of *B. dorsalis* wild populations. Also, sterile males were still trapped at least 1–2 months later, which is an indication of the efficiency of the SIT.

Moreover, the average FTD indicated that wild *B. dorsalis* were controlled at a level of less than 1 in the first year of the SIT approach, and for at least 5 years continuously, even when the SIT was not applied during the whole year. The treated area should be successful in becoming an area of low pest prevalence for fruit flies and could be declared as a low prevalence area for *B. dorsalis* following the ISPM No. 30 (FAO/IPPC, 2008) if it manages to minimize the spread of regulated fruit flies within the area.

The development of a genetic sexing strain (GSS) of *B. dorsalis*, which is under process, for the improvement of the SIT in Thailand is showing positive results. The process is being carried out in cooperation with the DOAE, the TINT, and the IAEA, using white pupae selected from the wild strain of the DOAE's mass-rearing facility and the white-striped back strain from the TINT.

23.4 CONCLUSIONS AND PERSPECTIVES

Our results indicate that environmental-friendly control techniques, orchard sanitation, alternative and wild host removal, mass trapping and interception traps, along with sterile male releases in an AW-IPM approach result in a reduction of *B. dorsalis* wild populations, a reduction of fruit infestation (from 30% to 2%), and an increase in fruit value. Overall, these results indicate a successful implementation of the SIT in AW-IPM and the establishment of a *B. dorsalis* low prevalence area in the Trok Nong subdistrict, with a positive impact in the country. Further research and national plant protection organization (NPPO) involvement are requested to fulfill standard international phytosanitary measures.

This positive result demonstrate that a sterile male release integrated with an AW-IPM approach allows a significant reduction of *B. dorsalis* wild populations in the area year by year. The DOAE implemented the SIT into the AW-IPM program as one of its key phytosanitary measures to control fruit flies, and growers in specific selected areas of 20 provinces joined the AW-IPM program.

As long as growers cooperate with each other and the SIT AW-IPM is effective, the program will continue to be implemented in the Trok Nong subdistrict. Also, as the CPMC grows, it will continue to be an important foundation for the future of the program. Nevertheless, growers will need an easy-to-use system based on consensus for buying irradiated pupae in case that the government stops subsidizing the program at some point.

In the near future, the GSS of *B. dorsalis*, provided by the three organizations of the IAEA, the TINT, and the DOAE, could be more effective in the SIT in AW-IPM for controlling fruit flies in Thailand because only males can be released. Using male-only strains would also provide a sense of confidence to fruit growers. Products from the treated area should also have access to a new niche of markets. However, further efforts should be made following the International Standards for Phytosanitary Measures (FAO/IPPC, 2012) to achieve the status of low prevalence area for *B. dorsalis* under the Thailand NPPO certification.

REFERENCES

Boonsirichai, K., S. Segsarnviriya, W. Limohpasmanee, T. Kongratarpon, T. Thannarin, and K. Sungsinleart. 2011. Genetic variation among the white-striped *Bactrocera dorsalis* (Hendel) in comparison with a Trok Nong-derived population. *12th Conference on Nuclear Science and Technology*, Thailand. In: INIS-TH-326. (https://inis.iaea.org/search/search.aspx?orig_q=RN:43095386).

FAO/IPPC. 2008. International standards for phytosanitary measures. Establishment of areas of low pest prevalence for fruit flies (Tephritidae) ISPM 30. 16 pp. (https://www.ippc.int/static/media/files/publication/en/2016/11/ISPM_30_2008_ALPP_fruit_flies_EN.pdf).

FAO/IPPC. 2012. International standards for phytosanitary measures. Systems approach for pest risk management of fruit flies (Tephritidae) ISPM 35. 10 pp. (http://www.fao.org/docrep/016/k6768e/k6768e.pdf).

FAO/IAEA/USDA. 2014. *Product Quality Control for Sterile Mass-Reared and Released Tephritid Fruit Flies, Version 6.0.* International Atomic Energy Agency, Vienna, Austria. 164 pp. (http://www-naweb.iaea.org/nafa/ipc/public/QualityControl.pdf).

Goergen, G., J. F. Vayssières, D. Gnanvossou, and M. Tindo. 2011. *Bactrocera invadens* (Diptera: Tephritidae), a new invasive fruit fly pest for the Afrotropical Region: Host plant range and distribution in West and Central Africa. *Environmental Entomology* 40: 844–854.

Salcedo Baca, D., J. R. Lomeli Flores, G. H. Terrazas Gonzalez, and W. Wnkerlin. 2010. Economic evaluation of the moscamed regional program in Mexico (1978–2008), pp. 179–188. In: B. Sabater-Muñoz, V. Navarro Llopis, and A. Urbaneja (Eds.), *Proceedings of the 8th International Symposium on Fruit Flies of Economic Importance.* Polytechnic University of Valencia Editorial, Valencia, Spain.

Schutze, M. K., N. Aketarawong, W. Amornsak, K. F. Armstrong, A. A. Augustinos, N. Barr, W. Bo, K. Bourtzis, L. M. Boykin, C. Cáceres et al. 2015. Synonymization of key pest species within the *Bactrocera dorsalis* species complex (Diptera: Tephritidae): Taxonomic changes based on a review of 20 years of integrative morphological, molecular, cytogenetic, behavioural and chemoecological data. *Systematic Entomology* 40: 456–471.

24 Implementation of an *Anastrepha* spp. Risk-Mitigation Protocol for the Mango Export Industry in Cuba

Mirtha Borges-Soto, Maylin Rodríguez Rubial, Evi R. Estévez Terrero, and Beatriz Sabater-Munoz*

CONTENTS

Abstract Mango (*Mangifera indica* L.) is one of the key products of the Cuban export market. This crop is threatened by a great number of pests and diseases, reaching between 10% and 50% of economic losses worldwide. Tephritid fruit flies are among the key pests of mangoes, deserving specific control programs in many countries. In the 1950s, Cuba established a risk mitigation protocol for *Ceratitis capitata* (Wiedemann) that has been updated regularly, including some other key tephritid species belonging to the genus *Anastrepha*. According to these programs, in late 2008, a study on tephritid invasions demonstrated that only two species of *Anastrepha*, namely *A. suspensa* (Loew) and *A. obliqua* (Macquart) are established in the island of Cuba, threatening the fruit export market. In 2015, the Plant Protection Cuban agency established the basis for the risk mitigation protocol for the mango export industry. This protocol is the objective of the present study. Four mango production areas were selected to survey the application of the Anastrepha *spp. Risk Mitigation Protocol*, following the systems approach indicated as the most appropriate for export. This protocol includes monitoring with a trap grid set at 0.3 McPhail baited traps per hectare, dissection of fruits (mango and guava), establishment of fruit traceability notebooks, training of local personnel, quarantine measures, and selection of orchards, among other measures.

* Corresponding author.

Fruit flies per trap per day (FTD) indexes were determined in each area. Trapping, inter-cropping of fruits and noncrop host fruit surveillance, orchard sanitation, and periodical data registry were set up. Only seven *Anastrepha* spp. were trapped throughout the whole study period (January 2016–June 2017), five *A. suspensa* females and two *A. obliqua* males, which were captured outside the studied commodity. A multicomponent systems approach has been established to reduce the risk of *Anastrepha* spp. in mango varieties destined for international export.

24.1 INTRODUCTION

Tephritid fruit flies are important pests of fruits and vegetables worldwide, with some species declared as threats for the worldwide trade of agricultural fresh products (Aluja and Rull, 2009). The natural distribution (Figure 24.1) of the species is being modified unintentionally by human worldwide trade and expanded due to climate change (Qin et al. 2015). In this sense, almost all fruit-producing countries are under menace of invasive species, especially those countries located in the border of species border boundaries. During the past decade, a number of models and approaches have been developed to determine the invasive risk of Tephritidae species, letting each country decide on the actions to prevent any invasion or establishment as part of the regular activities of their ongoing fruit fly management programs (Godefroid et al. 2015; Qin et al. 2015; David et al. 2017; Dias et al. 2018).

The Republic of Cuba, settled in the middle of the Caribbean sea, is threatened by several tephritid species in two ways: (1) by the risk of invasions from neighboring countries (Bahamas, United States, Mexico, Guatemala, Honduras, Costa Rica, Nicaragua, Panama, Colombia, Venezuela, Puerto Rico, Dominican Republic, and Haiti) during hurricane seasons, and (2) from transoceanic visitors (cruise ships or cargos with fresh fruits from other countries). Cuba has historical records of 30 Tephritidae species distributed in 15 genera, nearly all described from specimens from museum collections (from Cuba universities and research bodies, from the Natural History museum of Washington, DC, or from the Comparative Zoology museum of Harvard University) without a reference to their host plant

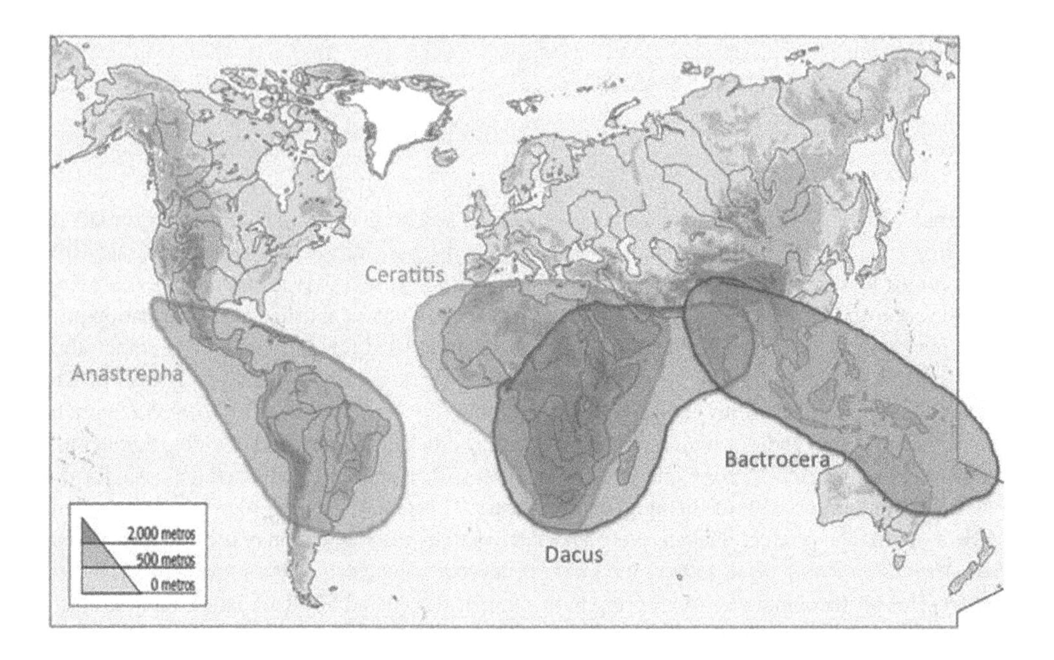

FIGURE 24.1 World atlas with the putative original biogeographical distribution of the four main Tephritidae genera. This distribution map was constructed from data from different papers and from the EPPO (European Plant Protection Organization).

or capture location (Rodríguez Velasquez et al. 2001). Of this Cuban Tephritidae species catalog, only six species belong to the genus *Anastrepha*, namely *Anastrepha suspensa* Loew, *Anastrepha obliqua* Macquart, *Anastrepha soroana* Fernandez y Rodríguez, *Anastrepha ocresia* Walker, *Anastrepha interrupta* Stone, and *Anastrepha insulae* Stone. The last two species have not been recorded in Cuba in the past 30 years, even if during these period some specimens that were caught in monitoring traps were assigned to the genus *Anastrepha*; however, they were not at all assignable to a specific species taxonomical descriptor, and the remaining, with a few specimens, were captured occasionally in minor crops (Borges-Soto et al. 2011, 2015). Only *A. suspensa* and *A. obliqua* were reported regularly with a detailed list of new hosts like pomarrosa (*Syzygium jambos* (L.) Alston, an invasive plant in Cuba), icaco (*Chrysobalanus icaco* [L.] L.) or caimito (*Chrysophyllum cainito* L.), none of which are an economically important crop in Cuba. These two species presented population dynamics in guava with peaks during the guava-maturation months (July–September), affecting up to 15% of guava fruits. Another tephritid species, the papaya fruit fly *Toxotrypana curvicauda* Gerstaecker, is present affecting mainly papaya (*Carica papaya* L.) and rarely affecting the mango production in Cuba. In addition to these species, *Ceratitis capitata* (Wiedemann) was also recorded as present from the specimens stored in the museum but was never found in any of the trapping systems established since early in the last century as part of the Cuba government's plant protection program (Vázquez et al. 1999; Rodríguez Velasquez et al. 2001; FAO, 2003; Drew, 2004; Borges-Soto et al. 2011).

Following the standard guidelines of the International Plant Protection Organization (IPPO n30), Cuba established its own operational procedures to control tephritid species outbreaks, reduce invasions, and determine the presence of these tephritid species in the island (Fernández et al. 1997; Rodríguez Velasquez et al. 2001; Armenteros 2005; Borges-Soto et al. 2011, 2015, 2016).

Mango is cultivated in several tropical and subtropical regions, with 13% of the global production concentrated in Latin America and the Caribbean countries (FAOSTAT 2018). Considered as an exotic rare fruit in Europe and North America, it has expanded its international trade as consumption increased among temperate-zone countries. Only in 2016, global production reached 46 million tons. The cultivars differ in size, shape, appearance, and physiological characteristics, including health-related antioxidant phenolic compounds, but they also differ in their susceptibility to diseases and pests. Tephritid fruit flies are considered key pests of mangoes, with 8 reported species of the genus *Anastrepha*, 30 of *Bactrocera*, 7 of *Ceratitis*, 2 of *Dirioxa*, and 1 of *Toxotrypana* (Yahia, 2011). However, in the Central American and Caribbean countries, only species from the genus *Anastrepha* have been reported to affect mangoes (Birke and Aluja 2011; Aluja et al. 2014). With a production of 420,191 tonnes, encompassing a crop surface of 38,307 ha, Cuba was ranked 17 out of 102 mango-production countries and third in the Central American and Caribbean region in 2016 (FAOSTAT 2018). Such significant position justifies the implementation of an *Anastrepha* spp. risk-mitigation program to protect the Cuban export market.

As previously indicated, surveillance, trapping, monitoring, control, and corrective action implementation procedures were established in several commodities throughout the whole island of Cuba (Rodríguez Velasquez et al. 2001; Borges-Soto et al. 2011, 2015, 2016). After this experience, the Cuba National Fruit Flies Control Program established an *Anastrepha* spp. *Risk Mitigation Program for Mango* following a "systems approach" as described previously for other species in other countries (Follet and Vargas 2009; Moore et al. 2016). Briefly (see Material and Methods for an in-depth description), it includes surveillance, trapping, monitoring, control, corrective actions, and postharvest regulation prior to exportation, all following Cuban national laws 731/98, 50/2008, and 435/94.

In this chapter, we present the results of this risk-mitigation program for mango in four selected areas of Cuba.

24.2 MATERIAL AND METHODS

During the period 2015–2017, 10 mango orchards from four different fruit-production enterprises were selected for the implementation of the risk-mitigation protocol (Table 24.1, Figure 24.2a). Some of these orchards are merged in an Unidad Economica de Base (UEB), the Cuban assignment

TABLE 24.1
Selected Production Areas with Indication of Their Assignment and Captures Obtained

Field ID	Geographical Area	Enterprise Name	Plantation Code	Total Surface (ha)	Traps (n)	Total Number of *Anastrepha* spp. Captured	FTD
1	Jaguey Grande, Matanzas	Agroindustrial "Victoria de Giron"	UEB[a] frutales-granja #4	226.04	62	0	0
2			UEB frutales-granja #5	69	23	1[b]	0[c]
3	Arimao, Cienfuegos	Citricos "Arimao"	UBPC "Breñas"	12	4	1[b]	0[c]
4			UBPC "Seibabo"	12	9	2[b]	0[c]
5			UBPC "La Cuchilla"	10	4	2[b]	0[c]
6	Caimito, Artemisia	Citricos "Ceiba del Agua"	UBPC[a] "24 de Febrero" – finca Ingenio Nuevo	10	4	0	0
7			UBPC "24 de Febrero" – finca Sandoval	12	10	1[b]	0[c]
8	Avila, Ceballos	Agroindustrial "Ciego de Avila"	UEB "Palmarito"	77	24	0	0
9			UEB "Colonia"	92	30	0	0
10			UEB "Nadales"	105	31	0	0

[a] UEB, Unidad Económica de Base; UBPC, Unidad Basica de Producción Cooperativa. Both UEB and UBPC indicate how the orchards are organized in economic units. Descriptions are given in Spanish because each country has a different economical organization of the production units.

[b] Some of the specimens were captured in traps located either at intercrop areas with avocados, guavas, coffee, or citrus, or in backyards, not considered for the fruit flies per trap per day (FTD) determination ([c]).

of crop surface for private economical administration, which will include more than one commodity (fruit fly–susceptible fruit species). Selection was based on the mango cultivars "AG-33 *cv* Tommy Atkins" and *cv* "*Super* Haden," the two varieties selected for this study.

The mango risk-mitigation protocol consisted of:

1. Surveillance of fruit fly populations throughout the year in an established grid across the targeted region;
2. Orchard sanitation (removal of wild noncrop hosts, isolated fruit trees, and ripe-fallen fruits);
3. Establishment of treatments and surveillance registry notebooks at each orchard;
4. Surveillance of any putative tephritid infested fruit by placing in-house designed development cages; and
5. Establishment of a training protocol in each new season.

In addition, postharvest quarantine measures (hot-bath thermal treatment) were also applied following Cuban laws 50/2008, 435/94, and 731/98. These directives allowed working with mango fruits from registered orchards for the export market, creating an "Export passport" that included traceability of origin, surveillance of quarantine species, quarantine postharvest treatment, and packing systems.

The traps used in this project were McPhail traps (IPS, International Pheromone Systems LTD, London, UK or from BIAGRO SL, Valencia, Spain) baited with a mixture of 3% Torula yeast

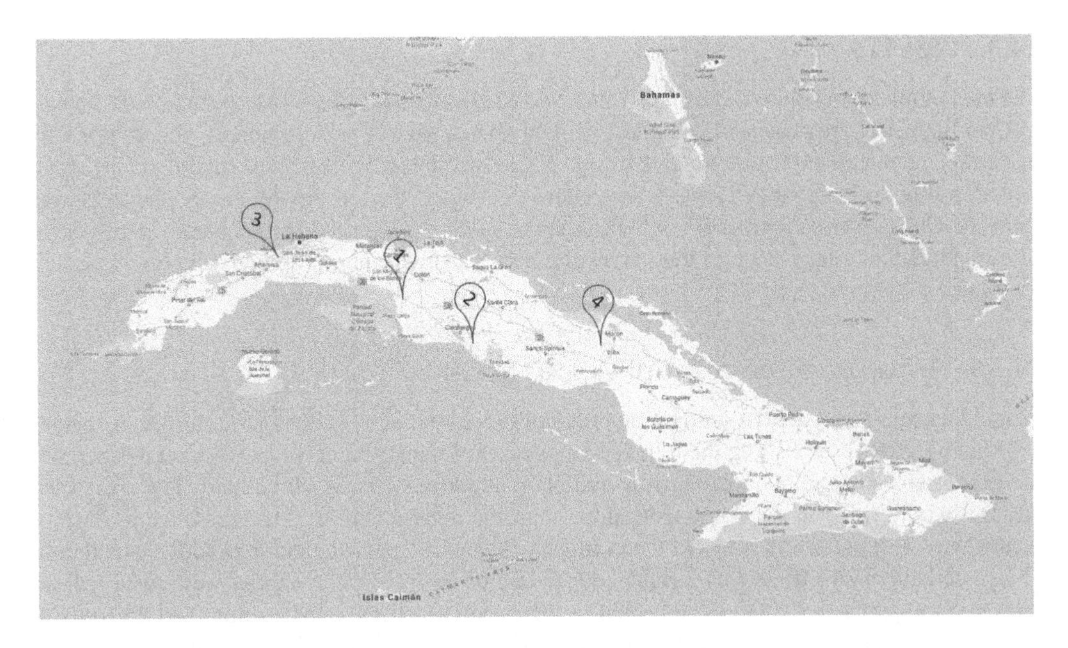

FIGURE 24.2 Geographical distribution of the areas under study 1: Jagüey Grande, Matanzas county; 2: Arimao, Cienfuego county; 3: Caimito, Artemisia county; and 4: Avila, Ceballos county.

(Fábrica de levadura de Torulas Alfredo Rafael Pérez, Central Azucarero Ciro Redondo, Ciego de Ávila, Cuba; https://www.ecured.cu/Fábrica_de_Levadura_de_Torulas_Alfredo_Rafael_Pérez) and 1%–3% borax (sodium tetraborate decahydrate from Empresa Laboratorios AICA, La Habana, Cuba).

From the geographical map and the plantation scheme (meaning the distribution of each mango tree within the plantation) of each plantation, a trapping grid was established in a one-by-one fashion. This method was adopted because the orchards were not regular and contained the selected mango varieties or mango plantations as well as other fruit fly host plantations, and houses with host plants in the backyards were crossed by service roads, train rails, or other vehicle pathways. Therefore, the trapping grid was composed of: (i) one McPhail trap set every 3 ha following the main diagonal of each mango plantation; (ii) another MacPhail trap was set in each cardinal direction (N, S, W, E) to the target trap per 10 ha of mango crops (as other varieties were established but not studied); (iii) a third trap every 33 ha of remaining mango crops (belonging to the Cuba National Fruit flies control program[1]); (iv) one McPhail trap every 5 ha of other fruit crops (like citrus or stone fruits, established as intercrop areas); (v) one trap per square kilometer in the closest town or inhabited area; and (vi) one trap in each backyard with putative host fruits, if houses were present within the plantation (see Figure 24.2).

At each plantation, a route was established allowing the service of all traps to be made in one inspection. All traps were serviced every 7 days, replacing the attractant solution (as reviewed in Epsky et al. 2014) and storing any trapped insects in 125-mL vials (recovering vials) with the corresponding trap number and collection date. Recovering vials were first evaluated in each enterprise, introducing all the data in their registry notebooks and then were retrieved to the Instituto de Investigaciones en Fruticultura Tropical (IIFT; Cuban Research Institute of Tropical Fruits) laboratory, and specimens belonging to the Tephritidae family were identified to species level under binoculars with the use of the corresponding taxonomic keys.

Infestation level was determined in all orchards as captured flies per trap per day (FTD), as previously determined (Borges-Soto et al. 2011, 2015, 2016).

Two to 5 days before the harvesting period, a sample of mango fruits (25 fruits per orchard, 5 per randomly selected tree) of variable size but nearly at the harvest stage were dissected to determine the presence of developing larvae. This equaled to approximately 13 to 20 kg of fruits per orchard per season.

24.3 RESULTS

The total number of captured *Anastrepha* flies and FTD values for all four study areas from January 2015 to June 2017 are presented in Table 24.1. Only seven *Anastrepha* specimens were trapped, five *A. suspensa* females and two *A. obliqua* males. These specimens were captured mainly in the mango intercrop zones or in inhabited areas where guava or avocado trees were present in backyards (traps from the Cuban National program, which in some cases were placed in nonmango tree species). The captures took place close to the harvesting period of these intercrop commodities, especially for guava.

24.3.1 Enterprise Agroindustrial "Victoria de Giron"

Within this enterprise, the risk-mitigation protocol started with an on-site visit, followed by personnel training. After establishing the trap grid, a new set of registry notebooks were established with the exact trap code and its location (row and plant number) within each orchard. The presence of all traps was verified in a second visit, along with the determination of the presence of development cages with putatively infested fruits (mangoes, guavas, papaya, and avocados). The number of assessed alternative fruits was variable, depending on the year, but mangoes were surveyed each season at the preharvest time, as indicated in the material and methods section, and 25 fruits per orchards were randomly selected from five trees (Figure 24.3). Some of the found isolated guava trees were removed as part of the orchard sanitation and risk-mitigation plan. All trap captures were submitted to the IIFT laboratory or to the Plant Protection national reference laboratory for species identification (see Table 24.1). Only one *Anastrepha* specimen was identified.

24.3.2 Enterprise Citricos "Arimao"

Within this enterprise, the mango-production area also included other fruits (mainly avocado and guava) and mangoes for the internal market. All mango-export orchards included field registry notebooks with all the applied treatments, including all steps performed for orchard sanitation, number and location of all types of traps, *Anastrepha* spp. monitoring, and fruit production. These notebooks also included the on-site visit routes from personnel of IIFT and from personnel of the quarantine department. Due to the presence of small guava orchards (sometimes used as intercropping systems), this area was under special surveillance, with traps also baited with Capilure® or Tridmelure® (Figure 24.4), as a part of the Cuban *C. capitata* management program. The presence of

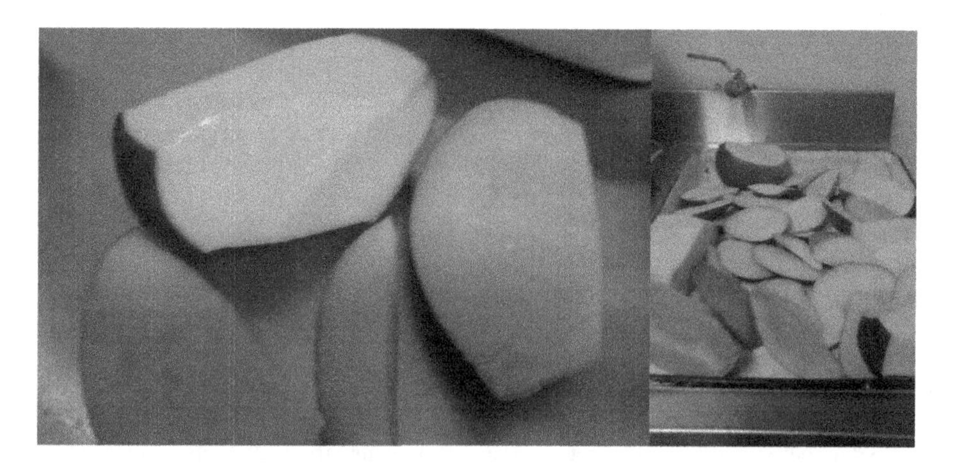

FIGURE 24.3 Enterprise Agroindustrial "Victoria de Giron." Sample of mangoes inspected for the presence of *Anastrepha* spp. larvae prior to the harvesting period.

FIGURE 24.4 Enterprise Citricos "Arimao." Detail of the young mango plantation (left) and a detail of a Rebell trap (right) set at the limit of guava orchards, with an *Anastrepha* spp. specimen.

three *Anastrepha* specimens (Table 24.1) jeopardized the inclusion of this enterprise in the export-targeted authorized list. To avoid this, the Cuban Plant Protection department has established that the guava orchards in this enterprise should be removed and replaced by others crops, such as citrus or mangoes. This replacement will take place in the near future.

24.3.3 Enterprise Citricos "Ceiba del Agua"

Within this enterprise, two different on-site visit routes were established to verify all the areas for export trade. All orchards within this enterprise included field registry notebooks, trap grids, and results. Some of the development cages were also surveyed in some of the field visits. From the same field visits, IIFT personnel noticed the presence of mango fruits with a great variability in size, probably due to the long-lasting blossom period in this enterprise. In this enterprise, only one *Anastrepha* specimen was reported (Table 24.1) in a trap located in a backyard, which contained one guava and several citrus trees for in-home consumption.

24.3.4 Enterprise Agroindustrial "Ciego de Avila"

Within this enterprise, and more precisely within the three selected UEBs, the mango-production area also included other fruits and mango varieties for the local market. All the mango-export orchards included field registry notebooks with all the applied treatments, including all steps required for orchard sanitation, number and location of all types of traps, *Anastrepha* spp. monitoring, and fruit production, which allowed for the record-keeping and traceability of all production from this enterprise. These notebooks also included the on-site visit routes from personnel of IIFT or the training days received. This enterprise was unable to include the established Torula-based attractant for the surveillance of McPhail traps, thus, the sugar cane molasses (3%) and borax (3%) mixture was kept during all the study period (Figure 24.5). Despite this constrain, this enterprise was the most successful in the application of the *Anastrepha* spp. risk-mitigation protocol in mango for export trade, as all the requirements (except for the type of attractant) were met. No *Anastrepha* specimens were recorded in any of the shriveled traps (Table 24.1).

FIGURE 24.5 Enterprise Agroindustrial "Ciego de Avila." Detail of blossoming mangoes (left) and a McPhail trap (right).

24.4 DISCUSSION

Due to current global warming and other climate alterations, along with unintentional man-driven dispersion, Tephritid species, irrespectively of their ancestral geographic origin, are becoming a global menace for many tropical fruits and vegetables (Godefroid et al. 2015; Qin et al. 2015; David et al. 2017). In the Caribbean Sea, the most noticeable invasive species belong to the genus *Anastrepha*, along with the worldwide distributed *C. capitata*. In Cuba, after several decades, a management program was established to control *C. capitata*, mainly in citrus species, which was used as a base program to establish the *Anastrepha* spp. phytosanitary surveillance program (reviewed in Borges-Soto et al. 2011, 2015, 2016). With this gained experience, the *Anastrepha* spp. *Risk Mitigation Plan* presented here for mango in Cuba was established with a detailed trapping network, surveillance methods, removal of alternative hosts, establishment of sanitation procedures, inspector on-site visits, in-field traceable fruit origins, and registry on the export-trade authorized orchard list. The results presented here allowed the re-assignation of the selected areas as areas with low prevalence of *Anastrepha*, making them suitable for fruit export to *Anastrepha* spp.-free countries as has occurred in other countries (Aluja and Rull 2009; Follet and Vargas 2009).

Historically, mango commodities were mainly subjected to postharvest quarantine treatments (hot-water baths) to reduce the risk of pest introduction into pest-free areas as part of the bilateral agreements between importing and exporting countries (reviewed in Yahia 2011), and the use of a systems approach to certify the "risk-mitigated status" for this commodity had not been considered. Hot-water postharvest quarantine treatments usually render the commodity with less nutritional value and shorter shelf half-life, thus threatening the mango trade without assuring a total "risk-mitigated status."

In the past 5–10 years, regulatory officials have embraced the use of *systems approaches*, within which the present work fits, by means of applying joint risk-mitigation processes with pre- and post-harvest quarantine procedures (Follet and Vargas 2009; Shelly 2014; Jang et al. 2015; Moore et al. 2016; reviewed in Dias et al. 2018). In this sense, this work provides for the first time the results of the implementation of the Anastrepha *spp. Risk-Mitigation Protocol for Mango* in Cuba, the third mango producer from the Caribbean countries, showing the cumulatively results of systems activities. These results will help the Cuban export market to grow as the systems approaches in course are mitigating the risk of invasion in the importing country by reducing the amount of putatively infested mango fruits that could contribute to invasive pest movements (Qin et al. 2015; David et al. 2017).

Similarly to what happened in other kinds of "push-and-pull" strategies or systems approaches for pest management (Cook et al. 2007; Aluja et al. 2009; ISPM 35 2012; Meats et al. 2012), the results of this work encourage the removal of intercropping tree plants and other *Anastrepha* spp. host fruits from the vicinity of the export-targeted production orchards. However, the benefits of these intercropping systems in the Anastrepha *spp. Risk Mitigation Plan* for mangoes should still be considered because these alternative hosts will attract fruit flies, which otherwise would forage for oviposition sites in mango plantations and would act as a reservoir for natural enemies (Deguine et al. 2015; David et al. 2017). In all, further research will contribute to improve our understanding on how *Anastrepha* fruit flies develop in this mango ecosystem.

24.5 CONCLUSIONS

In conclusion, Cuba has successfully developed and implemented a systems approach to reduce the risk of *Anastrepha* spp. infestations in mango varieties produced for export.

ACKNOWLEDGMENTS

The authors would like to acknowledge the growers for their help in establishing the trapping grids and for allowing the access to their home backyards to search for putatively infested fruits. We would also like to acknowledge the agricultural engineer N. Sellés from the quarantine department for assistance with the taxonomical identification of fruit flies. Last but not least, we would also like to acknowledge the two anonymous reviewers and the editorial staff of this book for their help in improving this manuscript.

NOTE

1. The Cuban national surveillance program is based on continuous year-round monitorization with three different types of traps (McPhail, Rebell, and Jackson as described in Borges-Soto et al., 2016) to verify the presence of several species of tephritid fruit flies. Traps are established in a triangle grid of 100 ha, setting one trap every 33 ha. Traps are switched in a counterclockwise fashion. In addition, all merchandise and people entry points (airports and ports) have each a complete set of traps, following the National law CNSV (2002).

REFERENCES

Aluja, M., and J. Rull. 2009. Managing pestiferous fruit flies (Diptera: Tephritidae) through environmental manipulation. In *Biorational Tree-Fruit Pest Management*, eds. M. Aluja, T.C. Leskey, and C. Vincent, pp. 171–213. Wallingford, UK: CAB international.

Aluja, M., J. Arredondo, F. Díaz-Fleischer, A. Birke, J. Rull, J. Niogret, and N. Epsky. 2014. Susceptibility of 15 Mango (Sapindales: Anacardiaceae) cultivars to the attack by *Anastrepha ludens* and *Anastrepha obliqua* (Diptera: Tephritidae) and the role of underdeveloped fruit as pest reservoirs: Management implications. *Journal of Economic Entomology* 107(1): 375–388.

Aluja, M., T. C. Leskey, and C. Vincent. 2009. *Biorational Tree-Fruit Pest Management.* Wallingford, UK: CAB International.

Armenteros, G. J. 2005. Reseña sobre la cuarentena exterior en Cuba. *Fitosanidad* 9(1): 47–64.

Birke, A., and M. Aluja. 2011. *Anastrepha ludens* and *Anastrepha serpentina* (Diptera: Tephritidae) do not infest *Psidium guajava* (Myrtaceae), but *Anastrepha obliqua* occasionally shares this resource with *Anastrepha striata* in nature. *Journal of Economic Entomology* 104(4): 1204–1211. doi: 10.1603/EC11042.

Borges-Soto, M., D. Rodriguez, M. Rodriguez Rubial, B. Sabater-Munoz, D. Hernandez Espinosa, and J. L. Rodriguez Tapial. 2016. A review on the Tephritid fruit flies of economic interest in Cuba: Species, plant hosts, surveillance methods and management programs implementation. In *Proceedings of the 9th International Symposium on Fruit Flies of Economic Importance*, eds. B. Sabater-Munoz, T. Vera, R. Pereira, and W. Orankanok, pp. 295–309. Bangkok, Thailand.

Borges-Soto, M., A. Beltrán, T. Mulkay, J. Rodriguez, D. Hernández, and A. Paunier. 2011. The Cuban experiences on monitoring and management of *Anastrepha* spp. (Diptera: Tephritidae) fruit flies in mango (*Mangifera indica*) and guava (*Psidium guajava*) orchards. In *Proceedings of the 8th International Symposium on Fruit Flies of Economic Importance*, eds. B. Sabater-Munoz, V. Navarro, and A. Urbaneja, pp. 206–211. Valencia, Spain: Editorial Polytechnic University of Valencia.

Borges-Soto, M., A. Beltran-Castillo, Y. Avalos-Rodriguez, B. Sabater-Munoz, D. Hernandez-Espinosa, and M. Rodriguez Rubial. 2015. Role of phytosanitary surveillance of *Anastrepha* spp. fruit flies (Diptera: Tephritidae) in the context of the citrus industry of Cuba. *Acta Horticulturae* 1065: 1027–1032.

CNSV. 2002. Programa de defensa contra moscas de la fruta. Republica de Cuba, Ministerio de la Agricultura, Centro Nacional de Sanidad Vegetal. 116 p. La Habana, Cuba.

Cook, S. M., Z. R. Khan, and J. A. Pickett. 2007. The use of push-pull strategies in integrated pest management. *Annual Reviews of Entomology* 52: 375–400.

David, P., E. Thebault, O. Anneville, P. F. Duyck, E. Chapuis, and N. Loeuille. 2017. Impacts of invasive species on food webs: A review of empirical data. *Advances in Ecological Research* 56: 1–60.

Deguine, J. P., T. Atiama-Nurbel, J. N. Aubertot, X. Augusseau, M. Atiama, M. Jacquot, and B. Reynaud. 2015. Agroecological management of cucurbit-infesting fruit fly: A review. *Agronomy for Sustainable Development* 35: 937–965.

Dias, N. P., M. J. Zotti, P. Montoya, I. R. Carvalho, D. E. Nava. 2018. Fruit fly management research: A systematic review of monitoring and control tactics in the world. *Crop Protection* 112: 187–200.

Drew, R. A. I. 2004. Biogeography and speciation in the Dacini (Diptera: Tephritidae: Dacinae). *Bishop Museum Bulletin in Entomology* 12: 165–178.

Epsky, N. D., P. E. Kendra, and E. Q. Schnell. 2014. History and development of food-based attractants. In *Trapping and the Detection, Control, and Regulation of Tephritid Fruit Flies. Lures, Area-Wide Programs, and Trade Implications*, eds. T. E. Shelly, N. Epsky, E. B. Jang, J. Reyes-Flores, and R. Vargas, pp. 75–118. Springer, Dordrecht, the Netherlands.

FAO. 2003. Cuba's citrus industry: Growth and trade prospects. Committee on commodity problems. Intergovernmental group on citrus fruit.ftp://ftp.fao.org/docrep/fao/meeting/006/y9316e.pdf (last accessed November 18, 2014).

FAOSTAT. 2018. Food and Agriculture Organization of the United Nations, Crop statistics (http://www.fao.org/faostat/en/#data/QC (last accessed on August 14, 2018).

Fernández, A. M., D. Rodriguez, and V. Hernández-Ortiz. 1997. Notas sobre el género *Anastrepha* Schiner en Cuba con descripción de una nueva especie (Diptera: Tephritidae). *Folia Entomológica Mexicana* 99: 29–36.

Follet, P. A. and R. I. Vargas. 2009. A systems approach to mitigate Oriental fruit fly risk in "Sharwil" avocados exported from Hawaii. *Acta Horticulturae* 880: 439–446.

Godefroid, M., A. Cruaud, J. P. Rossi, and J. Y. Rasplus. 2015. Assessing the risk of invasion by Tephritid fruit flies: Intraspecific divergence matters. *PLoS One* 10(8): e0135209.

ISPM 35 (international Standards for Phytosanitary Measures). 2012. Systems approach for pest management of fruit flies (Tephritidae), no. 35. Rome, IPPC, FAO. http://www.fao.org/docrep/016/k6768e/k6768e.pdf (last accessed August 14, 2018).

Jang, E., C. Miller, and B. Caton. 2015. Systems approaches for managing the risk of citrus fruit in Texas during a Mexican fruit fly outbreak. USDA. https://www.aphis.usda.gov/plant_health/plant_pest_info/fruit_flies/downloads/texas-citrus-systems-approach-risk-assesment.pdf (last accessed on August 2018).

Meats, A., A. Beattie, F. Ullah, and S. Bingham. 2012. To push, pull or push-pull? A behavioural strategy for protecting small tomato plots from tephritid fruit flies. *Crop Protection* 36: 1–6.

Moore, S. D., W. Kirkman, S. Albertyn, C. N. Love, J. A. Coetzee, and V. Hattingh. 2016. Partial cold treatment of citrus fruit for export risk mitigation for *Thaumatotibia leucotreta* (Lepidoptera: Tortricidae) as part of a systems approach. *Journal of Economic Entomology* 109(4): 1578–1585.

Qin, Y., D. R. Paini, C. Wang, Y. Fang, and Z. Li. 2015. Global establishment risk of economically important fruit fly species (Tephritidae). *PLoS One* 10(1): e0116424.

Rodríguez Velásquez D., A. M. Fernández, and V. Hernández-Ortiz. 2001. Catálogo de los Tefrítidos (Diptera: Tephritidae) de Cuba. *Fitosanidad* 5: 7–14.

Shelly, T. E. 2014. Fruit fly alphabets. In *Trapping and the Detection, Control, and Regulation of Tephritid Fruit Flies. Lures, Area-Wide Programs, and Trade Implications*, eds. T. E. Shelly, N. Epsky, E. B. Jang, J. Reyes-Flores, and R. Vargas, pp. 3–11. Springer, Dordrecht, the Netherlands.

Vázquez, L. L., I. Pérez, A. Navarro, and J. C. Casín. 1999. Occurrence and managing of the fruit flies in Cuba. *EPPO Bulletin* 29: 163–166.

Yahia, E. M. 2011. Mango (*Mangifera indica* L.). In *Postharvest Biology and Technology of Tropical and Subtropical Fruits: Cocona to Mango*, Vol. 3, ed. E.M. Yahia, pp. 492–567. Woodhead Publishing, Sawston, UK.

25 Fruit Fly Area-Wide Integrated Pest Management in Dragon Fruit in Binh Thuan Province, Viet Nam

Nguyen T.T. Hien, Vu T.T. Trang, Vu V. Thanh, Ha K. Lien, Dang Đ. Thang, Le T. Xuyen, and Rui Pereira*

CONTENTS

Abstract The area-wide integrated pest management (AW-IPM) to suppress fruit flies attacking dragon fruit was implemented in Ham Hiep village (Ham Thuan Bac district- Binh Thuan province, Viet Nam) since October 2016. The two targeted economically important tephritid fruit flies species were *Bactrocera dorsalis* (Hendel) and *Bactrocera correcta* (Bezzi). A pilot project consisting of a core zone (581 ha) and a buffer zone (986 ha) was implemented. Suppression strategies included both field sanitation and male annihilation technique (MAT) blocks in both zones. Additionally, in the core zone, protein bait spray was applied. A contiguous area under farmer suppression practice was used as a control. The average number of fruit flies per trap per day (FTD) was 1.8 and 2.2 in the core and buffer zones, respectively, compared to 11.6 in the farmers practice area. Another notable achievement was the involvement of the farmers in the surveillance activities, including trapping inspection, data collection, and sanitation by collecting and removing host fruits in the core and buffer zones. The results clearly indicated the advantage of integrating several methods in an AW-IPM approach. Further integration should include the sterile insect technique (SIT) in the overall suppression strategy.

25.1 BACKGROUND

Binh Thuan province is located in Southern Viet Nam. There are two seasons: wet (April–November) and dry (November–March). The temperature ranges from 20°C to 28°C during the year. The province has 28,000 ha of dragon fruit (*Hylocereus undatus*), which represent more than 70% of the total production in Viet Nam (Hien et al., 2012). Of these, 80% are for export. Farmers in 30%

* Corresponding author.

of the dragon fruit-growing areas adhere to the Vietnamese Good Agricultural Practices (VietGAP, 2008) and 8.7% to the Good Agricultural Practices (GlobalGAP) standards. However, most of them have limitations in the control of fruit flies (Diptera: Tephritidae), which are subject to strict quarantine measures and a barrier to fruit export for a large number of markets. Both *Bactrocera dorsalis* and *Bactrocera correcta* have been recorded to attack dragon fruit (Hien et al., 2011).

Since 2009, the Vietnamese government has been supporting the control of fruit flies; however, infestation is a limitation for fruit trade (Khanh et al., 2016). This is despite of fruit fly area-wide integrated pest management (AW-IPM), which is one of the most effective and environmentally friendly pest control strategies, already being applied successfully in many countries against Tephritid and other insect pests (Vreysen et al., 2007).

Since October 2016, an AW-IPM pilot project has been implemented in Binh Thuan province, Viet Nam in a 1,567-ha area, as a follow-up to a smaller-scale pilot project that was initiated in 2012 (Khanh et al., 2016). The objective of both trials was to suppress *B. dorsalis* and *B. correcta* tephritid fruit fly populations in selected dragon fruit-production areas by integrating different available control methods. A further goal would be the future integration of the sterile insect technique (SIT) into the control measures already taking place to aid sustainability to the program and to set areas of low fruit fly pest prevalence in the dragon fruit-production areas to reduce quarantine restrictions and facilitate trade.

25.2 METHODS

The pilot project area (1,567 ha) consisted of a 581-ha core zone where the full suite of available IPM control measures was implemented. The core zone was surrounded by a 986-ha buffer zone, which separated the core zone from the farmer zone (Figure 25.1). This farmer zone used existing farmer practices such as cover insecticide applications or lure traps and served as a control.

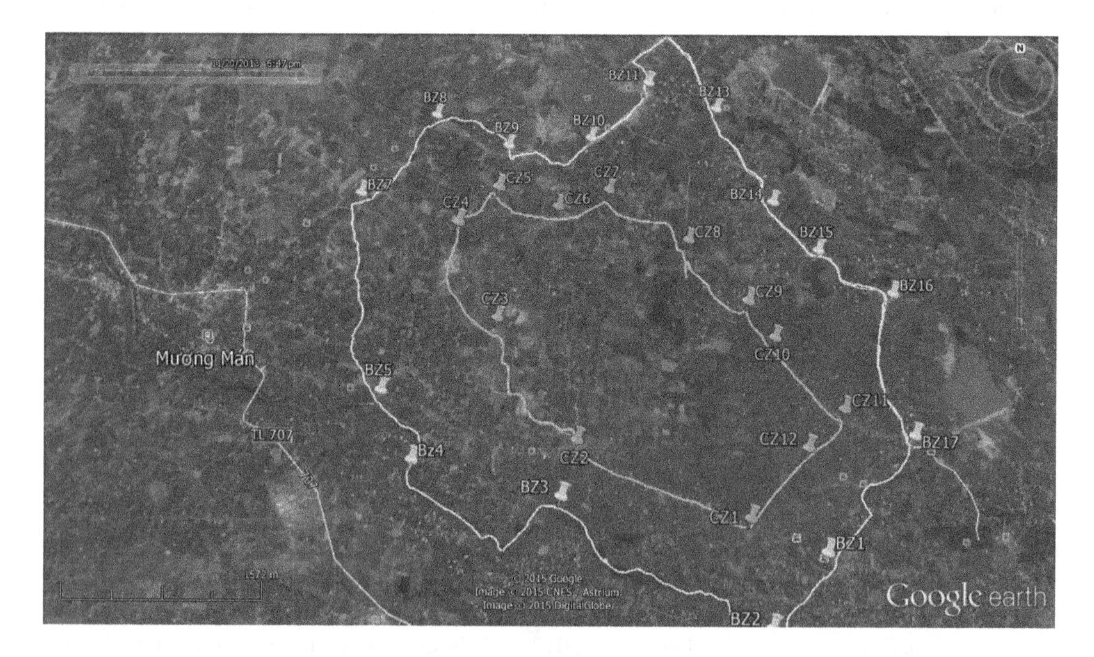

FIGURE 25.1 Map of fruit fly suppression in the dragon fruit-production area of Binh Thuan province. The core zone of 581 ha is inside the blue line, and the 986 ha between the blue and yellow line is the buffer zone. The area outside the yellow line is the farmer zone used as a control. Red, blue, and yellow letters refer to the location of the monitoring traps.

Three suppression methods were applied in the core zone. These included: (i) Field sanitation: Fallen and infested fruits were regularly collected and sealed into plastic bags that were exposed to direct sunlight to kill larvae in the fruit. Collected fruits were also burned or buried under the ground, at least 30 cm deep. Sanitation focused on dragon fruit plus fruits collected from backyards, such as mango (*Mangifera indica*), guava (*Psidium guajava*), and star fruit (*Averrhoa carambola*); (ii) Male Annihilation Technique (MAT): attract-and-kill blocks (containing 1 L of Methyl eugenol (ME) + 4 mL of fipronil) that were placed at 50-m intervals to suppress the population of males of both *B. dorsalis* and *B. correcta* (Hien et al. 2012, 2017). Blocks were replaced after 2–3 months (depending on the wet season); and (iii) Bait spray application targeting female fruit flies: Bait mixture (1 L of protein bait + 1 g of fipronil 800WG + 9 L of water) was applied every seven days from fruit maturation until harvest (Hien et al. 2012). Bait mixture was sprayed as spots (50 mL) under leaves or bushes (not applied directly on the fruits). Field sanitation and MAT block methods were also applied in combination in the buffer zone.

All information on host fruit maturation and infestation was recorded weekly during the implementation of the pilot project to obtain the status of the host (Khanh et al., 2016). Additionally, public information and training on AW-IPM for the farmers was conducted every week.

Adult populations were monitored during the full period of the pilot project (October 2016 to date) by using methyl eugenol (ME) traps (FAO/IAEA, 2018) inspected every 10 days and serviced every 2 months. A total of 72 traps were installed: 16 in the core zone, 48 in the buffer zone, and 8 in the control zone. All flies in each inspected trap were sent to the laboratory and the FTD was calculated.

To evaluate the impact of the suppression measures in the different zones and the control, weekly visual observations were conducted for tephritid damage on dragon fruits (FAO/IAEA, 2017). As of April 2017, a total of 300 dragon fruits in each zone were collected and observed for damage at the harvesting stage every month. They were then kept individually to allow larvae within the fruits to pupate and be counted, thus obtaining a percentage of fruit infestation. This study was initiated in April 2017 and is still in operation.

25.3 RESULTS

Flies per trap per day (FTD) varied from 0 to 5.43, from 1.31 to 14.97, and from 1.33 to 38.29 in the core zone, buffer zone, and farmer zone, respectively (for the period of October 2016–July 2018) (Figure 25.2). The number of fruit flies caught in all zones varied over the time period, with higher numbers being caught in the wet periods, from March to August/September and with population peaks in May/June. Fruit flies caught in traps in the core and buffer zones were significantly fewer

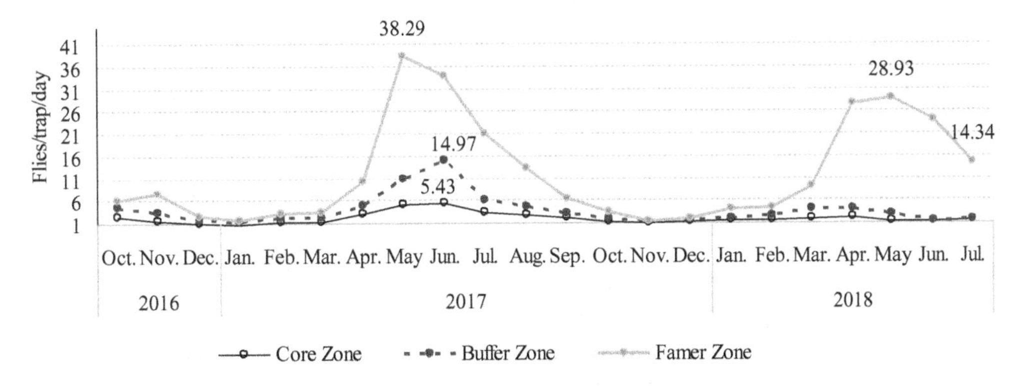

FIGURE 25.2 Mean population of fruit flies captured in the core, buffer, and farmer control zones in the area-wide integrated pest management (AW-IPM) pilot project (October 2016–July 2018) in Binh Thuan Province, Viet Nam.

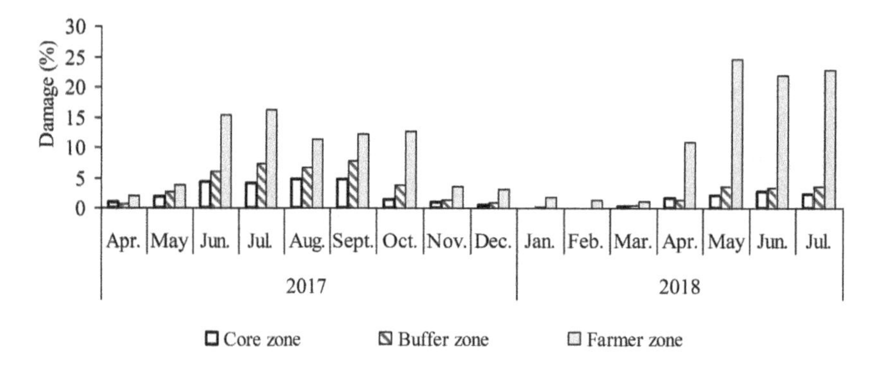

FIGURE 25.3 Mean percentage of damage in dragon fruits from the core, buffer, and farmer control zones in the area-wide integrated pest management (AW-IPM) pilot project (April 2017–July 2018) in Binh Thuan Province, Viet Nam.

than those in the farmer control zone (Figure 25.2). The reduction of the wild population was dramatic in both the core and the buffer zone compared to the control zone in the period of April–July 2018 (wet period). In the dry periods (October–March), fewer flies were caught, with no differences noted in trap catches between the three zones (Figure 25.2).

The average percentage of damaged dragon fruit ranged from 1% to 4.7% in the core zone, 0.7% to 7.7% in the buffer zone, and 2% to 24.7% in farmer's practice (control zone) during the period of April 2017–July 2018 (Figure 25.3).

At each sampling period, it can be clearly seen that fruits from the core area, where the AW-IPM treatments were applied, were far less damaged than in the farmers control fields, where no AW-IPM treatments were applied (Figure 25.3). This was markedly more evident in the wet months of June through October 2017 and even more so during the second year of the project, from April through July 2018, when fruit fly populations were generally higher than in the dry period (Figure 25.2).

25.4 CONCLUSIONS

Area-wide suppression methods using MAT with protein bait sprays and field sanitation were effective for controlling fruit fly populations in dragon fruit farms in the core area of the AW-IPM pilot trial in Binh Thuan province. However, for the continued implementation and maintenance of the AW-IPM program, a good knowledge base on alternative hosts and the effect of climate change is needed (Hien et al. 2012). In addition, further education for the continuation of monitoring and fruit damage evaluations is needed for farmers and stakeholders. The need of fruit sampling is very important, especially in situations where MAT blocks are used because adult males will be removed from the environment in these situations, making trapping data less reliable and requiring fruit sampling evaluations.

Another relevant achievement was the involvement of the farmers in the activities of surveillance, including trapping inspection and data collection, and implementing sanitation by removing host fruits in the core and buffer zones. This is the result of awareness and implementation in the field, which occurs first by leader farmers that attract others to use such methods.

The present results clearly indicate the impact and advantage of integrating several available methods in a controlled AW-IPM strategy. The implementation of the AW-IPM strategy in a planned and knowledge-based way and the lessons learnt during this pilot study should be a priority for the larger dragon fruit industries. The integration of SIT into this already established successful suppression strategy could be considered with additional suppression tools. Research on this integration is planned for 2019.

REFERENCES

FAO/IAEA. 2017. *Fruit Sampling Guidelines for Area-Wide Fruit Fly Programmes*, eds. W. R. Enkerlin, J. Reyes and G. Ortiz. Vienna, Austria: Food and Agriculture Organization of the United Nations.

FAO/IAEA. 2018. *Trapping Guidelines for Area-Wide Fruit Fly Programmes*, eds. W. R. Enkerlin and J. Reyes-Flores. Rome, Italy: IAEA.

Hien, N. T. T., L. D. Khanh, and L. Q. Khai. 2011. Thành phần loài ruồi hại quả (Tephritidae: Diptera) và ký chủ của chúng tại vùng thanh long BìnhThuận. *Journal of Vietnam Agricultural Science and Technology*, 9:41–44.

Hien, N. T. T., L. Đ. Khanh, L. Q. Khai et al. 2012. Nghiên cứu biện pháp quản lý ruồi hại quả thanh long trên diện rộng, nhằm góp phần nâng cao chất lượng sản phẩm quả xuất khẩu tại Bình Thuận. The National project report. Management fruit fly on Dragon fruit in Binh thuan province.

Hien, N. T. T, L. D. Khanh, V. V. Thanh et al. 2017. Influence of adult diet and exposure to methyl eugenol in the mating performance of *Bactrocera correcta. Nuclear Science and Technology* 7:42–48.

Khanh, L. D., L. Q. Khai, N. T. T. Hien, V. V. Thanh, T. V. R. Trang, and R. Pereira. 2016. Area-wide suppression of *Bactrocera* fruit flies in dragon fruit orchards in Binh Thuan, Viet Nam. In *Proceedings of the 9th International Symposium on Fruit Flies of Economic Importance*, eds. B. Sabater-Muñoz, T. Vera, R. Pereira and W. Orankanok, pp. 93–100. Bangkok, Thailand: International Fruit Fly Steering Committee.

VietGAP. 2008. Bộ Nông Nghiệp và PTNT. *Quyết định ban hành quy trình thực hành sản xuất nông nghiệp tốt cho rau, quả tươi an toàn*. Global Good Agricultural Practice.

Vreysen, M. J. B., A. S. Robinson, and J. Hendrichs. 2007. *Area-Wide Control of Insect Pests from Research to Field Implementation*. Dordrecht, the Netherlands: Springer.

26 Area-Wide Approach for the Control of Mango Fruit Flies in a Metropolis Containing Polycultures in Urban and Peri-Urban Areas in Nigeria

Vincent Umeh, Vivian Umeh, and John Thomas*

CONTENTS

Abstract Fruit flies impact the production of many fruit species and cause economic yield losses in all countries in West Africa. In such endemic areas, including metropolitan cities, fruit flies do not occur only in orchards but extend their infestation to trees in household backyards, private gardens, and stockpiled fruits for local and international markets. This scenario occurs in almost all towns and cities in Nigeria, contributing to fruit fly population explosions if left uncontrolled. We, therefore, attempted, for the first time in Nigeria, to implement a mass trapping technique over an area of about 20 km2 to capture mainly fruit flies infesting mango and other major alternative hosts. This study evaluated the population dynamics of the major mango fruit flies *Bactrocera dorsalis* (Hendel) and *Ceratitis cosyra* (Walker) during on and off season periods in Ibadan metropolis, Nigeria. The influence of environmental factors, such as temperature and humidity (rainfall), on the abundance of both species was also evaluated. Results of the implementation of fruit fly management techniques, which included orchard sanitation by picking dropped fruits, mass trapping using parapheromones, and the application of protein baits, are discussed. *B. dorsalis* dominated the trap catches, whereas the presence of *C. cosyra* on mango was very negligible throughout the study period. Although a higher number of *C. cosyra* was observed in the dry season months of January–March, it was totally absent in other months. The presence of *B. dorsalis* was recorded throughout the year, with higher populations occurring during the rainy season. The relative abundance of *B. dorsalis* across alternative hosts indicated that *Irvingia* harbored higher fly numbers compared to citrus. Fruits incubated during first- and second-year harvests showed a significant

* Corresponding author.

suppression of fruit fly populations by not less than 70%–82% for *B. dorsalis* and *C. cosyra*, respectively, in all fruit species compared to areas where no control was applied. Fruit fly population dynamics are influenced by environmental factors. Application of management strategies in a metropolis that is characterized by polycultures and diverse hosts can suppress populations. However, there is a need for awareness campaigns aimed at communities in the metropolis for their direct involvement in the control of fruit flies.

26.1 INTRODUCTION

Mango (*Magnifera indica*) and citrus are some of West Africa's most important crops and play a major role in local, national, regional, and international markets. They are also a major source of nutrition for rural populations in West Africa. In Nigeria, most of the fruit produced is consumed as fresh fruit, and ripe fruits can be made into juice and be preserved. Although Nigeria occupies the ninth position among the 10 leading mango-producing countries of the world, it does not feature among the 10 leading mango fruit exporters (FAOSTAT, 2007).

Pests and diseases are the primary constraints for fruit production in Nigeria. Although some insect pests are noted for contributing to the decline of citrus and mango (Umeh et al., 2000), some play an important role in reducing fruit yields and rendering them unacceptable to consumers (Drew et al., 2005; Umeh et al., 2008). Fruit flies are considered the most destructive insect pests of fruits (Ekesi et al., 2009; Vayssieres et al., 2007; Vayssieres et al., 2008). The oriental fruit fly, *Bactrocera dorsalis* (Hendel), is responsible for extensive economic losses of horticultural crops throughout West Africa, increasing the damages already caused by native fruit flies. This invasive species was recently identified in parts of Africa, which implies a further increase in yield losses.

The distribution and abundance of tephritids depend on several abiotic factors (e.g., temperature, relative humidity, rainfall) and several biotic factors (e.g., host plants, natural enemies) (Vayssières et al., 2008). Temperature and relative humidity have a significant effect on fruit flies, especially on their developmental stages. A decrease in temperature increases the duration of each stage. Rwomushana et al. (2008) reported high rates of survival for all immature stages in *B. invadens* (currently *B. dorsalis*) at 20°C–30°C. Similarly, Duyck et al. (2004) reported that a temperature ranging between 20°C and 30°C allows high survival rates of *B. zonata* (Saunders). Lower survival rates have been generally observed at extreme temperatures of 15°C–35°C for all developmental stages of tephritid fruit flies (Brévault and Quilici, 2000; Duyck and Quilici, 2002; Duyck et al., 2004; Rwomushana et al., 2008).

The main control methods employed in orchards are regular protein baiting of host trees and the implementation of the male annihilation technique (MAT). The bait application technique (BAT) is directed at killing both male and female flies, whereas MAT attracts and kills only male flies through the use of parapheromones. Presently, BAT has also been used in area-wide eradication programs on its own or in combination with other control methods. BAT is frequently used to eradicate exotic species entering into an area, and bait applications are used with sterile releases for the eradication of fruit flies (Permalloo et al., 1997). Methyl eugenol (4-allyl-1, 2 dimethoxy benzene-carboxylate) is used for the detection of the oriental fruit fly *B. dorsalis*. Trimedlure [t-Butyl-2-methyl-4-chlorocyclohexane carboxylate], a powerful lure for the Mediterranean fruit fly (*Ceratitis capitata* Wied), is used to detect incipient infestations of the destructive insect and is used in combination with an insecticide to reduce male populations to such low levels that mating does not occur. All these species have been introduced and have become severe pests of tropical fruits (Leblanc et al., 2011; Vargas et al., 2007).

MAT is aimed at reducing the number of male flies on an area-wide, long-term basis with the eventual effect of reducing female fertility due to the greatly reduced number of males available for mating. The male annihilation technique is a fruit fly control method that aims to remove male insects, thus reducing the male population. This affects the male-to-female ratio and reduces the insect's chances of mating, with females producing fewer progeny. Consequently, insect populations in target areas decline and insects can ultimately be eradicated (Stonehouse et al., 2008; Zaheeruddin, 2007). Lures in monitoring traps are used in MAT programs. The use of this method on incipient infestations of the oriental fruit fly should prevent the further development and spread of this species, with eradication being a definite possibility. Male attractants for other tropical fruit flies are strong enough to warrant consideration as possible male annihilation agents (Christenson, 2009). It has been reported that methyl eugenol and Cue-lure traps used in close proximity, about 3 m, to fruit trees show a high performance and are considered as the best attractants in mixed fruit orchards (Ullah et al., 2017).

The main objectives of this study were: (i) to evaluate the influence of environmental factors, such as temperature, relative humidity, and rainfall, on fruit fly population dynamics across a metropolis constituted of urban and peri-urban areas containing polycultures in Nigeria, and (ii) to assess the effect of mass trapping, using parapheromones and protein baits, and the cultural practice of picking dropped fruits, on the suppression of populations of the major mango fruit flies in polycultures.

26.2 MATERIALS AND METHODS

26.2.1 Study Site

High presence of mango was a determinant factor for the selection of the study site, which was located between N07° 30″, E003° 46″ and N07° 22″, E003° 53″. Other major economic fruit species common in the target area that are alternative hosts of *B. dorsalis* were also considered. The assessed alternative hosts were limited to citrus and bush mango *Irvingia* spp. The target area covered parts of urban and peri-urban areas in Ibadan metropolis, and it mostly comprised sole crops, polycultures, and homestead stands of mango, citrus, and *Irvingia* spp. Parts of the target area did not contain any fruit trees, whereas others had patches of small or big orchards ranging from 1–30 ha. Large orchards belong to the National Horticultural Research Institute (NIHORT) and the Forestry Research Institute of Nigeria (FRIN) (Figure 26.1).

26.2.2 Distribution and Placement of Traps

Trap layout (spatial distribution of traps) and trap density were influenced by various factors, including the sensitivity to the parapheromone of the fruit fly species associated with the host, type of survey (monitoring or control), trap efficiency, and assessed pest risk. Pest risk assessment was initially performed to identify the risk areas, with the lowest-risk areas requiring the lowest trap densities and the highest-risk areas requiring the highest trap densities. The identified and characterized risk factors (individually or as added effects) included the following:

- Host availability in the target area (number of species present, abundance, and distribution over space and time)
- Host preference (major and minor hosts)
- Human settlements (urban and peri-urban)
- Distance of host to infested areas
- Historical profile of pest occurrence in the area

A total of 330 traps were distributed in an area of about 20 km^2 to capture mainly fruit flies of mango and other alternative host plants, namely citrus and bush mango (*Irvingia* spp.), that are characteristic of the area. Tephri traps were used for the parapheromone baits. The parapheromones used were methyl eugenol and terpinyl acetate in a total of 60 traps each; whereas only methyl eugenol was used for citrus in a total of 60 traps. *Ceratitis capitata* populations in the target area were found to be negligible over the past 6 years. Thus, the parapheromone (Trimedlure) used for *C. capitata* associated with citrus was not included in this study. *Irvingia* stands were also supplied with a

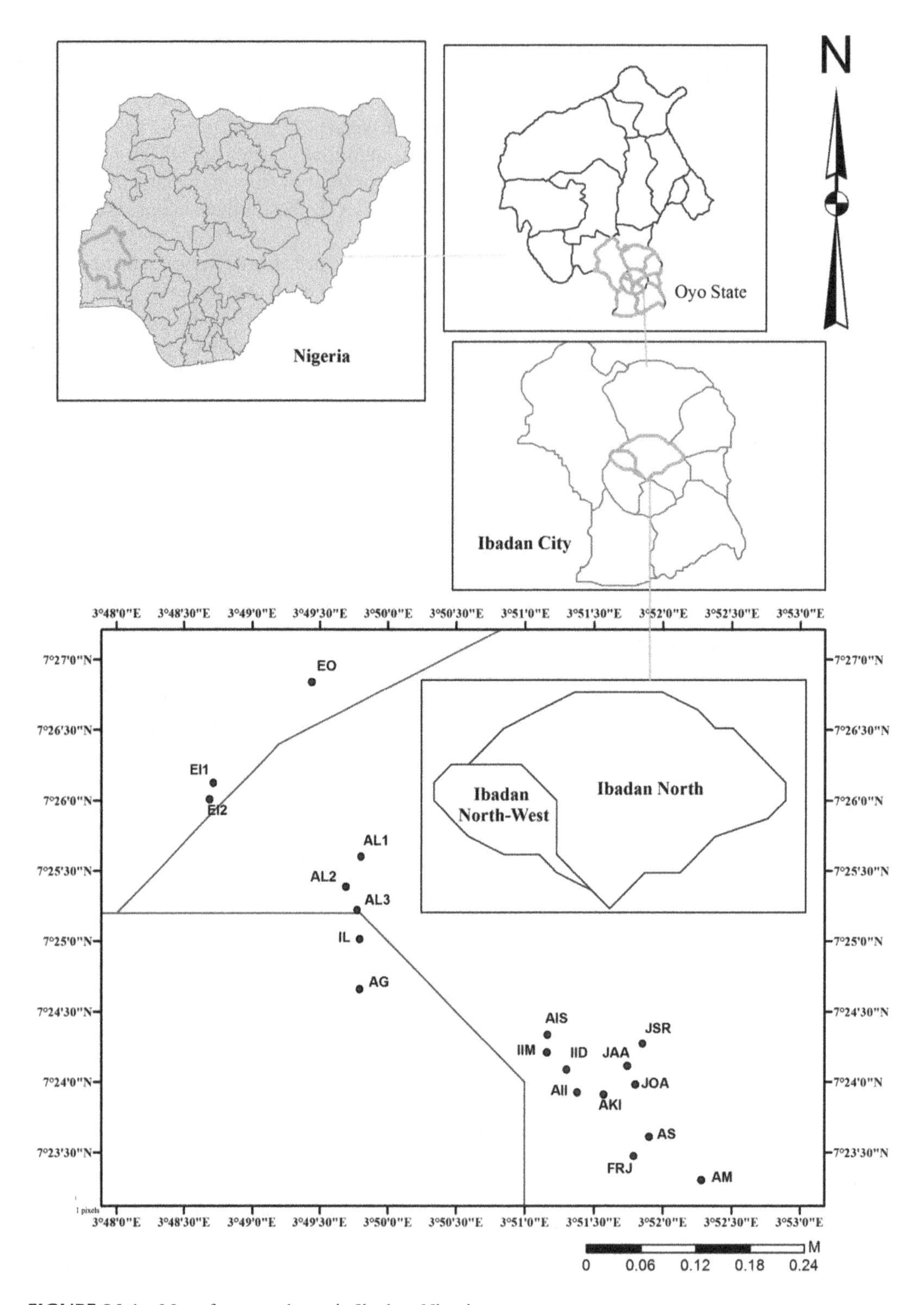

FIGURE 26.1 Map of surveyed area in Ibadan, Nigeria.

total of 60 traps baited with methyl eugenol and distributed in the target area. The fumigant DDVP was also placed in each trap device (i.e., parapheromone + fumigant). Traps were hung on mango, citrus, and *Irvingia* trees either in orchards or homestead stands. Protein baits, made of Torula® yeast pellets dissolved in warm water at 50°C, using 2 pellets per 150 mL of water, were also used. A total of 30 Torula yeast traps were placed in mango, citrus, and *Irvingia* trees in the target area.

The total number of trap devices for each plant type was divided into three batches and distributed according to the selected crop stands. Flies caught in the parapheromone traps were recorded and collected at weekly intervals, and the traps were repositioned. Parapheromone attractants were replaced at monthly intervals (IAEA, 2003). Torula yeast traps were replaced at weekly intervals after recording the number of trapped fruit flies. Three portions within the surveyed area were assigned as a control treatment and no traps were placed in those sites for any of the three fruit species (i.e., mango, citrus, and *Irvingia*). Picking and removal of dropped fruits was carried out throughout the target area.

We also carried out extension actions and publicity regularly to sensitize the stakeholders whose trees were included in the study by informing them about what to do to avoid disrupting the control activities. Fallen fruits in the control area were not picked. Monthly mean atmospheric temperature values and rainfall dates were obtained from a meteorological station belonging to NIHORT and located in the study area.

During harvest, 20 fruits of mango, citrus, and *Irvingia* each were collected in the experimental and control portions of the orchards for three consecutive harvesting regimes according to the maturation time of each of the fruit species. Collected fruits were incubated in small cages layered with sieved fine sand. Developed pupae were sieved and reared to adulthood in a cylindrical transparent plastic container covered with wire gauze.

The number of emerging fruit flies in the different treatments was recorded and the mean number of emerging fruit flies from each fruit batch was computed as mean number/fruit. Data were analyzed with an analysis of variance (ANOVA) and significant means were identified using the Student Newman Keuls (SNK) test. All tests were considered to be significant at $P = 0.05$.

26.3 RESULTS

Bactrocera dorsalis populations dominated the trap catches in mango. Its presence was recorded throughout the year. The population rose steadily from February 2017 and increased to a maximum mean of 14.5 fruit flies per trap per day (FTD) in June, which coincided with the rainy season (Figure 26.2).

The population remained relatively high during the rainy season but started to decline as the dry season began and stretched into the early part of 2018. After this time, the population started to increase again in a pattern similar to the trend observed in 2017 (Figure 26.2). The population of *C. cosyra* in mango was low throughout the study period compared to *B. dorsalis*. It ranged between 1 and 3 FTD in the dry season months of January–March, with a marked decrease to zero fruit flies recorded in the rainy season between August and November.

Although a relatively higher number of *C. cosyra* was observed during the harvest period of mango in the dry season, availability of early mango varieties also influenced its presence. Furthermore, the population decreased and became totally absent as the rainy season progressed and the mango season ended in the area (Figure 26.2). Observations made in traps baited with Torula yeast placed only in mango showed a dominance of female fruit flies from both *B. dorsalis* and *C. cosyra* (Figure 26.3) and a minimal number of *C. capitata* females and *B. dorsalis* males. These followed the same population trends that were observed in the parapheromone traps for male catches (i.e., a higher number of *B. dorsalis* females compared to *C. cosyra*).

Fluctuations in the number of FTD observed in the protein baits could be associated with the ripening period of mango. More visits to mango were made by female fruit flies during the ripening period in the months of March–June, when ripe fruits were still available. This agrees with the findings of other authors who observed similar trends (Manrakhan, 2016; Papadopoulos et al., 2003).

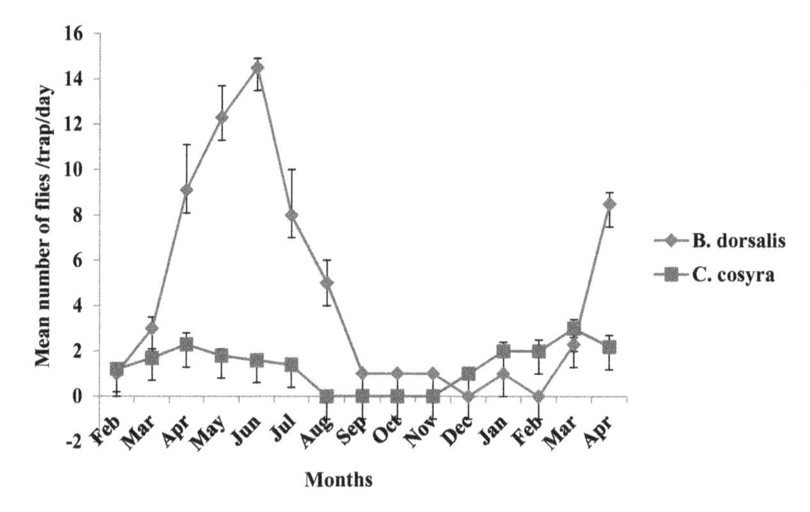

FIGURE 26.2 *Bactrocera dorsalis* and *Ceratitis cosyra* male population trends from trap catches in mango. Mean population of fruit flies from each of 30 traps replicated three times and separately baited with methyl eugenol and terpinyl acetate.

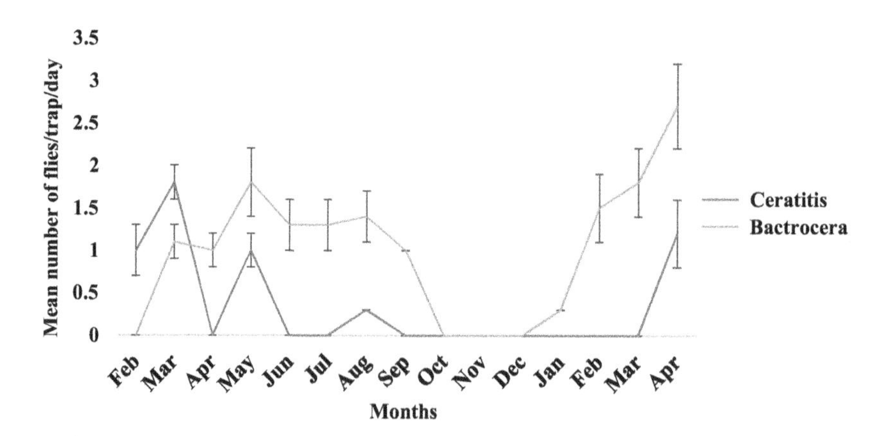

FIGURE 26.3 Population trends of *Ceratitis cosyra* and *Bactrocera dorsalis* females from protein bait catches in mango. Mean population of fruit flies from 10 torula yeast traps replicated three times.

The generally lower population of female fruit flies observed in this food-based attractant system may be attributed to the nature of the attractant, which can only be lethal if the flies drown in the liquid bait, unlike the parapheromone trap device, which has an insecticidal fumigant that kills the trapped male flies.

The relative abundance of *B. dorsalis* across other sampled alternative hosts, citrus and bush mango (*Irvingia* spp.), indicated that *Irvingia* attracted higher numbers of *B. dorsalis* compared to citrus (Figures 26.4 and 26.5). However, the population dynamics of *B. dorsalis* in citrus followed the same trend as that observed in mango, whereas the population of *B. dorsalis* in *Irvingia* differed slightly from the one in mango. The availability of *Irvingia* fruits in the months of January–March, which is usually a dry period when *B. dorsalis* populations are low, resulted in the unusual presence of *B. dorsalis* in traps in *Irvingia* earlier than in mango. Thus, this indicates that, apart from other environmental factors, the availability of preferred fruit types also influenced the population levels of this species.

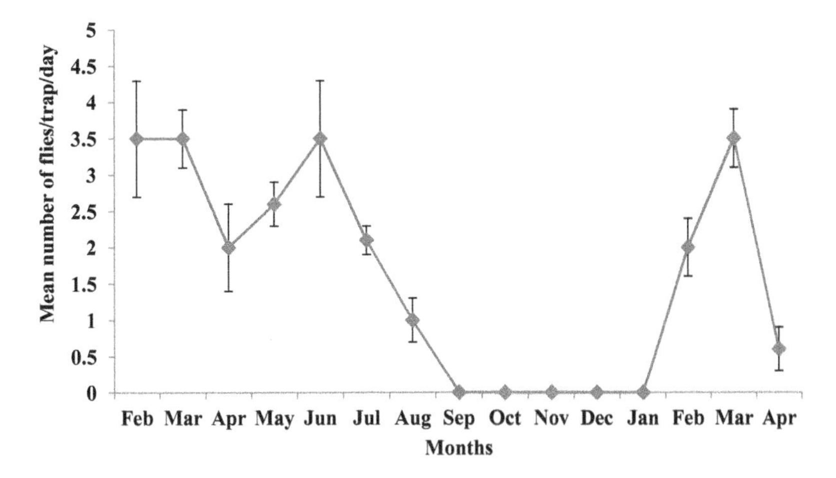

FIGURE 26.4 *Bactrocera dorsalis* male population trend from trap catches in citrus. Mean population of fruit flies from 30 traps replicated three times.

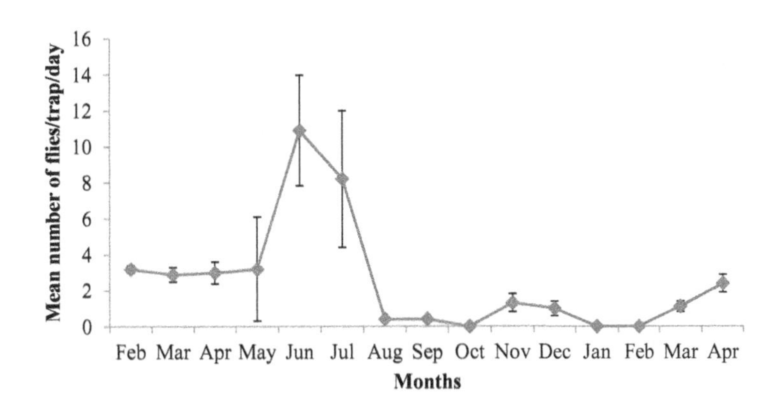

FIGURE 26.5 *Bactrocera dorsalis* male population trend from trap catches in *Irvingia*. Mean population of fruit flies from 60 traps replicated three times.

Population dynamics of *B. dorsalis* and *C. cosyra* in relation to temperature are shown in Figures 26.6 and 26.7 and are shown in relation to rainfall in Figures 26.8 and 26.9. Correlation analyses showed that the population of *B. dorsalis* was not significantly ($P > 0.05$) correlated with environmental temperature across the studied months, and the correlation coefficient was negative ($r = -0.037$; $n - 1 = 14$). On the other hand, the population of *C. cosyra* was positively correlated with environmental temperature ($r = 0.752$; $n - 1 = 14$; $P < 0.05$). These findings show that drier periods favor populations of *C. cosyra*, hence the larger population observed in the early part of the year until the beginning of the rainy season. In Ibadan, Nigeria, the highest annual temperatures are recorded during the months with a high abundance of *C. cosyra* (February–April) (Umeh and Onukwu, 2016). In the case of rainfall, there was a positive correlation with the population of *B. dorsalis* ($r = 0.434576$: $n - 1 = 14$; $P < 0.05$), whereas a weak negative correlation was observed between mean *C. cosyra* populations and rainfall levels ($r = -0.342$: $n - 1 = 14$; $P > 0.05$). These results are in line with findings by other authors in West Africa, especially the positive effect of humidity on *B. dorsalis* populations (Amice and Sales, 1997; Rwomushana et al., 2008; Sarada et al., 2001; Vayssières et al., 2009).

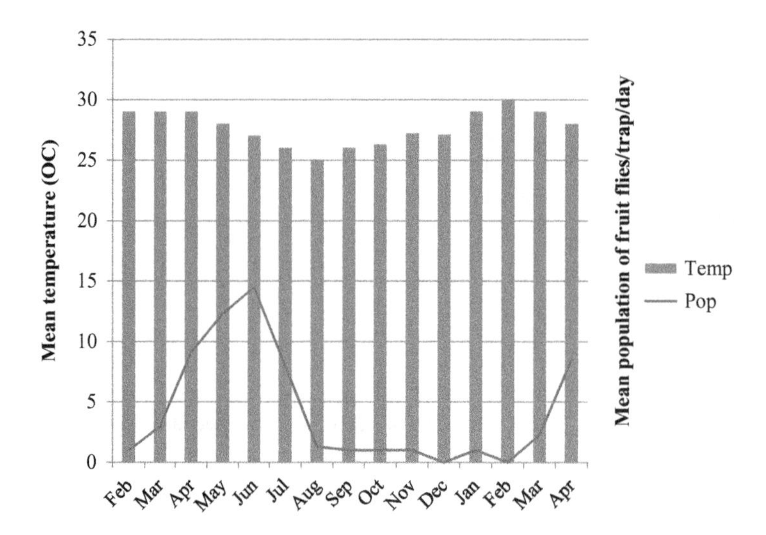

FIGURE 26.6 Relationship between *Bactrocera dorsalis* trap captures and temperature.

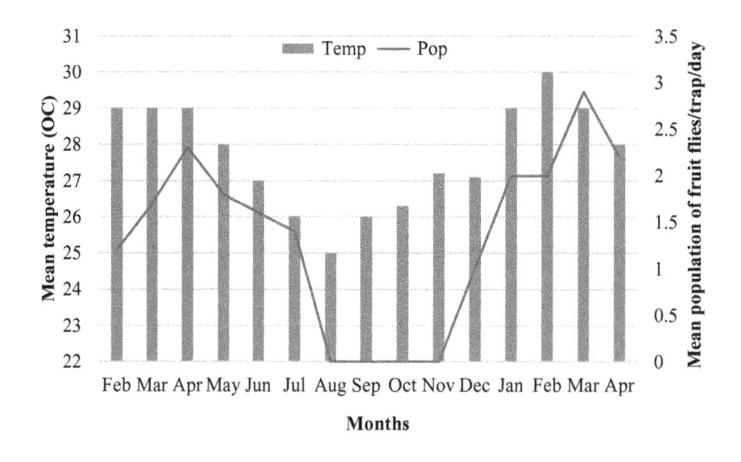

FIGURE 26.7 Relationship between *Ceratitis cosyra* trap captures and temperature.

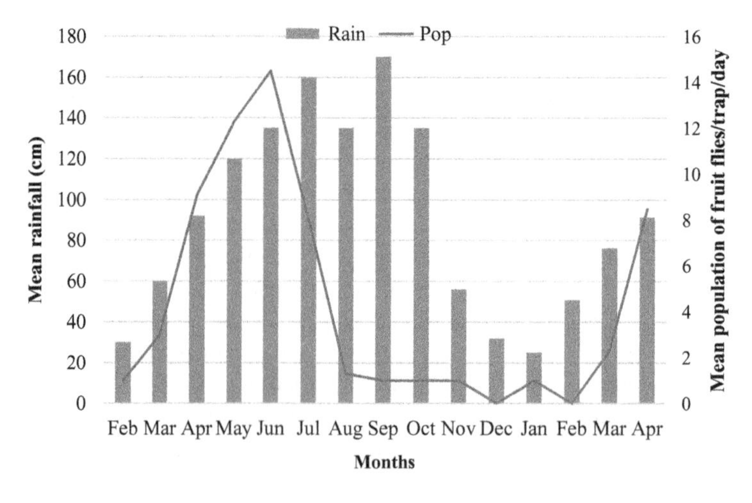

FIGURE 26.8 Relationship between *Bactrocera dorsalis* trap captures and rainfall.

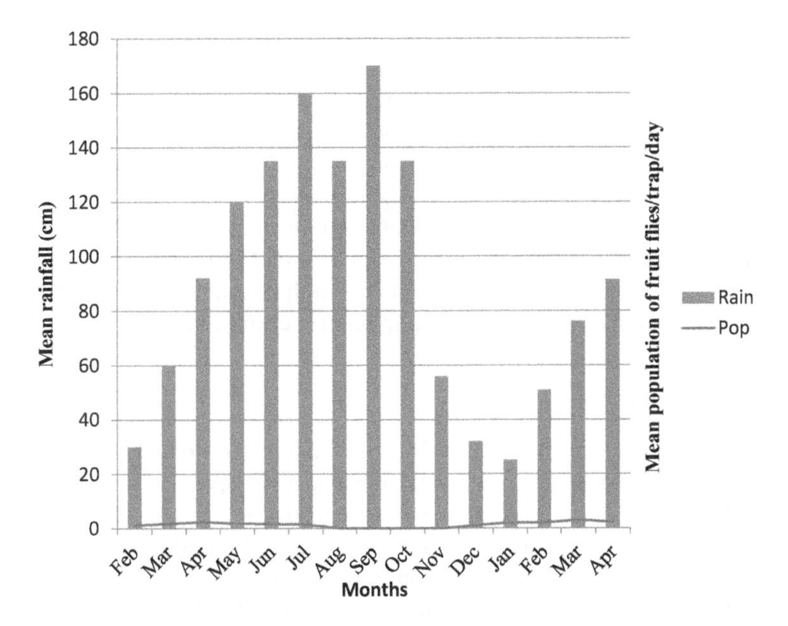

FIGURE 26.9 Relationship between *Ceratitis cosyra* trap captures and rainfall.

There seems to be an interplay of factors that affect the populations of these fruit fly species. These include the presence of preferred fruits, apart from temperature and humidity. Theron et al. (2017) reported, using a time series analyses, that adult populations of *B. dorsalis* increased 2 months after an increase in mean temperature in all sites of the study, 4 months after rainfall in natural and interface sites, and 1 and 3 months after fruit infestation in commercial and natural and interface sites, respectively.

Fruit flies obtained from fruits sampled in the different treatments showed various levels of fruit fly emergence per fruit. However, the mean number of *B. dorsalis* that emerged from fruits collected in areas where MAT traps were set up in mango, citrus, or *Irvingia* was significantly ($P < 0.05$) lower than that from those collected in the control area, indicating a decrease in the population of fruit flies due to mass trapping (Table 26.1).

TABLE 26.1
Relative Abundance of Fruit Fly Species Emerging from Fruits Collected from Different Treatment Plots in 2017 and 2018

	Mean Number of Fruit Flies/Fruit							
	Mango				Citrus		Bush Mango (*Irvingia*)	
Treatments	2017		2018		2017	2018	2017	2018
	Bd	Cc	Bd	Cc	Bd	Bd	Bd	Bd
Methyl eugenol	3.6 b	—	2.6 b	—	3.2 b	1.8 b	3.8 b	2.3 b
Terpinyl acetate	—	0.6 b	—	0.4 b	—	—	—	—
Control	10.2 a	3.8 a	7.8 a	2.0 a	7.2 a	5.6 a	9.4 a	6.4 a

Means in the same column followed by the same letter are not significantly different according to SNK ($P > 0.05$). Mean number of fruit flies obtained from 20 mango fruits (mean weight of 10.4 kg), 20 citrus fruits (mean weight 8.5 kg), and 20 Irvingia fruits (mean weight of 6.00 kg) are shown.
Bd, *Bactrocera dorsalis*; Cc, *Ceratitis cosyra*.

The number of *B. dorsalis* and *C. cosyra* per fruit recorded in the MAT fields ranged from 2.6 to 3.6/fruit and 0.4 to 0.6/fruit, respectively, compared to the control fields where recorded numbers were 7.8–10.2/fruit and 2–3.8/fruit, respectively. Similarly, the number of *B. dorsalis* that emerged per fruit in the alternative hosts citrus and *Irvingia* in the MAT fields ranged from 1.8 to 3.8/fruit compared to 5.6 to 9.4/fruit observed in the corresponding control fields. However, a lower *B. dorsalis* and *C. cosyra* adult emergence was generally observed in fruits collected in 2018 compared to those collected in 2017, which is probably due to the reduced population of fruit flies observed in 2018, possibly as a result of the effectiveness of MAT captures in 2017. Thus, fewer attacks resulted in 2018. A reduction of more than 70% in the population of fruit flies was obtained in the MAT fields, compared to the control area, in mango, citrus, and *Irvingia*.

26.4 CONCLUSIONS

Fruit fly population dynamics are influenced by environmental factors. The implementation of a combination of management strategies, such as the removal and disposal of dropped fruits, the application of the MAT, and the use of the BAT, in a metropolis that is characterized by polycultures and diverse hosts can suppress populations. However, there is a need for education campaigns aimed at communities in the cities for their direct involvement in fruit fly control using the tested techniques. Studies conducted by fruit fly experts indicate that the success of AW-IPM programs is highly dependent on the monitoring of fruit flies, appropriate and quick responses to incursions, and an active participation by all growers and the rest of the community in the area under the program. For a rapid population suppression and better results, the introduction of the sterile insect technique (SIT) will go a long way in achieving the desired result, especially in an area-wide approach.

REFERENCES

Amice, R., and F. Sales. 1997. Seasonal abundance of fruit flies in New Caledonia. In: *Management of Fruit Flies in the Pacific*, eds. A. J. Allwood and R. A. I. Drew, pp. 134–139. Camberra, Australia: ACIAR.

Brévault, T., and S. Quilici. 2000. Relationship between temperature, development and survival of different life stages of tomato fruit fly. *Neoceratitis Cyanescens. Entomologia Experimentalis et Applicata* 94: 25–30.

Christenson, L. D. 2009. The male annihilation technique in the control of fruit flies: New approaches to pest control and eradication. *Advances in Chemistry* 41: 31–35.

Drew, R. A. I, K. Tsuruta, and I. M. White. 2005. A new species of pest fruit fly from Sri Lanka and Africa. *African Entomol*ogy 13: 149–154.

Duyck P. F., and S. Quilici. 2002. Survival and development of different life stages of three *Ceratitis* spp. (Diptera: Tephritidae) reared at five constant temperatures. *Bulletin of Entomological Research* 92: 461–469.

Duyck, R. A. I., J. F. Sterlin, and S. Quilici. 2004. Survival and development of different life stages of *Bactrocera zonata* (Diptera: Tephritidae) reared at five constant temperatures compared to other fruit fly species. *Bulletin of Entomological Research* 94: 89–93.

Ekesi, S, M. K. Billah, P. W. Nderitu, S. Lux, and I. Rwomushana. 2009. Evidence for competitive displacement of *Ceratitis cosyra* by the invasive fruit fly *Bactrocera invadens* on mango and mechanisms contributing to the displacement. *Journal of Economic Entomology* 102 (3): 981–991.

FAOSTAT. 2007. *FAO Statistics*, Food and Agriculture Organization of the United Nations, Rome, Italy. http://faostat.fao.org/.

IAEA. 2003. *Trapping Guidelines for Area-Wide Fruit Fly Programmes*. Vienna, Austria: IAEA.

Leblanc, L., R. I. Vargas, B. MacKey, R. Putoa, and C. P. Jaime. 2011. Evaluation of cue-lure and methyl eugenol Solid Lure and Insecticide dispensers for fruit fly (Diptera: Tephritidae) monitoring and control in Tahiti. *Florida Entomologist* 94 (3): 510–516.

Manrakhan, A. 2016. Fruit fly. In: *Integrated Production Guidelines for Export Citrus. Integrated pest and Disease Management*, ed. T. G. Grout, pp. 1–10. Nelspruit, South Africa: Citrus Research International.

Papadopoulos, N. K., B. I. Katsoyannos, and D. Nestel. 2003. Spatial autocorrelation analysis of a *Ceratitis capitata* (Diptera: Tephritidae) adult population in a mixed deciduous fruit orchard in Northern Greece. *Environmental Entomology* 32: 319–326.

Permalloo, S., S. I. Seewooruthun, A. Soonnoo, B. Gungah, L. Unmole, and R. Boodram. 1997. An area wide control of fruit flies in Mauritius. In: *Proceedings of the Second Annual Meeting of Agricultural Scientists*, eds. J. A. Lalouette, D. Y. Bachraz, N. Sukurdeep, and B. D. Seebaluck, pp. 203–210. Reduit, Mauritius: Food and Research Council.

Rwomushana, I., S. Ekesi, C. K. P. O. Ogol, and I. Gordon. 2008. Effect of temperature on development and survival of immature stages of *Bactrocera invadens* (Diptera: Tephritidae). *Journal of Applied Entomology* 132: 832–839.

Sarada, G., T. U. Maheswari, and K. Purushotham. 2001. Seasonal incidence of population fluctuation of fruit flies of mango and guava. *Indian Journal of Entomology* 63: 272–276.

Stonehouse, J. M., J. D. Mumford, A. Verghese et al. 2008. Village-level area-wide fruit fly suppression in India: Bait application and male annihilation at village level and farm level. *Crop Protection* 26: 788–793.

Theron, C. D., A. Manrakhan, and C. W. Weldon. 2017. Host use of the oriental fruit fly, *Bactrocera dorsalis* (Hendel) (Diptera: Tephritidae), in South Africa. *Journal of Applied Entomology* 141: 810–816.

Ullah, F., M. W. Khan, F. Maula, M. Younus, and H. Badshah. 2017. Impact of close proximity of traps baited with various attractants on fruit fly catch. *Journal of Entomology and Zoology Studies* 5(6): 1843–1845.

Umeh, V. C., I. O. O. Aiyelaagbe, A. A. Kintomo, and M. B. Giginyu. 2000. Insect pest situation and farmers cultural practices in citrus orchards in southern Guinea savanna agro-ecological zone of Nigeria. *Nigerian Journal of Horticultural Science* 7: 26–32.

Umeh, V .C., L. E. Garcia, and M. De Meyer. 2008. Fruit flies of citrus in Nigeria: Species diversity relative abundance and spread in major producing areas. *Fruits* 63: 145–153.

Umeh, V. C. and D. Oukwu. 2016. Integrated management of fruit flies: Case studies from Nigeria. In: *Fruit Fly Research and Development in Africa—Towards a Sustainable Management Strategy to Improve Horticulture*, eds. S. Ekesi, S.A. Mohammed, and M. De Meyer, pp. 553–574. Cham, Switzerland: Springer International Publishing.

Vargas, R. I., L. Leblanc, R. Putoa, and A. Eitam. 2007. Impact of introduction of *Bactrocera dorsalis* (Diptera: Tephritidae) and classical biological control releases of *Fopius arisanus* (Hymenoptera: Braconidae) on economically important fruit flies in French Polynesia. *Journal of Economic Entomology* 100: 670–679.

Vayssières, J. F., Y. Carel, M. Coubes, and P. F. Duyck. 2008. Development of immature stages and comparative demography of two cucurbit-attacking fruit flies in Reunion Island: *Bactrocera cucurbitae* and *Dacus ciliatus* (Diptera Tephritidae). *Environmental Entomology* 37: 307–314.

Vayssières, J. F, S. Korie, and D. Ayegnon. 2009. Correlation of fruit fly (Diptera: Tephritidae) infestation of major mango cultivars in Borgou (Benin) with abiotic and biotic factors and assessment of damage. *Crop Protection* 28: 477–488.

Vayssieres, J. F, A. Sinzogan, and A. Adandonon. 2008. The new invasive fruit fly species, *Bactrocera invadens* Drew Tsuruta and white. IITA-CIRAD leaflet No. 2

Vayssieres, J. F., F. Sango, and M. Noussourou. 2007. Inventory of the fruit fly species (Diptera: Tephritidae) linked to the mango tree in Mali and tests of integrated control. *Fruits* 63: 329–341.

Zaheeruddin, M. 2007. Study of diffusion and adoption of male annihilation technique. *International Journal of Education and Development using Information and Communication Technology* 3: 89–99.

Section VIII

Social, Economic, and Policy Issues of Action Programs

27 Compendium of Fruit Fly Host Plant Information

The USDA Primary Reference in Establishing Fruit Fly Regulated Host Plants

Nicanor J. Liquido[*]*, Grant T. McQuate, Karl A. Suiter, Allen L. Norrbom, Wee L. Yee, and Chiou Ling Chang*

CONTENTS

Abstract The inherent ecological adaptiveness of fruit flies (Diptera: Tephritidae) ranks them among the worst invasive pest species, requiring vigilant detection, effective suppression, and regimented area-wide eradication. The US Department of Agriculture-Animal and Plant Health Inspection Service-Plant Protection and Quarantine (USDA-APHIS-PPQ) has a strategic goal to develop decision tools to prevent the entry and spread of quarantine-significant fruit flies posing threats to the health of US agriculture and natural resources. To achieve this strategic goal, USDA-APHIS-PPQ developed the *Compendium of Fruit Fly Host Information* (in short, CoFFHI: https://coffhi.cphst.org/), an interactive application integrating verified records of fruit fly infestations on their documented host plants, worldwide. Pertinent publications and manuscripts were acquired through the use of searchable online databases. Infestation data retrieved from the literature were classified as providing field infestation data, laboratory infestation data, interception data, or a mere listing of a fruit or vegetable as a host without providing any verifiable infestation data (i.e., listing only data). The taxonomy of recorded host plants was verified using the USDA-Agricultural Research Service (ARS) Germplasm Repository Information Network (GRIN, http://www.ars-grin.gov/) and other

[*] Corresponding author.

taxonomic resources. CoFFHI, Edition 4.0 has four integral components: (1) comprehensive fruit fly species-specific host plant databases of 24 select quarantine-significant fruit fly pests of horticultural commodities; (2) provisional host lists for the same 24 select fruit fly pests; (3) the *Tephritidae Databases*, which comprise name, host plant, and distribution data for all fruit fly species; and (4) infestation records of the Dacinae of the Pacific Islands. CoFFHI, Edition 4.0 is a vital USDA decision tool in achieving the core mission of APHIS-PPQ in preventing the introduction and establishment of exotic fruit flies into the United States and in facilitating safe domestic and international agricultural trade.

27.1 INTRODUCTION

Tephritid fruit flies exotic to the United States are regulated through the US Plant Protection Act of 2000 (7 U.S.C. 7701–7772) and relevant parts and subparts of the *Code of Federal Regulations (7 CFR – Agriculture).* Fruit fly infestations in host commodities impose enormous constraints on the diversification of agricultural production, emplace formidable trade barriers, and limit the expansion of safe agricultural commerce globally. The perennial detection and eradication of multiple species of fruit flies in the United States, especially in southern parts of the country, prompted the US Department of Agriculture-Animal and Plant Health Inspection Service-Plant Protection and Quarantine's (USDA-APHIS-PPQ) demand for up-to-date and readily accessible fruit fly host plant information. APHIS-PPQ has a strategic goal to develop decision tools to prevent the entry and spread of exotic fruit flies. To achieve this goal, one of the initiatives supported by APHIS-PPQ is the USDA Compendium of Fruit Fly Host Information Project. The project has the mandate to provide APHIS-PPQ with up-to-date, interactive, validated, and readily accessible information on suitable host plants of fruit flies of economic importance, as well as taxonomic and geographic information on fruit fly pests. The primary product of the project is the application *Compendium of Fruit Fly Host Information*, referred to in short as CoFFHI, and available online at https://coffhi.cphst.org/. Currently in its fourth edition, CoFFHI is interactive and integrates comprehensive botanical, geographic, and worldwide infestation biology data on reported host plants of quarantine-significant fruit flies. This scientific note presents the cataloged and managed databases in CoFFHI, Edition 4.0, and the impacts these databases have in achieving the core goals of APHIS-PPQ to strengthen fruit fly pest exclusion systems, optimize domestic fruit fly suppression and eradication programs, and promote safe domestic and global trade of fresh fruits and vegetables.

27.2 METHODS

Pertinent publications and manuscripts were acquired through the use of searchable online databases, as well as from searches of the USDA-APHIS-PPQ's pest interception databases. Infestation data retrieved from the literature were classified as providing field infestation data, laboratory infestation data, interception data, or a mere listing of a fruit or vegetable as a host without providing any verifiable supporting data (i.e., listing only data). Provisional host lists were prepared as lists of plant species ("suitable host plants") for which there are recorded infestations under natural field conditions. Each validated suitable host plant satisfies the definition and attributes of a fruit fly natural, suitable host plant consistent with the terms used in the International Plant Protection Convention (IPPC) International Standards for Phytosanitary Measures (ISPM) No. 37: "Determination of host status of fruit to fruit flies" (FAO, 2016) and the North American Plant Protection Organization (NAPPO) Regional Standard for Phytosanitary Management (RSPM) No. 30: "Guidelines for the determination and designation of host status of a fruit or vegetable for fruit flies '(Diptera: Tephritidae)'" (NAPPO, 2008). Lists of undetermined hosts, or hosts of uncertain regulatory status, were also prepared. The undetermined host category is conferred to a recorded host plant that has no validated record of infestation under natural field conditions, and its host association is based on reported laboratory infestation, interception at a port of entry, or a mere listing as a host without any accompanying verifiable data. The taxonomy of both suitable and undetermined host plants was verified according to current botanical classification using the USDA-Agricultural Research Service (ARS) Germplasm Repository Information Network (GRIN, http://www.ars-grin.gov/) and other taxonomic resources.

27.3 RESULTS AND DISCUSSION

CoFFHI has four integral components: (1) comprehensive fruit fly species-specific host plant databases of select quarantine-significant fruit fly pests of horticultural commodities, with summaries of field and laboratory infestation data, interceptions at ports of entry, and "listing only" host records; (2) provisional suitable host plant lists of select quarantine-significant fruit flies; (3) the *Tephritidae Databases* with name, distribution, and host plant data for all of the nearly 5,000 known tephritid species; and (4) host plants of the Dacinae of the Pacific Islands.

27.3.1 Comprehensive Fruit Fly Species-Specific Host Plant Databases and Provisional Host Lists

CoFFHI has provisional host lists for 24 tephritid fruit fly species of economic importance, with comprehensive documentation of host plant records for many of these species (see Table 27.1). The following species are included (in brackets, respectively, are the total number of recorded host plants [= the sum of suitable and undetermined host plants] and the total number of infestation records): Inga fruit fly, *Anastrepha distincta* Greene [73, 299]; South American fruit fly complex, *Anastrepha fraterculus* (Wiedemann) complex [267, 2133]; Mexican fruit fly, *Anastrepha ludens* (Loew) [95, 751]; West Indian fruit fly, *Anastrepha obliqua* (Macquart) [150, 924]; sapote fruit fly, *Anastrepha serpentina* (Wiedemann) [111, 729]; guava fruit fly, *Anastrepha striata* Schiner [100, 640]; white striped fruit fly, *Bactrocera albistrigata* (Meijere) [23, 137]; carambola fruit fly, *Bactrocera carambolae* Drew & Hancock [140, 257]; guava fruit fly, *Bactrocera correcta* (Bezzi) [73, 168]; Oriental fruit fly, *Bactrocera dorsalis* (Hendel) [647, 4363]; mango fruit fly, *Bactrocera frauenfeldi* (Schiner) [120, 605]; *Bactrocera kirki* (Froggatt) [62, 313]; Solanum fruit fly, *Bactrocera latifrons* (Hendel) [82, 425]; Chinese citrus fruit fly, *Bactrocera minax* (Enderlein) [20, 206]; *Bactrocera pedestris* (Bezzi) [28, 42]; peach fruit fly, *Bactrocera zonata* (Saunders) [134, 1384]; Mediterranean fruit fly, *Ceratitis capitata* (Wiedemann) [655, 8805]; greater pumpkin fruit fly, *Dacus bivittatus* (Bigot) [76, 311]; lesser pumpkin fly, *Dacus ciliatus* Loew [99, 758]; European cherry fruit fly, *Rhagoletis cerasi* (Linnaeus) [40, 485]; western cherry fruit fly, *Rhagoletis indifferens* Curran [15, 53]; apple maggot fly, *Rhagoletis pomonella* (Walsh) [73, 398]; melon fly, *Zeugodacus cucurbitae* (Coquillett) [273, 3953]; and *Zeugodacus tau* (Walker) complex [108, 297].

The CoFFHI team is in the process of adding comparable data for these additional fruit fly species of economic importance: papaya fruit fly, *Anastrepha curvicauda* (Gerstaecker); South American cucurbit fruit fly, *Anastrepha grandis* (Macquart); *Bactrocera occipitalis* (Bezzi); Japanese orange fly, *Bactrocera tsuneonis* (Miyaki); Queensland fruit fly, *Bactrocera tryoni* (Frogatt); Pacific fruit fly, *Bactrocera xanthodes* (Broun); mango fruit fly, *Ceratitis cosyra* (Walker); Natal fruit fly, *Ceratitis rosa* Karsch; eastern cherry fruit fly, *Rhagoletis cingulata* (Loew); walnut husk fly, *Rhagoletis completa* Cresson; blueberry maggot, *Rhagoletis mendax* Curran; *Zeugodacus caudatus* (Fabricius); three-striped fruit fly, *Zeugodacus diversus* (Coquillett); and striped fruit fly, *Zeugodacus scutellatus* (Hendel).

The fruit fly species-specific lists of provisional suitable host plants prepared by the CoFFHI team are reviewed by scientists and regulatory staff of APHIS-PPQ and State Plant Health Regulatory Officers (SPROs) of various states to establish the official USDA lists of fruit fly regulated host plants, which are published as federal orders. The vetting process follows a systematic procedure developed by the APHIS Fruit Fly Exclusion and Detection Working Group on host plants of quarantine-significant fruit flies.

TABLE 27.1
Tephritid Fruit Fly Species of Economic Importance Included in the USDA *Compendium of Fruit Fly Host Information*

	Suitable Hosts[a]			Undetermined Hosts[b]			
Fruit Fly Species	**Taxa**	**Genera**	**Families**	**Taxa**	**Genera**	**Families**	**No. Records[c]**
Anastrepha distincta	32	14	11	41	19	10	299
Anastrepha fraterculus	143	63	32	124	66	39	2133
Anastrepha ludens	45	24	17	50	32	18	751
Anastrepha obliqua	77	37	22	73	41	25	924
Anastrepha serpentina	52	27	16	59	38	20	729
Anastrepha striata	52	30	20	48	27	17	640
Bactrocera albistrigata	21	14	13	2	2	1	137
Bactrocera carambolae	100	58	38	40	29	16	257
Bactrocera correcta[d]	73	50	35	—	—	—	168
Bactrocera dorsalis	488	215	80	159	101	51	4363
Bactrocera frauenfeldi	94	51	33	26	20	15	605
Bactrocera kirki	42	28	26	20	15	9	313
Bactrocera latifrons	59	25	13	23	17	13	425
Bactrocera minax	15	2	1	5	3	1	206
Bactrocera pedestris	26	19	12	2	2	2	42
Bactrocera zonata	54	38	23	80	32	19	1384
Ceratitis capitata	408	179	68	247	148	62	8805
Dacus bivittatus	39	19	9	37	22	10	311
Dacus ciliatus	64	25	10	35	23	11	758
Rhagoletis cerasi[e]	15	5	4	25	8	5	485
Rhagoletis indifferens[d]	15	4	2	—	—	—	53
Rhagoletis pomonella	60	9	1	13	11	7	398
Zeugodacus cucurbitae	136	62	30	137	80	39	3953
Zeugodacus tau	77	44	23	31	21	16	297

Source: *Compendium of Fruit Fly Host Information* (CoFFHI) https://coffhi.cphst.org/.

Note: A provisional host list is included for each species, with comprehensive and annotated host infestation records for some of the species.

[a] Suitable hosts have validated records of field infestations under natural field conditions.

[b] The undetermined host category is conferred to a recorded host plant that has no validated record of infestation under natural field conditions, and its host association is based on reported laboratory infestation, interception at a port of entry, or a mere listing as a host without any accompanying verifiable data.

[c] No. records is the total number of infestation records documented in CoFFHI, Edition 4.0.

[d] Only host plants with field infestation records are recorded in CoFFHI, Edition 4.0.

[e] Includes infestation records in 87 cultivars of *Prunus avium* and 6 cultivars and varieties of *P. cerasus*.

27.3.2 Tephritidae Databases

The *Tephritidae Databases* compile taxonomic and host plant information for all recognized species in the family. Developed by Allen Norrbom and colleagues at the USDA-ARS Systematic Entomology Laboratory (SEL), earlier versions of the databases were searchable on the SEL website, which is no longer available. The *Tephritidae Databases* are now incorporated into CoFFHI, allowing integration of data and development of more efficient search capabilities. This makes records in the *Tephritidae Databases* available on a reliable server and more usable to scientists and regulators in conjunction with the other CoFFHI databases. The taxonomic database, originally developed as part of the *Biosystematic Database of World Diptera* (currently *Systema Dipterorum*, https://diptera.dk/) and published as a world catalog (see Thompson, 1999), now includes more than 10,000 valid and invalid scientific names for the nearly 5,000 currently recognized fruit fly species. The host plant database comprises over 36,000 records, and the distribution database more than 23,000 records. Although the host plant data are not comprehensive, the *Tephritidae Databases* document most of the known fruit fly/host plant relationships. Likewise, the distribution database is incomplete in regard to references documenting many records, but it provides the most comprehensive geographic distribution information available for all fruit fly species.

The *Tephritidae Databases* can be searched for information such as: (1) what fruit fly species have been reported to infest a particular host plant; (2) what are the reported hosts of a particular fruit fly species; (3) what are all of the names (valid or invalid) that have been used for a fruit fly species (i.e., to generate a list of associated synonyms and other invalid names, or to check the status of a name previously used in the literature); (4) what are all of the fruit fly species occurring in a particular country, or where does a particular fruit fly species occur; and (5) author and reference information pertaining to fruit fly taxonomy, distribution, and host plants. The name, host plant, and to a lesser extent, the distribution databases also provide citations to the references documenting each record. The *Tephritidae Databases* can be used to complement the fruit-fly-species-specific databases in CoFFHI by providing host plant data for the many fruit fly species for which comprehensive host plant databases have not been developed. It should be noted, however, that the components of the *Tephritidae Databases*, particularly the host and distribution databases, are working tools that are in a continuous state of development; thus, not all records have yet been fully verified and not all of the vast tephritid host and distribution literature has been incorporated.

27.3.3 Host Plants of the Dacinae of the Pacific Islands

Contributed by Luc Leblanc (University of Idaho), the Host Plants of the Dacinae of the Pacific Islands database provides records of infestation of 76 *Bactrocera* and *Zeugodacus* spp. and four *Dacus* spp. in 241 species of host plants; 31 of these fruit fly species are found only in Pacific Island countries and territories.

27.4 CONCLUSION

Using databases in CoFFHI, Edition 4.0, scientists and regulatory staff of APHIS-PPQ and SPROs of various states establish the official USDA lists of regulated host plants or regulated articles of select quarantine-significant fruit flies. As the USDA's primary reference on establishing fruit fly regulated articles, CoFFHI is designed to provide key information to regulatory scientists and regulatory officials to assess and mitigate the risk of fruit flies in fresh horticultural commodities and to serve as a decision tool in the design and implementation of effective fruit fly detection, monitoring, suppression, and eradication programs. CoFFHI is a vital USDA decision tool in achieving the core mission of APHIS-PPQ in preventing the introduction and establishment of exotic fruit flies that pose significant threats to US agriculture and natural resources.

ACKNOWLEDGMENTS

The development of CoFFHI would not be possible without the programming assistance of Sandra Sferrazza and technical support of Megan Hanlin, Amanda Birnbaum, Kelly Nakamichi, Kelly Ann Lee, Alexander Ching, Jessika Santamaria, and Melissa Seymour.

REFERENCES

FAO. 2016. International Standards for Phytosanitary Measures (ISPM) No. 37: Determination of host status of fruit to fruit flies. Rome, Italy, International Plant Protection Convention, Food and Agriculture Organization (FAO) of the United Nations. 18p.

NAPPO. 2008. *Regional Standard for Phytosanitary Measures (RSPM) No. 30: Guidelines for the Determination and Designation of Host Status of a Fruit or Vegetable for Fruit Flies (Diptera: Tephritidae).* Ottawa, ON: North American Plant Protection Organization. 19p.

Thompson, F. C. (ed.). 1999. *Fruit Fly Expert Identification System and Systematic Information Database.* Myia 9: 524p. and Diptera Data Dissemination Disk (CD-ROM) 1: \names\fruitfly\ffessib.pdf.

28 Tephritid-Related Databases
TWD, IDIDAS, IDCT, DIR-SIT

Abdeljelil Bakri[*], *Walther Enkerlin, Rui Pereira, Jorge Hendrichs, Emilia Bustos-Griffin, and Guy J. Hallman*

CONTENTS

Abstract The purpose of the databases developed by the Food and Agriculture Organization (FAO) and the International Atomic Energy Agency (IAEA) is to facilitate the collection and sharing of data among fruit fly workers and to provide access to information that details findings on doses required for phytosanitary irradiation (PI) and for the purpose of applying the sterile insect technique (SIT) as part of area-wide integrated pest management (AW-IPM) programs. These include: Tephritid Workers Database (TWD), the International Database on Insect Disinfestation and Sterilization (IDIDAS), and the World-Wide Directory of SIT Facilities (DIR-SIT). These databases have been continuously updated and populated with new data, including the TWD list of over 1500 members and more than 7000 literature references relevant to tephritid fruit flies. Furthermore, TWD hosts the web pages of the three regional tephritid worker groups and their respective Steering Committees: the Tephritid Workers of the Western Hemisphere (TWWH), the Tephritid Workers of Europe, Africa and the Middle East (TEAM) and the Tephritid Workers of Asia, Australia and Oceania (TAAO). IDIDAS includes 373 insect datasheets with radiation doses for sterilization and phytosanitary irradiation extracted from over 5400 references. DIR-SIT lists 38 mass-rearing facilities, including details about the insect species, the production capacity, and the irradiation sterilization parameters. The newly developed International Database on Commodity Tolerance (IDCT) helps to determine the tolerated PI dose for the disinfestation of fresh products. Up-to-now, data have been retrieved for IDCT from 243 references and have returned 156 different cultivars belonging to 89 fresh commodities (fruit, vegetables, and cut flowers). IDCT is an added value to IDIDAS and both share several common resources. With IDIDAS and IDCT data, food safety officers can select the optimum dose that balances between the insect/mite pest sterility or lethality and the commodity tolerance. In addition, technical resources, news,

[*] Corresponding author.

newsletters, event calendars, and photo galleries have been included in these databases. The monitoring and evaluation of the performance of these sites in terms of the audience and visits are tracked via Google Analytics.

With these four databases, TWD, IDIDAS, IDCT, and DIR-SIT, FAO and IAEA are offering a valuable repository of information and a comprehensive networking service to their member states. The objective of this chapter is to provide an overview of these resources to the community of tephritid fruit fly workers, including some information on their metrics.

28.1 INTRODUCTION

Four databases have been developed with the support of the Insect Pest Control Section (IPCS) of the Joint Food and Agriculture Organization/International Atomic Energy Agency (FAO/IAEA) Division of Nuclear Techniques in Food and Agriculture, which provide information related to tephritid fruit flies and area-wide integrated pest management (IPM), including the sterile insect technique (SIT) and phytosanitary irradiation (PI; disinfestation). These databases include: the Tephritid Workers Database (TWD), the International Database on Insect Disinfestation and Sterilization (IDIDAS), the World-Wide Directory of SIT Facilities (DIR-SIT), and the newly developed International Database on Commodity Tolerance (IDCT).

28.2 METHODS

To develop the databases, the first step was to design and set up an architecture suitable for the information we would like to convey. Information technology is a rapidly evolving science, thus, keeping up to date is a challenging endeavor. Since the development of the first database, the databases had to be migrated from a couple of systems not always compatible. Nonetheless, each time the architecture and data had to be adapted, and advantage was taken of the new functions available. These databases are continuously updated and populated with information and new resources. Extensive fine tuning has been carried out to ensure high-quality and user-friendly functions of the database platform based now on Microsoft SharePoint.

Analyzing scientific articles and technical documents and extracting the relevant data concerns mainly IDIDAS and IDCT. The taxonomy in general, either for insects or plants, is also an evolving science, and we had to take in consideration the changes in the names of species or their groups. The main IDIDAS data collected and assigned to the species datasheet were: the treated life stage, the irradiation conditions and doses, the quality control parameters either for PI or for insect sterilization, and the references. IDCT follows a similar procedure but for plant cultivars. The datasheet includes: the pre- and postharvest conditions, the irradiation doses with the tolerance aspects, and the references. For DIR-SIT, a standard form is sent to the focal points of all facilities worldwide to help collect data on the production of sterile insects and the irradiation process. TWD data are essentially publications on tephritid fruit flies and news on the same topic. Data were collected from various sources of academic databases and search engines such as the International Nuclear Information System (INIS), and from a number of specialized scientific journals in entomology, crop protection, PI, and related radiation biology.

28.3 RESULTS

28.3.1 Tephritid Workers Database (TWD)

This is a unique hub for tephritid fruit flies established 14 years ago (2004) by the Insect Pest Control Section of FAO/IAEA. The objective of the TWD is to provide a networking platform, news source, literature resource, a directory of fruit fly workers with information about their area of expertise, just to name a few (Figure 28.1). The most relevant news to tephritid workers

Tephritid Workers Database (TWD)

Organization
- Joint FAO/IAEA Programme
- Insect Pest Control Section
- Insect Pest Control Laboratory

- Tephritid Fruit Fly Steering Committees
- Regional groups
- News
- Newsletters
- International-Fruit-Fly-Symposia
- Entomology Events
- Search Publication
- Operational projects
- Resources
- Member Areas
- Photo Gallery
- Contact us

Dear fruit fly workers

Tephritid fruit flies are becoming increasingly important, due to the economic importance of many species and their threat to fruit and vegetable production and trade worldwide. Consequently, a tremendous amount of information is made available each year: development of new technologies, new information on their biology and ecology; new control methods made available, new species identified, new outbreaks recorded and new operational control programmes launched. Because things are evolving so rapidly in this field, it is of utmost importance to develop a site that collects and shares this information allowing each Tephritid fruit fly worker worldwide to keep up-to-date on the most recent developments.

International fruit fly symposia and other meetings, though providing a fantastic occasion to meet with colleagues, as well as exchange views and learn about new methodologies, do not provide an easily accessible and always available tool for a colleague in need of an e-mail address of a special field expert, or of information needed to solve a problem encountered in the field.

In the past, a very useful newsletter with information on fruit fly workers was issued annually by some fruit fly colleagues (mainly Boller and Liedo with others) under the auspices of the International Organisation of Biological Control (IOBC). Unfortunately this database is no longer being maintained. As part of its mandate to provide colleagues in Member States with mechanisms for increased collaboration, the Insect Pest control Section (IPCS) of the Joint FAO/IAEA Division of Nuclear Techniques in food and Agriculture has decided to re-establish the database on the web called Tephritid Workers. This database is available on the Internet to all fruit fly workers and other interested parties and is maintained by the participants, who gain access to their own pages by using their username and password. The goal of the initiative is to facilitate collection and sharing of data among fruitfly workers.

Follow us on Facebook

FIGURE 28.1 Home page of the Tephritid Workers Database (TWD).

are posted on TWD and its associated Facebook page (Figure 28.2) where users can freely add their comments and interact with other members. The Steering Committees (SC) page lists members of the International Fruit Fly Steering Committee (IFFSC) and those of the Tephritid Workers of the Western Hemisphere (TWWH), Tephritid Workers of Europe, Africa and the Middle East (TEAM), and Tephritid Workers of Asia, Australia and Oceania (TAAO). This page is regularly updated as new members (one third of the committee members) are elected every 4 years. In addition, the regional groups communicate with their members through newsletters and mailing lists by posting information about their activities such as meetings and ongoing fruit fly programs. The Fruit Fly News (FFN) newsletter (Figure 28.3), which is edited by a group of independent volunteer editors, is also distributed to the community and posted on the TWD and Facebook. Other information related to tephritids, such as events (e.g., meetings, symposium, workshops, and training courses), technical manuals/guidelines, and a photogallery, are regularly updated.

The TWD Facebook page (Figure 28.2) allows members to freely communicate, post comments, share findings, experience and expertise, exchange documents, inform about job opportunities, alerts, or other breaking news. Members can also get information about coming fruit fly events and express their wish and intention to participate. The result is shown on a dashboard indicating how many are planning to attend the event, which can be helpful for the meeting organizers.

Currently, 36 FFN have been issued since 1972, 16 TEAM newsletters since 2005, and six TAAO newsletters since 2015 (Figure 28.3).

As of July 2018, the TWD contains 1529 members from 120 countries. The top 10 countries in terms of the number of tephritid workers represented in the database in decreasing order are Mexico, United States, Thailand, Brazil, Spain, Australia, Argentina, South Africa, India, and China

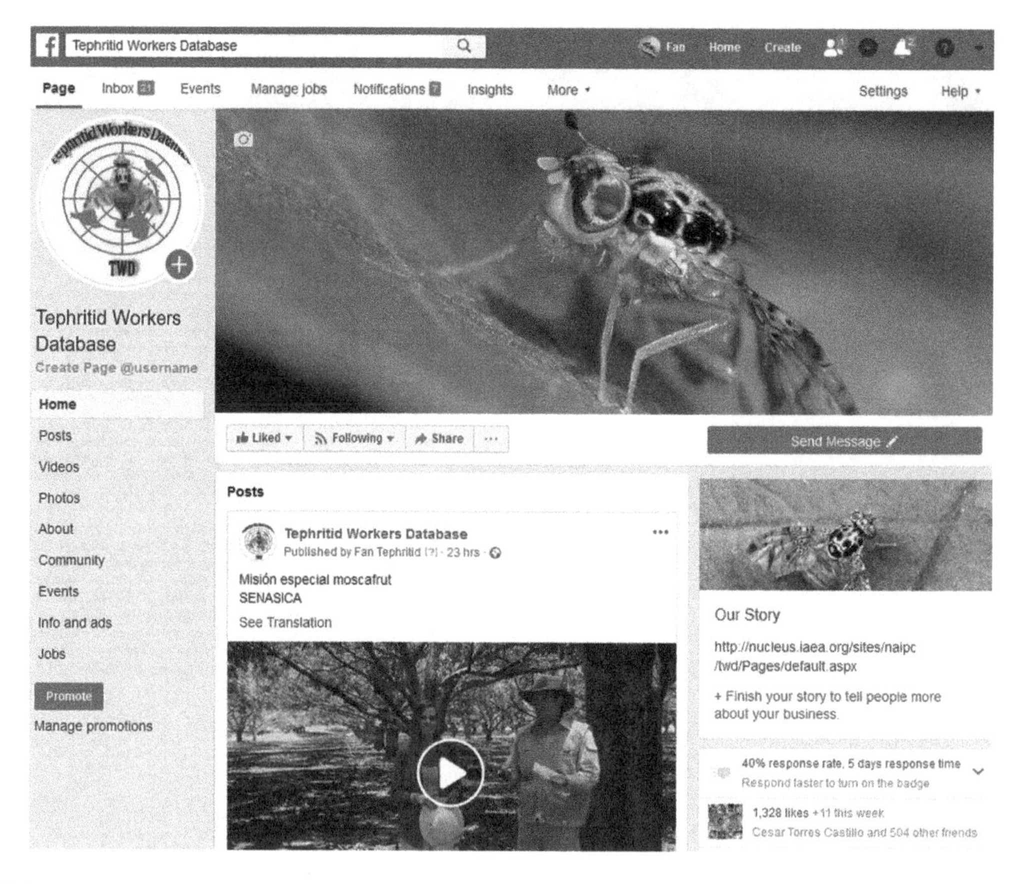

FIGURE 28.2 Facebook page of Tephritid Workers Database (fb-TWD).

FIGURE 28.3 Presentation of the latest eight issues of the Fruit Fly News (FFN) e-newsletters. There are 37 FFN issues since 1972.

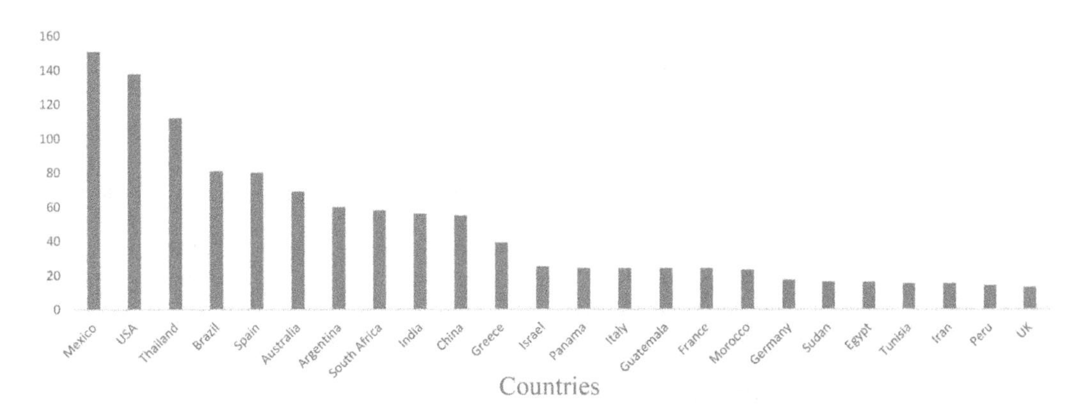

FIGURE 28.4 Top 25 countries based on the number of Tephritid Workers Database (TWD) members.

(Figure 28.4). Those countries most well represented are likely to have large tephritid fruit fly control programs and have hosted one of the past regional or international meetings.

Of the five continents, the Americas have the highest number of members (556), followed by Asia (368), Europe (261), Africa (234), and Oceania (89) (Figure 28.5).

These members are distributed in three regional groups: TWWH (Americas), TEAM (Europe, Africa, and the Middle East), and TAAO (Asia, Australia, and Oceania). Each regional group is likely to share similar challenges vis-a-vis the same fruit fly species present in their region (e.g., *Anastrepha* in the Americas and *Bactrocera* in Asia) and may have the same pest-management priorities. This makes the regional meetings more specific and relevant for the members of the regional group. Nonetheless, all these regional meetings remain open to all members from the other regional groups who might share their experience and learn from colleagues from the other geographical areas.

Three SCs were established to coordinate the activities within their respective regional groups. The IFFSC, however, coordinates the activities related to the International Symposium of Fruit Flies of Economic Importance (ISFFEI), such as receiving proposals and selecting the best proposals and venues to host the ISFFEI symposia that takes place every 4 years, providing support to the local organizing committee, editing and publishing proceedings, and other related tasks.

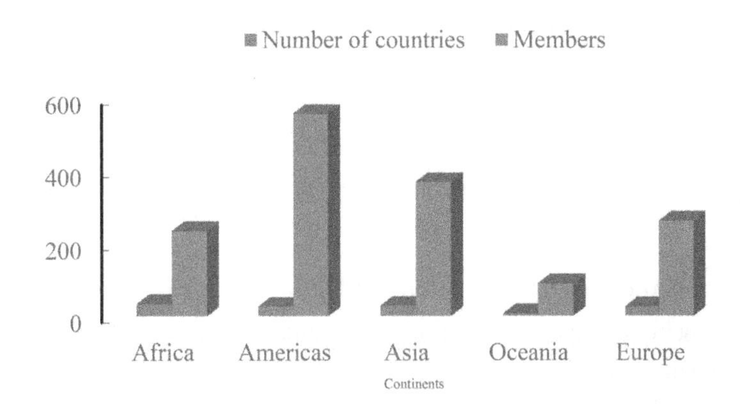

FIGURE 28.5 Distribution of Tephritid Workers Database (TWD) members by continents (2018).

TABLE 28.1
Meeting and Symposia on Tephritid Fruit Flies

ISFFEI	TWWH	TEAM	TAAO
1st ISFFEI (Greece 1982)	1st TWWH October 1992 (San José, Costa Rica)	1st TEAM April 2008 (Palma de Mallorca, Spain)	1st TAAO 15–18 August 2016 (Kuala Lumpur, Malaysia)
2nd ISFFEI (Greece 1986)	2nd TWWH August 1996 (Viña del Mar, Chile)	2nd TEAM July 2012 (Kolymbari, Crete, Greece)	2nd TAAO 18–21 August 2020 (Beijing, China)
3rd ISFFEI (Guatemala 1990)	3rd TWWH July 1999 (Guatemala City, Guatemala)	3rd TEAM April 2016 (Stellenbosch, South Africa)	
4th ISFFEI (United States 1994)	4th TWWH May 2001 (Mendoza, Argentina)	4th TEAM 2020 (La Grande Motte, France)	
5th ISFFEI (Malaysia 1998)	5th TWWH May 2004 (Fort Lauderdale, USA)		
6th ISFFEI (South Africa 2002)	6th TWWH September 2006 (Salvador, Brazil)		
7th ISFFEI (Brazil 2006)	7th TWWH November 2008 (Mazatlán, México)		
8th ISFFEI (Spain 2010)	8th TWWH July–August 2012 (Panama City, Panama)		
9th ISFFEI (Thailand 2014)	9th TWWH 16–22 October 2016 (Buenos Aires, Argentina)		
10th ISFFEI (Mexico 2018)	10th TWWH 16–20 March 2020 (Bogota, Colombia)		
11th ISFFEI (Australia 2022)			

Free proceedings are posted in TWD. (See Figure 28.5 for more details)

Up to now, 24 meetings specific to fruit fly tephritids have been organized, namely 10 ISFFEI symposia, 10 TWWH meetings, 3 TEAM meetings, and 1 TAAO meeting (Table 28.1). These meetings often include satellite meetings on a specific fruit fly topic, for example, Coordinated Research Meetings (CRP) or Consultants Group Meetings.

The global ISFFEI is the largest gathering of the tephritid fruit fly workers, and recent symposia can reach up to 400 attendees from all over the world (Figure 28.6).

Membership is open and freely available to all people working on tephritid fruit flies. For registration, one simply follows the steps indicated on the TWD website.

All the registration information required is about the fruit fly species being worked on, the subject of research, and how to reach the registrant in case colleagues need that persons' expertise and advice or wish to establish a collaborative project.

There are more than 7,100 relevant publications hosted on the TWD. Based on publication's search in TWD from the 1960s to 2018, the most widely represented genus are *Bactrocera* with 1034 publications, followed by *Ceratitis* with 934. The most well-represented single pest species is *Ceratitis capitata* (Wied.), with 844 publications, followed by *Bactrocera dorsalis* (Hendel), with 366.

28.3.2 The International Database on Insect Disinfestation and Sterilization (IDIDAS)

IDIDAS (Figure 28.7) compiles and analyzes information about insect and mites species that are subject to ionizing radiation mainly for reproduction sterilization (e.g., sterile insect technique [SIT]), phytosanitary disinfestation, sperm precedence studies, and host-parasitoid interaction studies. The information on irradiation doses (Gy) required for the various development stages

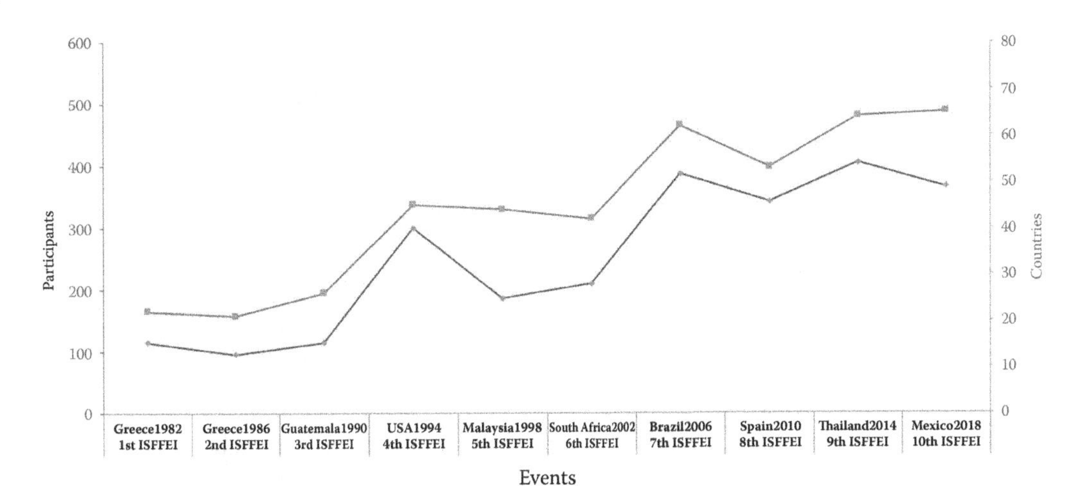

FIGURE 28.6 Trends in the International Symposium on Fruit Flies of Economic Importance (ISFFEI) attendance, 1982–2018.

(eggs, larvae, nymphs, pupae, and adults) were retrieved from more than 5,400 references. Data in IDIDAS also indicate the biotic and abiotic conditions of the irradiation treatment.

Three-hundred and seventy-three insect and mite pest species are recorded in IDIDAS, covering 13 orders, 82 families, and 211 genera. Tephritid fruit flies are the most widely represented in IDIDAS with 41 species from 7 genera included, namely *Anastrepha* (7 species), *Bactrocera* (17 species), *Ceratitis* (5 species), *Dacus* (1 species), *Myiopardalis* (1 species), *Rhagoletis* (7 species), *Toxotrypana* (1 species), and *Zeugodacus* (formerly Bactrocera) (2 species) (Table 28.2). In the

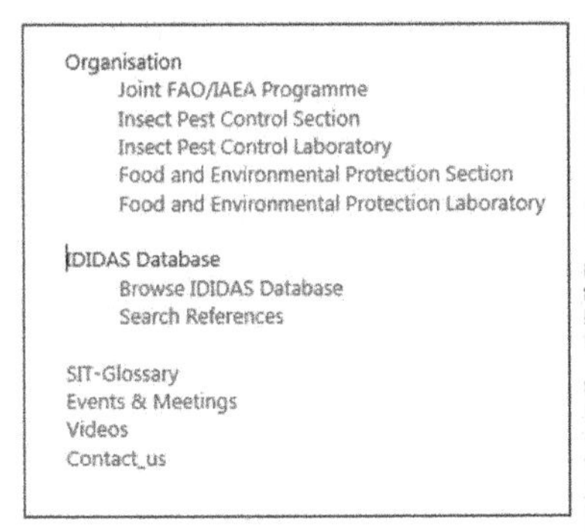

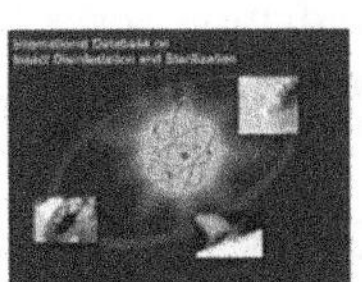

FIGURE 28.7 The International Database on Insect Disinfestation and Sterilization (IDIDAS) home page.

TABLE 28.2
Tephritidae Species Represented in the International Database on Insect Disinfestation and Sterilization (IDIDAS)

Anastrepha fraterculus	*Ceratitis capitata*
Anastrepha grandis	*Ceratitis cosyra*
Anastrepha ludens	*Ceratitis fasciventris*
Anastrepha obliqua	*Ceratitis quilicii*
Anastrepha serpentina	*Ceratitis rosa*
Anastrepha striata	*Dacus ciliatus*
Anastrepha suspensa	*Myiopardalis pardalina*
Bactrocera aquilonis	*Rhagoletis cerasi*
Bactrocera carambolae	*Rhagoletis cingulate*
Bactrocera correcta	*Rhagoletis completa*
Bactrocera dorsalis	*Rhagoletis fausta*
Bactrocera jarvisi	*Rhagoletis indifferens*
Bactrocera latifrons	*Rhagoletis mendax*
Bactrocera minax	*Rhagoletis pomonella*
Bactrocera occipitalis	*Toxotrypana curvicauda*
Bactrocera oleae	*Zeugodacus cucumis*
Bactrocera papayae (Syn *B. dorsalis*)	*Zeugodacus cucurbitae*
Bactrocera passiflorae	*Zeugodacus tau*
Bactrocera philippinensis (Syn *B. dorsalis*)	
Bactrocera tau	
Bactrocera tryoni	
Bactrocera tsuneonis	
Bactrocera zonata	

case of Tephritidae, sterilizing irradiation doses range, on average, from 83 Gy (low) to 85 Gy (mean) and to 108 Gy (high). These doses correspond to mean and 95% confidence limits (upper L2, lower L1) (Sokal and Rohlf 1995). For uniformity, the same irradiation conditions were considered to calculate the dose range. The data are for in-air irradiation of males treated mostly in late puparial stages. The ranges of the irradiation doses for each tephritid genus and species are reported in Bakri and Hendrichs (2004) and Bakri et al. (2005a, 2005b).

28.3.3 International Database on Commodity Tolerance (IDCT)

The IDCT assembles the responses of different cultivars to doses used PI. To date, the IDCT (Figure 28.8) includes the responses of 158 different cultivars belonging to 89 fresh commodities including 43 fruit (48%), 18 vegetables (20%), and 28 cut-flowers (32%) to radiation doses. The information was retrieved from 243 references.

The 158 cultivars belong to 22 families (Figure 28.9) and 28 genera (Figure 28.10). The top four commodities belong to Rosaceae (58 cultivars), Rutaceae (28 cultivars), Anacardiaceae (17 cultivars), and Sapindaceae (13 cultivars). The five top genera (Figure 28.10) are *Prunus* (37 cultivars), *Citrus* (28 cultivars), *Malus* (17 cultivars), *Mangifera* (17 cultivars), and *Litchi* (9 cultivars).

It is important to note that the doses (Gy) reported in the database correspond to the minimum and maximum dose range yielding acceptable marketability of the commodity, given the information presented in the reference cited. These doses are based on the data presented in the references and indicate the doses tolerated by the commodities in question. Pretreatment, treatment, and posttreatment conditions are described to help understand if the handling of the commodity is in line with current commercial marketing situations and if responses might be

The International Database on Commodity Tolerance (IDCT)

Organization
- Joint FAO/IAEA Programme
- Insect Pest Control Section
- Insect Pest Control Laboratory

IDCT Database
- Browse IDCT Database
- Search References
- Resources
- Events and Meetings
- Contact us

The objectives of the International Database on Commodity Tolerance (IDCT) are to gather and interpret the literature about commodity quality after phytosanitary irradiation treatment to aid stakeholders in Member States to identify the doses of radiation that are tolerated by different commodities including fresh fruits, vegetables and cut flowers in planning for commercial use of the technology. The information may also help users of the technology determine optimum methods of applying irradiation including pre- and post-treatment handling and if radiation may augment quality or prolong shelf life. It may furthermore be used to identify gaps and inconsistencies in knowledge that may be explored by researchers.

Important Note: For the best search results, please use the specific database search function at the left hand side of the page "Browse IDCT Database".

Acknowledgements: The content of the IDCT has been compiled by Emilia Bustos-Griffin in coordination with Guy Hallman, under the direction of the Insect Pest Control Section of the Joint FAO/IAEA Programme of Nuclear Techniques in Food and Agriculture. For questions about the content, please contact Walther Enkerlin (W.R.Enkerlin@iaea.org). For questions about access or functionality of the website, contact Abdeljelil Bakri (bakri@uca.ac.ma).

Disclaimer: The International Database on Commodity Tolerance (IDCT) is intended to provide access to scientific and other technical information which detail findings on dose tolerance required for phytosanitary irradiation of horticultural commodities. IDCT should be considered as a tool for first-cut analysis that interprets and summarizes existing information on commodity tolerance to radiation. However, it is the responsibility of the user to confirm the information and to verify it by way of the original publication and pilot testing as deemed necessary. Whilst every effort has been made to search all relevant publications, no assurance can be given that the reference search has been exhaustive.

FIGURE 28.8 The International Database on Commodity Tolerance (IDCT) home page.

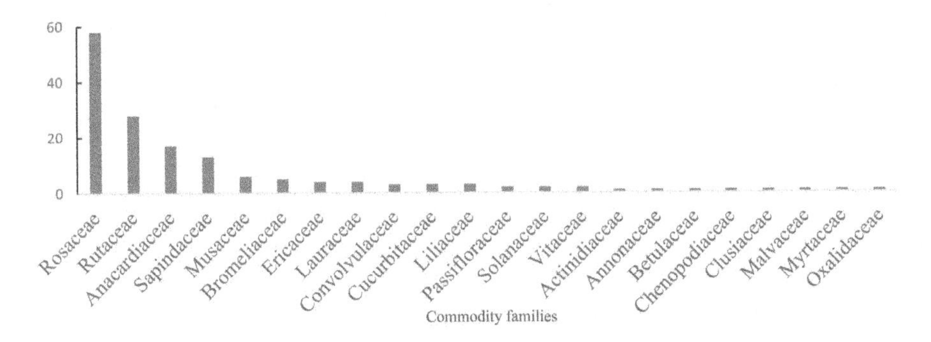

FIGURE 28.9 The number of cultivars, grouped per families, of fresh fruit, vegetables, and cut-flowers subject to phytosanitary irradiation. (From IDCT, https://nucleus.iaea.org/sites/naipc/IDCT/Pages/Browse-IDCT.aspx, 2018.)

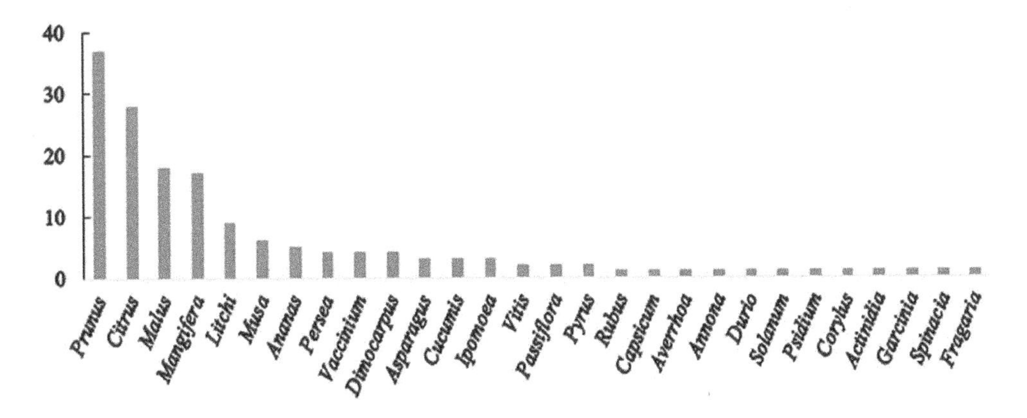

FIGURE 28.10 The number of cultivars, grouped per genera, of fresh fruit, vegetables, and cut-flowers subject to phytosanitary irradiation. (From IDCT, https://nucleus.iaea.org/sites/naipc/IDCT/Pages/Browse-IDCT.aspx, 2018.)

TABLE 28.3
Example of the Dose Range Variation for Phytosanitary Irradiation (PI) of Different Apple Cultivars

Family	Latin Name	Common Name	Cultivar	Dose (Gy)
Rosaceae	*Malus domestica*	Apple	Apple "?"[a]	1,000–1,500
			Apple "Ambri"	at least 500
			Apple "Boskoop"	Not estimated
			Apple "Cortland"	at least 288
			Apple "Fuji"	between 825 and 990
			Apple "Gala"	at least 440 and <880
			Apple "Golden Delicious"	Between 500 and 1,000
			Apple "Granny Smith"	at least 430 to <650
			Apple "Jonathan"	about 500
			Apple "Lobo"	Not estimated
			Apple "McIntosh"	at least 500 to <1000
			Apple "Red Delicious"	at least 600 to 1000
			Apple "Rhode Island Greening"	<384
			Apple "Rich-A-Red"	at least 600
			Apple "Rome Beauty"	at least 100 to <500
			Apple "Royal Delicious"	least 500
			Apple "Yellow Newton Pippin"	<750

[a] The question mark (?) indicates that the cultivar is unknown or not indicated by the author.

modified by handling or evaluation techniques. Users may check the original references for more details.

Considering PI at the cultivar level is very important as the example in Table 28.3 shows. For the same species, here apple, *Malus domestica* (Rosales: Rosaceae), the reported doses tolerated may vary considerably. Even for the same cultivar, the tolerated dose may vary widely according to the experimental conditions and the interpretations by the different researchers.

Beside the references to literature relevant to PI, the IDCT includes links to resources such as technical documents and e-learning courses about PI technology, as well as related meeting and event information.

28.3.4 The World-Wide Directory of Sit Facilities (DIR-SIT)

Up to now, DIR-SIT (Figure 28.11) includes data of 38 insect mass-rearing facilities from 25 countries. Out of these, 19 facilities from 15 countries produce the largest numbers of sterile tephritid fruit flies (at least 5 million/week) (Table 28.4). It is worth mentioning that the production capacity indicated in the table is the production when the program is running at its full capacity. For some facilities, the current production might be lower or nil depending on the current country program activity in managing fruit flies with SIT.

FIGURE 28.11 The World-Wide Directory of Sit Facilities (DIR-SIT) home page.

TABLE 28.4
Worldwide Mass-Rearing Facilities of Tephritid Fruit Flies (Diptera: Tephritidae), Their Production Capacity, Species and Strains, and the Radiosterilization Dose

Country	Facility Location and Name	Insect Reared	Strain	Production Capacity (million/week)	Dose (Gy)
Argentina	Mendoza, Bioplanta	*Ceratitis capitata*	TSL VIENNA 8	200	110
	San Juan, Bioplanta	*Ceratitis capitata*	TSL 2006	50	—
Australia	Camden (NSW), Queensland Fruit Fly Production Facility	*Bactrocera tryoni*	—	15	70
	Adelaide, National SITplus Facility	*Bactrocera tryoni*	—	50	70
	Perth, Sterile Medfly Production Facility	*Ceratitis capitata*	—	10	—
Austria	Seibersdorf, FAO/IAEA Insect Pest Control Laboratory	*Ceratitis capitata* *Bactrocera oleae* *Bactrocera dorsalis* *Anastrepha grandis* *Anastrepha ludens* *Anastrepha fraterculus* *Anastrepha obliqua* *Bactrocera aquilonis* *Bactrocera carambolae* *Bactrocera correcta* *Zeugodacus cucurbitae* *Bactrocera tryoni* *Bactrocera zonata* *Ceratitis quilicii* *Ceratitis rosa* *Zeugodacus tau*	A number of strains for each species were kept in small scale at Seibersdorf laboratory for research and development and to supply requests from member states	—	—
Brazil	Juazeiro, Bahia, Biofábrica Moscamed Brazil	*Ceratitis capitata*	TSL VIENNA 8/2004	100	100
Chile	Arica, Centro Produccion Insectos Esteriles	*Ceratitis capitata*	TSL VIENNA 8 Mix/2006	50	140

(*Continued*)

TABLE 28.4 (*Continued*)
Worldwide Mass-Rearing Facilities of Tephritid Fruit Flies (Diptera: Tephritidae), Their Production Capacity, Species and Strains, and the Radiosterilization Dose

Country	Facility Location and Name	Insect Reared	Strain	Production Capacity (million/week)	Dose (Gy)
Costa Rica	San José, Programa Nacional Moscas de la Fruta, Servicio Fitosanitario del Estado-MAG	*Ceratitis capitata*	Bisexual	5	150
Greece	Heraklion, University of Crete-Fruit Flies	*Bactrocera oleae*	Democritos/1966	5	95
		Ceratitis capitata	T(Y;5)1–61/1995	5	95
Guatemala	El Pino, Moscamed Guatemala	*Ceratitis capitata*	TSL strain/ Vienna7/Toliman 99/2.5 years	2000	100 Gy (local Program); 145 Gy (Exports)
		Anastrepha ludens	—	—	80
Israel	Sde-Eliyahu, Bio-Fly	*Ceratitis capitata*	TSL VIENNA 8	90	100
		Bactrocera oleae	Argov (2008) Yael (2010)	0.25	100
Japan	Naha, Okinawa Prefectural Plant Protection Center	*Zeugodacus cucurbitae*	Taiwan 6/2011	200	72
		Bactrocera latifrons	Yonaguni	1	70
Mexico	Metapa, Chiapas, Dr. Dieter Enkerlin Schallenmüller	*Anastrepha ludens*	Original strain with many refreshments (bisexual strain)	290	80
		Anastrepha ludens	Tapachula 7 (GSS color pupa)	10	80
		Anastrepha obliqua	Original strain with many refreshments (bisexual strain)	65	80
	Metapa, Chiapas, Moscamed, Jorge Gutiérrez Samperio	*Ceratitis capitata*	TSL VIENNA 7-Tol/October 2002	500	125
Peru	La Molina, Centro de Producción y Esterilización Mosca de la Fruta	*Ceratitis capitata*	GSS Vienna 8 TSL	300	120
Philippines	Quezon City, Philippine Fruit Fly Mass-Rearing Facility	*Bactrocera dorsalis* (Syn. *B. philippinensis*)	No strain	15	64–104
South Africa	Stellenbosch, Western Cape, FruitFly Africa (Pty) Ltd	*Ceratitis capitata*	V7-D53/Mix 2001 (December 2001) replaced with VIENNA 8	2	150

(*Continued*)

TABLE 28.4 (*Continued*)
Worldwide Mass-Rearing Facilities of Tephritid Fruit Flies (Diptera: Tephritidae), Their Production Capacity, Species and Strains, and the Radiosterilization Dose

Country	Facility Location and Name	Insect Reared	Strain	Production Capacity (million/week)	Dose (Gy)
Spain	Valencia, Bioplanta de Insectos Estériles	*Ceratitis capitata*	Vienna 8/refreshed 2015	500	95
Thailand	Pathumthani, Irradiation Center for Agricultural Development	*Bactrocera dorsalis*	Pakchong/July 30, 2000	30	90
		Bactrocera correcta	Thailand	40	80
Tunisia	Sidi Thabet, CNSTN, Medfly Facility	*Ceratitis capitata*	TSL VIENNA 8	5	100
USA	Hilo, Hawaii, USDA Pacific Basin Lab Research	*Bactrocera dorsalis*	1993	5	—
		Ceratitis capitata	1993	5	120
	Waimanalo, Hawaii, CDFA/USDA	*Ceratitis capitata*	—	—	150
	Texas, Mexican Fruit Fly Rearing Facility	*Anastrepha ludens*	Willacy County, 2010	200	70
	Gainesville, Florida Department of Agriculture	*Anastrepha suspensa*	—	50	70

Source: DIR-SIT, https://nucleus.iaea.org/sites/naipc/DIR-SIT/SitePages/All%20Facilities.aspx 2018.

CDFA, California Department of Food and Agriculture; CNSTN, Centre National des Sciences et Techniques Nucléaires; FAO/IAEA, Food and Agriculture Organization/International Atomic Energy Agency; USDA, US Department of Agriculture.

28.4 CONCLUSION

With these four open access databases, namely TWD, IDIDAS, IDCT, and DIR-SIT, FAO and IAEA are offering their member states a valuable repository of information and comprehensive networking services pertaining to tephritid fruit fly communities, as well as SIT and PI. With IDIDAS and IDCT data, food safety officers can select the optimum dose that balances between insect/mite pests' sterility or lethality and the commodity tolerance.

REFERENCES

Bakri, A., and J. Hendrichs. 2004. Radiation doses for sterilization of tephritid fruit flies. In *Proceedings, Symposium: 6th International Symposium on Fruit Flies of Economic Importance*, 6–10 May 2002, Stellenbosch, South Africa, eds. B. N. Barnes, pp. 475–479. Irene, South Africa: Isteg Scientific Publications.

Bakri, A., N. Heather, J. Hendrichs, and I. Ferris. 2005a. Fifty years of radiation biology in entomology: Lessons learned from IDIDAS. *Annals of the Entomological Society of America* 98: 1–12. doi:10.1603/0013-8746 (2005)098[0001:FYORBI]2.0.CO;2.

Bakri, A., K. Mehta, and D. R. Lance. 2005b. Sterilizing insects with ionizing radiation. In *Sterile Insect Technique. Principles and Practice in Area-Wide Integrated Pest Management*, eds. V. A. Dyck, J. Hendrichs, and A. S. Robinson, pp. 233–268. Dordrecht, the Netherlands: Springer.

Directory of the Sterile Insect Technique Facilities (DIR-SIT). 2018. Directory of the Sterile Insect Technique Facilities. https://nucleus.iaea.org/sites/naipc/DIRSIT/SitePages/All%20Facilities.aspx.

International Database on Commodity Tolerance (IDCT). 2018. International Database on Commodity Tolerance. https://nucleus.iaea.org/sites/naipc/IDCT/Pages/Browse-IDCT.aspx.

International Database on Insect Disinfestation and Sterilization (IDIDAS). 2018. International Database on Insect Disinfestation and Sterilization. https://nucleus.iaea.org/sites/naipc/ididas/Pages/Browse-IDIDAS.aspx.

Sokal, R. R., and F. J. Rohlf. 1995. *Biometry: The Principles and Practice of Statistics in Biological Research*, 3rd ed. New York: W. H. Freeman and Company.

Tephritid Workers Database (TWD). 2018. Tephritid Workers Database. https://nucleus.iaea.org/sites/naipc/twd/Pages/default.aspx.

29 Stewed Peaches, Fruit Flies, and STEM Professionals in Schools

Inspiring the Next Generation of Fruit Fly Entomologists

Carol Quashie-Williams[*]

CONTENTS

Abstract This chapter describes the science, technology, engineering and mathematics (STEM) Professionals in Schools volunteer program, and explains how STEM volunteers can use their experience and expertise to share agricultural and entomology skills with primary schools in Canberra, Australia, to inspire and engage students to consider careers in science in general and, entomology, in particular. STEM Professionals in Schools volunteers provide a valuable resource for teachers (e.g., using the fruit fly life cycle to demonstrate parts of the Australian Biological Sciences curriculum) and increase community engagement by involving entomologists with the wider community. The students and teachers learned about the Tephritidae fruit fly life cycle, which provided an alternative to the Lepidopteran life cycle, which is usually studied as part of the Australian biological sciences curriculum. The differences between true fruit flies and Drosophilidae flies were also observed and discussed. The school community also learned methods to reduce the incidence of fruit fly infestation in their gardens using environmentally friendly techniques.

29.1 INTRODUCTION

During the last few years, there has been a decline in the rate of students enrolling at universities in undergraduate science, technology, engineering, and mathematics (STEM) subjects in a number of countries, including the United States (Fairweather, 2008) and Australia (Figure 29.1) (PwC, 2014). Australia has one of the lowest rates of undergraduates enrolling in STEM subjects according to the Organization for Economic Cooperation and Development (OECD) (Singhal, 2017). This is of

[*] Corresponding author.

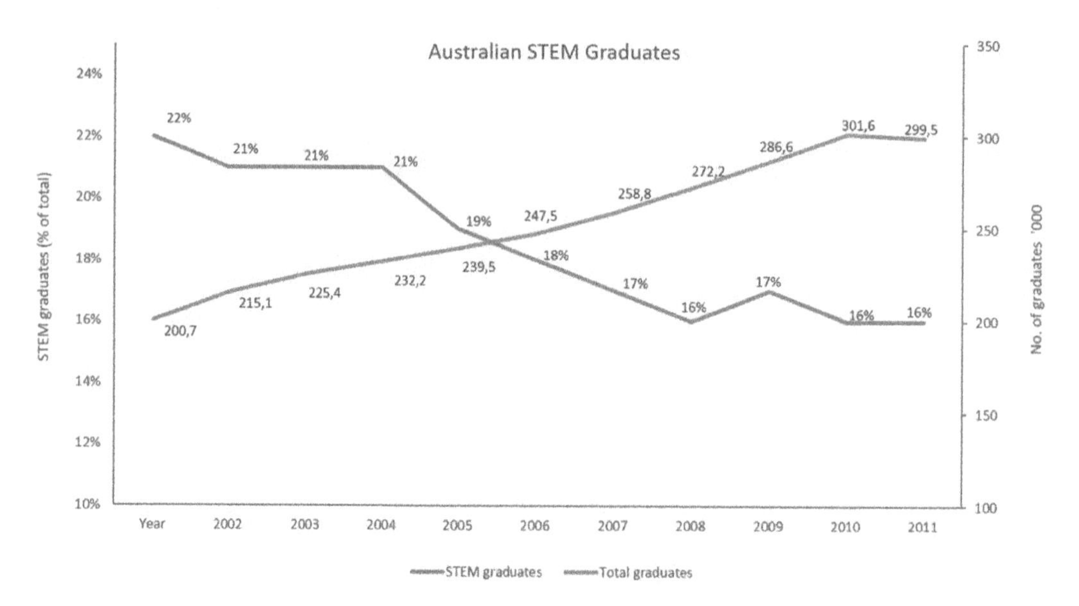

FIGURE 29.1 Australian STEM graduates as a percentage of total school graduates. (PwC, Fuelling NextGen digital innovation through education, http://www.digitalinnovation.pwc.com.au/education/, 2014)

concern when considering that 75% of the fastest-growing occupations require STEM skills and knowledge, and a lack of skilled personnel is cited as the number one barrier to industry innovation (PwC, 2014).

Under the National Science and Innovation Agenda (Department of the Prime Minister and Cabinet, 2015), the Australian government has invested almost A$100 million to inspire STEM literacy at all education levels and to help young Australians prepare for jobs in the future. The National STEM School Education Strategy 2016–2026 (Education Council, 2016) is endorsed by all Australian state and territory governments to invest in improving national STEM education through supporting the development of teachers STEM skills and increasing student engagement in STEM subjects. Australia's Chief Scientist's report *Science, Technology, Engineering and Mathematics: Australia's Future*, also focuses on STEM education in Australian schools to ensure young Australians are equipped with the necessary STEM skills for the future (Office of the Chief Scientist, 2014).

Science, Technology, Engineering and Maths Professionals in Schools (CSIRO, 2015) is Australia's leading STEM education volunteering program and is a major innovation in the national STEM education scene (CSIRO, 2015). *STEM Professionals in Schools* is an initiative of the Australian Government Department of Education and Training. It was established in 2007 as *Scientists in Schools* when the Department of Education and Training provided funding to the Commonwealth Scientific and Industrial Research Organisation (CSIRO) Education and Outreach to deliver the program through a national program team (Howitt and Rennie, 2008, Tytler et al., 2016). The program's establishment was supported by Australia's then Chief Scientist, Jim Peacock (Howitt and Rennie 2008), to address concerns with the decline in student enrolment in STEM subjects at tertiary level and the lack of support for teachers to teach contemporary scientific practices.

The general objective of the *STEM Professionals in Schools* program is to coordinate partnerships between primary and secondary school teachers and STEM professionals to facilitate real industry experience and encourage STEM learning skills in the classroom. Since 2007, the program has facilitated almost 6,000 partnerships across Australia. The aim is to enable teachers to build their knowledge and confidence in STEM subjects and inspire students to pursue STEM subjects and STEM-related careers.

Each partnership is unique, as the teacher and STEM professional determine what works best for them based on availability and location, with remote partnerships also encouraged. Activities can range from presentations (e.g., basic insect biology), classroom exercises, investigations and experiments (e.g., the science of popping corn, fruit preservation, fruit fly rearing, etc.), site visits and project mentoring (e.g., school projects for Science Week) to helping in their vegetable gardens (e.g., pest and disease diagnostics, crop rotations, basic horticulture), as well as after-school activities and participation in citizen-science projects.

The Australian Sciences curriculum includes *Living things grow, change and have offspring similar to themselves* as part of the primary school Year 2 curriculum (ACARA, 2018), and the life cycle of a butterfly is often studied in Year 2. Arthropods offer many opportunities as teaching tools when applied as part of inquiry teaching in primary and secondary education (Matthews et al., 1997), improving students' attitudes toward STEM subjects, enhancing their performance, and promoting scientific and environmental literacy (Golick and Heng-Moss, 2013). Tackling real-world problems is used to engage children and get them excited about what they are learning in STEM classes.

The *STEM Professionals in Schools* program aligns with the *Science Strategy 2013–2018* (DAFF, 2013) of the Australian Department of Agriculture (DA), which is committed to actively engaging DA scientists with schools and the community to increase STEM awareness. The department currently has more than 20 staff members volunteering in the *STEM Professionals in Schools* program throughout Australia.

This chapter describes how the STEM Professionals in Schools volunteer program uses the experience and expertise of STEM volunteers to share agricultural and entomology skills at a primary school in Canberra, Australia, to inspire and engage students to consider careers in science in general, and entomology in particular.

29.2 METHODS

Farrer Primary School in Canberra, Australia, has partnered with a DA *STEM Professionals in Schools* volunteer with more than 20 years' experience as an agricultural entomologist working in crop protection including Tephritidae fruit fly issues (e.g., biology, biosecurity, market access, and risk-mitigation management). The school has an environment center in which the students learn about sustainable agriculture. As well as animals, the environment center has raised garden beds for growing seasonal vegetables and a range of fruit trees. Working on solutions to real-world problems is the heart of any STEM investigation (Jolly, 2017), and while harvesting peaches, the presence of maggot-infested fruit allowed the students to study the fruit fly life cycle during *STEM Professionals in Schools* volunteer sessions.

Students collected infested peach fruit from the trees and the ground. The maggot stages were observed by cutting open the fruit. Potting mix was placed in the base of large glass jars or plastic ice cream containers, and the infested fruit was placed on top of the potting mix. The jars or containers were covered with mesh and secured with elastic bands to prevent emerging insects from escaping.

After a week, the fruit was checked for maggots and the potting mix was checked for the presence of pupae. All fruit without maggots were removed from the containers. After 2 weeks, adult flies began to emerge. They were fed by placing the following on the surface of the mesh: water-soaked pieces of sponge, checked daily to prevent the sponges from drying out, sugar for energy, and thin layers of VEGEMITE as a protein source for the fruit flies. VEGEMITE is a dark brown savory food spread, which is popular with Australian children. It is also one of the richest known natural sources in the vitamin B group. It is made from brewer's yeast similar to the product in protein bait sprays.

As the adult flies emerged, their colors and patterns were observed. Once they developed and matured, they started mating and laying eggs. Eggs were observed and collected from the mesh. Digital images were taken of all insect stages. The children, with the assistance of the *STEM*

Professionals in School volunteer, identified the fruit fly species as the Queensland fruit fly, *Bactrocera tryoni* (Froggatt). While learning about the life cycle of Tephritidae fruit flies, the students also learned to identify the differences between Tephritidae fruit flies and Drosophilidae flies because the latter also emerged from the infested peaches.

29.3 RESULTS

Approximately 20 Year-2 (7-year-old) students took part in this activity, which was carried out during the Australian summer in Term 1 of the school year (i.e., February). It was the end of a long drought period in Australia, and most of the children had not observed maggots in fruit before. Following the activity with that class, the environment teacher taught the whole school about the difference between true fruit flies (Tephritidae) and vinegar "fruit flies" (Drosophilidae).

Evidence that the students had retained knowledge of the differences between these flies occurred the following year when the school had a new environment teacher. When she called the small *Drosophila* flies buzzing around the compost heap "fruit flies," a number of children corrected her and told her they were "vinegar flies" and not fruit flies.

Although the Year-2 students had not seen maggoty fruit before, a number of their grandparents who lived near the school came into the environment center and reported that they had not had maggots in their backyard fruit (e.g., feijoa, apricots, peaches) since they had moved into the suburb in the early 1970s. They asked how they could prevent or reduce the incidence of maggoty fruit and were advised to pick up any rotting fruit and place it in plastic bags and expose the secured bags to the sun for 48 h and then dispose of the fruit through deep burial. The use of paper bags over young fruit to reduce the incidence of fruit fly attacks was also suggested. Infested fruit was also processed into jam to demonstrate sustainable uses of fruit once the infested sections were removed and disposed of.

In addition to the fruit fly life cycle, the author has also given entomological presentations on butterfly and moth life cycles, and a presentation on bees, their biology, life cycle, and pollination is also in development. The author has been advised by the teachers that these activities and presentations provide students and teachers alike with improved biological science education and awareness from an entomological perspective.

29.4 DISCUSSION AND CONCLUSION

STEM Professionals in Schools is a highly effective program that provides teachers, students, schools, the community, and STEM professionals with significant benefits. Benefits include raising the profile of STEM subjects in schools, increased opportunities for professional learning through communication with scientists and other teachers, inspiring and engaging students in science subjects and alerting them to science-related careers, aligning with DA workplace policy and improving professional scientific communication skills, and sharing a passion for science and information about entomology to increase community understanding of science (Tytler et al., 2016). The program has been evaluated four times (Howitt and Rennie, 2008, Rennie, 2012, Rennie and Howitt, 2009, Tytler et al., 2016), and the key strengths identified in the *STEM Professionals in Schools* program are that the partnerships between STEM professional and teacher are collaborative, flexible, and ongoing (Tytler et al., 2016) and that they have significant national reach with remote partnerships using social media and technology to communicate. The author has recently been partnered remotely with a school in the northernmost Torres Strait Islands where entomology as well as biosecurity knowledge will be shared.

Recommendations for the *STEM Professionals in Schools* program include expanding the program by recruiting more STEM professionals to be partnered with schools. For the author, expansion includes working with the teacher to identify parts of the Australian biological sciences curriculum where entomology can be used to further improve the teaching of biology. For example, discussing the differences in external features of insects from different orders (i.e., Diptera or Lepidoptera), different life stages of insects and how they feed (i.e., chewing or sucking mouthparts) and different life cycles (i.e., complete metamorphosis [Diptera] and incomplete metamorphosis [Hemiptera]), and undertaking a mini-beast excursion to identify insects found on plants grown in the environment center gardens.

Similar STEM programs are run in the United States (AAAS, 2018; STEM-H Center, 2018; Scientist in the Classroom, 2018), Mexico (STEM Movimiento, 2018), the United Kingdom (STEM Ambassadors, 2018), the European Union (STEM Alliance, 2018), Cambodia (STEM Cambodia, 2018), Malaysia (National STEM Movement, 2018), Ghana (STEM Bees, 2018), and many other countries.

These real-world STEM community engagement programs combined with national STEM policy initiatives (e.g., provide specialist STEM schools, update the STEM curriculum, develop smart monitoring, early intervention and access for all, regardless of gender and socioeconomic backgrounds [Timms et al., 2018]) should result in an overall improvement and participation in STEM subjects in schools in Australia and throughout the world.

ACKNOWLEDGMENTS

The success of this program couldn't have been achieved without the commitment of the CSIRO STEM Professions in School team and the Principal, environment center teachers and students of Farrer Primary School.

REFERENCES

AAAS (American Association for the Advancement of Science). 2018. STEM Volunteer Program. http://www.stemvolunteers.org/ (accessed September 29, 2018).

ACARA (Australian Curriculum, Assessment and Reporting Authority). 2018. Science—Foundation to Year 2 ACSSU030. https://www.australiancurriculum.edu.au/Search/?q=ACSSU030 (accessed September 29, 2018).

CSIRO (Commonwealth Scientific and Industrial Research Organisation). 2015. STEM Professionals in Schools. https://www.csiro.au/en/Education/Programs/STEM-Professionals-in-Schools (accessed September 27, 2018).

DAFF. 2013. DAFF Science Strategy, Department of Agriculture, Fisheries and Forestry, Canberra. http://www.agriculture.gov.au/Style%20Library/Images/DAFF/__data/assets/pdffile/0009/2338947/science-strategy-2013.pdf (accessed September 27, 2018).

Department of Education. 2012. Award course completions. https://docs.education.gov.au/documents/2012-award-course-completions (accessed February 24, 2019).

Department of the Prime Minister and Cabinet. 2015. National Innovation and Science Agenda. https://www.industry.gov.au/sites/g/files/net3906/f/July%202018/document/pdf/national-innovation-and-science-agenda-report.pdf (accessed September 27, 2018).

Education Council. 2016. National STEM School Education Strategy, 2016–2026. http://www.educationcouncil.edu.au/site/DefaultSite/filesystem/documents/National%20STEM%20School%20Education%20Strategy.pdf (accessed September 29, 2018).

Fairweather, J. 2008. Linking Evidence and Promising Practices in Science, Technology, Engineering, and Mathematics (STEM) Undergraduate Education: A Status Report for The National Academies National Research Council Board of Science Education. http://otl.wayne.edu/wider/linking_evidence--fairweather.pdf (accessed September 29, 2018).

Golick, D. A., and T. M. Heng-Moss. 2013. Insects as Educational Tools: An Online Course Teaching the Use of Insects as Instructional Tools. *American Entomologist* 59 (3): 183–187.

Howitt, C., and L. J. Rennie. 2008. Evaluation of the Scientists in Schools Pilot Project. Perth: Curtin University of Technology. https://www.csiro.au/en/Education/Programs/STEM-Professionals-in-Schools/How-the-program-works/Program-evaluation (accessed September 27, 2018).

Jolly, A. 2017. The Search for Real-World STEM Problems. Education Week, July 17, 2017. https://www.edweek.org/tm/articles/2017/07/17/the-search-for-real-world-stem-problems.html (accessed September 29, 2018).

Matthews, R. W., L. R. Flage, and J. R. Matthews. 1997. Insects as Teaching Tools in Primary and Secondary Education. *Annual Review of Entomology* 42: 269–289.

National STEM Movement. 2018. Ministry of Higher Education, Ministry of Education, and Ministry of Science, Technology & Innovation of Malaysia. http://www.stem-malaysia.com/ (accessed September 29, 2018).

Office of the Chief Scientist. 2014. Science, Technology, Engineering and Mathematics: Australia's Future. Australian Government, Canberra. https://www.chiefscientist.gov.au/wp-content/uploads/STEM_AustraliasFuture_Sept2014_Web.pdf (accessed September 29, 2018).

PwC (PricewaterhouseCoopers). 2014. Fuelling NextGen digital innovation through education. http://www.digitalinnovation.pwc.com.au/education/ (accessed September 29, 2018).

Rennie, L. J., and C. Howitt. 2009. "Science has changed my life!" Evaluation of the Scientists in Schools Project 2008–2009. Perth: Curtin University of Technology. https://www.csiro.au/en/Education/Programs/STEM-Professionals-in-Schools/How-the-program-works/Program-evaluation (accessed September 27, 2018).

Rennie, L. J. 2012. "A very valuable partnership" Evaluation of the Scientists in Schools Project 2011–2012. Dickson, ACT: CSIRO Education. http://www.scientistsinschools.edu.au/downloads/SiSEvaluationReport 2011–2012.pdf (accessed September 27, 2018).

Scientist in the Classroom. 2018. National Center for Science Education, Inc. Oakland, California, USA. https://ncse.com/scientistinclassroom (accessed September 29, 2018).

Singhal, P. 2017. Australia falling behind in science graduates and public funding: OECD report. Sydney Morning Herald. https://www.smh.com.au/education/australia-falling-behind-in-science-graduates-and-public-funding-oecd-report-20170912-gyfigs.html (accessed September 29, 2018).

STEM Alliance. 2018. European Schoolnet, Bringing Industry and Education Together. http://www.stemalliance.eu/ (accessed September 29, 2018).

STEM Ambassadors. 2018. National STEM Learning Centre, STEM Learning Ltd. University of York, Heslington, UK. https://www.stem.org.uk/stem-ambassadors (accessed September 29, 2018).

STEM Bees. 2018. http://www.stembees.org/ (accessed September 29, 2018).

STEM Cambodia. 2018. Ministry of Education, Youth and Sport. http://www.stemcambodia.org/ (accessed September 29, 2018).

STEM-H Center. 2018. University of New Mexico STEM-H Center for Outreach, Research & Education. Albuquerque, New Mexico, USA. http://stemed.unm.edu/volunteer (accessed September 29, 2018).

STEM Movimiento. 2018. Movimiento STEM A. C., Mexico City, Mexico. http://movimientostem.org/ (accessed September 29, 2018).

Timms, M., K. Moyle, P. R. Weldon, and P. Mitchell. 2018. Challenges in STEM learning in Australian schools: Literature and policy review. https://research.acer.edu.au/cgi/viewcontent.cgi?article=1028&context=policy_analysis_misc (accessed September 30, 2018).

Tytler, R., D. Symington, G. Williams, P. White, C. Campbell, G. Chittleborough, G. Upstill, E. Roper, and N. Dziadkiewicz. 2016. Building productive partnerships for STEM education: Evaluating the model and outcomes of the Scientists and Mathematicians in Schools program 2015, Deakin University, Burwood, Australia. Website: www.csiro.au/~/media/Education-media/Files/STEM-Prof-Schools/Productive-Partnerships-STEM-Education-PDF.pdf (accessed September 29, 2018).

30 Phytosanitary Education
An Essential Component of Eradication Actions for the Carambola Fruit Fly, Bactrocera carambolae, *in the Marajo Archipelago, Para State, Brazil*

Maria Julia S. Godoy[*], *Gabriela Costa de Sousa Cunha, Luzia Picanço, and Wilda S. Pinto*

CONTENTS

Abstract This chapter presents phytosanitary education actions carried out in support of official control actions implemented to eradicate outbreaks of the quarantine pest *Bactrocera carambolae* (Drew and Hancock) (carambola fruit fly) in the municipalities of Curralinho, Portel, Gurupa, and Breves, Marajo Archipelago, State of Para, Brazil. All actions were carried out from an Emergency Action Plan of Phytosanitary Education, including household visits, meetings with local authorities, technical meetings, lectures, training courses for multiplier agents based on the SOMA Method, radio and TV interviews, notes for websites, participation in local social and agricultural events, workshops, and puppet theatre. The phytosanitary education actions reached 24,750 people (3,058 people in 2014, 6,543 people in 2015, 4,128 people in 2016, and 11,021 people in 2017) in the aforementioned municipalities and also in neighboring ones considered to be at high risk of pest dispersal. Even after pest outbreaks have been declared officially eradicated, phytosanitary education actions must be continued to support passenger baggage transit control that prevents pest host fruit smuggling from the state of Amapa to the state to Para through the Marajo Archipelago boat route to maintain the eradicated area.

[*] Corresponding author.

30.1 INTRODUCTION

In Brazil, especially in the State of Amapa, the carambola fruit fly (CFF), *Bactrocera carambolae* (Drew and Hancock), is considered a quarantine pest that, although present, is not widely distributed and officially controlled. This pest is of great economic importance for Brazilian agribusiness exports. In the states of Para and Roraima, it is considered a transient pest according to the National Programme for Eradication of *Bactrocera carambolae* (PBC) of the Ministry of Agriculture, Livestock and Food Supply (MAPA). Para's Agrihealth State Agency (ADEPARA), under the coordination of the Brazilian National Plant Protection Organization (NPPO), Department of Plant Health (DSV), MAPA, carried out phytosanitary education activities to support the eradication of *B. carambolae* in the municipalities of Curralinho, Portel, Gurupa, and Breves located in the Marajo Archipelago, state of Para, Brazil, from 2014 to 2017 (Figure 30.1).

In these municipalities, there is a great risk of entry and spread through water, land, and air transportation of contaminated fruits due to the proximity to the Marajo Archipelago and the state of Amapa, where *B. carambolae* is being controlled. The cultural habits, especially of the riverine population, are to consume fresh fruits during river journeys, especially of host plants like *Mangifera indica* L., *Averrhoa carambola* L., *Malpighia emarginata* DC., *Syzygium malaccense* L., *Psidium guajava* L., *Citrus x sinensis* Osb., *Solanum lycopersicum* L., and *Capsicum annuum* L. Although

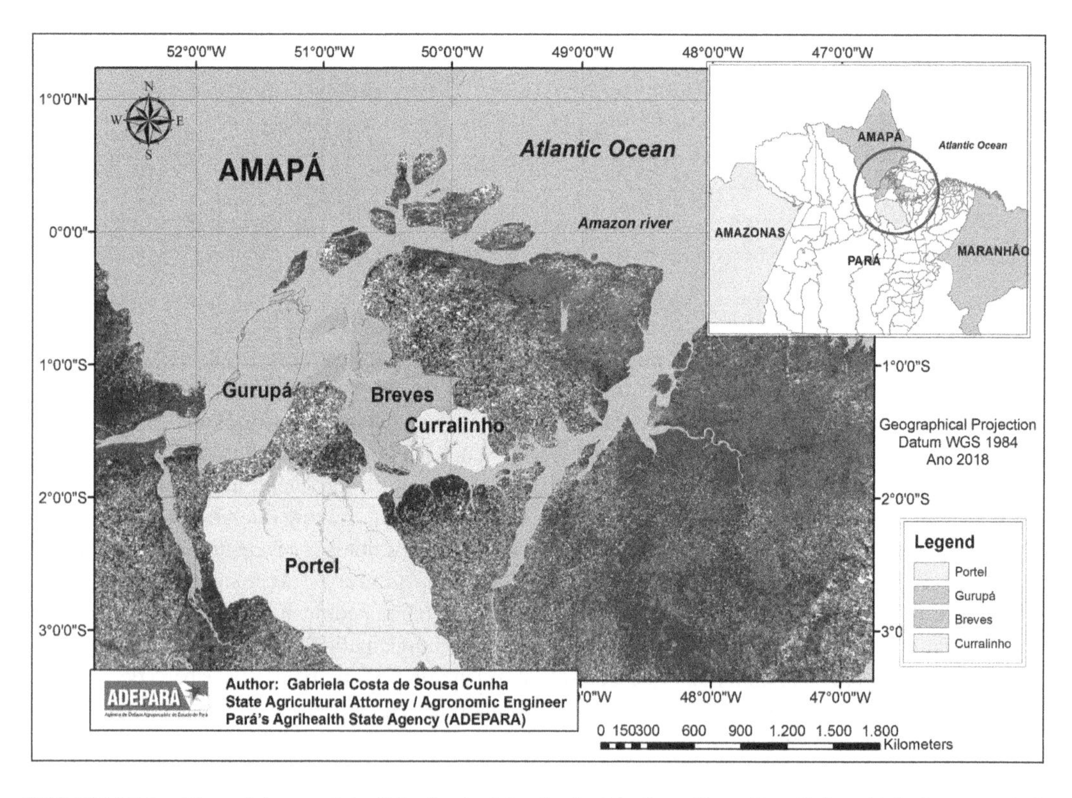

FIGURE 30.1 Map of the municipalities in the Marajo Archipelago, Para, Brazil, in which the carambola fruit fly, *Bactrocera carambolae*, was eradicated.

regulations were immediately enforced and measures were taken when outbreaks were detected, including the prohibition of host fruit transportation and commercialization from infested areas to pest-free areas, it is necessary to raise community awareness about the danger that the movement of this fruit imposes. Raising community awareness regarding the risk of transporting host fruits from areas where CFF is known to occur to CFF-free areas, and also about the importance of eradication of the pest for Brazilian fruit exports, is a key component of any action program. Phytosanitary education supports pest inspection, control and eradication actions, as well as activities aiming to inform and encourage the change of habits in communities and farmers. This is achieved through the development of educational campaigns and community awareness about agriculture and agro-industry activity projects.

The objective of this contribution was to describe the activities undertaken to increase the knowledge of the local people about the *B. carambolae* control program during the 2014–2016 outbreaks in the municipalities of Curralinho, Portel, Gurupa, and Breves located in the Marajo Archipelago, state of Para, Brazil.

30.2 MATERIALS AND METHODS

Based on the Phytosanitary Education Emergency Action Plan, prepared by the team of the National Program for the Eradication of *Bactrocera carambolae* (PBC), and considering the local cultural habitats of the riverine population, phytosanitary education actions are implemented at the time of notification of an outbreak. PBC procedures establish that phytosanitary education and control measures should be implemented together and within 48 h of the notification of the outbreak. Activities involve visits to municipal authorities and state and federal agencies present in the municipality with the purpose of providing official information on the phytosanitary condition of the location of the outbreak. This is followed by radio and TV interviews, presentations of the topic to primary and secondary schools, courses for training multiplier agents using the SOMA method (education tool whose acronym means systemic [S], objective [O], monitoring [M], and evaluation [A]), technical lectures and participation in social and agricultural events, and workshops and puppet plays. According to Albuquerque (2000), the SOMA method allows to quantify the students' knowledge before and after the technical lecture and to identify the learning efficiency of each objective, indicating to the teacher the need to clarify the presented subject. Teachers, rural extension agents, health agents, high school and university students, public servants, community leaders, among others, are invited to participate in the training of multipliers based on the SOMA method. After contacting the municipal authorities, visits are made to the community explaining the detection of the pest, identifying hosts for trapping in backyard orchards as support for the actions of the pest control team, as well as on-site visits to commercial establishments, waterways, and homes in both urban and rural communities. With the accomplishment of a training course for multiplier agents, Municipal Phytosanitary Education Nuclei are implemented with representatives of community agents such as teachers, health agents, and others. The community is informed of the activities that are being carried out, and after the eradication of the outbreak, a Post-Eradication Plan of Phytosanitary Education is implemented to continue carrying out activities in the municipality as shown in Figure 30.2.

During 2014–2017, actions were carried out in partnership with local agricultural and related institutions, as well as with the communities, considering interinstitutional integration and local knowledge. All educational activities were carried out in a continuous way, with alternation of

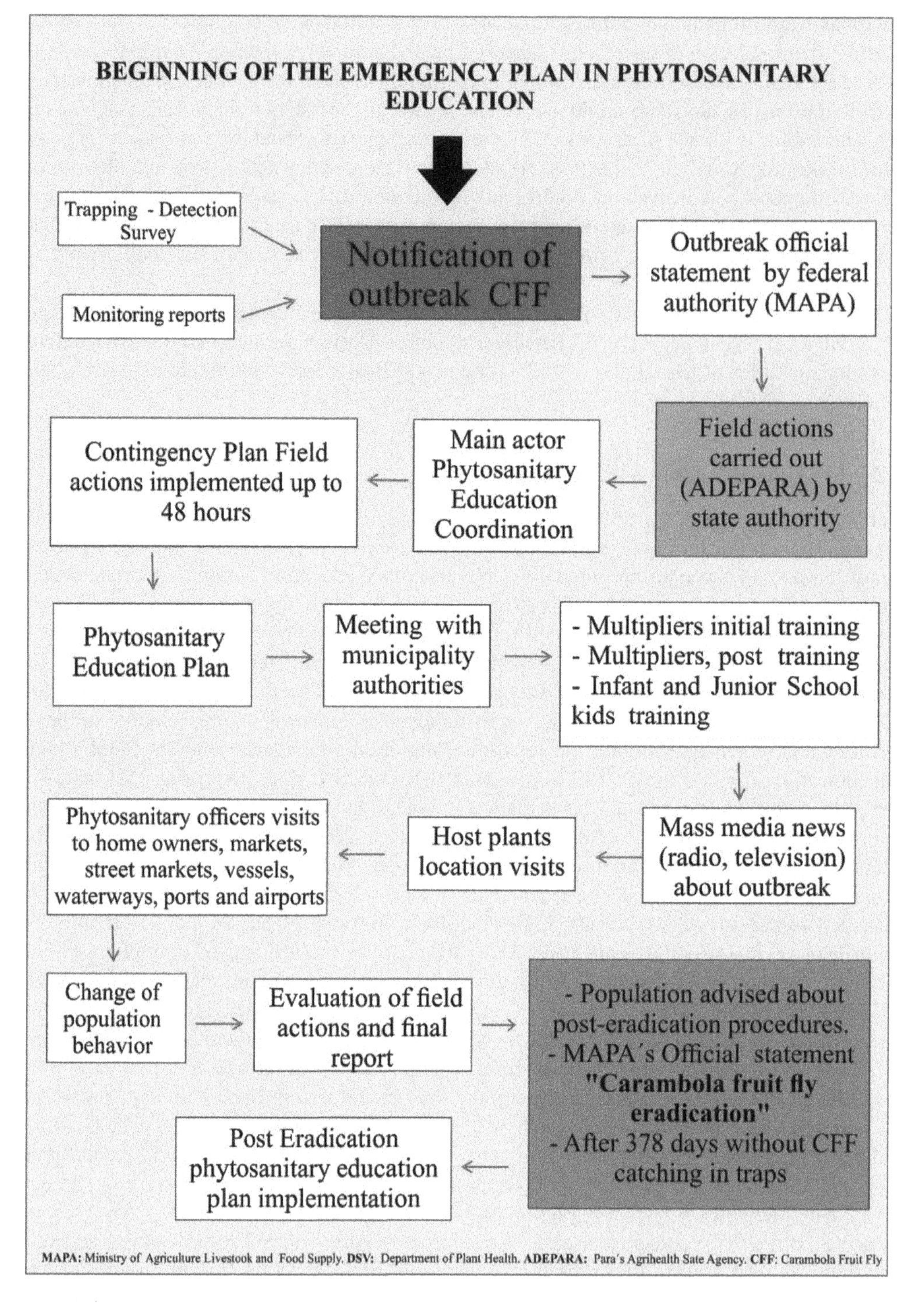

FIGURE 30.2 Flowchart of the Emergency Phytosanitary Education Plan of the National Carambola Fruit Fly Eradication Program. CFF, carambola fruit fly.

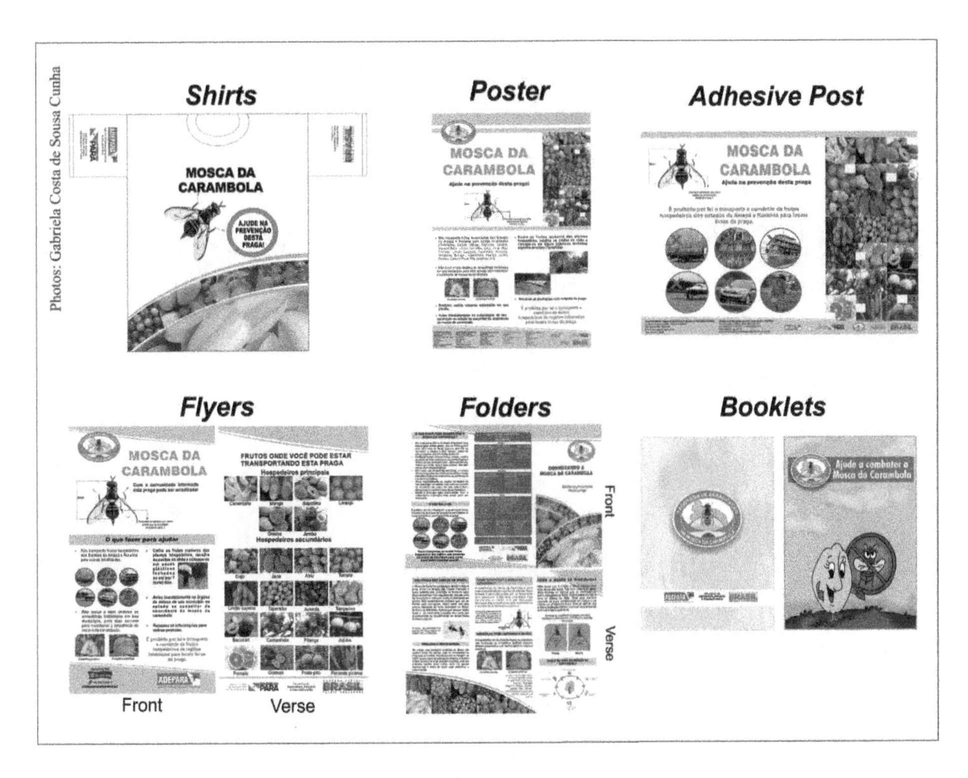

FIGURE 30.3 Educational material used in the campaign to eradicate the carambola fruit fly in the Marajo Archipelago, Para, Brazil.

teams in the field and with the minimum interval between them. The crew and passengers of vessels are considered strong allies in actions to prevent the dispersion of the pest. After they receive guidelines, they become co-participative by spreading the acquired knowledge, mainly in relation to host fruit transit restriction and pest identification. Educational materials were used to support all activities, such as banners, stickers for SOMA method courses, flyers, folders, booklets for children, and other printed material for places without electricity (Figure 30.3).

30.3 RESULTS AND DISCUSSION

In compliance with the actions contained in the Emergency Plan for Health Education Actions, technical lectures for children and adolescents, interviews on radio and TV, visits, and courses were held. From 2014 to 2017, a total of 24,750 people participated in the educational activities (3,058 people in 2014; 6,543 people in 2015; 4,128 people in 2016; and 11,021 people in 2017), both in the municipalities of the Marajo Archipelago and in others considered of high risk for pest dispersal (Figures 30.4 and 30.5).

During this period, four outbreaks were detected in the Marajo Archipelago. Restrictions related to movement of host fruits from infested sites to pest-free areas were immediately published. In each municipality, a group of multiplier agents of the program was formed, with a total of 186 multiplier

FIGURE 30.4 Educational activities in fairs and residences and lectures at schools carried out by field teams aimed at supporting the eradication of the carambola fruit fly.

FIGURE 30.5 Interviews on radio and television to clarify carambola fruit fly's outbreak detection.

agents: 14 in Curralinho, 35 in Gurupa, 66 in Portel, and 53 in Breves. Also, another group with 18 students was formed in the municipality of Melgaço (Figure 30.6).

Because SOMA is a method that does not require large audiovisual aids, it can be used in areas without much infrastructure. In addition, because it uses repetition as part of the learning process, it can be implemented in an audience with any level of schooling, including those who are illiterate. Training using this methodology improved the educational tools the community received by improving questionnaires, manual tabulation of data, calculations of average efficiency and learning improvement, and by identifying the weaknesses of the training. The use of this method contributed

FIGURE 30.6 Class of multiplier agents trained through the SOMA method in the municipalities of Curralinho, Portel, Gurupa, and Breves. SOMA, systemic, objective, monitoring, evaluation.

significantly to the immediate efficiency of the teaching-learning process, whose diagnosis facilitated the planning and continuity of future actions.

The Marajo Archipelago, with 104,606.90 km², is divided into 16 municipalities and is the main route of entry of the pest into the Amazon, where the main road network is fluvial. Vessels leave the state of Amapa to Belem, capital of the state of Para, Manaus in Amazonas, and other cities, distributing freight and passengers, thereby becoming a pathway for the distribution of the pest. During the program, health education teams intensified activities with passengers and crew on a daily basis with approaches before boarding and after landing, explaining to the public the restriction of transit of all host fruits, in any quantity, as well as with the distribution of informative material to reinforce the information (Figure 30.7). It is important to emphasize that the control teams,

FIGURE 30.7 Approach carried out in vessels and waterways with passengers and crew in the municipalities of Curralinho, Portel, Gurupa, and Breves.

monitoring agents, phytosanitary educators, and general coordinators were always interconnected and motivated. This was supported through meetings destined to update about new situations found in the field activities, which also contributed to the harmony and success of the activities. The first approach with the local population, including meetings with leaders and city hall authorities, was carried out in an enlightening, convincing, and respectful manner, which favored educational activities aimed at supporting pest eradication. Therefore, control activities became a community action, and the community was not afraid of the program, and the acquisition of knowledge contributed to the good progress of the work.

The communities of the municipalities of Curralinho, Portel, Gurupa, and Breves played a fundamental role in the CFF-eradication process. This was evidenced by behavioral changes, mainly in relation to host fruit transportation, fruit collection, and contribution to the technical staff of the control team to carry out trapping surveys and sprayings around homes. They also provided permission to display posters in commercial areas, looked after the traps, participated in lectures and events promoted by the program, and provided valuable support by reporting houses with host fruits and potential outbreaks.

As a result of control actions, supported by the education program, outbreaks were declared as eradicated in Curralinho on April 24, 2015; in Portel on October 16, 2016; in Gurupa on September 17, 2016; and in Breves on July 2, 2017. Nevertheless, local activities, including the control of passengers moving from the state of Amapa, continued after the declaration of eradication of each outbreak.

30.4 CONCLUSIONS

The local population, through awareness activities, understood the dangers of pest dispersal, as well as the economic and social costs that occur when CFF spreads to production areas. The program also resulted in community participation and in a strengthened partnership between the community and the PBC team. Phytosanitary education activities were found to be essential for the success of the eradication programs against fruit flies. Therefore, such programs should be part of each contingency plan of each federal state and should be carried out jointly with control actions.

REFERENCE

Albuquerque, C. 2000. *Método SOMA: Capacitação de agricultores, educação sanitária, educação ambiental.* Gráfica e Editora Bandeirante, Bandeirante, Brazil.

31 Phytosanitary Education as a Component of Eradication Actions of the Carambola Fruit Fly (CFF) *Bactrocera carambolae* in the Raposa Serra Do Sol Native Reserve, State of Roraima, Brazil

Maria Julia S. Godoy, Gabriela Costa de Sousa Cunha, Elindinalva Antônia Nascimento, Maria Eliana Queiroz, Luzia Picanço, Luiz Carlos Trassato, and Wilda S. Pinto*

CONTENTS

Abstract This chapter presents the results obtained through the phytosanitary education methodology used by the Carambola Fruit Fly (CFF) Eradication Program based on the SOMA Method. Since the initial detection of the quarantine pest *Bactrocera carambolae* (Drew & Hancock) in the Raposa Serra do Sol native reserve, Roraima, Brazil, this program has contributed to the eradication of the CFF and the maintenance of a protected area, which is the minimum area necessary for the effective protection of an endangered area found in the extreme north of Roraima.

* Corresponding author.

31.1 BACKGROUND

Two initial outbreaks of the carambola fruit fly (CFF), *Bractocera carambolae*, in the state of Roraima, Brazil, were detected on December 19, 2010, and February 2, 2011, respectively, in the municipality of Uiramutã, located in the northeast of Roraima, Raposa Serra do Sol. After this detection, the outbreaks were kept under control without dispersion outside the Raposa Serra do Sol region because of the promptness of control actions and phytosanitary education.

The state of Roraima, located in the northern region of Brazil, borders to the north and northwest with Venezuela, to the east with Guyana, to the southeast with the Brazilian state of Para, and to the south and west with the Brazilian state of Amazonas. The municipality of Uiramutã is located at 04° 35′ 45″ N and 60° 10′ 04″ W, border with Venezuela and Guyana. It has a total area of 8,066 km² and an estimated population of 8,375. It houses one national park and part of the Raposa Serra do Sol native reserve, and it exhibits tropical savanna climate (Aw) according to the Koppen climate classification. Large plains have savanna lowbush and grass vegetation, and mountains are covered with tropical rainforest. The reserve is located between the Tacutu, Mau, Surumu, and Miang Rivers, and it is occupied by the indigenous groups of ingaricos, macuxis, patamonas, taurepangues, and uapixanas. (https://pt.wikipedia.org).

The municipality of Normandia is located at 3° 52′51″ N and 59° 37′ 22″ W, with a border to the north with Uiramutã and the Co-operative Republic of Guyana, to the south with Bonfim, to the east with the Co-operative Republic of Guyana, and to the west with Boa Vista and Pacaraima. It has a total area of 6,967 km² (https://en.wikipedia.org) (Figure 31.1).

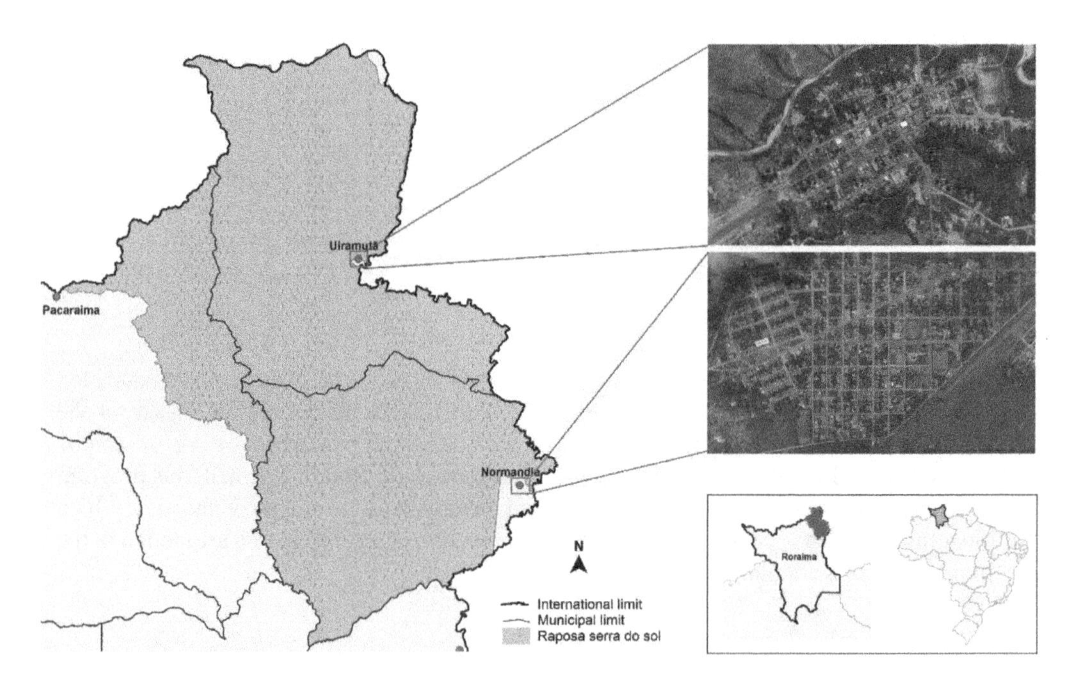

FIGURE 31.1 Location and aerial view of Uiramutã and Normandia, Raposa Serra do Sol, Roraima, Brazil.

Although this region is not an important producer of CFF host fruits, since the occurrence of the outbreaks in 2010, Roraima's host fruit production has not been allowed to be marketed out of the state. For instance, the most important host crop of Roraima is mango, with 860 ha planted and 4,214 tons of fruit harvested in 2017 (Instituto Brasileiro de Geografia e Estatística - IBGE 2017), and it has its main market in the neighboring state of Amazonas. The restriction on host fruit traffic prevents the spread of CFF to Amazonas and from there to the main Brazilian fruit-producing states (Figure 31.2).

Factors that helped the dispersion of CFF into Brazil were the uncontrolled presence of CFF in Guyana; the lack of control actions; continuous and regular informal commercial exchange between native people living in Guyana's regions 7 (Cuyuni-Mazaruni), 8 (Potaro-Siparuni), and 9 (Alto Takutu-Alto Essequibo) and the northeast of Roraima, Brazil; gold prospecting routes; and strong winds (Ezilon Maps, 2018) (Figure 31.3).

According to the Brazilian legislation, emergency plans for CFF eradication must be implemented no later than 48 h after one specimen of the pest has been detected (Normative Instruction Nº. 28, of July 20, 2017). In the case of native people, previous community authorization is mandatory before any control action takes place, which leads to the prioritization of phytosanitary education actions. The Raposa Serra do Sol native reserve has many villages, with Maturuca being the main one, and Willimon, Formoso, Caraparu, Morro, Pedra Branca, and Serra do Sol being other important villages.

FIGURE 31.2 The Mau River, border between Karasabai town, the Co-operative Republic of Guyana, and the state of Roraima, Brazil.

FIGURE 31.3 Administrative regions of Guyana (Ezilon Maps, 2018).

31.2 MATERIALS AND METHODS

The SOMA method, an education tool whose acronym means systemic (S), objective (O), monitoring (M), and evaluation (A) (Albuquerque 2000), was applied in all the communities of the Raposa Serra do Sol native reserve. The SOMA method can be understood as systemic: results are essential, as they guide the system; objectives: need to be well defined and clearly measurable; monitoring: needs to be continuous with the capacity of building a process to allow trainee evolution and evaluation, and to adjust the system to reach expected results; evaluation: all work is done under continuous evaluation, allowing system improvement along the process (Albuquerque 2000).

This method can be used in areas without infrastructure and for a public with any level of education because it uses repetition in the learning process and does not require any specific audiovisual resources. It allows the establishment of results through the improvement of questionnaires based on feedback; manual tabulation of data; average calculations, efficiency, and increase of learning; and identifying the weaknesses of the training. It also contributes significantly to the evaluation and immediate effectiveness of the teaching-learning process, whose diagnosis facilitates the planning/

continuity of future actions. The first contact is with the health secretary of the city hall for the approval of the participation of community health agents, servers linked to the education area and community leaders.

Subsequently, these servers become multipliers of the National Carambola Fly Eradication Program, becoming a focal point in these municipalities. A series of visits are made by the multiplier agents, who then provide feedback of the situation that allows to guide actions in the communities.

During training, the instructors emphasize to multipliers, students, and the general public the importance of actions to control the eradication of CFF and the risks related to the transport of the most common host fruits in the region; these are fruits such as mango, carambola, acerola, lemon cayenne, and chili pepper from the infested areas of Uiramutã and Normandia to cities without the occurrence of the pest within the state of Roraima, as well as to other states where CFF is not present in Brazil. The Raposa Serra do Sol native reserve comprises Uiramutã city, Pacaraima, and Normandia city.

In Uiramutã city, the distance between these communities varies from 500 m to 10 km, with the largest distance being between the Maturuca village and the Mutum and Willimon villages, reaching 60 km. It is important to note that to carry out visits to all the native villages, 700 km have to be covered.

The indigenous population has the habit of carrying host fruits from one locality to another, either to offer them as gifts or to consume during journeys and, in the case of peppers, during festivities in which the indigenous population of Guyana and Brazil take part. Therefore, all involved must be alerted about the risks of uncontrolled transit of host fruits from infested areas to CFF-free areas and about the economic loss of exports of Brazilian fruit to other countries. This method has certainly contributed to sensitizing the population about the risks of dispersion of the pest within the communities, thus promoting an awareness through effective change in behavior.

31.3 RESULTS

The initial activity is the training of multipliers, comprised preferentially of leaders from native villages and municipalities, school teachers, community health agents, and civil servants based in the region. The multipliers integrate the Municipality Phytosanitary Education Group which supports actions carried out by the CFF Eradication Program teams. Through community and school lectures and host plant product transit control, the local population is informed about economic and social losses in the case of CFF dispersion to other states in Brazil, as well as about legal responsibilities assumed by those who disrespect legislations forbidding transit and sale of CFF host fruits. Interviews are regularly held in locally and statewide broadcasted radio programs, which also reach towns bordering neighbor countries. Roraima's team is composed of eight members from the Ministry of Agriculture, Livestock and Food Supply, and the Agri-health State Agency, all of whom have already been trained through the SOMA method (Albuquerque 2000), which is a mandatory condition for membership to the phytosanitary education team.

From 2011 to 2017, the phytosanitary education team in Roraima presented 32 technical lectures for a total audience of 1,533 people, 72 presentations in junior schools for 2,472 students, 168 meetings with native leaders reaching 1,170 people, 24 phytosanitary education blitzes in the borders reaching 9,413 people, and 12 training courses attended by 2,472 CFF Eradication Program multiplying agents (Figure 31.4). The education team also carried out 221 educational activities reaching 8,454 people, including radio programs.

The Raposa Serra do Sol native reserve is partially located in the territories of the Pacaraima, Normandia, and Uiramutã municipalities. Normandia comprises 68 native villages, Pacaraima 60 villages, and Uiramutã 85 villages. Each one of the 213 villages has its own leader and maximum authority named "Tuxaua," which is chosen by village members to rule during a 2-year term. As the main authority, the "Tuxaua" must be the first one to be consulted about any issue related to the village. Thus, the activities related to phytosanitary education and CFF pest control actions could only be carried out after his authorization.

FIGURE 31.4 Training of people in indigenous communities as multipliers of the carambola fruit fly (CFF) Eradication Program, Raposa Serra do Sol, Uiramutã, Roraima, Brazil.

The first activity performed by the phytosanitary education team is usually a lecture given to the community explaining about the CFF, its biological cycle, why it is considered a pest, how it can spread, and the risks it presents to the domestic and export fruit industries. At the end of the lecture, the team members and the community organize priority activities to be carried out, taking into account the outbreak, the public to be worked with, and the physical structure and access routes.

The phytosanitary education team performs activities such as school and community lectures, puppet theater, training courses for multipliers (SOMA method), individual home surveys to locate host plants and control actions, teaching, control actions in commercial establishments, radio and TV interviews, and transit control related to host fruit transportation. Trained multipliers have been working on a regular basis with phytosanitary education teams in education and control actions, resulting in the successful eradication of CFF outbreaks (Figure 31.5).

FIGURE 31.5 SOMA training meeting with indigenous people learning to identify the carambola fruit fly (CFF). SOMA, systemic, objective, monitoring, and evaluation.

FIGURE 31.6 Children trained as trap guardians from the native groups of ingaricos, macuxis, patamonas, taurepangues, and uapixanas.

Children also play an important role in the community; thus, they were trained to become trap guardians (Figure 31.6). However, sporadic CFF outbreaks can occur in the border because of pest pressure, demanding continuous phytosanitary education activities.

31.4 CONCLUSIONS

The first detection of CFF in the state of Roraima occurred in the Raposa Serra do Sol native reserve located close to the border between Brazil and Guyana. This specific situation required prioritization of phytosanitary education actions before control actions could take place due to the need of previous authorization from native community leaders.

Phytosanitary education actions must be performed in a continuous, respectful, and clear manner, taking into account the particularities of the communities and the organization of the people, as well as the education of the indigenous people related to the way of life and beliefs of the community. In this way, the inclusion of indigenous groups led to a successful CFF-eradication process.

The SOMA method was selected because it is understandable for people with different levels of school education, and it uses repetition as a learning basis, allowing previous identification of specific objectives and the creation of local groups supporting phytosanitary education teams. The education activities related to control actions, especially CFF trap maintenance, were key for the successful eradication of CFF and the maintenance of the protected area based on the current Brazilian legislation.

REFERENCES

Albuquerque, C. A. 2000. *Método Soma: Capacitação de agricultores, educação sanitária e educação ambiental*. Gráfica e Editora Bandeirante, Bandeirante, Brazil.

Ezilon Maps. Guyana Map—Political Map Of Guyana. https://www.ezilon.com/maps/south-america/guyana-maps.html. (accessed December 20, 2018).

Instituto Brasileiro De Geografia E Estatística – Ibge. Banco De Tabelas Estatísticas. Produção Agrícola Municipal. https://sidra.ibge.gov.br/pesquisa/pam/tabelas. (accessed December 18, 2018).

Normative Instruction N°. 28. 2017. The operational procedures for the prevention, containment, suppression and eradication of the quarantine pest present *Bactrocera carambolae* (Carambola Fruit Fly). Ministry of Agriculture and the Agrihealth State Agency. Official Journal of the Union. Brasília, Brazil, July 26, 2014, No. 142, Section 1, p. 8.

Wikipedia The Free Encyclopedia. https://pt.wikipedia.org. (accessed December 18, 2018).

Index

Note: Page numbers in italic and bold refer to figures and tables, respectively.

C

D

J

L

M

N

O

P

Q

R

S

T

U

V

W

Z